Engineering Design, Planning, and Management

Engineering Design, Planning, and Management

Second Edition

Hugh Jack

Academic Press is an imprint of Elsevier
125 London Wall, London EC2Y 5AS, United Kingdom
525 B Street, Suite 1650, San Diego, CA 92101, United States
50 Hampshire Street, 5th Floor, Cambridge, MA 02139, United States
The Boulevard, Langford Lane, Kidlington, Oxford OX5 1GB, United Kingdom

Library of Congress Cataloging-in-Publication Data
A catalog record for this book is available from the Library of Congress

British Library Cataloguing-in-Publication Data
A catalogue record for this book is available from the British Library

ISBN: 978-0-12-821055-0

Publisher: Katey Birtcher
Acquisition Editor: Stephen Merken
Editorial Project Manager: Chris Hockaday
Production Project Manager: Rukmani Krishnan
Cover Designer: Christian J. Bilbow

Typeset by TNQ Technologies

Printed in the United States of America
Last digit is the print number: 9 8 7 6 5 4 3 2 1

Contents

Preface

Engineers are professional inventors, researchers, and developers. Education imbues each engineer with discipline-specific knowledge. Combining the different disciplines allows engineers to solve more complex problems.

A design project has a fixed time frame, allocated resources, and defined outcomes. There are many books about project management without the engineering context. And, most project management books are for working professionals who have a few years of project management experience. Most design books pay minimal attention to project management. To make engineering students more effective, an integrated approach to project management and creative design is necessary.

This book represents a compilation of essential resources, methods, materials, and knowledge developed and used over 2 decades of teaching project-oriented courses. It is for engineering students taking courses with technical design projects. Students who do project work in parallel with the book can benefit greatly. Reading chapters out of order, or omitting some chapters entirely, can accommodate unique curricula. Readers can find technical examples specific to their disciplines and to other forms of engineering. This book uses methods and knowledge that are applicable to all disciplines. A mixture of cross-discipline and discipline-specific examples relates application knowledge to the multidisciplinary field of engineering design projects.

The best approach to design project education is to actually do projects; students or professionals should work on technical problems at the same time they are reading the book. Read Chapter 1 for an overview of the design project process. The remaining chapters can be read in any order, suited to the course or project. For example, some instructors might choose to omit the "People and teams," "Communication, meetings, and presentations," or "Customer requirements and specifications" chapters. Some chapters are for smaller audiences, such as Chapter 10, "General design topics," which is aimed at all engineering disciplines.

The construction of the book supports the comprehension and use of engineering theory in applied practice.

Notable features include the following:

- There are many figures and clear procedural steps, which support learning and application.
- Abstract and concrete learning styles are accommodated with parallel text and/or figures for each concept.
- Visual models provide a foundation for knowledge. Flowcharts illustrate decision-making examples, office procedures, and human relations.
- Many methods are illustrated with tables so that they can be done using spreadsheet software.
- Some critical topics are discussed in depth. Other topics provide enough description to understand the strategic importance of the methods, and prepare the reader to quickly locate, and use, learning resources.
- Chapters are concise and focus on design project skills (Fig. 1).
- The sequence of the chapters supports a relatively generic project sequence for just-in-time learning. However, instructors may change the chapter sequence as necessary.
- The book is intended for a multidisciplinary audience. There is an assumption that readers have a strong grounding in their chosen discipline and are in need of an integrative experience.

Management	Engineering Judgment	Professional Effectiveness
Design Projects Chapter 1	Concepts and Technical Specs. Chapter 4	People and Teams Chapter 5
Planning and Managing Projects Chapter 2	Decision-Making Chapter 6	Communication, Meetings, and Presentations Chapter 9
Customer Reqs. and Specs. Chapters 3	Reliability and System Design Chapter 8	
Finance and Budgets Chapters 7	General Design Topics Chapter 10	

FIGURE 1

The big-picture outcomes.

- Short single-sentence axioms highlight key concepts.
- Cases and examples bring the concepts to life.
- Simpler problems are placed throughout the chapters to allow readers to test their knowledge as they read. Problem solutions are available on the support website: http://www.engineeringdesignprojects.com.
- Professional engineering topics in most chapters include human factors, law, ethics, and communication. These topics prepare engineers to collaborate with peers and eventually manage employees, budgets, customers, and more.
- Accredited and certified engineering programs can benefit from topics ill suited to traditional courses. The chapters of the book correlate to engineering and technology accreditation criteria. Appendix C provides a map to accreditation criteria for a few bodies such as ABET and CEAB.
- The book uses système internationale (SI) units. Imperial (English) units appear when appropriate.
- A tutorial on using Microsoft Project is provided for project management tasks.

This book is multidisciplinary in nature and benefits from the author's extensive background, which includes a bachelor's degree in electrical engineering and master's and PhD in mechanical engineering. Professional work in manufacturing, automation, and robotics brings a solid understanding of manufacturing engineering. In addition, the author has had over 2 decades of teaching experience with industrial, academic, course, thesis, and capstone projects. These have resulted in hundreds of projects for industrial sponsors at the undergraduate and graduate levels. The project types have included new product designs, test equipment, production equipment, and applied research.

ANCILLARIES

Instructors using this text for a course can find useful support materials, including electronic images from the text, recommended schedules for projects, and other resources, by registering at textbooks.elsevier.com.

For readers of this text, additional materials such as forms, checklists, and spreadsheets discussed in the book are available at https://www.elsevier.com/books-and-journals/book-companion/9780128210550. In addition, the author maintains a book-related website with selected materials at www.engineeringdesignprojects.com.

ACKNOWLEDGMENTS

This book has benefited from constructive comments and suggestions from the following reviewers, as well as several additional anonymous reviewers. The author takes this opportunity to thank them for their contributions.

- Kamal Amin, Florida State University
- Samuel Bechara, Colorado State University
- Peter Childs, Imperial College
- Shabbir Choudhuri, Grand Valley State University
- Irena N. Ciobanescu Husanu, Drexel University
- Ed Cydzik, San Jose State University
- Dan Dolan, South Dakota School of Mines and Technology
- Jan Gou, University of Central Florida
- John Farris, Grand Valley State University
- Sebastien Feve, Iowa State University
- Alicia Jack, Haywood Community Collete
- Anthony D. Johnson, University of Huddersfield
- J. Carson Meredith, Georgia Tech
- Curtis O'Malley, New Mexico Institute of Mining and Technology
- Christopher Pung, Grand Valley State University
- Ramin Sedaghati, Concordia University
- Joshua D. Summers, Clemson University
- X. Jack Xin, Kansas State University
- Robert T. Balmer, University of Wisconsin–Milwaukee (Emeritus)
- Tariq Tashtoush, Texas A&M International University

The author would also like to thank Steve Merken, Senior Acquisitions Editor; Rukmani Krishnan, Project Manager; Ali Afzal-Khan and Chris Hockaday, Editorial Project Managers; and Valerie Koval, Copy Editor, for helping guide the project through to publication.

Engineering Design, Planning, and Management

CHAPTER 1

Design projects

Think-Plan-Do-Repeat

1.1 Introduction

The five key steps in design projects are to:

(1) Develop a clear detailed plan to reach the goal.
(2) Work to control cost and reduce uncertainty and risk.
(3) Continually monitor the progress of the plan and update when needed.
(4) Initiate, facilitate, and end activities.
(5) Assess.

Projects are finite activities that seek to develop new and improved tools, methods, products, and equipment. The Project Management Institute describes the typical phases of an engineering design project using the waterfall model shown in Fig. 1.1. The important theme of this model is that a plan is carefully created to address the customer needs and then used to control the project to a point on completion, called *closing*.

Much of the design project terminology is self-evident, while other terms require some explanation. The terms below appear repeatedly throughout this book and technical project work in general:

- Deliverables: things of value, such as:
  - Things: items, products, machines, prototypes, software, production facilities, documents, drawings;

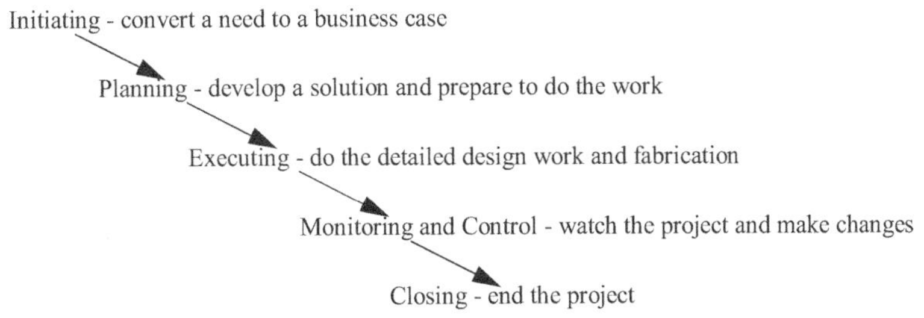

Initiating - convert a need to a business case

Planning - develop a solution and prepare to do the work

Executing - do the detailed design work and fabrication

Monitoring and Control - watch the project and make changes

Closing - end the project

FIGURE 1.1

The Project Management Institute waterfall model of projects.

Engineering Design, Planning, and Management. https://doi.org/10.1016/B978-0-12-821055-0.00001-3

    - Reports and analysis: key terms, parameters, state of the art, financial, test results, recommendations;
    - Plans: evaluation, maintenance, business, phase-out, renovation, implementation, specifications;
    - Procedures/methods: assessment, training, reporting, corrective action, documentation, process models;
    - Services.
- Tasks: work with definite outcomes:
    - Subtasks: a task that is subdivided into a number of smaller tasks;
    - Event: something that occurs to start and/or end a task;
    - Milestone: a major event at the start or end of a major task.
- Project management (PM): those items that go beyond the technical work of the project:
    - Manager: somebody appointed to help organization, planning, communication, and administration;
    - Budget: a planned amount of money, including contingencies;
    - Stakeholder: anybody with some direct or indirect interest in the project;
    - Customer or sponsor: the individual or group that is requesting the project work; a project sponsor may be a person in the same department, a paying customer, or a design that anticipates customer needs (note: customer and sponsor will be used interchangeably throughout the book);
    - Scope: a limit for what is, and is not, part of the project;
    - Resources: people, equipment, materials, money, and other things of value required for the tasks.

How do you eat a whale? One bite at a time.

PM is based on reducing the work to smaller and simpler tasks. The result is reduced complexity, greater coordination, better progress tracking, and a greater chance of success. Without the formal PM approaches, large projects collapse under the weight of details. A summary of design project activities is given in the following lists. As engineers we tend to focus on the technical and creative tasks. The professional, management, and strategic tasks are critical to the success of a project in a business sense. Technical task examples:

- implement, build, or fabricate
- test, analyze, certify, and troubleshoot
- document
- design for testing and manufacturing
- develop procedures for customer use, maintenance, and end-of-life disposal
- system-level integration and testing
- maintain quality standards

Creative task examples:

- need-driven problem solving
- design using creative and systematic problem-solving methods
- intangibles such as aesthetics, user interfaces, and personal preferences
- using competitive product design practices

Management task examples:

- gather and distribute resources
- maintain a plan to achieve the goals
- constantly assess progress
- revise the plan as needed
- coordinate the plan activities
- procurement/purchasing
- budget development and management

Professional task examples:

- people skills, including meetings, correspondence, and site visits
- critical thinking skills for decision-making
- managing employees and delegating
- communicating with stakeholders

Strategic task examples:

- develop needs, constraints, functional requirements, and customer specifications
- define a project scope, time frame, and the outcomes for all work
- risk management and decision-making

Fig. 1.2 shows a reasonable start to a project proposed by a customer, Ian. He identifies the need for the project, an entertaining gift. Ian suggests a cost specification and robot concept. At this point, Hugh begins thinking through the project plan and anticipating customer specifications for Ian. The

From: Ian B. Wantin
Date: Wed., Sep 3, 2015 at 4:24pm
To: Hugh Jack <hugh.jack@supplier.com>
Subject: Meeting to discuss a low cost robot souvenir

Hugh,

We were meeting today and the sales department said they would like to find a distinctive gift for our customers. The idea that seemed to get the most interest was a small robot kit that could be made for a few dollars. I would like to sit and talk about it. I am open Thursday morning and Friday 10am-1pm.

Ian

From: Hugh Jack
Date: Thur., Sep 4, 2015 at 7:24am
To: Ian B. Wantin <ian@customer.com>
Subject: Re: Meeting to discuss a low cost robot souvenir

Hi Ian,

We will drop over tomorrow at 11am.

Hugh

FIGURE 1.2

The robot project begins with an email exchange.

subsequent steps include concepts, technical specifications, and a quote for Ian. Hugh is also thinking of the detailed design and manufacturing options. Hugh wants to sell something that is easy to design, build, and deliver, hence, a small robot with a plastic body and inexpensive microcontroller.

PROBLEMS

1.1 How are project tasks and milestones related?

1.2 Which parts of the Project Management Institute waterfall model use a project plan?

1.3 Make a list of 10 different stakeholders in a project.

1.4 Develop a set of five questions that Hugh can ask Ian at their first meeting; see Fig. 1.2.

1.2 Projects and design

There are a number of models for the engineering design process, but they all contain the same basic activities, as illustrated in Fig. 1.3. During the needs and customer specification steps (1 and 2) the objective is to get a clear picture of the design problem. A very clear set of customer specifications focuses the design process, leading to clear and measurable success. The customer specifications drive the conceptual and technical specifications steps (3 and 4). The outcome of this is a technical specification that includes key components, algorithms, mechanical sketches, and process plans. The detailed design process (5) turns these into concrete plans that should satisfy the technical specifications and meet quality requirements. The build-and-test phase (6) varies by discipline, but should lead toward the final project deliverable. Meeting the customer and technical specifications is critical for project closure (7). This model suggests that the project work occurs in clean sequential steps. In practice a good designer thinks ahead to the harder tasks to reduce backtracking iterations and unnecessary problems. The phases in the model can define objectives and deadlines that clearly define the progress through project phases.

To illustrate the different possible interpretations of the design models, Fig. 1.4 shows an example of a civil or mechanical-structural engineering problem. The needs and customer specifications phases define a force loading case. The conceptual design phase is primarily selecting the member type and basic fittings. The detailed design phase involves selecting the size of the components and then doing an analysis for suitability. When the detailed design is complete, other companies can fabricate and install it. Mounting the structure to a building is the final test. The emphasis on various design steps varies with the type of design. For a company that is developing a variant of an existing product, the emphasis is on the later steps. For example, a medical device may need a multiyear testing process mandated by the US Food and Drug Administration (FDA).

From a business perspective, the design process is framed with contractual obligations and exchanges of money. Fig. 1.5 shows one such model for a new company. The company starts by developing the business plan, getting money, and organizing. The next task is to find a customer who is willing to issue a purchase order, essentially a legal contract to buy something. The company then goes about design work until there is something to deliver. Delivery of a project normally includes transferring deliverables to the customer, verification, and acceptance. After delivery, the supplier sends an invoice and the

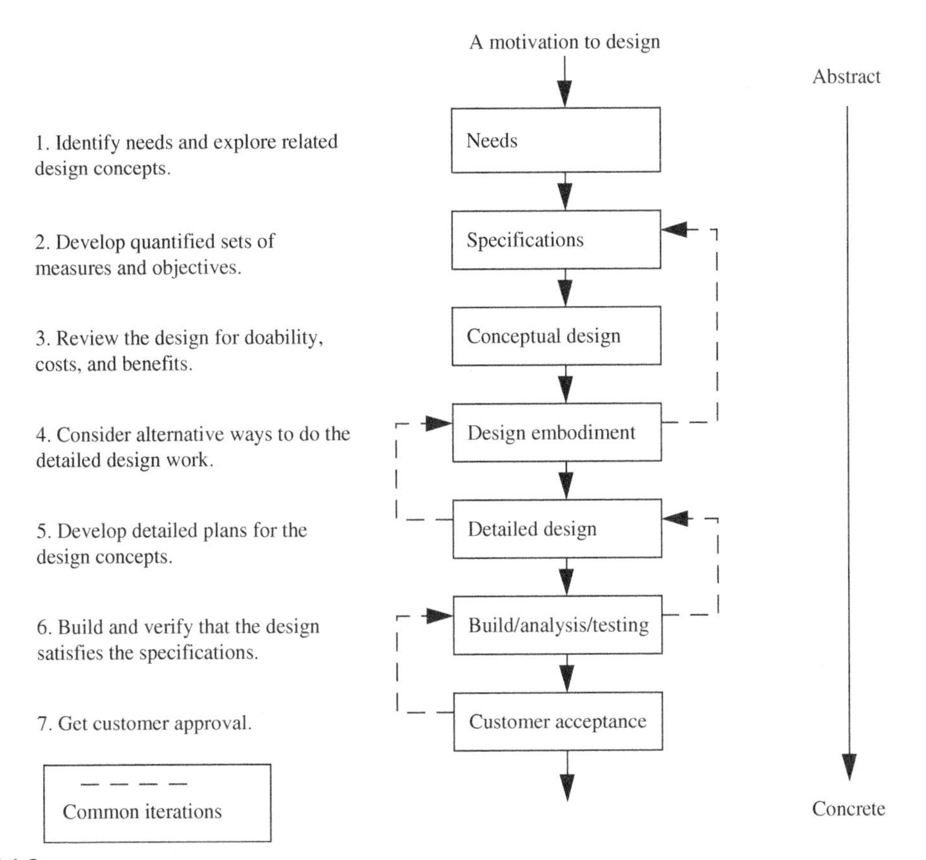

FIGURE 1.3

A seven-step model of project work.

customer responds with a payment. The number of people, and administrative steps, increases with the company size. A single entrepreneur may do all parts of the process. Doing project work without a specific customer is speculative, or "spec." In these projects, the designers make assumptions about the needs of the customer. Most consumer projects are "done on spec" using market studies and projected sales.

Smaller, or more entrepreneurial, design businesses prefer a less structured approach, as shown in Fig. 1.6. In this example there are three individuals or groups that take care of securing customers, do design work, and then build the project. The diagonal lines between tasks indicate that project phases may not end cleanly; they overlap. Overlapping tasks could include critical component selection while the customer specifications are still incomplete. This model allows fewer people to flexibly handle multiple phases of a design project. The counterpoint is that every design project is a special case, and the failure of one person can have an impact on multiple stages of the process.

The project scope is a document or clear understanding of a project's objectives, stakeholders, constraints, resources, etc. A practical project scope is a reasonable compromise between everything and nothing. A utopian project scope has no cost and satisfies all needs. A horrific project scope does not meet needs and incurs high costs. A practical project scope has a set of boundaries. At a minimum the scope should define the supplier resources and authority required to meet the customer needs.

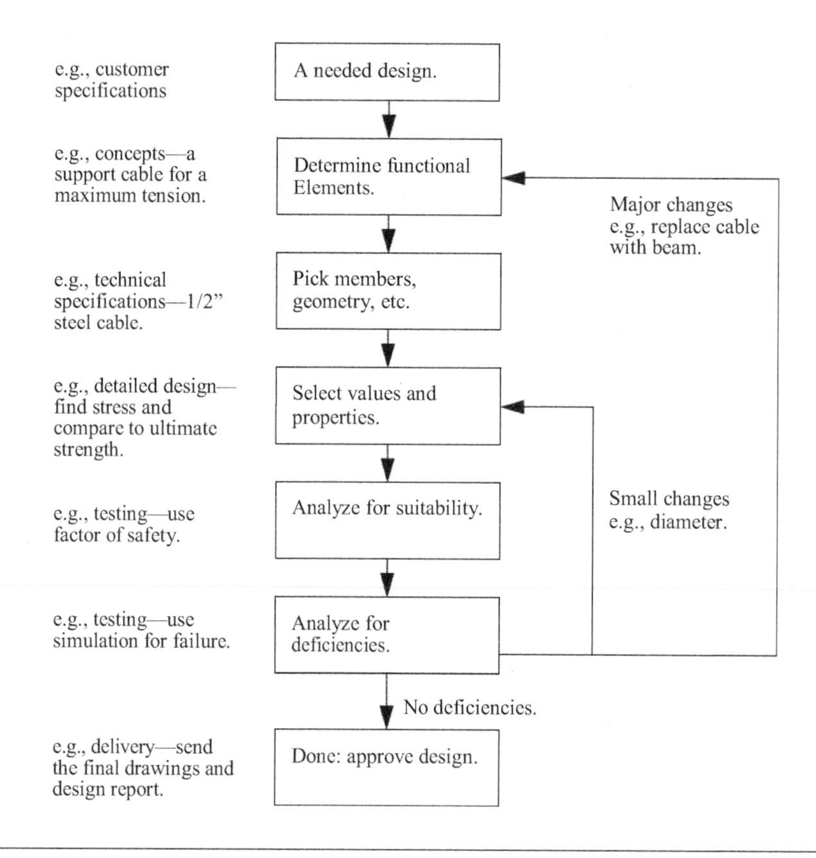

FIGURE 1.4

An example civil engineering design project.

The idealistic goals shown in the first of the following lists apply to generic projects. Although these are impractical and do not always agree with business needs, they are very useful when questioning assumptions and conventions.

A perfect design project:

- satisfies every need and solves every problem;
- assumes immediate and infinite supply;
- does not need maintenance, oversight, upkeep, or consumable supplies;
- does not require training or changes in behavior;
- does not require materials, labor, or equipment.

A poor design project:

- requires extreme manufacturing precision and complexity;
- involves untested solutions and methods;
- has excessive, unnecessary, or limited functionality;
- requires major effort and change by the user;
- has high costs and limited availability.

Returning to the robot design example, Hugh met with Ian and then sent a postmeeting email, Fig. 1.7. It is a normal practice to send a postmeeting summary to the other participants, to create a

Big Picture	Functions	Tasks and Documents
Approve business plan	Create business functions	Gather people, knowledge, resources, market info, advertising/sales material
	Project recruiting (sales)	Networking, cold calls, advertising, entrepreneurial
Issue quote and receive purchase order	Project screening	Early analysis of risk, financial, analysis of customer, match to mission/values
	Pre-quote work	Patents and competition
	Conceptual	Make case for incremental improvement (lower risk)
Approve spending	Detailed design	Set the team toward a detailed design and budget
	Order components	Materials list to purchasing, accounting ready to release budget
	Build	Rotate design into production schedule
	Test	Test labs and quality control
	Review	Sign-off meeting
Issue invoice	Deliver	Contractual acceptance and shipping
Service contract	Maintain	Service plan and warranty

FIGURE 1.5

A business view of the design process.

written record and correct any misunderstandings. Hugh's email outlines the project milestones and suggests a few more customer specifications. Looking ahead, the next project steps will be developing concepts and technical specifications.

The technical specifications will be used to create a price quote. The email includes some discussion of delivery, showing that Hugh was developing a project plan to estimate equipment needs and anticipating scheduling problems.

PROBLEMS

1.5 Which of the seven design steps answer the following questions: a) What should the design do? b) How can the design do it? c) What is the product? d) Is it acceptable?

1.6 The seven-step model shows backward loops for design improvement. Suggest methods for reducing the number of backward iterations.

1.7 The seven-step model shows very clear steps. Explain if it is possible to combine steps. Explain
if it is possible to skip steps.

1.8 Of the seven design steps, list those that are used in the processes of a) sales, b) quoting, c) receiving orders, d) delivery, and e) issuing invoices.

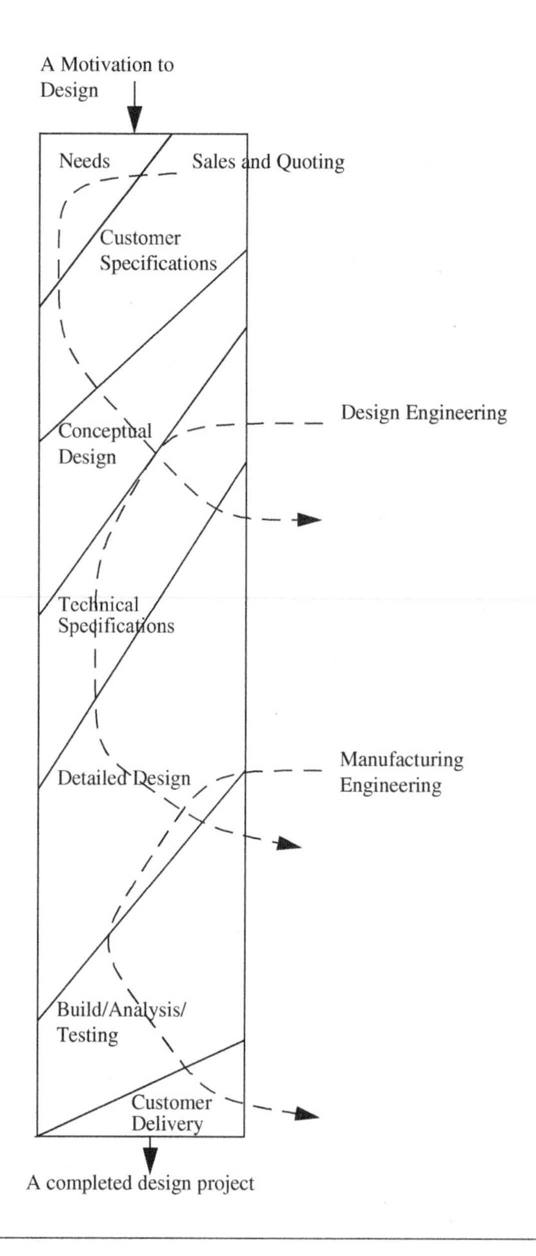

Speculation—The sales team finds a business opportunity and a quote is generated with some input from the design engineering team. The design team does not have a complete set of customer specifications, but has done some conceptual design and selected some major technical specifications.

Approval—The designers create a detailed design, write programs, order parts, create part drawings, etc. Eventually this process is shifted to the manufacturing or fabrication group.

Completion—The manufacturing group finally gets the last design work and builds the last parts. The first half of the order already been shipped to the customer.

FIGURE 1.6

An overlapping model of design projects.

1.3 Needs identification and customer specifications

If it is untestable or unprovable, do not put it in the specifications.

In a strict sense, we must know what a problem is before we can solve it. For design projects we require a clear reason for solving a problem, the needs, and a clear measure of when the solution is satisfactory,

From: Hugh Jack
Date: Mon., Sep 8, 2015 at 11:21am
To: Ian B. Wantin <ian@customer.com>
Subject: Re: Meeting to discuss a low cost robot souvenir

Hi Ian,

We have had a chance to discuss the project and I think we can do enough design work so that we can give you a quote in a couple of weeks. If you can get us a P.O. quickly we should be able to review the design with you before the end of October. If all goes well we could start production in December. It will take a while to deliver all 2000 units, but we should be able to give you 500 by January.

Hugh

FIGURE 1.7

A follow-up email to the customer.

the specifications. If you cannot understand the needs, it is almost impossible to develop a useful set of specifications. If any specification is vague and undefined, the end of the project is a moving target that may never be reached. The common mistake at this phase is to expedite the process by leaving a few things loosely defined, which you will "figure out later." Do not leave unresolved specifications. The first set of specifications is based on the customer input; we will call these the customer specifications. As the design moves forward these will be replaced by the technical specifications. In some cases, as with knowledgeable customers, the customer and technical specifications may be the same. Conceptual design is the critical step between customer and technical specifications.

Customer specifications are created using needs and wants. These often arise from a number of sources, including:

- incremental improvements to existing designs
- replacements for existing designs
- speculative design

Existing designs are the result of imperfect processes. Each design includes imperfect decisions that allow future improvement. A design in use may be adequate or need changes, providing a motivation for subsequent design projects. Speculative designs use an open-ended process of discovery for estimating design needs, as shown in Fig. 1.8. Many of the formal tools used in this process are within the domain of other disciplines, such as marketing and sales. Reduce the needs to a clear list of testable items, regardless of the source of the need. Develop the detailed technical specifications to satisfy the customer needs. Acceptance testing verifies that the end product meets the customer and technical specifications, signaling the end of the project. Unclear specifications often delay the project completion.

It can be helpful to draft the customer specifications by putting the needs and wants list on a spreadsheet with the intention of refining these to specific values later. Needs typically fall into one of three categories: (1) minimum needs, (2) assumed needs (wants), or (3) unrecognized needs/wants. Stated, or minimum, needs are standard, and wants are new features that the customer considers important. For a car, needs would mean seating and dashboard electronics. Assumed needs include basic operation, safety, and reliability. This would mean the ability to drive. Customers have unrecognized needs, such as voltage levels or compression ratios that would normally be captured in the technical specifications. Sometimes customer needs include technical specifications. Customers can provide numbers and details used directly, or indirectly, in customer specification development. Customers commonly suggest specification values that exceed the minimum required because they do not trust the design

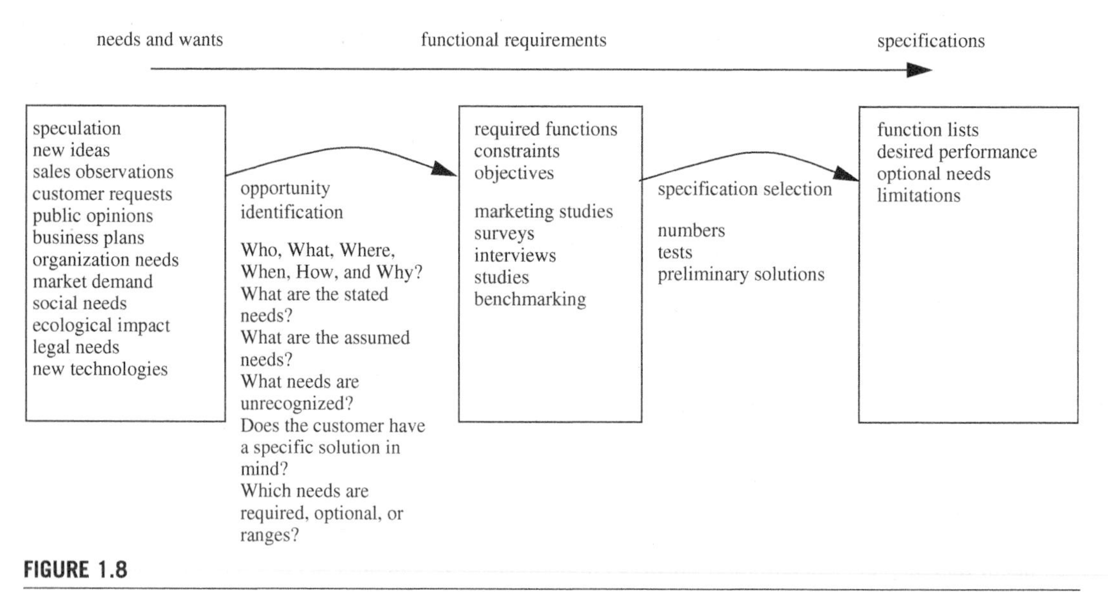

FIGURE 1.8

Needs clarify the problems.

process. However, without care these inflated values could become the requirements for the project. Consider inflated values to be wants.

PROBLEMS

1.9 What are the advantages and disadvantages of overlapping steps in the seven-step design model?

1.10 Is it important to enforce the end of each design step when they overlap?

1.11 Briefly describe the seven design process steps.

1.12 Do the steps in the design process need to be cleanly divided?

1.13 List the details now known for the toy robot design.

Compare the needs and customer specifications to verify that each need is addressed. Each of the needs should lead to a detailed customer and technical specification. Eliminate or modify customer specifications that do not address a need. For unquantifiable needs, the designers should ask, "How will I know when I have satisfied the need?" This will be very important when you want the customer to accept the design and pay for the work. The project expenses and effort will not end cleanly without a clear specification-based acceptance process. Frame the design details as (1) required, (2) constraints, and (3) objectives (wants). When the process is complete, each of these should result in a numerical value for specific acceptance tests. If it does not, the result should be a clear binary yes/no test. Vague items, like aesthetics, require some sort of test protocol for acceptance.

Customer wants are harder to resolve. It is common to have many more wants than is reasonable. It is the job of the sales/design team to eliminate, prioritize, or convert these to needs. When prioritizing wants, they can be combined into an objective function. The objective function will have an overall limit, like a cost, that helps increase or decrease the rank of various wants. Individual customer objectives can be quantified as a set of basic requirements and hopeful outcomes, subject to realistic constraints. Design objectives and functional requirements are what the technical specifications should strive for. It is best to define the ideal and acceptable values. The danger is that when not clearly defined, the design specifications will be below or above the requirements. Examples of well-defined objectives include the following:

- cost between $100.00 and $120.00
- provide a maximum power from 50 to 75 kW
- a maximum speed of 100–150 m/s
- heating time between 40 and 60 s

Design constraints are firm limits that must be met for the design to be acceptable. Examples of these are listed below. These can be specified by the customer or by the nature of the problem, limitations of the fabrication processes, or technology:

- must be delivered by June 15
- carcinogen levels below 10 ppm
- surface temperature below 50°C
- cycle times below 3 s
- power levels above 5 W
- a data transmission rate above 5 megabauds

The needs worksheet shown in Fig. 1.9 can be used to guide the customer and technical specification development process. These columns can be added to a specifications spreadsheet to consolidate documents. The best source of specification details is existing designs. When these don't exist, it may be necessary to do some early conceptual design. When converting needs to specification, approaches to consider include the following:

- Find similar designs from the past, present, or future.
- Examine each of the objectives and constraints and find methods for implementation.
- Identify components or OEM (original equipment manufacturer) modules.
- Find sources for standard components.
- Determine what capabilities exist for fabrication and skills.
- Ask, "Who can do things not possible in-house?"
- Meet with customers, users, or others.

Functional Requirements: Toy Robot

Needs	Ideas in the market	Patents	Easy or obvious solution	Alternatives	Difficulty
Cost < $6 basic motion avoid walls	toys < $2				

FIGURE 1.9

A needs worksheet example.

- Study target groups.
- Locate applicable standards, e.g., SAE, IEEE, ASTM, FCC, FDA, UL, CSA, CE, BIFMA, AGMA, JIC.
- Research regulations and governing bodies, e.g., NHTSA, FCC, FDA.
- Perform benchmarking with competitive designs.
- Identify technologically challenging features, e.g., Is it new? Have others used it? For how long?

The process of defining the specifications is critical to the success of the project. A clear set of specifications that can be measured and tested without ambiguity is ideal; anything less may lead to problems. The following are just some of the issues that can arise when the specifications are less than precise:

- "That is not what we asked for." The customer rejects the final design. This occurs when specific acceptance tests are not defined.
- "We asked for feature X." A customer asks for additional features after the design work begins. If not detailed in the customer specifications, it should not be added without renegotiation. If a feature is added after the customer/technical specifications are set, it is called a *feature creep.*
- "It doesn't meet quality standards." Unclear measurements can be contested and delay the final acceptance.
- "We are willing to compromise." To end projects sooner, the customer might be willing to compromise on agreed specifications.

For each of the specifications, you should consider the questions in the following list. If it is impossible to clarify a specification, then it must be managed carefully. A good strategy is to review the arbitrary specifications often with the customer and seek early approval, before the conclusion of the project. This is especially true for requirements such as aesthetics, user interfaces, manufacturing processes, reliability, and packaging. Consider the design of a new computer mouse. Some of the specifications are exact, while the "comfort" specification is hard to quantify. The designers should try to get the customer to give final approval for the "comfort" of the mouse early in the detailed design process. The following questions can be used to explore the specifications:

- Is the specification measurable?
- How will the specification be measured?
- Are there formal tests that can be used?
- How will the customer see that the design specifications will be met?
- How many ways can the specifications be interpreted?
- What are the failure criteria?
- Are the numerical values and ranges clear, with units?
- What are the tolerances?
- How will acceptance criteria be measured (safety, usability, etc.)?
- How can we manage feature creep?
- Can some of the specifications be made optional? If not, consider using a performance bonus.
- Are there standard tests available for a specification?

A sample spreadsheet for final specifications is shown in Fig. 1.10. It emphasizes a concrete list of specifications that will ultimately be approved by the project team and the customer. Once approved, the list will act as a contractual agreement about acceptable deliverables. As mentioned before, anything not clearly written will lead to confusion, disagreement, delays, and extra costs.

Specification	Required or optional	Value or yes/no	Units	Acceptance test or method	Accepted

FIGURE 1.10

Specification worksheet example.

Once the customer specifications are set, the design team can move forward with the selection of design concepts and detailed design. Ideally the following will apply: (1) the specifications are accepted by the project team and customer, (2) the project team will have a clear understanding of the work required, and (3) the customer will have a clear understanding of what they will receive. If anyone wants to change the customer specification, a formal process can be used to review the changes and adjust the time and budget as needed.

Failure to adhere to the customer specifications results in feature creep, a common and critical problem that occurs when specifications are poorly written or don't exist. After this phase, the addition or change of any specification will result in delays and increased costs. So regardless of how small a change is, it will normally take longer than required, add complexity to debugging and testing, and have unintended consequences. In other words, agreeing to add features is like giving away profit and delaying delivery. Look for a win−win situation in which the customer gets what they want in exchange for fair compensation.

The robot design example continues in Fig. 1.11, when Hugh sends a draft of the robot specifications to Ian. The customer specifications were developed by comparing robot kits available on the Internet. The specifications include two motors, rubber tires, tank steering, plastic body, domed top,

From: Hugh Jack
Date: Wed., Sep 10, 2015 at 12:57pm
To: Ian B. Wantin <ian@customer.com>
Subject: Re: Meeting to discuss a low cost robot souvenir

Hi Ian,

We have spent some time and think we have a reasonable solution. Give me a call and we can work through any details. At this point our target specifications are:
- cost $6.25
- two motors and rubber tires
- tank steering
- traces your logo but avoids the walls
- it has a plastic body with your logo on two sides.
- a domed top and round body with a 10-cm diameter.

Hugh

FIGURE 1.11

An email outlining the specifications for the customer.

and a 10-cm-diameter body. The vague specifications are to avoid the walls, to trace a logo, and to have logos printed on the sides. The specifications do not contain much detail and may lead to problems later. The customer might want "tasteful" logos, but Hugh envisions something large and bright. The logo it traces on the floor might be an out-of-date version. These would be excellent reasons to refuse delivery of the product. Listing the "target price" of $6.25 is a risky commitment before the technical specifications have been drafted.

PROBLEMS

1.14 Can a need and a specification be the same? Explain.

1.15 Should the customer needs and the supplier specifications overlap? Explain.

1.16 Specifications should be numerical and/or testable. Why?

1.17 Rewrite these customer needs as specifications:
- **(a)** a comfortable room temperature
- **(b)** will fit in a pocket
- **(c)** light enough to carry all day

1.18 What is the difference between a design objective and a constraint?

1.19 The government mandates that a class of recreational vehicles do not carry more than 10 L of methane. Is this an objective or constraint?

1.20 For office furniture locate the following:
- **(a)** standards and regulations
- **(b)** trade groups
- **(c)** ten major suppliers
- **(d)** international trade shows
- **(e)** Internet retailers
- **(f)** a local retailer with a showroom

1.21 A project can become stalled if the customer and the supplier don't agree on a specification definition. One strategy is to leave the description vague and return to it later. What are the advantages and disadvantages of this approach?

1.22 Should optional, that is not required, items be listed in the specifications?

1.23 How can a design team be encouraged to address optional specifications?

1.24 Use the specification research worksheet to develop a simple set of specifications for a coffee cup lid.

1.25 Develop a simple set of technical specifications for an ice cube.

1.4 Concept generation and technical specifications

Design work falls across a spectrum that can be loosely defined as having three categories: (1) standard, (2) evolutionary, and (3) revolutionary. Sorting the design problems into these categories allows more focused effort. The standard design problems do not require much or any conceptual design work. The evolutionary design problems will require some conceptual work with a practical focus. Revolutionary design allows the designer to set aside previous work. In any design there will be some combination of these. And, naturally, we want to begin by generating concepts for the revolutionary problems, hopefully breaking these into standard and evolutionary design problems. Evolutionary design concepts are then addressed to find ways to improve design components and, it is hoped, reduce everything to a standard design type. Once the design has been reduced to a set of standard design problems the detailed design work can begin:

- Revolutionary: The design is completely open ended. Concepts are system wide, e.g., number of wings.
- Evolutionary: An existing design is to be refined. Some new concepts are required, e.g., wing geometry.
- Standard: Well-understood approaches to design are used. The focus is on the details, .

Use the customer specifications to begin conceptual design. The technical specifications with obvious design solutions do not need conceptual design work and can be added directly to the technical specifications. The others will require some level of concept generation and selection. Naturally, this process may be iterated until all of the customer specifications have been addressed with standard design concepts. When there are choices, it is better to deal with the revolutionary design work first, as this tends to have an impact on more of the technical specifications. This should then be followed by more focused concepts for design components or subdesigns:

(1) Identify technical specifications that have single, or obvious, design solutions. These standard designs should not need any conceptual work.
(2) Look for specifications that have a few reasonable alternative designs. These evolutionary designs will require some investigation of concepts and eventual selection.
(3) If there are specifications that do not have a clear design approach, new concepts will need to be generated and then one selected.
(4) Repeat this process until all of the technical specifications are addressed.

A reasonable approach to concept generation and selection is shown in Fig. 1.12. To generate concepts, we start with the customer specifications and prior knowledge. The outcome of this process should be a set of concepts. Depending on the generation method, these may range from overconstrained to wildly improbable.

The selection process is normally applied after generation, so that the generation process can work outside conventional wisdom. In selection, the unreasonable concepts are eliminated, and then we rank the remaining.

Generating concepts is a fun process that allows for a measure of creativity. It is a chance to think outside the box for evolutionary designs and a chance to create a new box for revolutionary designs.

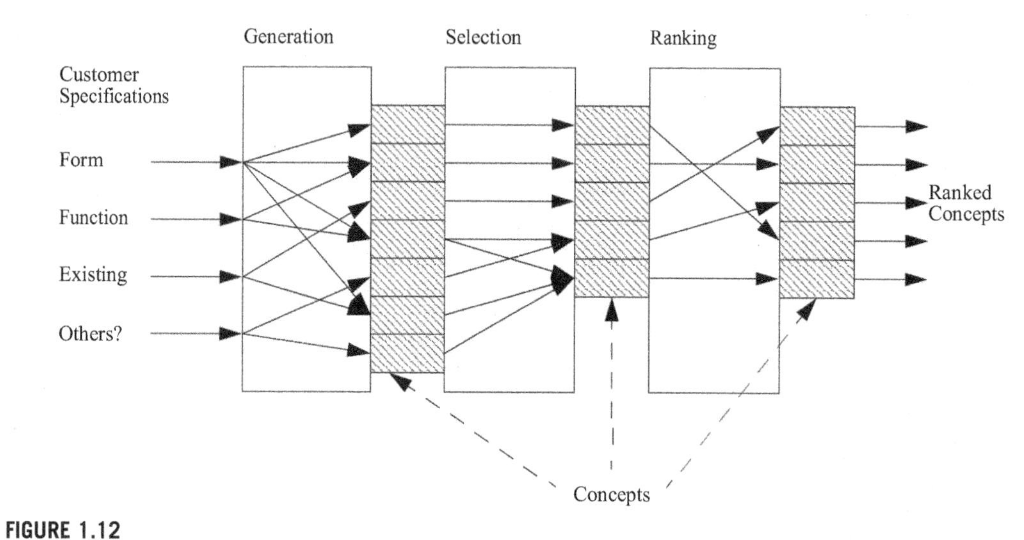

FIGURE 1.12

Mapping customer specifications to concepts.

The focus is on creating new concepts using knowledge we already have. In the absence of inherent creativity, a few of the many techniques are as follows:

- Brainstorming: Generate many concepts at high and low levels.
- Prototyping: Build physical and virtual models to test the unknown approaches in the design.
- Pilot testing: Run limited tests of a method, procedure, or product for assessment.
- Simulation: This allows basic trial-and-error "what-ifs" without physical construction.
- Axiomatic: Use general design rules to generate new concepts.
- Derivation: Extending or reapplying existing design solutions.
- Survey: Talk to people to get their ideas.

After the concepts have been generated, it is necessary to screen them for viability. Some concepts are easily eliminated, such as nuclear-powered mousetraps. Likewise, some concepts will have obvious merit. Concepts in the gray zone will require better analysis. It is a major point of concern if this stage generates only a single viable solution because (1) one solution means there is no backup, and (2) good ideas have been excluded. Typical questions to ask yourself are:

- Is the basic physics/chemistry/technology sound?
- Do the engineering approaches exist?
- Could the concept meet the cost and time budget?
- Is the concept consistent with other parts of the design?
- How well will it satisfy the customer specifications?

To rank the concepts, we should objectively assess how well each addresses the customer specifications and at what cost. This is also the point at which the objective functions for the wants come into play. Consider that the best set of concepts will satisfy all of the customer specifications at no cost. Unless there is a bonus for exceeding the customer specifications, the ranking often focuses on reducing the project budget and time. The best situation is to have many excellent alternatives. This indicates a greater chance of success and alternatives if problems arise.

When the conceptual design is complete, there should be a set of standard design problems to address. If any conceptual design problems remain beyond this stage, then they indicate uncertainty about a solution. This means that a solution may not be found or may not be sufficient later. If this occurs, it may be necessary to backtrack and make larger changes to the design. In other words, if you move forward without ending the conceptual design, you are very likely to add variability and risk to the project. If it is absolutely necessary, find a way to split the project into completely independent subprojects. Obviously, this will require additional planning and tracking for the project. This stage should produce deliverable items that define the detailed design problems. Examples of these are:

- system block diagrams
- critical component lists or bills of materials
- software architecture, data structures, pseudocode, high-level code
- mechanical sketches, frame designs, component layouts
- process flow diagrams, plant layouts, material flows

The major mistakes made in this process are (1) rushing through to "get to the design work," (2) finding only a few or a single viable design, (3) leaving customer specifications unaddressed, (4) generating poor designs to justify another, (5) taking design decisions personally, (6) not receiving approval for the design concepts before moving to detailed design.

PROBLEMS

1.26 What is the difference between revolutionary and incremental (evolutionary) design?

1.27 Are concepts required for all parts of the design? Explain.

1.28 Concept generation must refer to the customer specifications and indirectly to the needs. Some specifications have obvious solutions and do not need new concepts. Do these obvious specifications need to be considered during concept generation?

1.29 If there are three alternative concepts, does each one need to satisfy the same customer specifications? Explain.

1.30 After concept selection, some customer specifications may not be addressed. How should this be resolved?

1.31 (a) Develop a set of 10 concepts for moving water between two containers. b) Eliminate unreasonable concepts. If there are fewer than five, then generate additional concepts. c) Order the concepts from best to poorest.

1.32 How can a large system concept include lesser concepts for subsystems and components?

1.33 Develop a list of materials for the toy robot design problem. The major parts include the plastic housing, wheels, dome, motors, and controller.

1.5 Detailed design

At the end of detailed design, we should be ready to build and test the system. Typically, the detailed design phase ends with a formal agreement between the customer and the design team. The design team uses the outcome of the conceptual design process as the basis for the technical specifications and detailed design phases. Once the technical specifications are approved, it is assumed that there will be very few changes. The project team uses the inputs, outputs, and controls of the detail design process, as shown in Fig. 1.13. At the end of detailed design the team will have devices ready to produce and test, often in the form of drawings, parts lists, and work orders. The detailed design process is one of refining the design until it is buildable. During the detailed design stage, it is normal for the design team to involve suppliers, customers, and the manufacturing department.

Complexity increases problem-solving difficulty and slows progress. What we want is a set of very small, easy-to-do steps. Therefore the general approach to detailed design is to (1) reduce problems to standard approaches, including free body diagrams, schematics, power, and architecture; (2) solve the problems; (3) integrate the solutions; and (4) resolve the differences. Each engineering discipline has tools for simplifying and solving standard technical problems. The goal for the detailed design process is approval of a detailed design that will meet the specifications. A good approach is to have design review meetings within the project team and with the customer. The desired outcome is an unconditional approval to continue. However, minor and major design changes are normally required. The key to concluding the detailed design phase is to push for one or more of the following decisions:

- agreement that the design meets the technical specifications;
- approval to release the budget;
- a list of items to be addressed, including required and optional categories;
- an agreement for final delivery date;
- recognition of risks and backup plans;
- if necessary, agreement on any revisions to the technical specifications.

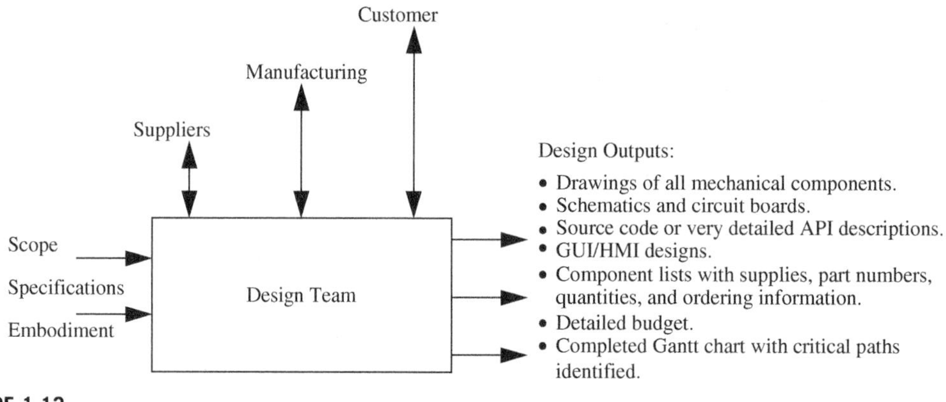

FIGURE 1.13

A black box model of product design. *API*, application programming interface; *GUI*, graphical user interface; *HMI*, human–machine interface.

Some of the mistakes that occur during this phase include (1) leaving the difficult or time-consuming work for later, (2) delaying decisions until "the solution becomes clear," (3) poorly documenting design work, (4) assuming that things will work without preliminary testing, (5) unnecessarily exceeding the customer and technical specifications, and (6) not recognizing and planning for risks.

PROBLEMS

1.34 Breaking one large design problem into multiple problems appears to increase the total work. How do the smaller problems change the complexity?

1.35 What is the relationship between the specifications and the design?

1.36 Is detailed design a process of creating new concepts?

1.6 Building and testing

In most circumstances the designer and the builder are not the same person. The detailed design documents must clearly communicate the design intent. Others can easily produce a design that is correctly documented. For example, an electrical engineer might design a complex surface-mount PCB assembly that is built in another country. A mechanical engineer might design a metal frame that is welded by somebody at another plant. Even when the design team will build and test the device, the detailed design should be well documented. Relationships between designers and fabricators are as follows:

- Separated: One group designs and the other fabricates.
- Oversight: The group that designs oversees the construction.
- Combined: The same group is responsible for design and fabrication.

During the build and test phases, problems will always arise. These should be dealt with as soon as possible, including updating the detailed designs. An approach that will simplify this build–test process is to iteratively build a piece, test the piece, and then integrate the piece. This repeated cycle finds small problems before they become large problems and provides an early opportunity to revise the detailed design or project plan. When testing occurs less often it is much harder to find a problem in a larger system. In a good build-and-test approach the final test is a formality. The best approach is as shown in the following list:

(1) Identify a small system piece to build and test.
(2) Gather needed resources and parts.
(3) Build the piece.
(4) Test the piece.
(5) Revise the piece if it doesn't meet the technical specifications.
(6) Integrate the piece into the larger system and test the integrated system.
(7) Review the system level design if needed.
(8) Go back to step 1, until the build-and-testing work is done.

At the conclusion of the build–test stage, the result should be a design that is accepted by the customer. This will occur if the design has been satisfactorily tested to meet the technical and customer specifications. Anything less will require negotiation and the possibility of rejection. Ideally, the customer has been involved and knows exactly what to expect at the acceptance meeting. It is advisable to have the customer visit for midpoint reviews during building and testing. These extra reviews avoid "surprises" at the end. Surprises may be fun for parties, but they are not fun in a technical environment. The final acceptance for a project should strive for one of the following outcomes, even if it is negative:

- Accept: The project outcomes are acceptable.
- Conditional: A limited set of changes is required.
- Reject: The customer is not willing to accept the design without major changes.
- Failure: The customer will not continue the project.

Each customer, industry, and sector handles project acceptance differently. A very relaxed customer may simply accept your assurances. Others may require exhaustive testing before approval. The various testing options include the following:

- Accept delivery without testing.
- Use an independent testing lab.
- Conduct a simple usage trial.
- Perform exhaustive testing.
- Use a product in service for a period of time.
- Use a holdback, whereby a portion of the final payment is kept for a period of time.

Problems that occur during this phase are (1) the design has not been tested against the specifications, (2) changes have been made but not documented, (3) changes have been made without telling the customer, (4) the project budget or schedule has not been revised to reflect changes, (5) the build-and-test results have been hidden as a surprise for the customer, and (6) details have not been provided to "meddling" customers.

PROBLEMS

1.37 List three build problems that might occur if a design is missing details.

1.38 How do the specifications relate to predelivery testing procedures?

1.39 What is the objective for testing?

1.7 Project closure

The project closure process starts once a customer has accepted the deliverables. Essentially it is the conclusion of all the contractual obligations and preparation to move on to other projects. In simple terms, the customer should have everything they need (a successful design) and the supplier should have everything they need (payment and profit). Project closure will often contain one or more of the following elements:

- delivery
- installation
- formal approval documents
- additional testing
- certifications
- maintenance plans:
  - software patches and updates
  - recalls for consumer products
  - regular maintenance of production, calibration, testing, etc.
  - safety inspections
- documentation
- training

When done, the project team is disbanded and assigned to other projects, all of the financial accounts are settled and closed, and the managers take the project off the regular review lists.

PROBLEMS

1.40 List three activities a detailed designer could perform at the project closure.

1.8 Project planning and management

Pick the task you don't want to do, and do it first.

The old saying is, "They won the battle, but lost the war." The same relationship exists with project details and progress. On a daily basis, designers become engrossed in specific activities. These specific successes contribute to the project progress. However, the project may be failing because of neglected details and tasks. A poor manager may not even be aware that the project is failing. A planning process is used in advance to choose activities that will reach the project goals. The plan will include task start and end times so that the team will "do the right things at the right time." Necessary resources are identified and scheduled so that they are available when needed.

The project manager is responsible for creating, tracking, and revising the plan. However, the process of planning should include multiple groups and people, typically called stakeholders, or sometimes the audience. The stakeholders should be involved with creating the plan, be participants in tracking the project, and assess the outcomes. The project manager should develop a plan that indicates when each of the stakeholders should be invited to participate, asked to make decisions, or assess the outputs. Stakeholders can be identified as those who:

- have something to gain, or lose; e.g., money, time, resources, etc.;
- may, or will, participate; e.g., employee, bank, customer, government, management, etc.;
- have an interest in the past, present, and/or future.

The planning process has multiple steps, including (1) breaking the project into a set of tasks, (2) developing a list of approvals and milestones, (3) itemizing required resources, and (4) checking the plan for consistency.

The plan often includes a schedule for tasks and time, a budget for money, and a communication list. Fig. 1.14 shows the first and second drafts of a project timeline. In this example, the process begins with a request for quotes from a customer. The completion date remains the same for the initial and revised project plans, but delays in the quoting process and purchase order from the customer have compressed the remainder of the project activities. The time available for design and shipping remains the same, but the purchasing, production, testing, redesign, and rebuilding time is compressed. This plan reveals a problem whereby a schedule delay early in the project could have an impact on all of the groups doing succeeding tasks.

Once the plan is approved, the project can move forward. Management is the process of using the plan to initiate actions, track the progress, adjust the plan as needed, and take actions to keep the project on track. In simple terms, a cycle of continuous improvement follows a simple sequence of planning–action–evaluation–assessment of objectives, as illustrated in Fig. 1.15. Project resources and tasks are tracked formally with the objective of achieving project success. The best project objectives are to complete the project on time and on budget, or maybe a little sooner at a slightly lower cost. Major variations in the schedule and budget, even if good, indicate planning problems. If the progress of each project step is not completely clear, the project can be said to be "out of control."

Every project has risks. The risks will be the greatest at the beginning of the project, but these should be reduced to negligible levels for final delivery. A good manager will be able to identify a risk and minimize it. Sample risk factors include:

- technical limitations
- time constraints
- resource/people constraints
- purchasing limitations

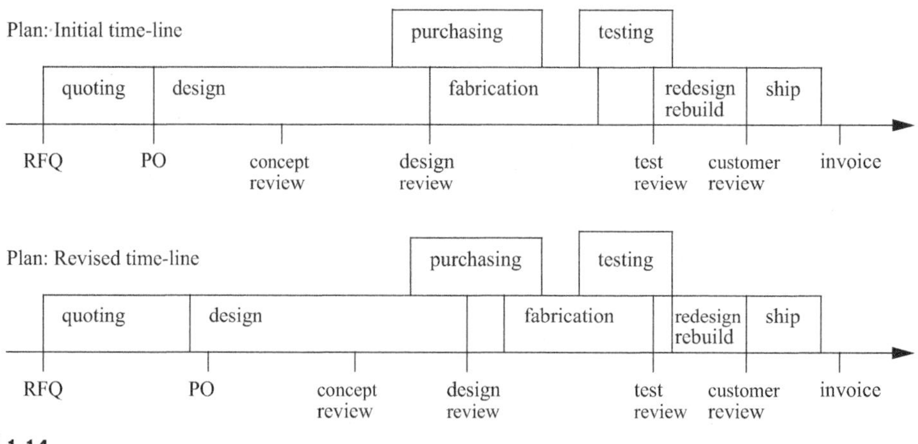

FIGURE 1.14

Project timeline planning. *PO*, purchase order; *RFQ*, request for quotes.

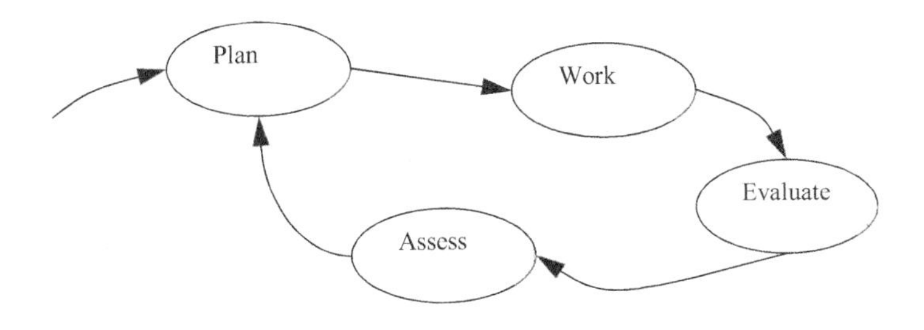

FIGURE 1.15

Another work approach similar to think–plan–do–assess.

PROBLEMS

1.41 Each meal at a restaurant is a small project. List four project stakeholders.

1.42 What would happen if a project did not have a planned timeline?

1.43 Why do some project activities overlap?

1.44 Plans are developed using estimates. How should a plan accommodate the differences between actual and planned details?

1.45 What is the difference between evaluation and assessment?

1.46 Is it reasonable to say that without assessment we would repeat the same mistakes? Explain.

1.47 Is a project manager responsible for doing daily project tracking or replanning? Explain.

1.9 Project problems and disasters

Problematic design project plans are illustrated in Fig. 1.16. With careful planning and assessment these problems can be avoided. When these problems occur, there are a few possible effects, including (1) mild, with no substantial impact on the project, but some extra work and/or planning required; (2)

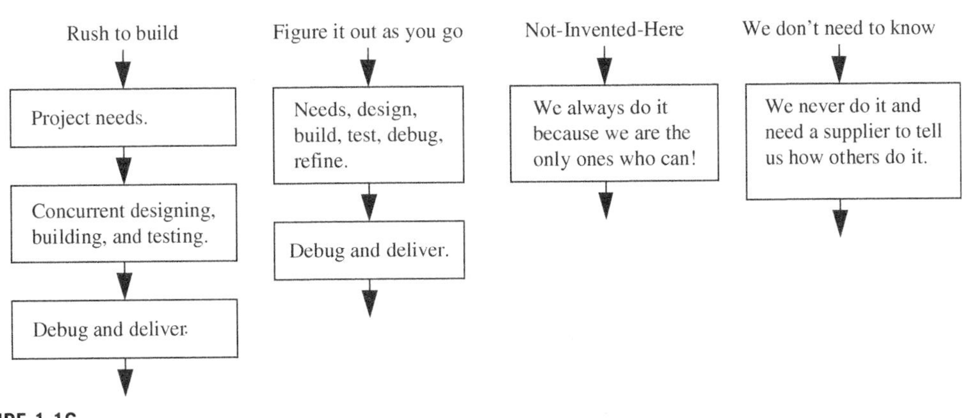

FIGURE 1.16

Some troublesome project approaches.

minor, requiring a major change to one or more tasks and the plan; or (3) major, putting the project at risk of failure and requiring substantial effort to correct. Some possible disaster scenarios are as follows:

- *Rush to build.* Movie and television characters often skip the design work and focus on the building and testing. This hides all of that "unexciting thought, design, and planning." Early on, the project seems to make progress, but slows quickly once the neglected details emerge. At the conclusion, there is always a mismatch between customer and supplier expectations.
- *Figure it out as you go.* This approach really helps move forward quickly, although a substantial amount of time is spent refining the original motivations for the project. Goals often start out as lofty, but eventually the hard tasks are abandoned. This approach is great for personal hobbies and "exploration." These projects are difficult to assess because the lack of an original goal means that the goal is adjusted to match what was done.
- *Not invented here.* Overconfidence results in trying to do everything yourself. When you have the expertise, and ample time, this can be fine.
- *We don't need to know.* A complete lack of confidence results in a complete dependence on outside suppliers. Sometimes this will work, but other times there will be problems that cannot be corrected. When you do not have enough knowledge to assess a supplier's products, you need to invest time to learn enough to make decisions.

Project problems will happen. With experience these will be easier to predict and control. Although you may not be able to deal with every problem, such as the loss of a key employee, awareness will give more control.

More elaborate methods for identifying, and minimizing, project risks are provided in following chapters. The following list outlines some common problems:

- People issues
  - Loss of trust or loyalty between the project team and customer
  - Lack of interest or too much interest
  - Key people are removed, to service other projects
  - Serving two masters: split loyalties and rewards
- Technical
  - Unexpected failures
  - Technology limits are reached
  - Tasks take much longer than expected
  - A new technology does not work as described by the supplier
  - A major flaw is exposed in testing
- Business
  - Budget and schedule setbacks
  - Loss of key people
  - No control of project
  - Goals and milestones missed
  - Unclear or inconsistent business practices, objectives, or structures
  - Business goals change, making a project irrelevant or more important, e.g., a change in oil prices
  - Price and supply fluctuations, e.g., agricultural supplies may drop if there is a frost

- Scope
  - Project specifications are added or expanded, also called feature creep
  - Project specifications are modified or replaced
  - The deliverables don't match the customer expectations
  - Project resources are reduced without a reduction in deliverables

PROBLEMS

1.48 It can be frustrating to design for a few months before touching any parts. List three excuses that could be used to start building earlier.

1.49 When do engineers use trial and error in design?

1.50 How could a feeling of ownership of ideas create team problems?

1.51 Provide an example of feature creep.

1.52 Is it better to be a technical optimist or pessimist?

1.53 What can be done if business changes eliminate the need for a project?

1.10 Businesses

In practice, a company operates as a collection of individuals working together. To make these people effective, some sort of coordination of activity and motivation is required. In a very small company this could be an owner working with a couple of employees. In a larger company there are more alternatives, but the hierarchical structure has become ubiquitous. In such organizations, the business uses a high-level mission and vision for coordination. The mission and vision are then shared, from the chief executive officer out to the hourly employees.

A functional organization focuses on specialized departments (see Fig. 1.17 for a sample organization chart). Senior executives direct company-wide projects by coordinating functional managers. In this example, the marketing, design, and manufacturing departments would need to cooperate. This organization structure is larger but more efficient because of the functional specialization. However, this organization is not well suited to overlapping project phases. In this structure, project work is completed by a department and then "thrown over the wall" to the next group. This "over the wall" approach allows departments to focus on each task, with minimal distractions. But this approach results in many decisions that generate inefficiencies in other departments. For example, the design department may specify a part coated in gold. If the part was coated with silver, it could be fabricated in-house at a much lower cost. In this structure the choices are to (1) make the gold part at a higher cost or (2) have the design department change the design. If the part were not sent back to design, they would continue to specify gold-coated parts.

A project-oriented structure, like that in Fig. 1.18, can appear to be multiple smaller companies inside a larger structure. This organization can develop substructures that specialize and/or compete. The risk is inefficiency resulting from duplicated and uncoordinated capabilities. Examples include large automakers that have separate design and manufacturing divisions for each major vehicle line. The benefit of these groups is that they share expertise from multiple departments. For example, the design group can involve the manufacturing group in a decision on part coatings during the detailed design. So instead of specifying an expensive gold coating, they can specify silver and eliminate unnecessary iterations and cost.

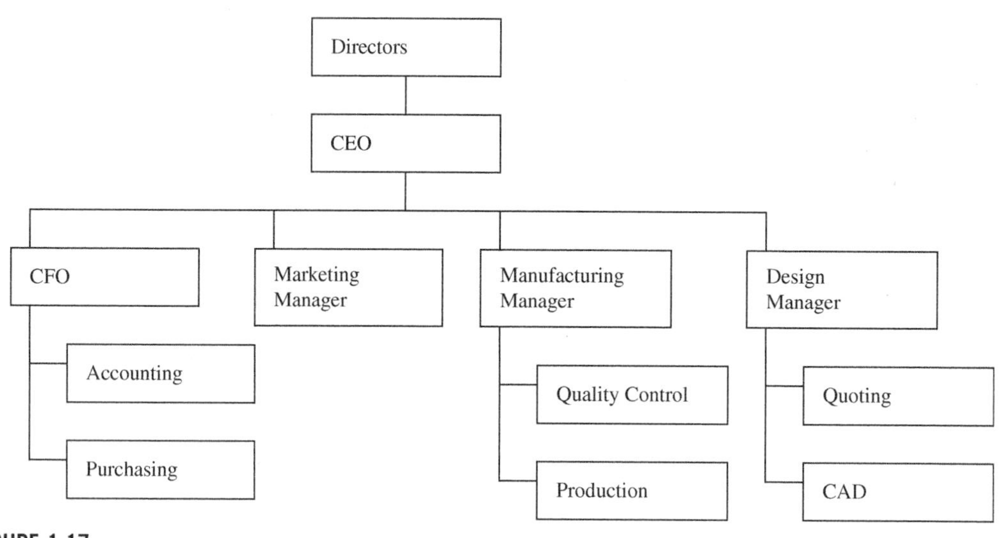

FIGURE 1.17

Example organization chart for a corporate structure. *CAD*, computer-aided design; *CEO*, chief executive officer; *CFO*, chief financial officer.

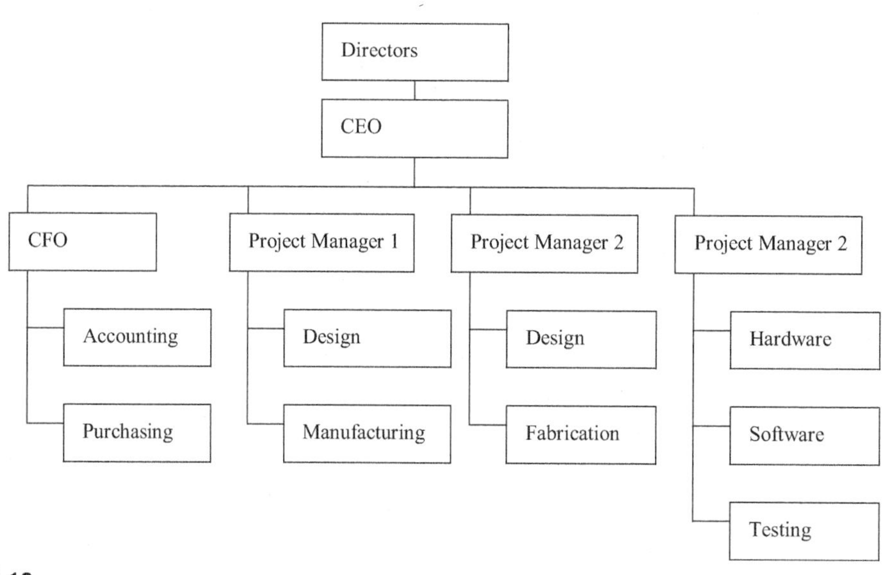

FIGURE 1.18

Example of a project-oriented organization chart. *CEO*, chief executive officer; *CFO*, chief financial officer.

In a matrix organization, like that in Fig. 1.19, the project managers are independent of the functional managers. Employees answer to both the functional manager and the project manager. Needless to say, without care this can lead to conflicting requests, priorities, and evaluations for employees. This

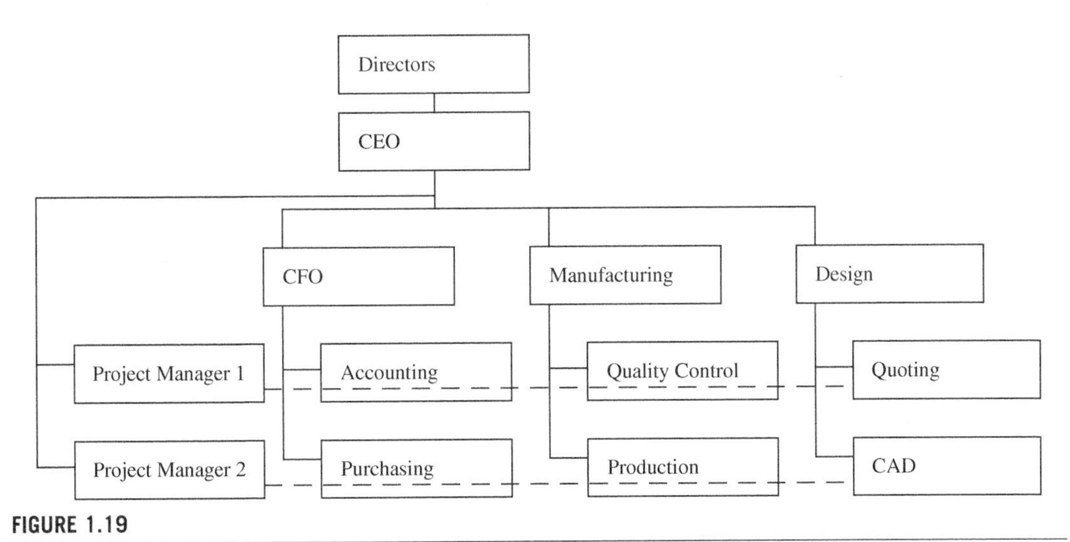

FIGURE 1.19

Example of a project matrix organization chart. *CAD*, computer-aided design; *CEO*, chief executive officer; *CFO*, chief financial officer.

organization structure allows company-wide projects that efficiently share the resources. The cost is that employee expectations and activities must be more carefully controlled. An example of this organization is a construction company. The project managers oversee single building construction but they share the corporate resources for doing concrete, steel, electrical, and excavation work.

Businesses approach design projects differently, as shown in the following list. Technical craftsmanship is used for priority when setting technical specifications and detailed design. For example, a family car design might emphasize durability, quantity, and cost. A sports car might emphasize performance, quality, and precision. Markets determine how the customer needs are addressed in initial design and eventual use of the product. For example, a fast-food restaurant company would emphasize commodity, service, end products, and stability. A roller coaster maker would emphasize innovation, service, equipment, and expansion. The business type determines the complexity and size of projects and market reach available to the company:

- Technical craftsmanship
  - Performance or durability: higher performance with service or longer life with less performance
  - Quantity or quality: large volume production or fewer products with individual attention
  - Cost or precision: lower precision costs less
- Market focus
  - Innovation or commodity: the rate of new feature introduction relative to competitors
  - Service or distribution: service-oriented companies work directly with the customers, distribution companies delegate the role to resellers
  - Parts, equipment, or end products: parts are sold to other companies, equipment is used by other companies, and end products are used by consumers
  - Expansion or stability: stability is growing a market share, while expanding is entering new markets
- Business type
  - Entrepreneurial: born small and grows (tolerates high risk, less to lose, more to gain)
  - Small: a few employees

- Medium: tens or hundreds of employees
- Large: hundreds or thousands of employees
- National: multiple locations in a country
- Multinational: a worldwide presence (conservative and stable, less to gain, more to lose)
- Conglomerate: a large collection of related company types that are under the same corporate body

PROBLEMS

1.54 Find an organization chart for your employer or school.

1.55 In an organization chart the financial and manufacturing functions are always separated. What does this mean?

1.56 Apply the PM organization chart to a restaurant with a grilling, deep-frying, and fresh food section.

1.57 If a project matrix organization has three managers, how would any technical department expect to divide its time?

1.58 Consider five major electronics companies and rank them for a) sales versus technical approach, b) cost versus precision, c) innovation versus commodity.

1.11 Decision-making

You can do it, but should you?

In 1986, engineers at Morton-Thiokol were asked to approve an overdue launch for the space shuttle.

Initially, the team turned down the request because the solid fuel rocket engines were not rated for the freezing temperatures. There was heavy pressure to approve the launch and the company eventually gave in and agreed to approve the launch. One of the O rings in the engines did not deform and seal; it failed, and the resulting flame caused the vehicle to explode. This resulted in lost lives and shut down the program for years. In hindsight, the team should have stuck with their original decision, even if it was very disappointing. Similar problems occur daily in project work. Decisions should always be categorized as (1) reject, (2) delay, (3) modify, or (4) accept. Examples of design project decisions include the following:

- the addition of a new design feature
- decreasing the cost or time to delivery
- acceptance of designs or test results
- requests for additional meetings and appointments
- competing suppliers offering similar parts
- enough time to do only one of two projects
- multiple concepts to produce the same design
- submitting a bid for a project that carries higher risk and lower benefit

When the benefits clearly outweigh the costs, it is often easy to say "yes" (Fig. 1.20). When the benefits are the same, or less than the costs, a negative decision is more suitable. When the benefits or costs are not clear, it is better to decline the decision, delay it, or reject it outright. If possible, the decision may be modified to increase the benefits and/or decrease the costs. Sometimes, high benefit decisions will be rejected because the risk is too high, such as bankruptcy. The most important benefit in most decisions is money. The best projects will use existing abilities and resources, satisfying the customer. When a project team chooses a project, they commit resources, time, equipment, people, and more. These resources will not be available for other projects that may have greater benefits; this is known as the *opportunity cost*. As the project team goes through the project stages, there are risks that problems may arise, increasing the cost of the project. If a risk leads to some sort of problem, there will be an associated cost. Only a few of the possible risks will develop into problems during any project. Adding a contingency fund, a form of internal insurance, to each project will increase the project cost, but it will also produce a more accurate project budget.

Accept: Benefits > Costs
Reject: Costs > Benefits
Reject: There is a "Killer"
Delay or Reject: Benefits > ?? < Costs
Modify: Benefits close to Costs

Benefits may include:

Financial income
Use of existing capabilities
Supports company mission
Uses free resources
Uses current expertise
Opportunities
New skills and knowledge
Reputation and brand value
New customers
Builds customer loyalty and trust
New or expanded markets

Costs may include:

Financial expenses
Opportunity cost for money, people, and resources
Company resources:
Employee time, knowledge, and skills
Equipment and materials
Marketing and sales time
Support from management
Risk
Company health
Brand value
Unclear objectives
Customer relationship style
Unreasonable project expectations
Requires new skills and equipment
Depending on a single customer
Long projects and market changes
Competitors
Customer stability
New technology is needed
Legal, regulation, standards, government
Insurance

Killers:

Not enough resources or time available
Risk is too high and failure would be catastrophic
Does not meet project objectives
Technical failure is very likely
Component availability and delays are uncertain
The customer or project definition is unreliable and poorly defined

FIGURE 1.20

Cost versus benefit factors.

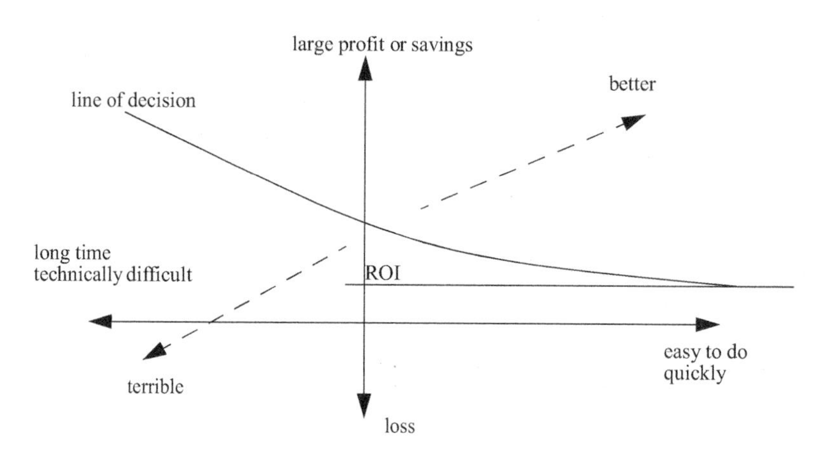

FIGURE 1.21

Project decision space. *ROI*, return on investment.

Change the task to suit your methods, not the other way around.

A simplified decision space is provided in Fig. 1.21. The best projects result in a large financial benefit and can be done quickly and with little effort. The worst projects are long, complicated, and lead to loss. Given that companies and people need some sort of motivation, there should be benefit in each decision. In the figure, the line of decision is a reasonable relationship. At the lowest level the project should be at or above a *return on investment* (ROI) as dictated by management. For example, an ROI of 15% would generally mean that benefit $= 1.15 \times$ cost. To do more intensive, or longer duration, projects there should be greater benefits to offset the risks.

Poor decisions will be presented as urgent; meet haste with delay.

When making urgent decisions, there are a few simple ways to understand the options. In particular, consider "who, what, where, when, and how?" with respect to money, time, resources, people, and knowledge. If you can imagine a clear path to the outcome of the decision, then you have enough knowledge to decide. If you cannot picture the outcomes, then ask for more information or delay until you can weigh the value of the decision. Recognize that your intuition and personal feelings are involved in hasty decisions. If you must make hasty decisions use a "devil's advocate" approach to compare your options. For example, an optimist might say the decision will "help you move forward faster," but the pessimist might say it would "help you to start losing money sooner."

PROBLEMS

1.59 Weigh the benefits and costs of making and freezing 20 sandwiches for daily work lunches for the next 4 weeks.
- **(a)** List five obvious costs and benefits.
- **(b)** List at least one killing factor.
- **(c)** For each cost and killing issue, provide a strategy to eliminate it or convert it to a benefit.

1.60 When should a project decision not produce a positive benefit or profit?

1.61 Mini-case: "I thought I asked …"

Bev was reviewing her daily email and found a request from her assistant Mario that read, "We want to have a department party and want you to select a date." Bev replied, "We don't have anything scheduled on June 4 and 5." The following day an email was sent to the division that said "Please join Bev's group for a summer barbecue on June 4." The date arrived, the party was held, and everybody returned to work.

Three weeks later, Bev was reviewing the expenses and saw that $300 had been used for "entertaining." After visiting the accounting department, she found that Mario had submitted a claim for food on June 6. During the resulting discussion between Mario and Bev, it became obvious that Mario thought he was asking for permission. Bev had not considered food and beverages and did not think she was being asked for approval. Mario said he would be clearer when asking for decisions in the future. What could Bev do differently next time?

1.62 How can the need, specification, and concept development stages be combined?

1.63 What is the difference between tasks and deliverables?

1.64 Briefly describe the following terms:
(a) Function
(b) Deliverable
(c) Decision
(d) Organization chart

1.65 Draw a graph that shows the level of design detail development over the life of the project. Mark the major milestones.

1.66 A small rural road connects two cities that are 10 km apart. The traffic volume on the road has grown enough to expand the road to two lanes in both directions. The project does not require any new land, but two bridges will need to be widened. Making reasonable assumptions:
(a) Write a project scope.
(b) Write a list of stakeholders.
(c) Develop a table of needs.
(d) Discuss the need for concept development and selection.
(e) Prepare a two-year schedule for the project.

1.67 Illustrate the seven-step design process for a new flashlight design. For each step write a one-sentence description of the work and outcomes.

1.68 How does the volume of product detail influence the need for PM?

1.69 Develop a 10-step process for customer acceptance of a new automobile.

1.70 Develop a process flow for a project you have done recently. If you are a student this could begin with a project assignment from a professor.

1.71 Use the seven-step process to design a toy robot.

Further reading

Baxter, M., 1995. Product Design: Practical Methods for the Systematic Development of New Products. Chapman and Hall.

Bennett, F.L., 2003. The Management of Construction: A Project Life Cycle Approach. Butterworth-Heinemann.

Bryan, W.J. (Ed.), 1996. Management Skills Handbook: A Practical, Comprehensive Guide to Management. American Society of Mechanical Engineers, Region V.

Cagle, R.B., 2003. Blueprint for Project Recovery: A Project Management Guide. AMACON Books.

Charvat, J., 2002. Project Management Nation: Tools, Techniques, and Goals for the New and Practicing IT Project Manager. Wiley.

Davidson, J., 2000. 10 Minute Guide to Project Management. Alpha.

Dhillon, B.S., 1996. Engineering Design: A Modern Approach. Irwin.

Fioravanti, F., 2006. Skills for Managing Rapidly Changing IT Projects. IRM Press.

Gido, J., Clements, J.P., 2003. Successful Project Management, second ed. Thompson South-Western.

Heldman, K., 2009. PMP: Project Management Professional Exam Study Guide, fifth ed. Sybex.

Heerkens, G.R., 2002. Project Management. McGraw-Hill.

Kerzner, H., 2000. Applied Project Management: Best Practices on Implementation. John Wiley and Sons.

Lewis, J.P., 1995. Fundamentals of Project Management. AMACON.

Marchewka, J.T., 2002. Information Technology Project Management: Providing Measurable Organizational Value. Wiley.

McCormack, M., 1984. What They Don't Teach You at Harvard Business School. Bantam.

Portney, S.E., 2007. Project Management for Dummies, second ed. Wiley.

Rothman, J., 2007. Manage It! Your Guide to Modern Pragmatic Project Management. The Pragmatic Bookshelf.

Salliers, R.D., Leidner, D.E., 2003. Strategic Information Management: Challenges and Strategies in Managing Information Systems. Butterworth-Heinemann.

Tinnirello, P.C., 2002. New Directions in Project Management. CRC Press.

Ullman, D.G., 1997. The Mechanical Design Process. McGraw-Hill.

Verzuh, E., 2003. The Portable MBA in Project Management. Wiley.

Verzuh, E., 2005. The Fast Forward MBA in Project Management, second ed. John Wiley and Sons.

Wysocki, R.K., 2004. Project Management Process Improvement. Artech House.

CHAPTER

Planning and managing projects

2

2.1 Introduction

A project plan is based on a schedule and a budget. The first major goal in developing a plan is the identification of the required work. The project constitutes a set of high-level tasks. Each of those tasks can be reduced to smaller tasks. Regardless of the level in an organization, tasks share similar attributes. A task must lead to a useful outcome and can be done by employees, suppliers, customers, and so on. The time required for each task is used to set a budget for the project. Combining the time and sequence for multiple tasks provides an overall time frame for the project. As a minimum, each task will require labor, but should also require tools, equipment, and expenses. From an administrative perspective, there must be enough information to track each task from start to end and assess effectiveness. The administrator is also responsible for communicating with outside groups and managing the team:

Activity:

- Internal actions
- Outside suppliers
- Inputs to the task
- Outputs and outcomes of the task

Time:

- Earliest start
- Latest end
- Duration
- Variability

Sequence:

- Preceding tasks
- Following tasks
- Conflicting tasks
- Concurrent tasks

Resources:

- Specific people, expertise, manual labor, etc.
- Equipment
- Expenses

Engineering Design, Planning, and Management. https://doi.org/10.1016/B978-0-12-821055-0.00002-5

Administration:

- Initiation: What actions start each task?
- Conclusion: What actions end each task?
- Risks of failure or change?
- Who needs to approve actions?
- Who needs to be informed?
- Who is responsible for each task?

PROBLEM

2.1 What is a task?

2.2 Chunking the project

The key to planning is the task list, or work breakdown structure (WBS). For a new plan, this list is very general, has many details missing, and has crude estimates. The planner, or planning group, will work to refine the task list until there is adequate detail and certainty to move ahead. Eventually the task list will become a schedule, budget, work plan, and more. The process is shown graphically in Fig. 2.1, and described as the following steps:

(1) Lay out the tasks.
  (a) Start with a spreadsheet for the overall project tasks.
  (b) Break the major tasks into subtasks, focusing on what different groups will do.
  (c) Further divide the tasks to the individual level for larger assignments.
(2) Add details and estimates to the spreadsheet.
  (a) Times: Add estimated step times and variances (e.g., standard deviation). If you are uncertain use a larger variance.
  (b) Cost: Estimate the cost for components, services, and labor. Use a variance for uncertainty (e.g., standard deviation).
  (c) Resources: Indicate what special resources will be needed. This is critical if they need to be formally scheduled.
  (d) People: Identify people who are critical to project phases.
  (e) Other details: You can add additional details as they arise, to reduce the work later.
(3) Sum the values to get an estimate of the overall project time and cost.
  (a) Times: Sum the nominal times for overall project time. Sum the variances for time risk.
  (b) Costs: Sum the costs for an overall projected cost. Sum the variances for cost risk.
(4) Add sequence details.
  (a) Inputs: Determine what each task needs, such as solid models, schematics, alpha stage source code.
  (b) Outputs: What is the tangible outcome from the task, such as parts, test reports, simulation models, etc.?
  (c) Preceding tasks: What tasks must come before?
  (d) Following tasks: What tasks must come after?

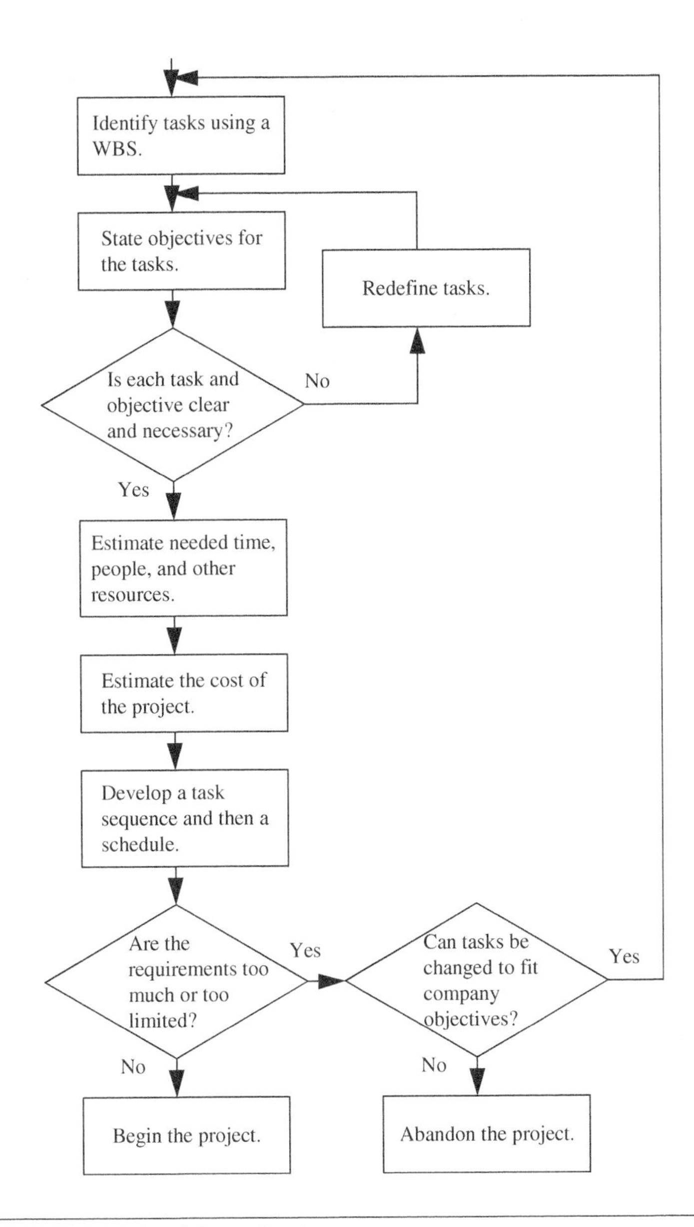

FIGURE 2.1

Pragmatic planning. *WBS*, work breakdown structure.

(5) Review the task list details for sequencing.

- **(a)** Conflicts: If one resource, such as an engineer, is needed for two steps, then they conflict and cannot overlap.
- **(b)** Concurrence: Some tasks share resources and will work best in parallel, such as software and hardware design.

(6) Put the entire plan in a time sequence.
   (a) Use the preceding, following, conflicting, and concurrent tasks to create an overall task sequence.
   (b) Use the critical path method (CPM) or program evaluation and review technique (PERT) (Section 2.7) to find the time limits for the plan.

(7) Review the plan and address the variance.
   (a) Time: Large variances on critical tasks will require better estimates or alternate plans.
   (b) Cost: The budget will need to include the cost variance for the estimates. Improving estimates will reduce the variance.
   (c) Resources and people: Prepare backup and alternate plans for overused or critical resources.

(8) Prepare project documents for the different groups.
   (a) Gantt charts for project schedules;
   (b) budgets and reporting schedules for accounting;
   (c) resource and people schedules for managers;
   (d) a to-do list for yourself and/or the project team.

Developing a formal plan for a simple project may not be necessary. However, a professional will have a plan in mind. For example, it would be unusual to develop a formal schedule for a week long project to be done by a couple of people. Project plans become essential as a project stretches into hundreds of hours of work, multiple expenses, multiple people, suppliers, shared resources, and competing projects. Failing to develop a plan will result in financial, time, and opportunity losses. The alternative to planning is to order all of the time-critical parts first, and start the critical processes as soon as possible; this is easily said but very difficult to do.

The advantages of planning are the following:

- There is a good estimate of time, money, and resource needs.
- Critical parts and tasks can be started early.
- Multiple projects can share resources with few conflicts.
- Problems can be detected and solved earlier.
- The plan keeps everybody informed about what is happening.
- There is clear responsibility for all tasks.

The disadvantages of planning are the following:

- Developing a plan takes time.
- Tracking a plan takes time at all levels of the organization.
- Planning delays the start of the project.
- A plan appears to reduce flexibility and threaten individual autonomy.

PROBLEM

2.2 Why should each task have a specified starting point and outcome?

2.3 What is a subtask?

2.4 Every task must be assigned to a person, even if it is to be done by a team. Why?

2.5 What can be done if the schedule does not fit the necessary time?

2.6 Give five examples of tasks that must be in a sequence.

2.7 Give examples of five tasks that cannot be done in parallel.

2.8 What are three options when a project schedule will not fit the constraints?

2.9 Develop a WBS spreadsheet for a project that uses the seven-step project model. The deliverable is a cheese sandwich.

2.3 Task identification

A plan is a sequence of steps leading to a predictable end.

A common trick for athletes is to break big training and competition activities into smaller goals. Focusing on the next smaller, and reachable, goal improves focus and gives an ongoing sense of accomplishment. The same approach is very important for projects, to keep us focused on the next deliverable goal and improve morale with frequent achievements. To this end we want to clearly separate the work into phases so that we can put our best efforts into the current task. Some of the main features of tasks are listed as follows.

Milestones

- Milestones are major points in the overall project.
- There is a clear end to one phase of the project and start of the next.
- They should involve some major approval or permission.
- A reasonable range for milestones is weeks to months.

Tasks

- Include all the critical tasks (e.g., ordering, testing, meeting, etc.).
- Have a clear start and end. Some form of review should occur at the end of the task.
- Have a very well-defined outcome or deliverable. Measurable, reviewable, or testable is best.
- Have a lead person responsible for delivery. If a required person or resource changes, create another task.
- Consider other resources needed: equipment, departments, suppliers, consultants, facilities, weather, etc.
- Subtasks can be used to break up bigger tasks with less clearly defined outcomes and times.
- A reasonable range for tasks is weeks, and for subtasks it is days.

Constraints

- Constraints may be due to relationships between tasks, such as sequence.
- There may also be specific dates for availability.

Milestones are normally selected using business objectives. Fig. 2.2 shows a couple of project timelines, one for an established business product line, the other for a new venture. Consider the top timeline for a design project. It would be normal for the milestones to include (1) project launch, (2) design review, (3) external testing, and (4) hands-off. Other milestones might include system integration, individual component completion, or progress review. Avoid having milestones that are simply placeholders. Use them to obtain approval for work completed and approval to move ahead.

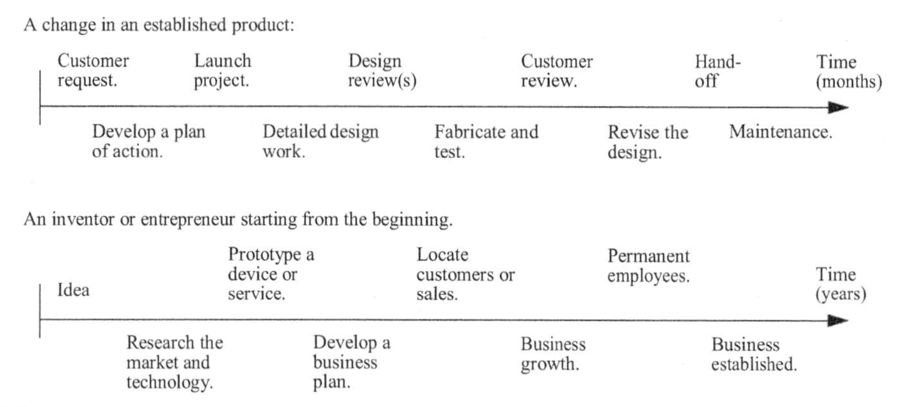

FIGURE 2.2

Example project timelines.

Every company is different, but the project phases are similar (Fig. 2.3). Project phases sometimes combine, overlap, or are broken into smaller steps. Some organizations tolerate or encourage fuzzy milestones, although it is better to close phases cleanly at milestones to help the project move forward. Even though the project work happens in stages it can be valuable to anticipate or work ahead of the

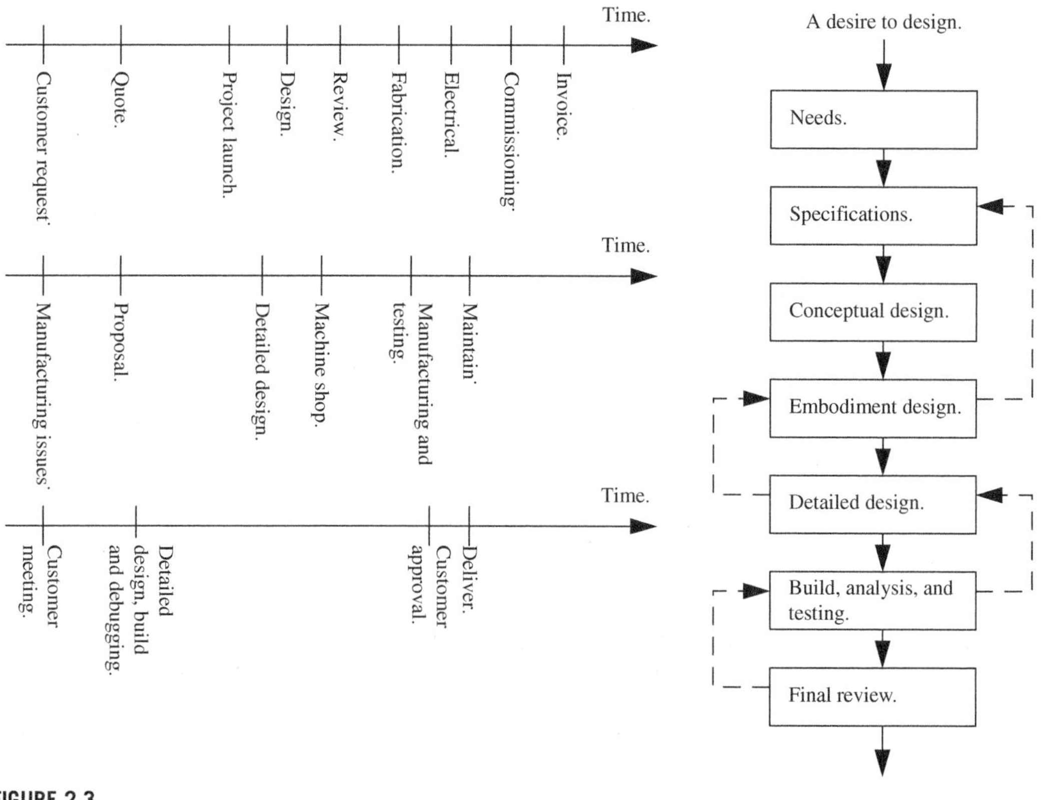

FIGURE 2.3

Time-wise views of three different projects.

next deadline, but only if the current project work is on track. The names for project phases will vary between organizations and business models. Regardless of these variations, always work toward the goals of (1) customer specifications or a clear project definition, (2) a preliminary design with concepts and technical specifications, (3) the end of detailed design, (4) the end of building and testing, and (5) the end of the project. An amateur mistake is to leave details for later; professionals will solve the difficult problems first.

Most designs begin with a perceived need from a project customer. The customer will play a driving role in defining the needs and objectives, providing resources, and approving project steps and outcomes. (Note: When the customer is also the designer, there is a concern about making poor decisions.) The project needs are sometimes documented in a project charter or sometimes as a set of draft customer specifications. The team will work toward detailed specification to drive the design work and provide a clean conclusion to the end of the project. In more challenging technical designs it may be necessary to do some of the prototyping, conceptual, and technical specification work to prove the viability of some customer specifications. Concepts and technical specifications are done to identify the technical and detailed design work. The conceptual and technical specification development phases are when the detailed project plan gains many details. By the start of the detailed design most of the major design choices have been set by the technical specifications, the major budget is set, and the work for each project phase has been defined. In detailed design, the work should be steady and predictable, leading to the build phases of the project. The testing phases need to verify that the design meets the technical and customer specifications. If there are problems it is wise to leave time for limited redesign and rebuilding work. When the project meets the specifications it is time to close the project work and deliver the results to the customer.

If you are unsure about the project phases, use the standard design process of needs, customer specifications, concepts, technical specifications, details, build and test, and deliver. You can then modify terms and phases to match customer and organization expectations. Again, always have a crisp end to each project phase. Failure to close the project phases will waste endless time as you redo work. At all points in the project, remember that the objective of the project is to satisfy the customer by meeting the customer specifications. At the end of any project the following actions should be anticipated. If there is an action that does not lead to one of the project outcomes it should be questioned. Other items to consider when planning include the following:

- Review all of the specifications and the final design.
- Obtain and document testing results.
- Review and deliver final documentation.
- Train the customer.
- Create maintenance plans.
- Review budgets.
- Reassign teams to new projects.
- Report on the project to internal management.
- Resolve all open issues.
- Review all contractual obligations.
- Facilitate customer hand-off.
- Prepare project documents and materials for long-term storage, disposal, or transfer to other projects.
- Undertake a final assessment of the work.

Who	What form	Due	Content
With: Sponsor	Review meetings	1, 4, 6 months	design details and updates
To: Manager	Progress reports	weekly	progress summary, deviations, budget
From: Design group	Hours log	weekly	time logs for job tracking
From: Production	Schedule	5 months	a list of available production times

FIGURE 2.4

A communication plan example.

Customers should also ask for internal meetings and reviews during the project. Sometimes this is a few phone calls or emails to check on progress, or it could be more detailed meetings, reports, and approvals. It can be helpful to develop a communication plan to share with all the project stakeholders (Fig. 2.4). This plan should indicate who initiates the communication, what is being communicated, how it will be communicated, who gets it, and when should it occur. Sharing this plan will avoid confusion and help customer–designer relationships. The communication plan should be aligned with the start and end of major project phases.

PROBLEM

2.10 List 10 types of meetings that should be included as a task.

2.11 List six activities that could occur at the end of the project, including customer training.

2.12 Multiple groups are normally involved in projects. For example, the project leadership will eventually pass from design to manufacturing. If there is overlap, should some manufacturing tasks be occurring while the detailed design task is incomplete?

2.13 Give five examples of planning constraints.

2.4 Work breakdown structure

Cut the big problems into smaller problems.

There are multiple ways to divide work into stages. In a modular system this might be done by major system components, as illustrated in the following list. In a process-oriented environment this may be a sequence of operations such as foundations, structure, inspection, electrical, inspection, drywall, and so on. For the remainder of this section we will assume that you have already (1) developed a list of milestones and tasks using the appropriate methods, (2) identified rough sequences or time, and (3) developed rough estimates for tasks. A WBS is often numbered to organize the tasks into a hierarchy (tree structure). Level 0 is the main task/project.

Level 1 is the next division of tasks, and so on until a reasonable level of detail has been reached:

(1) Microcontroller hardware
- **(1.1)** Component selection
- **(1.2)** Circuit design
- **(1.3)** Printed circuit board (PCB) layout
- **(1.4)** Review
- **(1.5)** Order parts
- **(1.6)** Assemble and test hardware

(2) Software, etc.

The contents of a WBS are very well suited to a spreadsheet. The columns for a reasonable spreadsheet are shown in the following list. Many of these fields, such as progress, can be calculated automatically. It can also be helpful to use colors to indicate the status of tasks, including current, upcoming, and overdue:

- Task number (e.g., 1, 3.2, 4.2.6)
- Milestone name
- Task name
- Subtask name
- Next tasks: any tasks that must wait until another one is done
- Start date
- End date
- Required work $\leq$ end $-$ start
- Actual work/work done to date
- Progress: normally percentage $= 100 \times$ (required $-$ actual)/required
- Status: ahead, behind, on track, late
- Lead person
- Other resources, people, etc.

The baseline is a reasonable estimate of the time to complete the plan. This is often based on a critical path through the network diagram: the most time-constrained tasks. Tasks that are not on the critical path typically have some variable, or slack time, for when they start.

PROBLEM

2.14 How would a task, subtask, and subsubtask be numbered?

2.15 What is a milestone?

2.16 Mini-case: Shifting goals
Automotive parts that are visible cannot have visible defects, including fingerprints. A manufacturer, Cleanco, was using a process to coat one side of decorative transparent plastic pieces. Cleanco had managed to redesign the process so that the pieces needed to be touched only once by workers wearing gloves. In the process, two parts were joined and a heating process was used to cure the glue. The Supplyco design team was contacted by Todd, a project manager at

Cleanco. The design team was asked to design a replacement process that would eliminate the last human handling operation. The Supplyco team, led by Angus, began the process of determining the needs and specifications for the project. After a few weeks the Supplyco design team received an email from Todd about their specifications that said, "They look good." The email was sent to Ranjiv, who approved the work, and the Supplyco team began the detailed design work. Two weeks later, Todd called to say that he had talked to his manager, Bailey, and they decided to "go in a different direction." A meeting was called with Bailey, Todd, Angus, and Ranjiv. After some discussion, it was obvious that Todd and Bailey had different ideas about the project. When Todd sent the email, he thought that Angus's team was on the right track but did not consider it to be a formal approval. Given the absence of a formal acceptance, Ranjiv decided to discuss and pursue the project as Bailey saw it.

Bailey's vision for the project was research into a different method using paper strips to hold the plastic parts as they pass through an oven. The original process developed by Angus's team used routine fixtures and methods to hold the mirrors. Bailey's method had great potential but had never been tried before. Suddenly the team project had expanded to include research into new materials and development of new processes. After the meeting Ranjiv asked Angus's team to develop a new plan for the project. The specifications were initially written for the final machine, assuming that the research and development process went smoothly. After some consideration Ranjiv asked the team to break the project into separate phases for the research and development steps. This plan included backups in case any of the research and development steps failed. The worst-case scenario was that the team would design and build the originally proposed fixture-based machine. Another meeting was held with Ranjiv, Angus, Bailey, and Todd to discuss the new project plan. The meeting outcome was an agreement that the project plan include contingencies for failure. After this the team moved ahead with the research plans and was able to find paper strips that would hold the parts in the heat but did not have time to incorporate these into a machine. Having found a solution the team was able to deliver a solution to the problem, but not a constructed machine. The customer was happy with the end result of the project.

(a) What are the pros and cons of discussing the chance of failure in a project?

(b) How does research and development change the design process?

(c) How could the Supplyco project process be changed to avoid similar problems?

2.5 Resources and people

Estimate when and how much employee time and resources are needed for a project. Consider that employee time, equipment, purchases, and facilities cost money. This needs to be added into project costs, but this is possible only if you know how much time you will need. Most companies do not have unused equipment and idle workers, and what they do have is shared between multiple projects and regular operation. Tell employees ahead of time what you need them to do. They will work to get other tasks done and not book vacations in the middle of your project. If you do not schedule equipment, somebody else will be using it when you need it. The same is true for outside suppliers, shippers, customers, and others.

To estimate people and equipment needs you can use a combination of top-down or bottom-up approaches. In top-down estimating, the budget grand total needs are estimated first. These are gradually divided into more detailed estimates. In bottom-up estimation, also called roll-up, the estimation begins at specific tasks and then is added up to get the grand totals. Practically, it makes sense to work from the top and bottom to refine estimates, putting emphasis on the larger tasks and items first.

Fig. 2.5 shows a couple of working lists to compile and estimate people and resource needs. The top of Fig. 2.5A shows people, space, software, and suppliers. Moreover it shows who is the primary person responsible for task completion; the other people on the task are also listed. Including information and approvals helps to schedule time. The bottom of Fig. 2.5B includes cost, lead time, and other details. It is also possible to put this information directly into a WBS spreadsheet if additional columns are added for the resources and specific employee names.

Consider a complex airplane project with components made by 12 different teams at 9 companies. One company makes the wings and landing gear, another makes the fuselage, and another makes the engines. These separate projects must individually complete and deliver components for testing. Each of these project groups can work independently, but deliver collectively. For early planning it can be helpful to create a sketch such as that in Fig. 2.6. In this context each of the design companies is treated like a resource to be scheduled and managed. The nodes (circles) in the diagram represent different physical components. The lines indicate parts of a larger system. This graph can be used in constructing the schedule.

Writing a plan is like writing a script for a play.

For a truly complex set of resources it can be helpful to develop a story line graph to track people and resources across multiple tasks and projects (Fig. 2.7). In this graph each line represents an important person or resource. Over time people move in and out of the project work, indicated by the dashed lines. At the beginning of this project Peter and the production department are busy with another project, but Vijay and Jane are working with the customer to prepare a quote for the customer. Peter and the production department have a chance to review the quote before it is sent out. After that, Jane is no

WBS Item (task)	Person 1	Person 2	Person 3	Supplier 1	Machine 1	Software 1	Room 1	Sponsor
CAD work	P (5hr)	S (1hr)				(5hr)		I
CNC machining		S (2hr)	P (1hr)		(4hr)			A
PCB assembly				I (6hr)			(6hr)	I

Legend:
P = Primary
S = Secondary
I = Information
A = Approval

Name	details	lead time	cost	hours
designer				
production				
supplier				
machine				
etc....				

FIGURE 2.5

Sample stakeholder lists. *CAD*, computer-aided design; *CNC*, computer numerical control; *PCB*, printed circuit board; *WBS*, work breakdown structure.

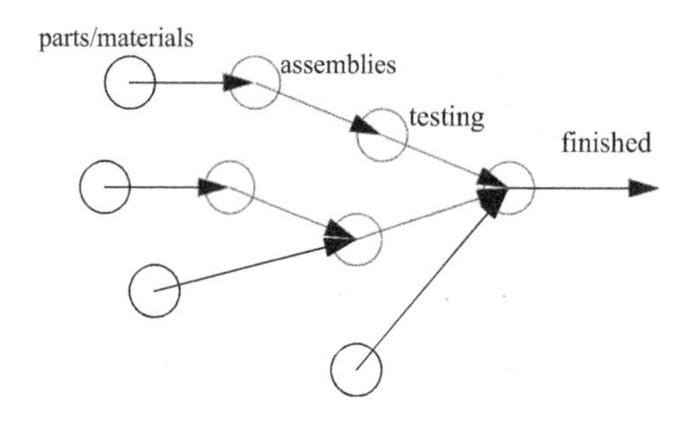

FIGURE 2.6

A component assembly and testing view of design tasks.

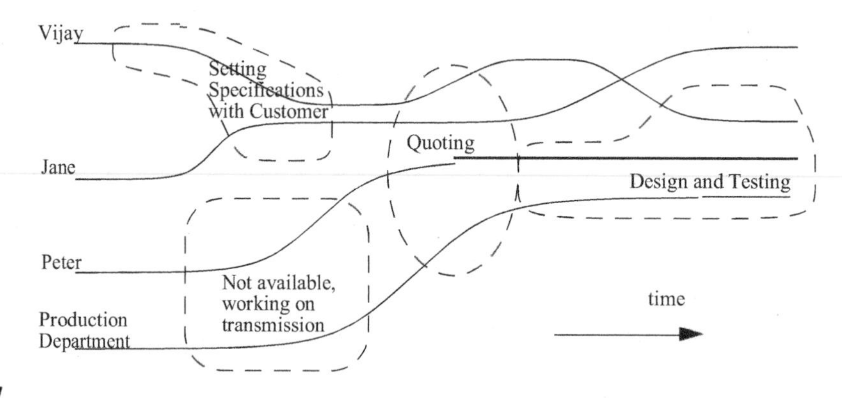

FIGURE 2.7

A story line graph for multirole people and resources in a project.

longer involved. Peter and the production department begin the detailed design and testing and then Vijay joins them halfway through the design and testing. This is not a formal method, but it can be very helpful for organizing thoughts.

Parkinson's law: Work will expand to fill the allotted time.

The time and costs are estimated while, or after, the people and resources are being identified. There is no single source or method to collect the planning information. Educated guesses are a good place to start, but these should be replaced with better estimates as the plan evolves. It is better to consider an estimate as a mean time to completion, plus a probability distribution to compensate for uncertainty, natural deviations, and estimating errors. A normal probability distribution would be a simple choice for estimating variations. Some pointers are provided in the following list to help the estimating process; however, a more detailed list can be found in the appendix:

- Use estimates and final totals from similar projects.
- Consult people with experience.
- Ask employers and suppliers for rough estimates.
- The accounting department can sometimes provide estimates for similar jobs.
- Consider other company jobs.

- If there is new equipment required, consider the costs and training.
- If parts of the project are new, use larger time and cost variations.
- People over- and underestimate time needs. Until you know a person, use a larger variance.

PROBLEM

2.17 People, equipment, and other resources are often heavily used and shared between multiple projects. Describe two ways that resource and people availability can be tracked.

2.18 One of the designers was reviewing the budget and noticed that there were two options for a critical part. Normally, the part would cost $20,000 and take 7 months to deliver. For a cost of $25,000 the supplier would expedite the order and deliver it in 3 months. Discuss the advantages and disadvantages of both of the purchasing options.

2.6 Microsoft Project tutorial: setup and work breakdown structure

Projects can be planned and tracked with many tools, including paper, spreadsheets, websites, and paid for/free software. A common tool for project managers is Microsoft Project. A minimal use of Project is capturing a WBS and outputting a Gantt chart (these charts are covered later in this chapter). The software has many advanced functions that include tracking resources, people, work completion, etc. At various points in this book there is discussion of some of the common uses for the software. This section will be tutorial in nature and focus on creating a basic WBS. The example is based on a multi-month design project for a class that meets on Mondays.

1. Start the "Project" software. When it starts you should as screen with templates. Select or create a blank project. What appears should look like the screen in Fig. 2.8.
2. Notice that the "Task" tab is selected. On the left-hand side of the work area there is an area to enter tasks; we will enter the WBS here.
3. To begin, enter the following tasks on their own line under "Task Name": Project Definition, Customer Specifications, Conceptual Design, Technical Specifications, Detailed Design, Building, Testing, Delivery. These will form the major phases of the project as shown in Fig. 2.9.
4. Next, we will add details for each of the project phases. Task properties can be changed by clicking on the number at the far left of the task line. This opens a "Task Information" dialog box with multiple tabs, as seen in Fig. 2.10. For the first task, "Project Definition," we will select a start date of January 20, 2020; we will leave the end date blank. For the "Delivery" task, we set a date of April 20, 2020. Notice that these add date lines to the "Timeline" bar near the top of the screen.
5. Now it is time to break the major tasks into smaller tasks. There are various ways to insert tasks. The simplest is to right click on a line and pick "Insert Task." The selected line will be pushed down, leaving a new blank line. If a task is in the wrong line, the entire line can be moved by clicking on the line number at the far left. When the up/down/left/right arrows appear, the task line can be dragged to a new location. Add the subtasks as shown in Fig. 2.11. Subtasks are indented by selecting the line and then clicking on the "Indent Task" icon on the toolbar. Notice

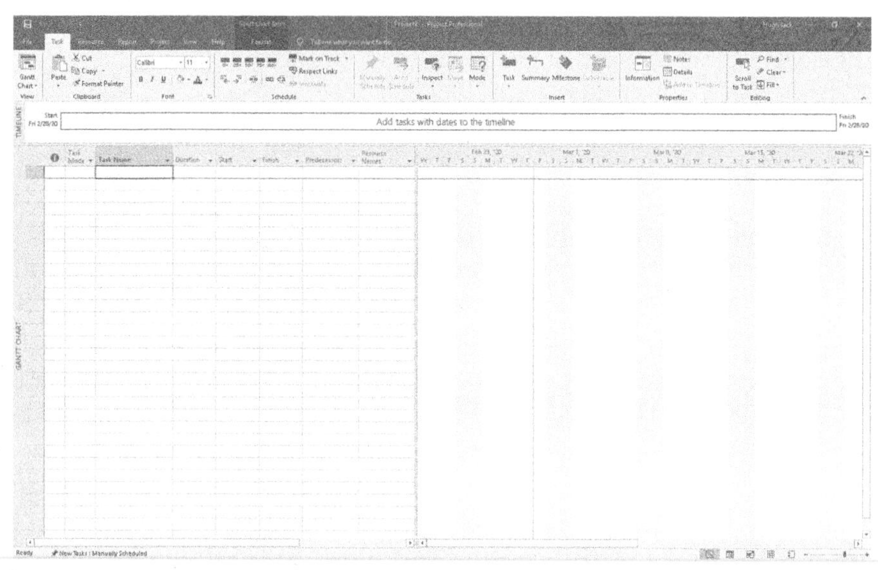

FIGURE 2.8

A blank Microsoft Project template.

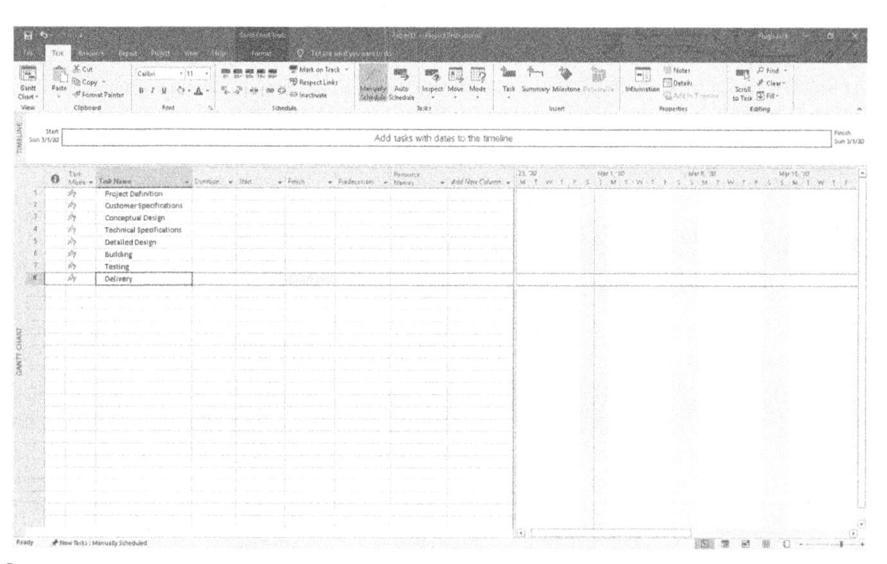

FIGURE 2.9

The task list is populated with higher-level activities.

that adding subtasks adds an arrow beside the task name. This allows us to collapse detail when not needed.

6. Next we must estimate the time required for tasks. If we set a duration of zero, it will be marked as a milestone. A milestone is a major project goal. For the "Project Scope Definition" set a

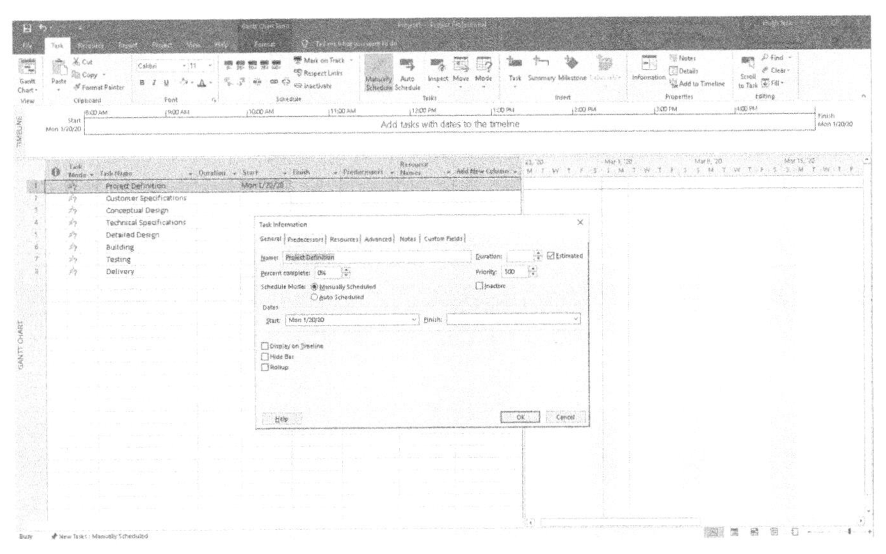

FIGURE 2.10

The task information dialog box was opened by clicking on the "1" at the far left.

	Task Mode	Task Name	Duration	Start	Finish	Predecessors	Resource Names	Add New Column
1		Project Definition		Mon 1/20/20				
2		Launch Meeting						
3		Project Scope Definition						
4		Stakeholder List						
5		Stakeholder Approval						
6		Customer Specifications						
7		Conceptual Design						
8		Technical Specifications						
9		Detailed Design						
10		Building						
11		Testing						
12		Delivery			Mon 4/20/20			

FIGURE 2.11

Subtasks are added to the Project Definition task.

duration of "6 h" for 6 hours. The Stakeholder list should take "3 hrs." Finally we add a "Stakeholder Approval" period of "3 d" to allow feedback and approval before launching the project. See Fig. 2.12 for additional subtasks and times. If start and finish dates are added, these can be deleted.

7. Next, we need to add a sequence to tasks. This can be done in the "Task Information" window, or by clicking in the predecessor cell and then clicking on the down arrow. Fig. 2.13 shows predecessors selected for a number of subtasks. The last entry for "Customer Approval" on line 12 shows "11 Instructor Approval" as something that must occur before the customer is consulted. There is an estimate that it will take 3 days for the response.
8. At this point the plan is estimating dates based on task time and sequence. The plan is currently set on "Manually Scheduled," as indicated by the thumbtacks on the left side. Clicking on the item allows you to select "Auto Scheduled," as is shown in Fig. 2.14. If the instructor takes "1 d"

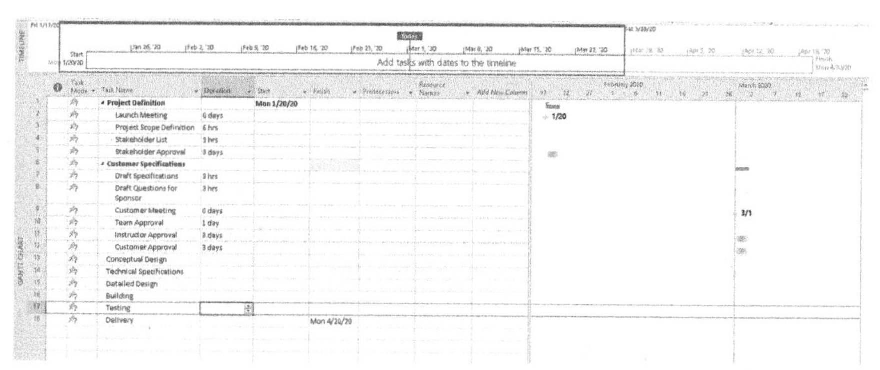

FIGURE 2.12

Task durations are added to the subtasks.

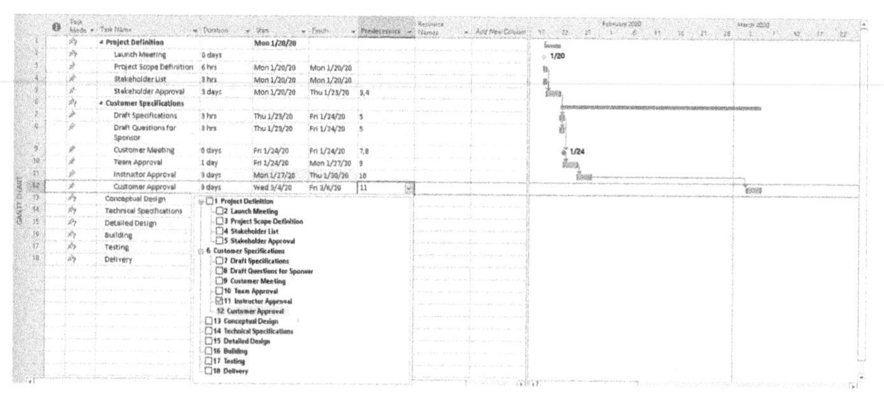

FIGURE 2.13

Task sequence is defined using predecessors.

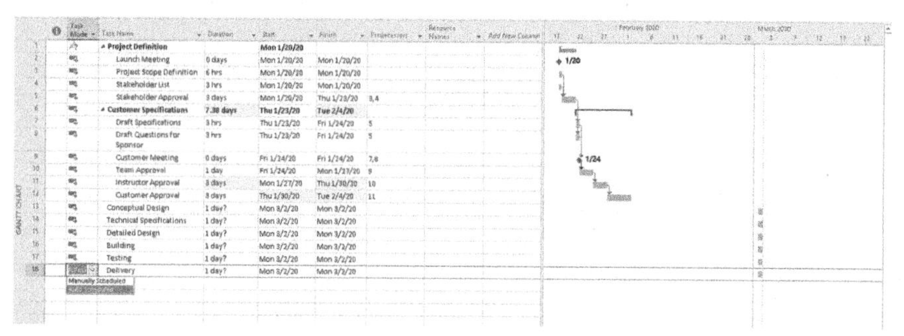

FIGURE 2.14

Timing can be adjusted dynamically by selecting automatic scheduling.

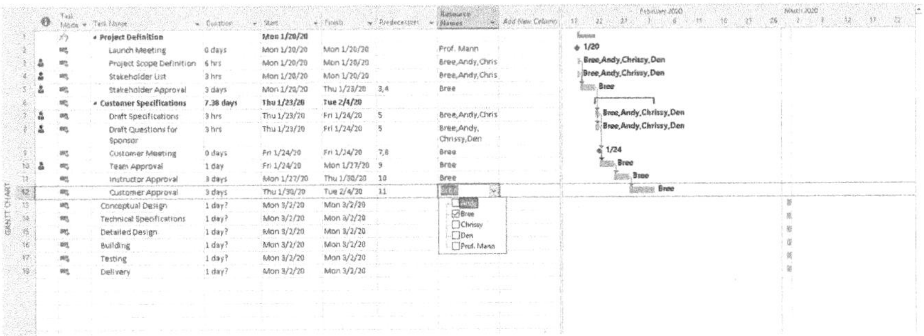

FIGURE 2.15

People are assigned to subtasks as resources.

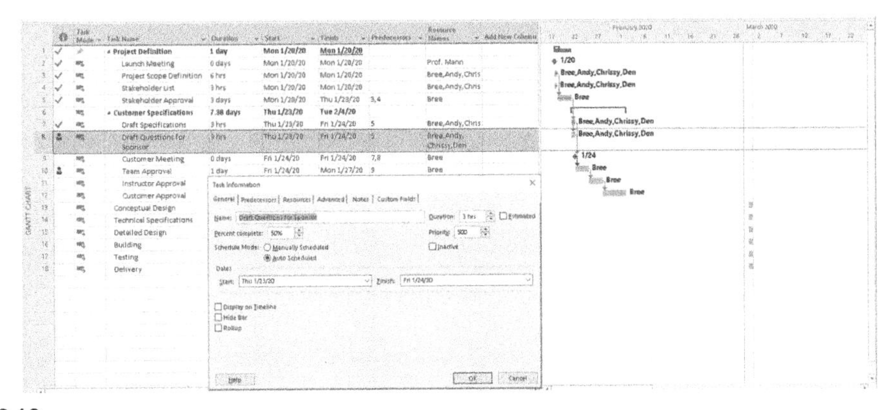

FIGURE 2.16

Tracking project progress by updating the completion of subtasks.

instead of "3 days," we can change the "Instructor Approval" duration, and the "Customer Approval" dates move forward. Or, if the instructor takes longer, the schedule will adjust to move dates to later.

9. Any task that is not assigned to a person has a very good chance of being forgotten. We can resolve this problem by adding people as resources. Note: This is also done for equipment, facilities, services, etc. For "Launch Meeting" we will assign the task to the course instructor, "Professor Mann." We do this by clicking on the resource cell for the row and typing in the resource name. After this we can select them from a list as shown in Fig. 2.15. For the remainder we add the team members Andy, Bree, Chrissy, and Den. Bree is the team leader and is assigned most of the organizational tasks, but is often backed up by other team members. Notice that the names also appear on the graph to the right. We now have a simple and visual plan for tracking the project.
10. We can also keep track of work that has been completed and is underway using the Task Information popup. In Fig. 2.16 the tasks that have been marked "100% complete" have a check mark in the information column. An estimate of "50% complete" was entered for line 8.

2.7 Schedule synthesis and analysis

Time is money.

Scheduling is simple when all of the tasks must be done in a sequential order and there is no option to reduce the project duration. In this case any delay for a single task delays the entire project. The other extreme is many tasks that can be done in parallel, and in any order. These are also very easy to schedule, and the project can be completed in less time. The project length is dictated by the task that is completed last, a combination of task start date and duration. Real projects have a mix of sequential and parallel tasks. When tasks must follow some sequence we need to indicate the preceding tasks. This relationship can be included in the WBS as a "next tasks" or "previous tasks" column.

The steps for scheduling are listed as follows. Schedule development can be done manually using a spreadsheet. For complex projects, the planning process will require a number of iterations to find a plan that will work:

(1) List all of the major project phases and milestones. Milestones must have measurable outcomes to end and start project phases.
(2) Divide the WBS tasks under each project phase. The tasks need to identify work, time, and resources.
(3) Identify the task sequences, i.e., can tasks be done in parallel or must they be done in series?
(4) Set start and end dates using the task sequence and duration.
(5) Verify that resources do not overlap.
(6) Adjust start times to meet all of the constraints and objectives.

2.7.1 Critical path method

The scheduling constraints and times for the tasks can be represented graphically using (network) diagrams like the example in Fig. 2.17. Each of the circles (nodes) indicates some sort of event, like the start and end of a task or a milestone. In this example, "p" may be the launch of the project. Three activity

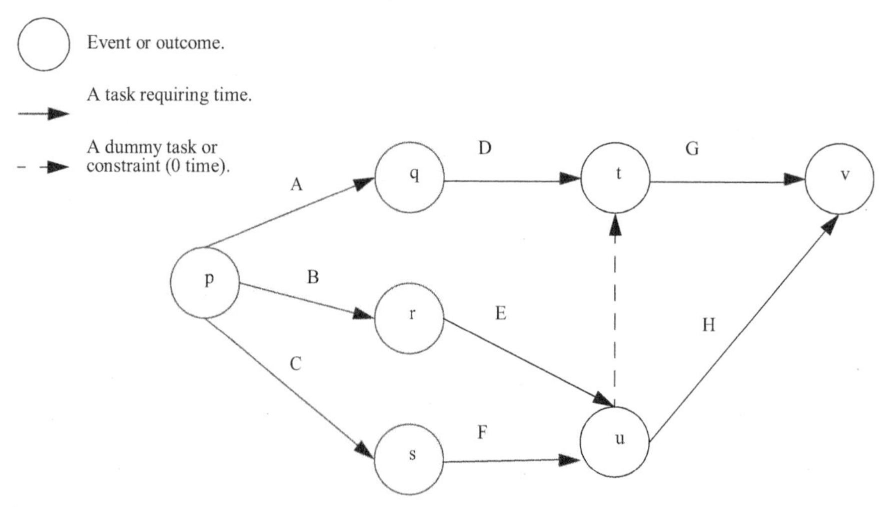

FIGURE 2.17

Arrow diagramming method.

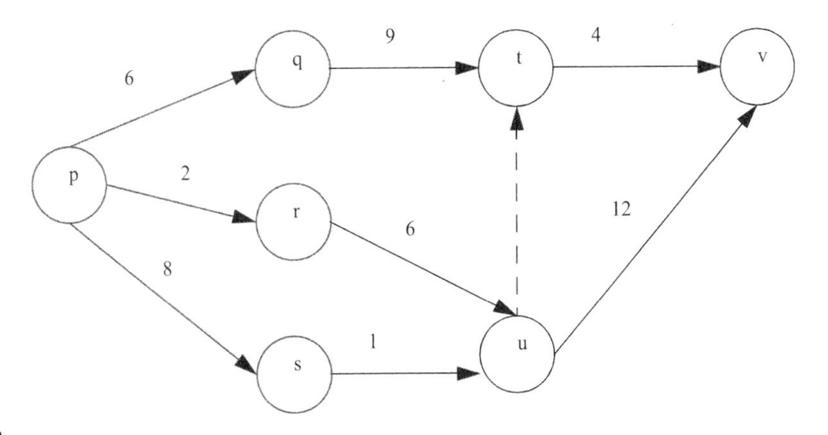

FIGURE 2.18

An arrow diagram with time estimates.

arrows (arcs) start in parallel (A, B, and C). Activity D cannot start until A is complete. Activity E must wait for B, and F must wait for C. Activity H cannot start until E and F are both complete. This diagram also has a "dummy constraint" that is used for coordination so that G requires the completion of D, E, and F, but only uses the results from D. In concrete terms, activities A to F may be mechanical, electrical, and software design. "t" is the complete mechanism, and "u" is the complete computer systems with software. A customer review (i.e., dashed line) must occur before integration and testing. The project is complete at "v." (Note: If some or all of the subtasks are included, this diagram might become unreasonably complex. Select a level of detail that captures parallel and serial scheduling issues.)

Once the diagram is complete we can add estimated times for each task. It is understood that these are estimates and will be somewhat inaccurate. If in doubt, it is wise to use slightly pessimistic, longer, estimates. Let's assume that the numbers shown in Fig. 2.18 are the estimated number of days to complete a task. It is very reasonable for the tasks to take less time than estimated if things go well, or longer if problems occur. Also, a 3-day task might be done over 21 days if somebody is working on multiple projects. So in simple terms, we can stretch out task times easily, but to compress them is costly, difficult, or impossible.

Begin the analysis by determining the shortest possible time to complete the example in Fig. 2.19.

Consider the sequence of tasks p—s—u—v that would take at least $8 + 1 + 12 = 21$ days. The task sequence p—r—u—v would take at least $2 + 6 + 12 = 20$ days. Likewise, the task sequence p—q—t—v would take $6 + 9 + 4 = 19$ days. The dummy task, the dashed line, indicates that p—r—u—t—v would require at least $2 + 6 + 4 = 12$ days. The longest path time is 21 days and dictates that the project must take at least 21 days to complete. In other words, p—s—u—v is the critical path. As the critical path, any increase in task time will increase the project duration. For example, if task p—s increases to 9 days, the minimum project time is 22 days. However, if task p—r increases to 3, then the minimum project time is unchanged. The critical path is used when developing schedules. When tracking progress, a manager will focus on the tasks on the critical path.

In the first pass of the CPM we went through the diagram from beginning to end to determine the minimum project duration possible. In the second pass, Fig. 2.20, we can work backward from the end once an overall project duration has been selected. In this case we can estimate how much flexibility we have in the schedule. Let's assume that the approved project length was 40 days. It is possible to review the schedule to determine variability. In this case, we work backward from the latest project completion at 40 days. So task u must start no later than 12 days before, or no later than 28 days after,

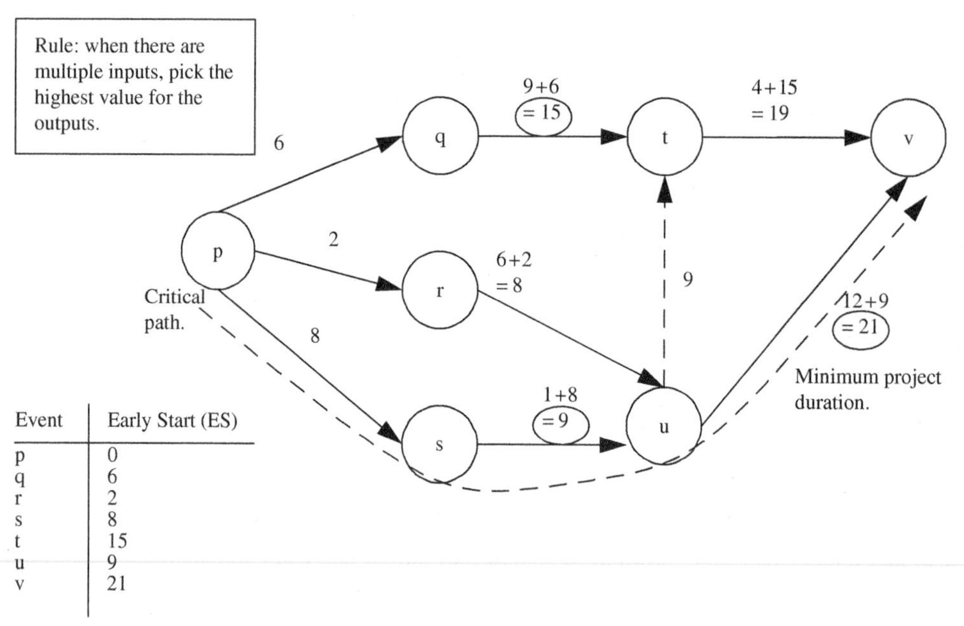

FIGURE 2.19

Critical path and minimum project duration example calculations.

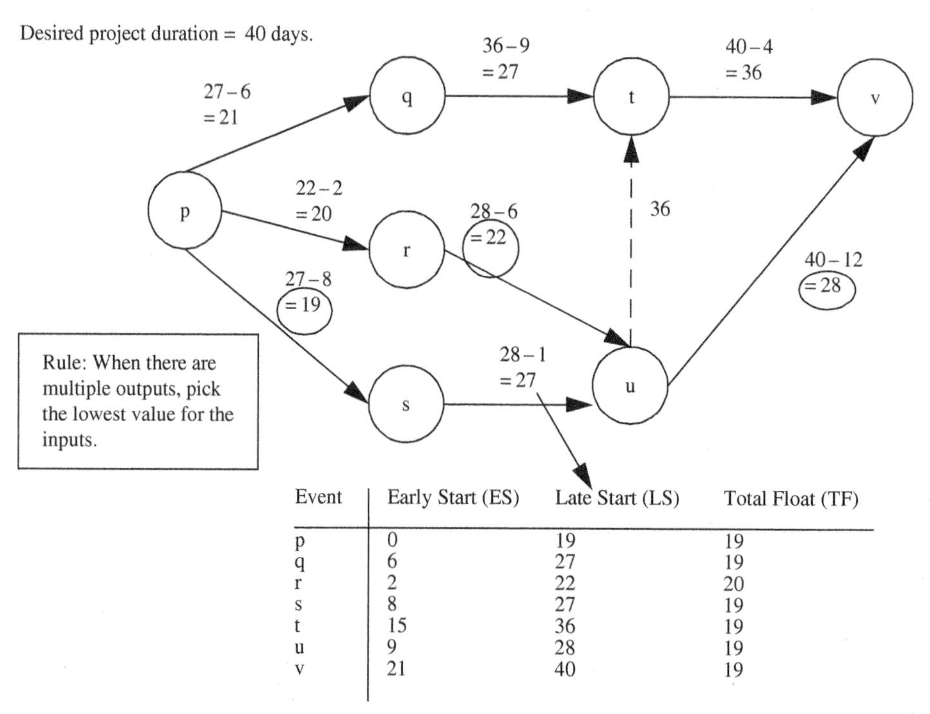

Event	Early Start (ES)	Late Start (LS)	Total Float (TF)
p	0	19	19
q	6	27	19
r	2	22	20
s	8	27	19
t	15	36	19
u	9	28	19
v	21	40	19

FIGURE 2.20

Critical path method variability.

the start of the project. The process is repeated through the diagram to find the latest start for all of the tasks. Finally the total float is calculated for each of the tasks. In this case all of the tasks have 19 days of float, except for task r, which has an extra day of permitted variability.

Once the critical path analysis is complete, the numbers can be used to adjust, or accept, the plan. Some general rules include the following:

- The early and late start dates will indicate if the project can fit the schedule.
- Tasks with the lowest float (on the critical path) should be scheduled as early as possible.
- Tasks with more risk or variability should be scheduled near the earliest start.
- Use the late start dates to create escalation points for reviewing the schedule and project plan.
- Use the critical path to schedule project review meetings.

It is possible that the initial schedule cannot meet the required delivery date. In these cases it is possible to use a process called crashing to pick tasks on the critical path and attempt to reduce the required time. In many cases additional money, resources, or people can be used to reduce the time required for a project step. Clearly this adds inefficiencies and cost and should be done strategically—hence the numbers.

PROBLEM

2.19 What are the circles in the CPM diagram?

2.20 What are the arrows in the CPM diagram?

2.21 What is the critical path?

2.22 What is "crashing"?

2.23 Why are dummy tasks added to CPM diagrams?

2.24 Consider the CPM network diagram shown here. Find the late start, early start, and float for each task, if the project is 50 weeks. Identify the critical path.

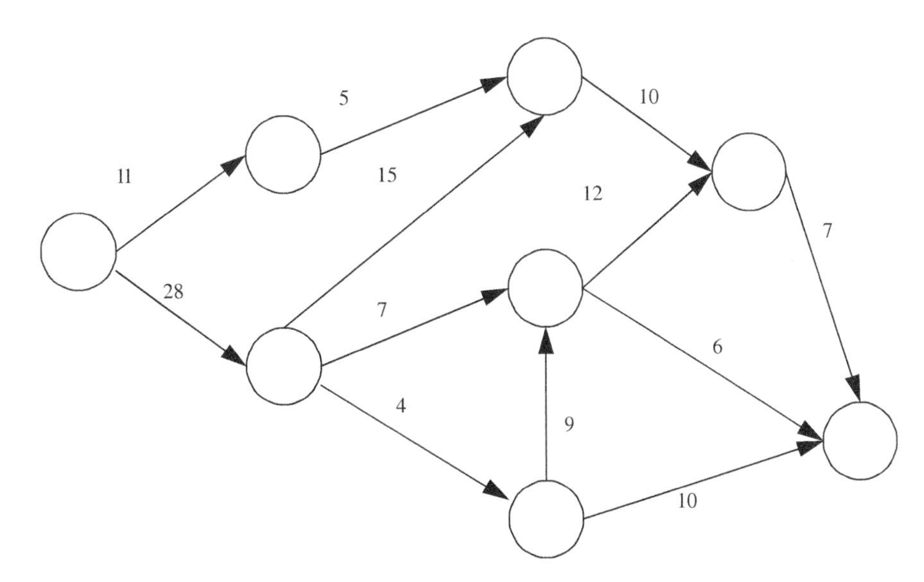

2.8 Program evaluation and review technique

Murphy's law: Anything that can go wrong will go wrong.

In CPM, a single task time value is used for each process, even though, in truth, it has been estimated using a specific value and a range of uncertainty. In mathematical terms, we can describe these using statistics. Consider the beta distribution shown in Fig. 2.21. Based on research, experience, and discussions, we see or guess that similar tasks have ranged from the optimistic times (very fast) to the pessimistic times (slowest), but the time required generally falls near the most likely value. With less knowledge and experience the range should be much larger. It is also worth considering your personal tendencies: do you normally overestimate or underestimate time for tasks?

Task times are expressed with three numbers separated by dashes $t_o-t_m-t_p$ to represent task times. These can be used in place of the normal CPM diagram task times, as shown in Fig. 2.22. For each of these we could perform the calculations in Fig. 2.23 to get an estimated time. For example, the task from p to r would have an expected time of $(2 + 4 \times 2 + 4)/6$ and a variability of $(4 - 2)/6$. The numerical values can be used to do a CPM analysis of a network diagram. Once the critical path is identified, the overall task time and variance can be calculated using T_e and σ_T (Fig. 2.23). To find the

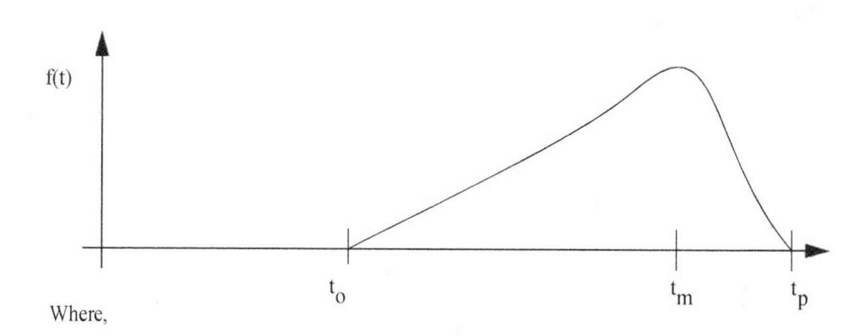

t_o = optimistic time estimate

t_m = most likely time estimate

t_p = pessimistic time estimate

t_e = effective, or mean, time

σ_i = approximate standard deviation for each task

For each task,

$$t_e = \frac{t_o + 4t_m + t_p}{6} \qquad \text{eqn. 2.1}$$

$$\sigma_i = \left| \frac{t_{pi} - t_{oi}}{6} \right| \qquad \text{eqn. 2.2}$$

FIGURE 2.21

Statistical time estimation.

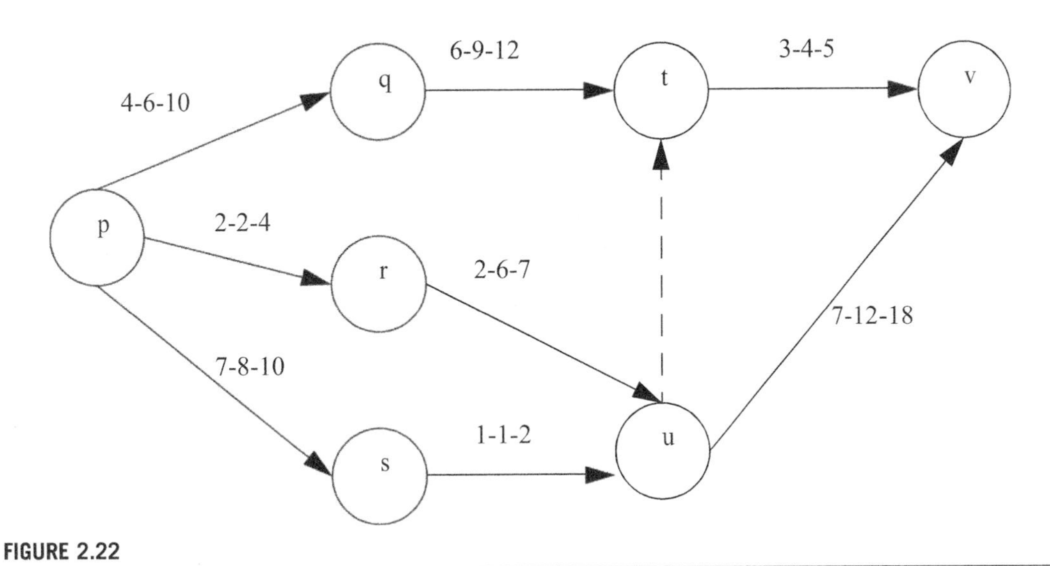

FIGURE 2.22

A network diagram with time estimates.

Where;

T_e = overall time estimate.

σ_T = overall variability estimate.

T_s = sample project completion time.

z = the probability that a task will be complete at time T_s.

$$T_e = \sum t_{e_i} \qquad \text{eqn. 2.3}$$

$$\sigma_T = \sqrt{\sum \sigma_i^2} \qquad \text{eqn. 2.4}$$

$$z = \frac{T_s - T_e}{\sigma_T} \qquad \text{eqn. 2.5}$$

FIGURE 2.23

Statistical time estimation.

chance that the process will be done by the time T_S, the z value can be calculated. The z value can then be used to find the probability of completion using the cumulative normal distribution function.

PROBLEM

2.25 What are the three numbers for each task in PERT?

2.26 Where do the pessimistic and optimistic values come from?

2.27 Why does the method include a standard deviation?

2.28 Consider the PERT network diagram shown here.

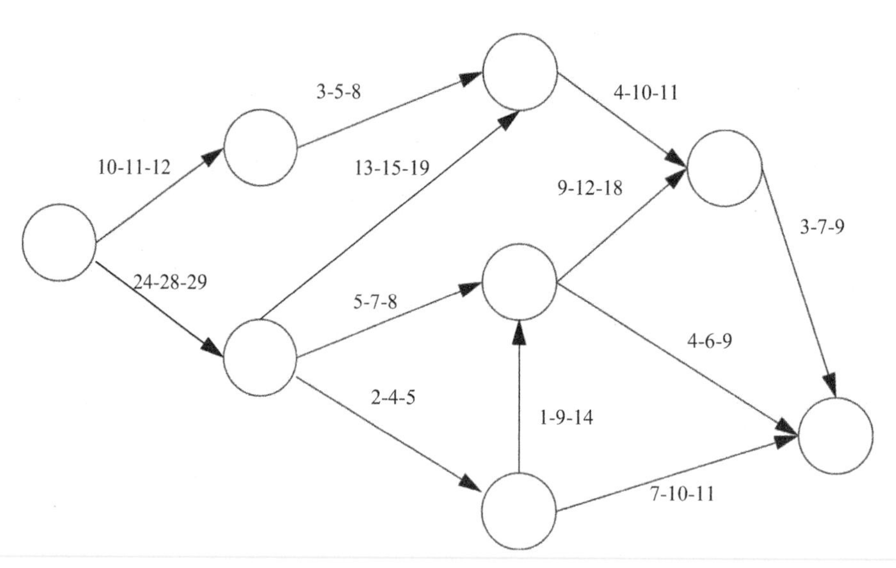

(a) Find the likelihood that the project will be complete in 40 days.

(b) Find a target completion date for the project that will make it 50% likely that it will be complete.

2.9 Plan review and documentation

Very few people will want to see all of the details in a project plan. Documentation is derived from the project plan and shared with different groups. For simpler projects, the WBS may be suitable for scheduling. When a WBS for a project becomes larger, it is very helpful to use it in a graphical form; the standard visual form is the Gantt chart. A simple example is shown in Fig. 2.24. On the

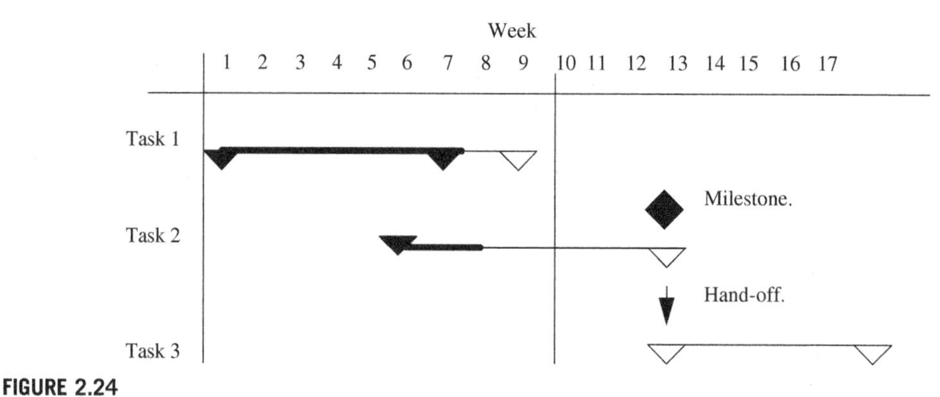

FIGURE 2.24

Gantt chart layout.

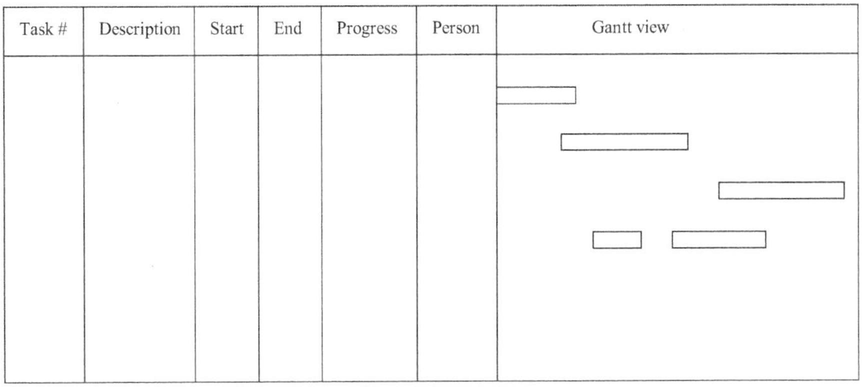

Task #	Description	Start	End	Progress	Person	Gantt view

FIGURE 2.25

Tabular Gantt chart.

left axis is a list of all of the milestone/task/subtask numbers and names. The horizontal axis is time. There are two main ways of drawing the task length.

The traditional method uses triangles to indicate the start and end dates, connected by a thin line. The other method is to use narrow unshaded boxes from the start to the end date. The actual work progress is shown by making the line thicker or adding a shaded box. Empty triangles indicate a task that has not been started or completed. Once the task is started or finished the triangles are shaded. In this example, task 1 started at week 1 and was scheduled to finish 8 weeks later in week 9. Task 1 was completed early, in just over 6 weeks. Task 2 has started but is almost 2 weeks behind schedule. There is a milestone in week 12 that will end task 2 and begin task 3.

There are a large number of free and commercial project-tracking software packages. For scheduling, they allow a WBS to be entered, and a Gantt chart is generated automatically. On a regular basis the progress is updated and a new Gantt chart is printed. (Note: Normally Gantt charts are printed on wide-bed printers or plotters normally used for large drawings.) The chart is posted in a public place so that the team can easily assess the progress of the project. Spreadsheets are an attractive option for midsized projects. Fig. 2.25 shows an example that could be added to a WBS stored in a spreadsheet, as described in the previous section. Examples for specific software packages are available on the Internet. Web-based spreadsheets can be shared by multiple users for live task tracking and Gantt chart generation. Given that a number of other project items will probably be stored in a spreadsheet (on multiple tabs) it is very useful to have a single shared spreadsheet to coordinate all project activities. The flexibility of the spreadsheet permits customization of processes and data specific to the project.

Resource 2.1 1 You can see a Gantt chart on this book's website: www.engineeringdesignprojects.com/home/content/resources.

A project team is most likely to coordinate activities with a Gantt chart. For clarity and accountability each task should have a lead person identified and a list of other people working on the task. Other details that might be communicated with the Gantt chart or separately include:

- Who will execute the task?
- Who needs to be informed?
- Who needs to be consulted?
- Who has approval authority?

As with any engineering problem solution, it is critical to look at the methodology and results to see if they are rational and reasonable. A rational plan will have a set of times and costs that "sounds right." Some problem indicators are as follows:

- There are task times that are extremely long or short.
- There are tasks that do not have a lead person.
- Resources and people are overscheduled.
- There is inadequate buffer time to compensate for time variations.
- Holidays, shutdowns, and other projects are not considered.
- The plan does not match the project contract or proposal.
- Special equipment is not available.
- Outside supplier lead times, and late times, are not considered.
- Some tasks are too long. At the team level this may be weeks; at the individual level this could be a week or less.
- There are no clear starts and ends to tasks.
- Some low-level tasks are not aligned with major milestones and project level tasks.
- The critical path cannot be identified.
- There are time periods in which no tasks are being done.
- There are too many concurrent tasks.

PROBLEM

2.29 What are the cash-flow projections used for?

2.30 If a spreadsheet is used to track the tasks, why is a Gantt chart used also?

2.31 Does the Gantt chart need to be graphical?

2.32 Can a bill of materials be used to track when parts should be ordered and when they arrive?

2.33 A new building is being constructed and the following tasks are required. The normal workdays are 7:00 a.m. to 3:00 p.m., Monday to Friday. Overtime is possible; however, the costs make it highly undesirable. Expected task times are listed.
Site preparation: 1 month
Foundations: 1 month after site preparation
Framing: 2 weeks after foundations
Plumbing: 5 weeks after framing
Electrical: 6 weeks after framing
Inspection: 1 week after plumbing and electrical
Drywall: 2 weeks after inspection
Painting: 1 week after drywall
Hardware: 1 week after painting
Carpet: 3 days after painting

(a) Write a Gantt chart for completion of the job in 3 months.

(b) Develop a project activity network.

(c) Identify the critical path.

2.10 Project tracking and control

Surprises in projects are bad, even when they are good.

A project plan is a best guess for how a project will be guided to success. Ideally the project will follow the plan exactly; practically it should be adjusted somewhat. If the plan and actual progress are relatively close we will say the project is in control. By control we mean that we can predict the outcome. When a project is in control the project team will be well prepared. When a project is not in control there are real and unmanaged threats to the project success. A good engineer and/or manager will recognize the signs of a project that is out of control. One particular fallacy is that saving time and money is always good. A project task that is finished quickly could mean that the job was done poorly or that work was missed. Budget savings could indicate that poorer quality components were used, the wrong parts were purchased, or that the original budget was incorrect. Some general warning signs, not necessarily problems, include the following:

- Tasks are done too quickly or are taking too long.
- Major budget items are not close to budget estimates.
- Details, such as expenses, are not clear.
- The specifications are being questioned or changed.
- Questions receive "fuzzy" or no answers.
- There are issues and questions that were not expected.
- Significant meetings are rescheduled.
- Parts of the work have started to blend together or overlap, or are forgotten.
- People have differing views about the project plan details, especially tasks.
- There are very strong emotions.

The progress of the team should be tracked by looking at details, large and small, on a regular basis. The review includes comparing actual with planned expenses, task progress, task completion, customer interaction, scheduling resources, and reporting to senior management. A good project manager will track projects informally every day and formally every week while asking questions such as, "Does the project match the plan?" When the plan and the project agree, the project will be deemed in control. To track the project formally, actual progress should be compared with the plan details. The schedule can be assessed by looking at the WBS or Gantt chart for tasks that are far behind or ahead of schedule. Likewise, the budget should be close to the actual expenses or money and time. Some of the tasks performed by a project manager are outlined in the following list.

Categories

- Budget: any point that marks the beginning or end of a major budget item
- Performance tests: when values are available
- Component completion: when major subsystems are complete

- Specification changes: if details in the specifications are changed or approved
- Preliminary: before major reviews to reduce questions and issues

Routine tasks

- Reviewing and approving customer invoices
- Reviewing and approving purchase requests
- Reviewing and approving budget items
- Tracking the project budget
- Arranging for shipping
- Routine correspondence
- Arranging travel and meetings
- Scheduling and approving employee work hours and overtime for payroll
- Working with suppliers for samples, technical support, and components
- Communicating with other corporate departments: manufacturing, HR, accounting, security, IT, etc.
- Maintaining corporate policies and practices
- Making commitments on behalf of the team
- Personnel reviews

Reflective questions

- How does this help the project?
- Is it good for the company?
- Is it good for me and my people?

Plan, act, assess, repeat.

At a high level, project tracking should be objective. A formal tracking tool can be used to aid in this process (Fig. 2.26). The worksheet can be updated at regular intervals and then reviewed to identify issues. Budget estimates and task times are relatively easy to assess. Other elements, such as risk, involve some consideration of larger issues. For example, a project that has large energy costs might need to monitor crude oil prices. Another concern would be a change in company ownership that might result in project cancellation.

Make decisions, get support, move forward.

Don't confuse apparent and actual progress.

Fig. 2.27 shows a stepwise procedure for reviewing plans. If the plan matches the expected progress, then it can move forward, possibly with some minor variations. Higher levels of uncertainty or threats to the plan may require escalation. Escalation is a process whereby upper level management is informed or asked to consider the project status. These sudden and/or major changes increase the chances for mistakes and losses. Some companies will set escalation procedures for project managers. For example, if the project expenses reach more than $10,000 above or $20,000 below the budget it may be necessary to inform the director of engineering. The worst situation, firefighting, is to have a project problem that must be addressed quickly and/or change the project in a substantial way.

The two extremes of control are hands-off and micromanaging. In a hands-off environment there is very little effort put into tracking and reporting, thus saving money. However, there are more lapses that lead to financial losses and time overruns. Micromanagers put extraordinary attention on every

Element:	Current status	Action items
budget spent? budget committed? design risk (worst case)? matches business mission? cost risk time risk projected completion plan is still relevant? unknowns? new or modified specs/details performance test results		

FIGURE 2.26

Project review worksheet.

detail and the reporting takes away from the value-added project work. The two curves in Fig. 2.28 show that with the level of control there is also an increase in control costs but a decrease in losses. Adding these two reveals a minimum in the overall cost. A project manager must learn to find the optimal point, accepting that there should be a reasonable level of losses and delays. We can say that a project that is "in control" has a predictable and acceptable level of losses and delays.

PROBLEM

2.34 What does "out of control" mean?

2.35 Is being under budget and ahead of schedule always good?

2.36 What should a project manager do when the engineers are saying inconsistent things?

2.37 Who tracks budgets?

2.38 List six items that a manager should track for each project.

2.39 When should a project plan be updated immediately?

2.40 If a project requires firefighting, is it out of control?

2.41 What can occur if a project plan is too specific, or the plan is tracked too eagerly?

2.42 How should a manager select a reasonable level of project control?

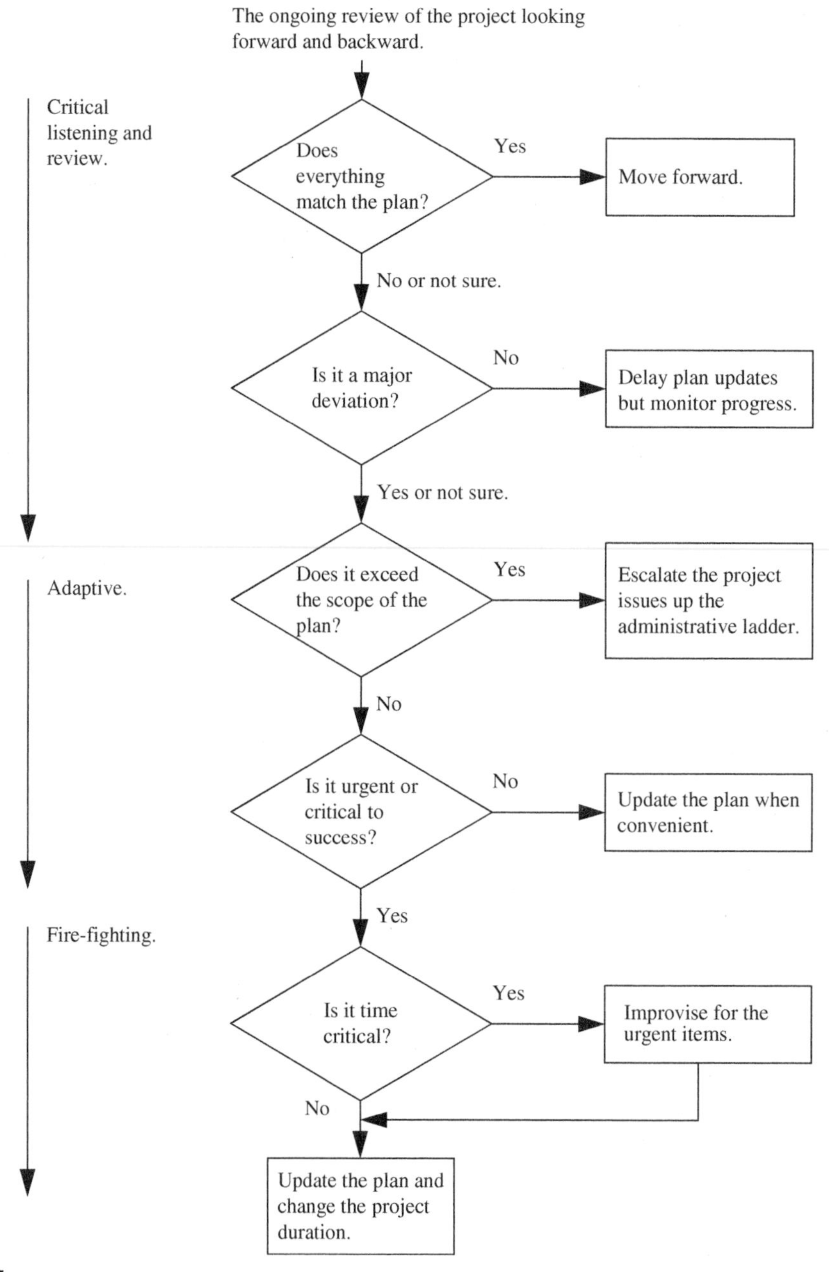

FIGURE 2.27

Situation review.

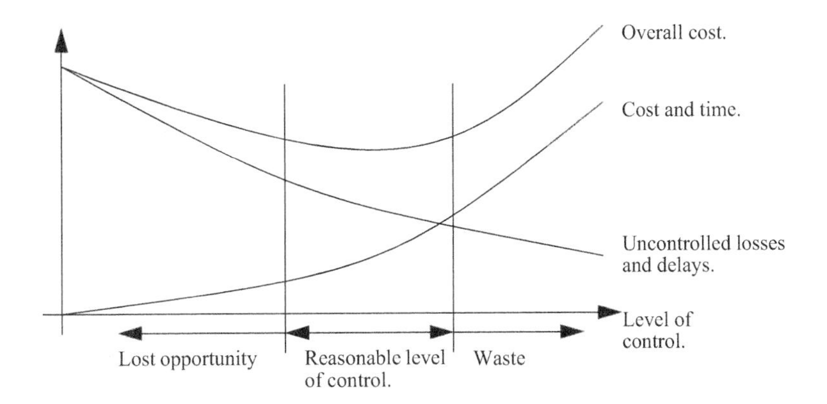

FIGURE 2.28

The cost of control.

2.11 Assessment

Learn from mistakes

Assessment comes after the routine business of the day. The assumptions and approaches used in project work will sometimes fail, or worse. When these things happen, consider what led to the issue and how it can be avoided or minimized the next time. Thinking or talking about these issues can provide insight into how the problems occurred and possible solutions. Another approach is to use a simple assessment table for the assumptions and new approaches for each project or design (Fig. 2.29).

Plan–do–check–act (PDCA) is a four-step management method used for continuous improvement. The focus is on the assessment process for a specific task. The basic steps are as follows:

- Plan: Define the target outcomes and the processes and methods to achieve them.
- Do: Do the work and collect data for assessment in the "check" and "act" steps.
- Check: Review the actual work and the plan while looking for deviations. If possible, use numerical metrics for long-term tracking.
- Act: Make changes to the "plan" and "do" steps to improve the process.

PROBLEM

2.43 Why is assessment used?

2.44 What will happen if assessment is not used?

2.45 Draw a flowchart illustrating the PDCA approach.

2.46 Develop a reasonable plan for a one-semester school project to build a robot.

2.47 What is the objective of a design review?

	Good	Impartial/not sure.	Bad
budget estimates			
resources / people			
sequences			
important and unimportant tasks			
value added vs. non-value added			
accurate estimates of task time (early, late, variations)			
personal style			
management methods			
	What should we continue to do or duplicate?	Opportunities	What must be changed or anticipated?

FIGURE 2.29

Assessment table.

2.48 Mini-case: Progress

Anetha and Bjorn are given a programming challenge. The program must generate 50 random numbers and then find the median, average, standard deviation, and 90th percentile. Anetha decides to sit and write the entire program quickly and then debug. Bjorn decides to break the program into five parts that he will write and then debug.

Anetha makes fast progress and within an hour has the program written and begins to debug. As expected, there are a few minor syntax errors when she compiles the program but they are fixed quickly. The first time her program runs completely, the program output is a mix of numbers and characters. She is not sure if the problem is in the random number generator or the calculations, so she decides to change the random number generator. At 1 h and 20 min, her 140-line program was 100% done but 0% working.

Bjorn took 20 min to write the 30-line random number generator. It took him another 15 min to debug the program. He then spent 15 min calculating the average and 10 min debugging his 30-line program. In another 20 min he had added and tested 20 lines to calculate the standard deviation. At 1 h and 20 min, Bjorn had written only 80 lines of code that were working.

Anetha continued debugging by changing parts of her program, and saw the numbers begin to look more correct. By 2 h, her average was near the middle of the range and her standard deviations numbers were around a sixth of the range, both good signs. However, her median and 90th percentile numbers were clearly wrong.

Bjorn spent 20 min writing a number sort routine and 10 min debugging the 30 lines of code. In another 10 min, he had written and tested the 10 lines of median code. At 2 h, Bjorn had 120 lines of code working. He spent another 10 min to complete and test the 90th percentile calculation and was complete at 2 h and 10 min.

Anetha continued debugging until 2 h and 40 min, at which point the numbers looked correct, and she declared her program working.

In total Anetha spent 1 h and 20 min programming, compared with Bjorn, who spent 1 h and 35 min. However, Anetha spent 1 h and 20 min debugging, compared with less than an hour for Bjorn. In simple terms, Bjorn was looking for 1 problem in a 30-line program, while Anetha was looking for 1 line in a 140-line program. Assuming the same number of errors, the probability of finding the error for Bjorn was roughly 1/30, but for Anetha it was 1/140. Anetha had made more apparent progress but Bjorn had made more actual progress. Draw timelines for Anetha and Bjorn. How would a WBS differ between Anetha and Bjorn?

2.49 Inexperienced designers will often become stuck trying to make the right decision. For example, an engineer knows that the wrong decision will cost a day of wasted time. The engineer then spends 3 days looking for the right solution. Write a simple rule to avoid the trap.

2.50 Mini-case: Purchasing expertise

Cutco needed a workcell to cut metal plates. The manager had found a similar design; it did not meet their specific needs but illustrated the main design features. The design team at Designtec was asked to develop a similar design. The main components of the design were a robot, a plasma cutter, a protective enclosure, and a material feeder. The specifications were easily determined using the similar design and the requirements from Cutco. After specification approval, Designtec began the detailed design and kept Cutco informed of its progress. It became obvious that Designtec did not have enough expertise or time to design and build a reliable material feeder.

Cutco and Designtec met to discuss alternatives, and decided that they would find a supplier to make the feeder to specifications. Within 2 weeks a subcontractor, Rollerzip, had been found and a purchase order written for the equipment to be delivered in 6 weeks at a relatively low cost. The other stages of the design were done, with the exception of the mechanical and electrical connections to the feeder. Designtec began to purchase the components for the cell and construction started. During the later stages of the detailed design, and while the cell was being built, the team contacted Rollerzip a number of times to ask for interface details. After numerous calls and messages, Designtec was told that Rollerzip was still working on the design, but it was proceeding well.

The 6-week delivery date came, and passed, without delivery of the material feeder. Designtec completed the building and testing work and then stopped, waiting for Rollerzip to deliver the material feeder. Over the next 2 months there were numerous phone calls and emails sent to Rollerzip by Designtec requesting technical details and by Cutco requesting delivery. Most calls and emails went unanswered. A couple of times they reached the owner's wife, who assured them the owner was working on the material feeder. Two months after the delivery date, Rollerzip began to provide technical details about the electrical and mechanical interface. By the third month, the material feeder was delivered and the work on the cell was completed, 2 months late.

(a) Suggest methods that could have been used to avoid the material feeder delays.

(b) If Designtec had done the design work itself, it would have spent more time building and debugging but at a comparable cost. What was the risk trade-off for using a subcontractor?

Further reading

Baxter, M., 1995. Product Design; Practical Methods for the Systematic Development of New Products. Chapman and Hall.

Bennett, F.L., 2003. The Management of Construction: A Project Life Cycle Approach. Butterworth- Heinemann.

Bryan, W.J. (Ed.), 1996. Management Skills Handbook; A Practical, Comprehensive Guide to Management. American Society of Mechanical Engineers, Region V.

Cagle, R.B., 2003. Blueprint for Project Recovery—A Project Management Guide. AMCON Books.

Charvat, J., 2002. Project Management Nation: Tools, Techniques, and Goals for the New and Practicing IT Project Manager. Wiley.

Davidson, J., 2000. 10 Minute Guide to Project Management. Alpha.

Fioravanti, F., 2006. Skills for Managing Rapidly Changing IT Projects. IRM Press.

Gido, J., Clements, J.P., 2003. Successful Project Management, second ed. Thompson South- Western.

Heldman, K., 2009. PMP: Project Management Professional Exam Study Guide, fifth ed. Sybex.

Heerkens, G.R., 2002. Project Management. McGraw-Hill.

Kerzner, H., 2000. Applied Project Management: Best Practices on Implementation. John Wiley and Sons.

Lewis, J.P., 1995. Fundamentals of Project Management. AMACON.

Marchewka, J.T., 2002. Information Technology Project Management: Providing Measurable Organizational Value. Wiley.

Portney, S.E., 2007. Project Management for Dummies, second ed. Wiley.

Rothman, J., 2007. Manage It! Your Guide to Modern Pragmatic Project Management. The Pragmatic Bookshelf.

Salliers, R.D., Leidner, D.E., 2003. Strategic Information Management: Challenges and Strategies in Managing Information Systems. Butterworth-Heinemann.

Tinnirello, P.C., 2002. New Directions in Project Management. CRC Press.

Verzuh, E., 2003. The Portable MBA in Project Management. Wiley.

Verzuh, E., 2005. The Fast Forward MBA in Project Management, second ed. John Wiley and Sons.

Wysocki, R.K., 2004. Project Management Process Improvement. Artech House.

CHAPTER

Customer requirements and specifications

3

3.1 Introduction

Needs, customer specifications, concepts, and technical specifications are a compromise.

A design project normally starts with the process of discovery during the initiation stages of a project (Fig. 3.1). The first step begins with (1) defining the need for the project, (2) developing detailed customer specifications to guide the design, and (3) accepting the specifications. This process goes by different names, including framing, problem definition, and scoping.

All of the topics in this chapter support the single goal of developing customer specifications. Well-written specifications help to define the project and guide the work. Customer specifications are developed from the needs. In simple terms, the needs are a mixture of quantitative, qualitative, and intangible factors. Needs come from a variety of sources with different motivations and expectations. Examples include:

- inventors: a perceived need;
- entrepreneurs: a project essential to establishing a new business;
- sponsors/customers: a group that comes with a previously established need; they may also provide specifications;
- yourself: a self-identified project that has some value to solve your own needs;
- social: a humanitarian project motivated by helping people in need;
- competition: a design objective constrained by contest rules.

It is essential to have a clear understanding of needs, to establish expectations for final deliverables. Without clear needs every solution can be accepted or rejected on a whim. With less experienced customers, such as inventors, part of the job will involve clarifying the needs. Once the needs are clearly established the customer specifications can be developed. Detailed specifications are beneficial to the customer and project team because they (1) ensure a clear understanding of deliverables throughout the project and (2) control the work, budget, and delivery date for the project. Agreement between the customer specifications and deliverables is required for a successful project. Developing detailed specifications is not meant to be an adversarial process; the enemy is ill-defined specifications. Always work toward a win—win set of specifications so that you know what to deliver and your customer knows what to expect.

When it is not possible to establish a clear set of needs or customer specifications for a project, it is unwise to advance to the conceptual design phase. A wise approach is to create a pilot project that has a goal such as developing a draft set of needs and technical specifications, creating a testable prototype, identifying problems, or refining user interfaces.

Engineering Design, Planning, and Management. https://doi.org/10.1016/B978-0-12-821055-0.00003-7

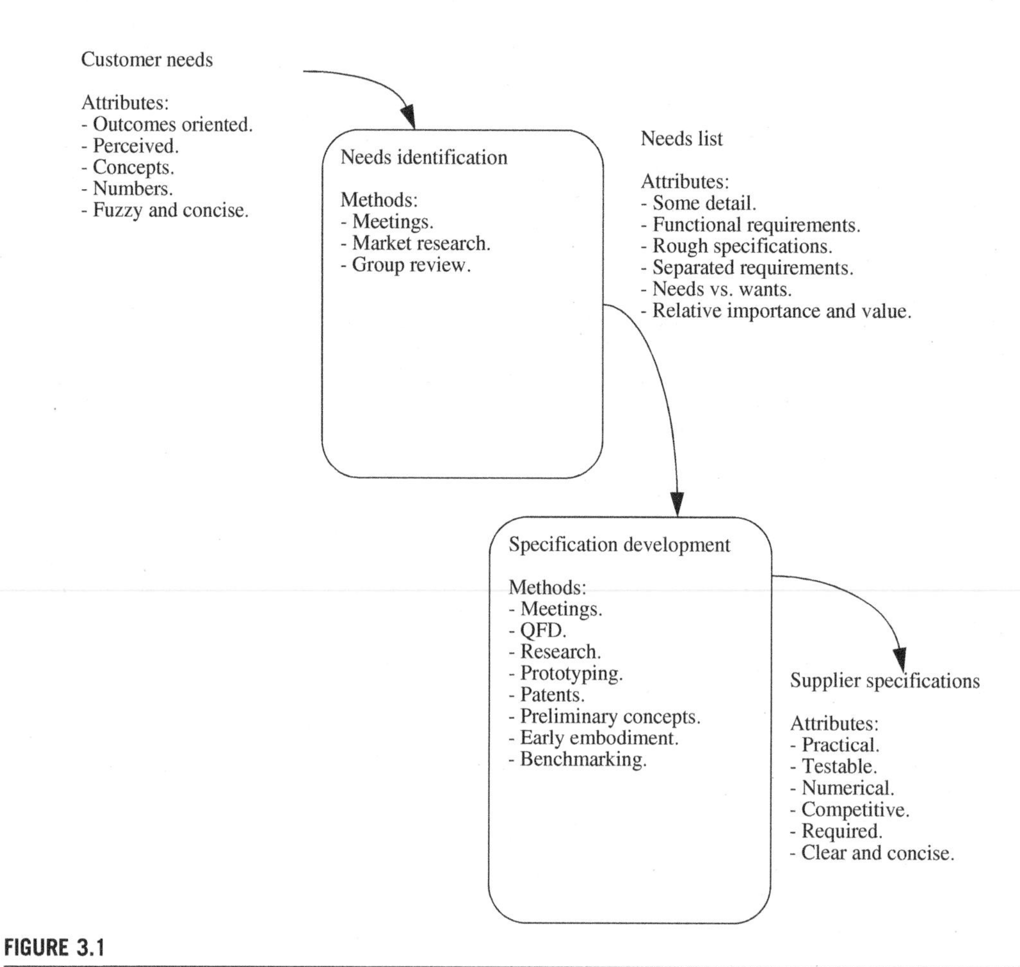

FIGURE 3.1

The initiation stages of a project. *QFD*, quality functional deployment.

In commercial design, there is typically a customer and a supplier. The customer expresses needs and the supplier develops customer specifications. This process finishes successfully when a customer places an order for the design. A quotation, or quote, is a formal document prepared for a customer by a supplier. It outlines the work to be done and the cost. A customer will accept a quote with another formal document, such as a purchase request. In a process in which multiple suppliers are competing, the quotes are called bids. The process of developing quotations varies by industry, business, and project type. The request for quotes (RFQ) bidding process is used when an experienced customer performs substantial planning before talking to suppliers (Fig. 3.2). These customers have already considered the design needs, how much they are willing to pay, and detailed technical specifications for the design. An RFQ is created for suppliers, who examine the needs and specifications and prepare quotes. The three critical business decisions for the customer are releasing an RFQ, selecting a bid/quote, and accepting the final deliverables. The critical business decisions for the supplier are deciding to prepare a bid, submitting a bid, and issuing an invoice. This is the preferred approach for large projects.

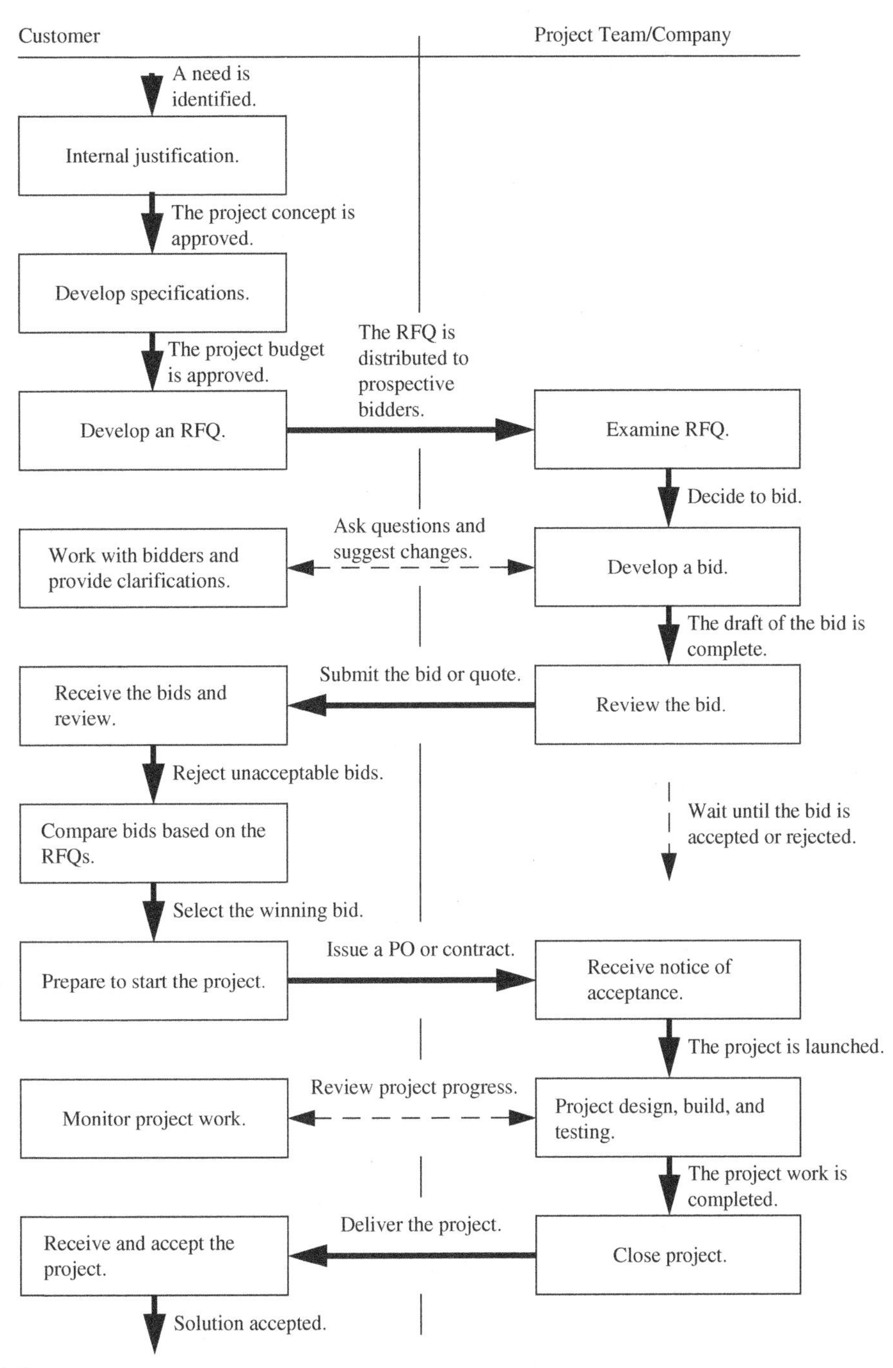

FIGURE 3.2

A request for quotes (*RFQ*) competitive proposal process. *PO*, purchase order.

A faster and simpler quoting process is used for a single supplier, as shown in Fig. 3.3. A single supplier is selected to do the design work, but a quote is still used as a decision point for both customer and supplier. This approach can be used between business divisions in a company or with separate companies. For example, a manufacturing department could be a customer looking for a new test station. The project team is the engineering department, acting as an internal supplier. The quote is a budget request and

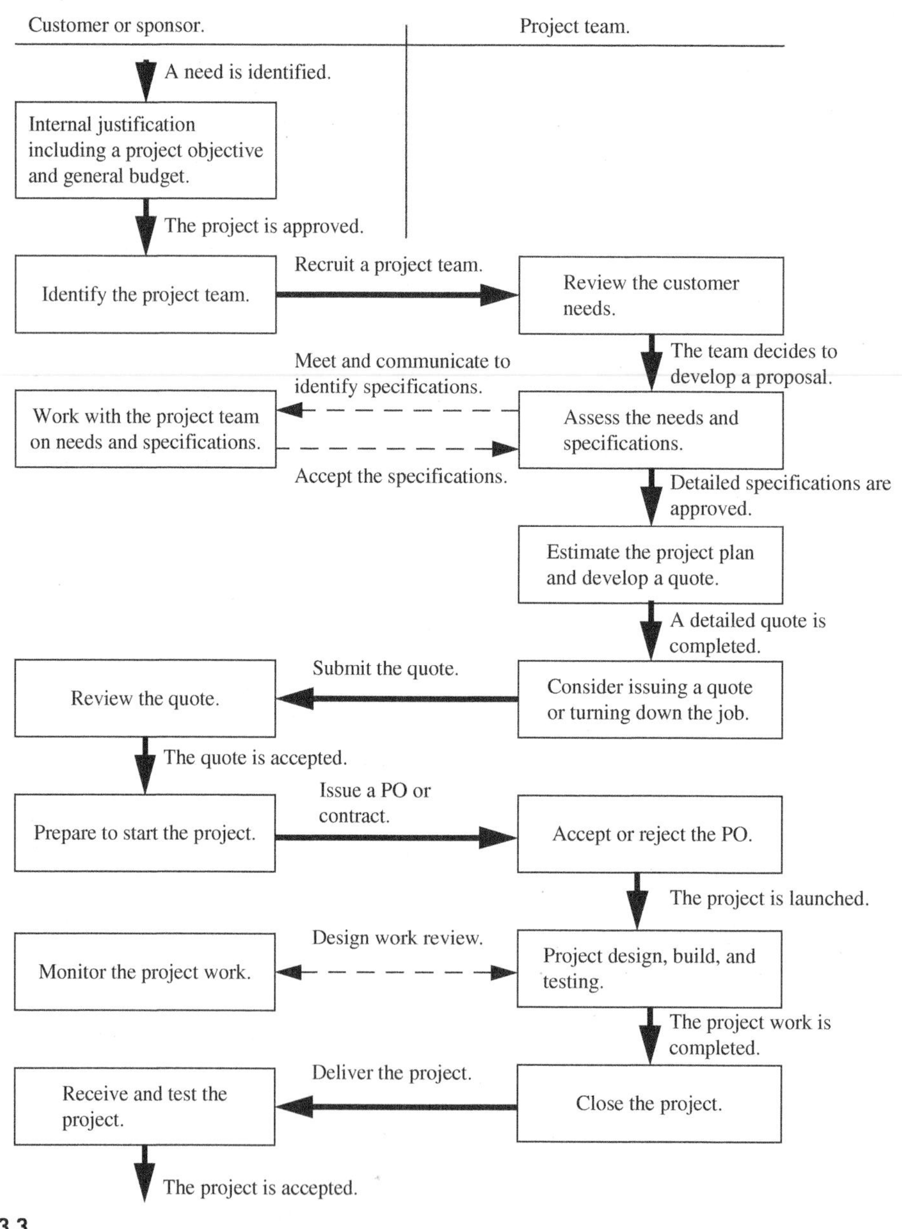

FIGURE 3.3

Preferred supplier quote development. *PO*, purchase order.

payment is a transfer of money between accounts in the company. Another example is a homeowner looking for a new backup generator installation. A supplier is contacted to generate a quote and do the work. In these approaches the supplier develops the customer specifications from the customer needs. This approach is used when a project requires specialized skills and knowledge. Often the project customer and supplier are the same person or group. Internal projects are less formal but still follow the process of proposals and approvals, as illustrated in Fig. 3.4. These projects require a person, often called a champion, who develops the project from the needs to the closure phase. At some point early on, the champion will develop a proposal for management and request a budget for the work. At all points the champion is considering the progress of the project work, as illustrated by the rightward arrows.

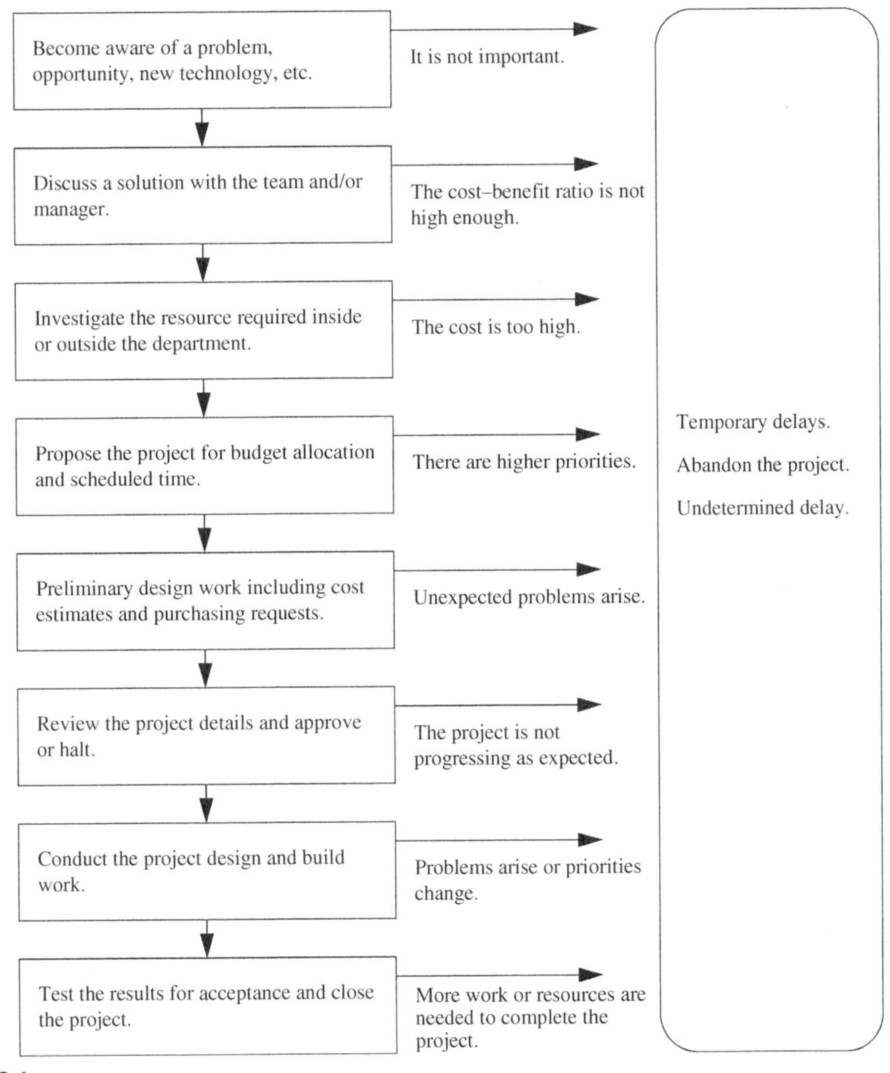

FIGURE 3.4

Internal department or individual project.

Given the arbitrary nature of the specification selection process, the process will need to iterate until a set of specifications is acceptable to the project team and the customer. Only when both are satisfied that the specifications are reasonable should they be approved and the project move forward to the conceptual design phase. Fig. 3.5 shows a sample procedure.

From an abstract perspective, perceived customer specifications should converge to a final set of detail specifications (Fig. 3.6). If the detailed specifications are developed too quickly, there are probably misunderstandings and rushed decisions. Eventually, the perceived specifications will converge on an acceptable set of detailed specifications. Good specifications are an acceptable compromise between the customer and the supplier objectives. A small but recognizable difference between ideal and accepted customer specifications is normal and indicates a healthy process. If the refining process takes too long, it probably means that more compromise is needed.

The most common and troublesome project issue is called feature creep. Once the project needs and specifications are accepted, the customer often asks for modifications or addition of other functions and specifications. Normally, these are presented as trivial additions to the design, but they usually increase the overall cost, delay the project, add complexity, require backtracking, and increase the risks. After the customer specifications have been accepted, changes should require negotiation. The absolutely critical steps at this point are (1) freeze the detailed specifications by agreement and (2) provide a mechanism for considering specification change requests, including schedule, budget, and deliverable requirement modifications. In simple terms, if the specifications are suitably detailed, a customer may request changes, but he or she can be asked to accept budget and timeline changes. Other problems that arise during specification development include the following:

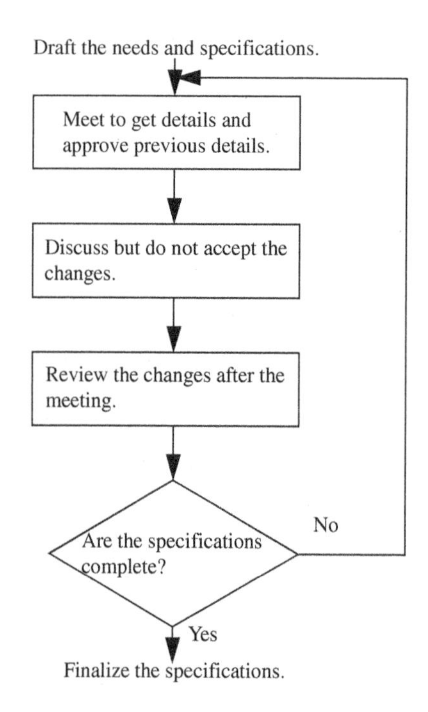

FIGURE 3.5

Refining project needs.

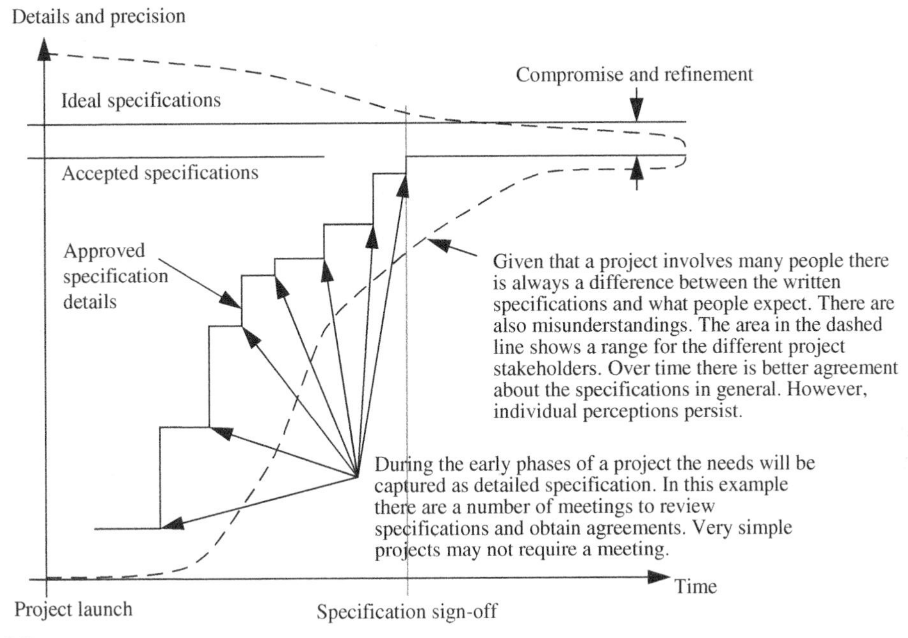
Details and precision
Compromise and refinement
Ideal specifications
Accepted specifications
Approved specification details
Given that a project involves many people there is always a difference between the written specifications and what people expect. There are also misunderstandings. The area in the dashed line shows a range for the different project stakeholders. Over time there is better agreement about the specifications in general. However, individual perceptions persist.
During the early phases of a project the needs will be captured as detailed specification. In this example there are a number of meetings to review specifications and obtain agreements. Very simple projects may not require a meeting.
Time
Project launch
Specification sign-off

FIGURE 3.6

Refining customer specifications.

- Details are "left for later."
- Specifications are vague.
- Specifications cannot be tested for project closure.
- Performance measures are not clear and measurable.
- Critical details are omitted, such as training and documentation.
- Details intimidate and delay decisions.

PROBLEMS

3.1 Can you be a project customer and supplier?

3.2 When a supplier is developing a quote, what can a customer do?

3.3 What is the difference between an RFQ and a quote?

3.4 Describe a multistep process for developing customer specifications.

3.5 Propose a numerical approach to measuring the progress of specification development.

3.6 Assume that a draft specification is "a comfortable weight." Brian wants to wait until a prototype is designed, built, and tested with a customer. Fernando wants to use a target of 650 g and adjust it later. Who is right? Why?

3.7 Customers sometimes request changes after a quote is accepted. Can the supplier object if the quote does not include the specifications? How can the problem be solved?

3.8 List three options if a customer makes a request that exceeds the specifications.

3.9 Is "aesthetically pleasing" acceptable in specifications? Explain.

3.10 Mini-case: I already knew what to do

Engineers are educated to invent. The inventive spark is often ignited when observing some daily inconvenience. The hero of our story, Ahmed, was first touched by the flame of invention in a forest. As a hiker he would travel long distances. On one particularly hot day, he stopped to quell his growing thirst with water. Reaching for his canteen he realized that he had filled it but forgotten to clip it to his pack. During the remainder of the trip, Ahmed's burning thirst brought him to the realization that a solution would help him and many other outdoor sports enthusiasts. He imagined many possible solutions and eventually settled on a special hook for a canteen. If shaken the hook would beep unless a canteen or something similar was attached. The clip would remain attached to the backpack, waiting for the next trip with the canteen. With a little thought he conceptualized something in the shape of a carabiner with a solar panel to recharge the battery. He even went as far as thinking about a sensor to detect the bottle, an accelerometer for motion detection, a microprocessor, and a milled aluminum frame to hold it together. By the end of the hike, Ahmed had a very good idea about the device he would build.

At home, Ahmed calculated that he needed $15,000 to apply for a patent, purchase components, and do some sales work. He had $5000 and approached the bank for a $10,000 loan. The loan manager asked to see a set of the specifications and a patent search. Ahmed developed a set of specifications that outlined his design that included the following: (1) accelerometer, (2) microprocessor, (3) 5-cm hook for canteen, (4) solar cell and battery, (5) contact switch, (6) software, and (7) cost $20 to make; $80 to sell. (Note: This is a poor example of specifications.)

Ahmed took the specifications and ideas to a patent lawyer and paid $1600 for a patent search. The search did not find any similar systems for canteens, but there were a few patents for similar construction equipment. Ahmed used the search results and specifications to obtain the loan. He then applied for a patent, built a prototype, and approached outdoor sports companies. Each company told him that the device was too expensive.

After some disappointment Ahmed consulted with a product engineer, Saed, who explained that his problem was fundamental: his specifications were for the solution, not the problem. Saed helped Ahmed rewrite the specifications to read (1) hold a 2-cm canteen hook, (2) detect accelerations over 30 m/s^2, (3) detect when a weight of 200 g to 5 kg is attached, and (4) provide an audible 80-dB alarm when it is in motion with no weight attached. Saed then led Ahmed through a new concept development process in which they developed a mechanical-only solution with a small metal piece that would bounce with a loud noise when the canteen was not holding it in place. The new design could be added to existing clips and cost $4 or less in retail stores.

Ahmed made the mistake of planning ahead in the project and then making the other project steps fit his plan. He should have still performed a seven-step, or equivalent, design method. Expand the list of detailed specifications to 10 items. Be careful to avoid suggesting the solution.

3.2 Needs

Fuzzy design objectives will mean more work later.

Needs start the design process. The sources of the initial needs will vary wildly, but at the end of the needs identification stage we must have something that will drive the design process. Given that the needs will be used to develop the customer specifications it is a good idea to identify all of the necessary, assumed, and desired needs. It is even better if the needs are expressed as measurable, or testable, qualities. In fact, the needs could be a draft version of the specifications. However, these must be reviewed technically before finalizing them as specifications. One common process error is to assume a solution and then select the needs and specifications for that solution, hence constraining you to a single design. To determine needs you should:

(1) Form a general idea of the problem and the motivation for a solution.
(2) Further define the problem and need.
(3) Check the need for completeness and consistency.
(4) Iterate as necessary.

Customer needs can be captured using a form, like the example in Fig. 3.7. The form provides a few prompts that are often found in needs statements. These needs can be captured as free-form bullet lists, sketches, diagrams, photographs, and so on. Sometimes a customer will be able to describe their needs in detail.

Sometimes a customer may not be able to express a clear set of needs, and the supplier will need to develop these, too. A detailed description of customer needs should focus on what the design needs to do, not how to do it.

Restated, the needs should avoid requiring a specific design implementation. Examples of needs include:

- be able to withstand hurricane/typhoon winds
- use international electrical outlets
- be easy to carry
- last 10 years
- have storage for 10 boxes
- use an engine from the same manufacturer
- be fun
- be similar to a competitor's design, but avoid a patent
- copy an existing product or design (also called reverse engineering)

After the needs have been captured, the process needs to work toward customer specifications. Naturally some of these will be established during the needs-capturing phase, but after this point the other needs must be converted to testable, designable, and buildable functions and values. Some of the needs are easily translated to technical specifications, while others are vague and hard to define. For example, "be fun" could mean many things, and the designers and customers will interpret this differently. This need has to be refined before trying to develop technical specifications.

Design Project Need Definition and Assessment	
INFORMATION COLLECTION	
Need Description:	Who, what, where, when, why, how?
Importance:	critical, important, useful, optional
Function:	Operation Numerical performance measures Testable values Minimums, maximums, ranges, and ideals Other constraints Mass and dimensions
Requirements:	Timeline and critical dates Cost constraints Legal Published standards or regulations (e.g., ASTM, IEEE, SAE, OSHA, BIFMA, NFPA) Available power, utilities, and facilities
Complex Needs:	Aesthetics Usability Feel Documentation Safety approval
NEED ASSESSMENT	
Design Type:	Commodity Reverse engineered Incremental Revolutionary Niche/specialty Consumer Similar designs
Conceptual Stage	Design type unknown or vague A specific design type has already been chosen
Quality of information and stage of understanding	The provided problem description is vague. You are not ready to ask the sponsor questions yet. Read the description in detail and break it down into small "requirements."
Acceptance and testing	Written acceptances including tests, rates, limits, counts, measurements, etc. Yes/no checklist items Tolerances Standard tests - if they do not exist specify your own
Possible Implementation:	technology, method, existing solution

FIGURE 3.7

A needs worksheet.

PROBLEMS

3.11 Is it acceptable to have specification numbers in a needs list?

3.12 Is it fair to say that needs come from the customer and the specifications come from the supplier? Explain.

3.13 Write a reasonable list of at least five customer needs for a package of pasta.

3.14 Use a needs worksheet to define the need for a package of pasta.

Design project need definition and assessment

Information collection

Need Description:
1 A package of pasta that is fresh
2 Easy to store and handle
3 Appeals to customers

Importance:
1 Important: older pasta loses taste and changes texture
2 Important: a package that breaks results in waste
3 Critical: customers must want to have it

Function:
1 Air seal
2 Contain the contents
3 Aesthetic

Requirements:
1 Resist flow of oxygen and humidity
2 Must not fail when pushed or pulled
3 A careful choice of materials, colors, and shapes

Complex Needs:
The product packaging must be easily recognized and understood by the customer. At the same time it must look distinctive enough to catch attention.

Need assessment

Design Type:
The design should be commodity in nature with incremental improvements on markets and technologies. However, the aesthetic requirement of customer appeal will create many problems and should be addressed before detailed design.
Conceptual Stage:
The standard forms are boxes and bags with windows to make the pasta visible. There are a variety of methods for opening and resealing the package, including flaps and plastic zipping seals.
Quality of Information and Stage of Understanding:
The general information is enough for a routine design. Some level of complexity can be achieved by studying competitors designs.
Acceptance and Testing:
The design can be tested by using variable temperature and humidity test chambers. Drop tests will

Design project need definition and assessment

determine package toughness. Aesthetic design can be assessed with market surveys.
Possible Implementation:
Simple A: A printed card-stock box with a transparent plastic window.
Simple B: A heat-sealed plastic bag with printed packaging.
Complex: A blow-molded plastic bottle with a resealable cap.

3.3 Research

Reinventing the wheel is not useful because many others have probably identified needs and associated customer specifications. They have also done the work to find the blind alleys and successes. Tricks that can be used to gather this information are listed below.

Competitors' products

- Searching—Use the Internet to search for similar designs. For consumer products, search sales and auction websites (e.g., Amazon or eBay). For industrial components and processes, search in general or use industrial search resources (e.g., ThomasNet.com or GlobalSpec.com).
- Shopping—For a number of products, you can drop by sales outlets and look at the alternatives. In some cases you may buy the product to look at later.
- Catalogs—Standard suppliers will often have exhaustive lists of parts, data, and costs.
- Contact—Phone or email some of the known suppliers and ask questions. If you are planning to purchase their systems a supplier may be very helpful. You might even be able to get references to other experts.
- Manufacturers—Many manufacturers maintain and freely distribute product data sheets, manuals, application notes, product brochures, and more.

Informal

- Crowd sourcing—Use public groups to develop ideas.
- Internet—There are many professional websites where you will be able to find opinions, technical reviews, group discussions, etc. These can be very valuable sources of unfiltered information. Even inaccurate opinions can provide value if reviewed critically.

Technical

- Consultants—Paying for advice and knowledge is an option if suitable consultants are available.
- Library—Look for references in public and private libraries. Buying books is always an option.
- Internal—Find internal people to talk to who have similar experiences.
- Network—Find people you know who may have advice or suggestions.

Requirements

- Legal issues—Liability
- Intellectual property—Patents, trade secrets, ownership
- Testing—Acceptance testing
- Standards—UL, CSA, CE, SAE, IEEE, NIST, ASTM, BIFMA, ANSI, etc.
- Regulations—NEC, NFPA, FAA, FDA, NHTSA, FCC, etc.

- Additional supplier requirements—FMEA, ISO 9000/14000/26000, supply-chain management, etc.
- Safety

PROBLEMS

3.15 Find five sources of information and a standard for breakage forces for residential window glass. Provide details for the resources, including URLs, references, or paper/electronic copies.

3.16 What are the CE and ASTM standards for wax crayons?

3.4 Benchmarking and surveys

People don't always know what they want.

Industrial customers generally understand what they want and why they need it. Public consumers have needs that are less defined. Larger companies have marketing departments that identify market segments and customer needs. In smaller companies marketing is performed by engineering, sales, and management.

Consumers state their needs in opinions and with purchases. Current and past user needs can be inferred from sales numbers and current product features. The common name for this method is benchmarking:

- Look at desired and proposed features for existing products.
- Select products that are current or may be future competitors.
- Use customer feedback, such as surveys, to select the most important new product features.
- Prioritize the features for engineering development.
- Benchmarking outputs
  - a list of competitive products
  - a list of features of features
  - consumer perceptions of the device features
  - an engineering analysis of key components
  - your devices tested against the same criteria

Opinions capture future need. Initial consumer opinions state what they think they want, and these opinions serve as an excellent starting point for investigation. Market surveys and tests are used to develop hypotheses and then test them statistically. Market surveys begin with a statement of a perceived market segment and need. If the needs statement proves true, it is refined and the process is repeated. The eventual outcome is a detailed picture of the range of customers and range of needs. The greatest marketing mistakes are caused by trying to make the target market too broad or narrow. Broad marketing plans try to be everything to everyone and fail to more focused products. Narrow marketing plans prepare a design for one consumer and assume that others will decide to adopt it. Marketing tools include the following:

- Survey opinions
  - actively seek, or passively review, data
  - customers

    - paid study groups
    - review of public information such as Internet discussion forums
- Multiple surveys and testing to refine the detail
- Use of scientific methods for hypothesis testing
- Developing critical questions such as
    - How often will this be used?
    - What feature is the most important?
    - Do you own something similar?
    - What would help you select one product over another?
- Market study outputs
    - estimated market size(s)
    - minimum features (this can include numbers)
    - desired features
    - relative feature importance
    - intangibles and observations
    - basic price and feature values
    - competitors

A formal method for incorporating consumer needs into the design process is quality functional deployment (QFD). This method, discussed later, is used to rank product features by using customer demands and features found in competitors' products. The outcome is a relative ranking of customer specifications that can be used to focus engineering efforts where they will have the most market impact. Less formal methods include specification review using customer opinions and benchmarking data.

PROBLEM

3.17 Find three companies that produce 10-kW audio amplifiers and identify similar products from each. Identify a website that discusses and reviews the products. Read reviews and find 10 features/specifications that are mentioned as advantages and disadvantages for each of the amplifiers. Estimate customer importance and satisfaction for each. Use a scale that ranges from 0.0 ("don't care" or "don't like") to 1.0 ("must have" or "very happy").

3.5 Market-driven design

Revolutionary or evolutionary change?

A market is a collection of individuals who want different solutions to their own particular needs. The narrowest extreme is a market of one individual. This is relatively easy to identify, but limited in a commercial sense. The broadest extreme is all of the people who could be in the market, requiring that any design "be all things to all people." Broad designs are remarkably difficult to define because each person has design features that he or she requires, some that he or she wants, and some that are less important. Consequently, broad designs usually end up satisfying no one fully.

Identifying a market requires that a group of customers be defined by the features or specifications that they require, they want, and are relevant. The task of identifying the market, product features, and specifications is often done by sales and marketing professionals. For our purposes the descriptors markets and customers will be used interchangeably, assuming that they have been grouped by needs. Typical categories of customers include the following:

- a single customer who is well known
- a large market base that must be characterized statistically
- a complex market that has many different identifiable groups
- an unknown customer base

Whether designing consumer products or a single piece of industrial equipment, it is valuable to recognize some of the factors that influence individual customers. Simple examples of these variations include floor cleaner dust collectors that are transparent in North America but opaque in Japan so that dirt is not visible.

Computer mice for computer gamers may be "technology black" and have over a dozen buttons, but a mouse for young children may have a single button that is shaped like a cartoon character. A piece of equipment designed for the Canadian marketplace may use 120 V 60 Hz AC for the controls and provide user interfaces in English and French. A similar piece of equipment for Indonesia may use 220 V 50 Hz AC for the controls and have a user interface that uses colored pictures that are language neutral. Some of the factors used when defining customers are:

- region
- culture
- personal history
- interests
- ethics/morals/religion
- financial standing

Most designs have some sort of predecessor. When a design is a major departure from previous approaches we call this a revolutionary design. However, almost all design work makes evolutionary, or incremental, improvements on existing designs. One way to consider designs is as a set of features. Examples of features for a laptop computer include cameras, CD drives, serial ports, USB ports, printer ports, wireless networking, and so on. Over time the value of each feature will change. A new feature can be expensive to add and appeal to only a small number of consumers. Over time the new feature becomes expected, and eventually becomes obsolete. For example, CD drives were new and expensive features in the 1990s. These were often expensive options for new computers. Over the subsequent 2 decades these were replaced with faster and higher-capacity drives at lower prices. As of 2010, many computers began to use other forms of storage with other advantages, and many expensive laptop computers are now offered without any CD drive option. Older features cost money to include, but do not add much value to the consumer. These legacy features are eventually removed from the standard product design.

A designer must decide how many evolutionary changes, or new features, should be added. If there are too many, the user will end up paying a premium for features he or she does not need. If there are too few features, then the design will be stale and probably obsolete. The Kano curve shows this trend (Fig. 3.8). Over time the new features become expected features and the customer looks for newer features. Techniques such as benchmarking and the house of quality techniques can be useful (see Section 3.2). Some important points are:

- Over time new features are added; these are novel at first, but become standard, and then obsolete.
- Over time the number of features will increase, but the consumer no longer considers older features to be special.
- A product with a large number of new features often exceeds customers' needs and maximum purchase price.

The Kano model concept is valuable for setting an expectation for design. Primarily, the market rewards incremental improvements, but punishes too little or too much change. A fast approach to evaluating a feature is the neighbor test. Imagine purchasing a new product, taking it home, and meeting your neighbor on the way. You excitedly show them your new purchase. How do you reply when he or she asks, "Why did you need a new one?" If you can answer in one clear sentence the improvement is "understandable." If it costs less and does more it is "acceptable." Is it "exciting"? Consider a new car with an advanced braking system. You might tell the neighbor, "The new brakes cost a little more, but they are safer." Most neighbors would ask, "Were the old brakes not working?"

The neighbor test: How would you reply if a neighbor asked why you needed a new product?

Fig. 3.9 shows the life cycle of a new design. Early in the design life there are high development costs required for each copy of the design. However, at this point, the number of features and durability are low. Over time the design evolves, more are sold, the cost per copy drops, and the quality and number of features improve. People who buy designs in the early stages are called early adopters and they pay a

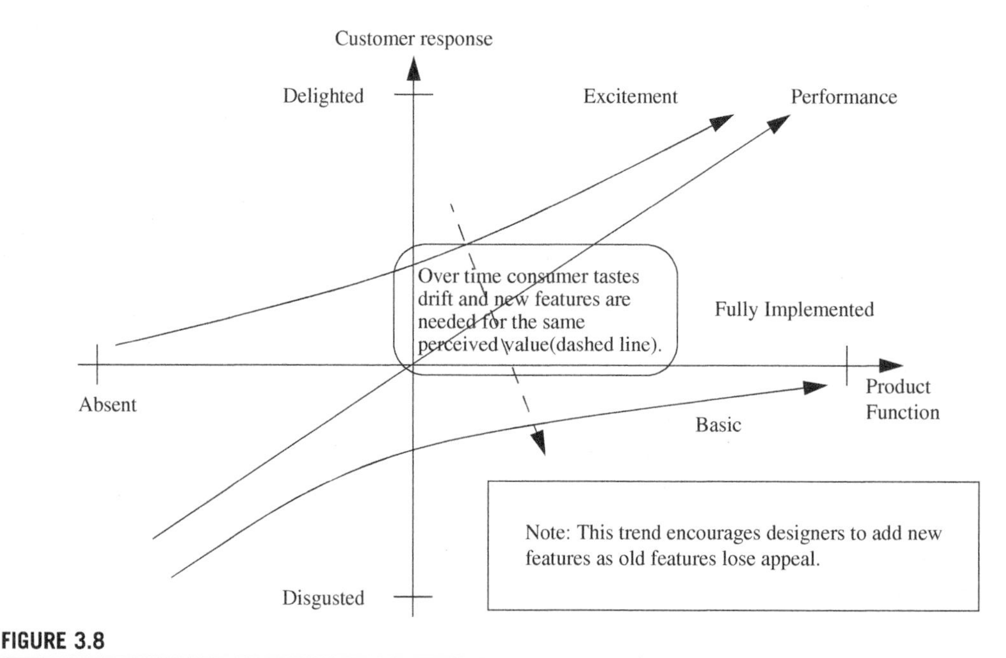

FIGURE 3.8

The Kano model of design features.

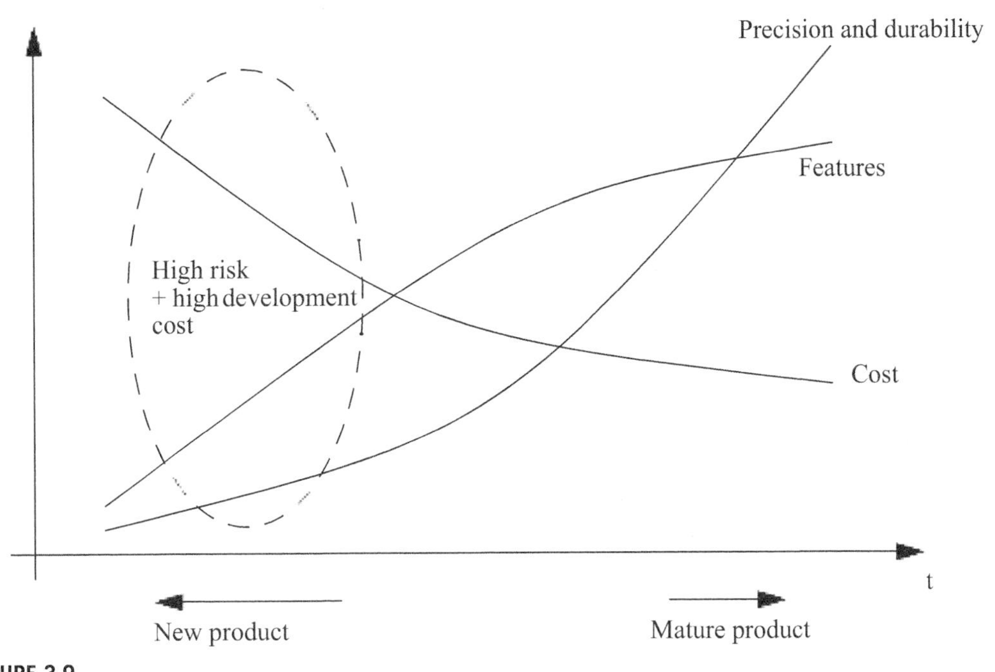

FIGURE 3.9

Design features.

premium for the newer design. A good example of this trend is the semiconductor industry. Moore's law suggests that the cost halves and transistor quantity doubles every 18 months. We see evidence of these trends in digital electronics such as phones and computers.

To put both of the previous graphs in a simpler context, a designer needs to select the optimum number of design features (Fig. 3.10). The customer, sales, or marketing groups often indicate the range of interest in specific features and which features have the most value. As with most things, we want to add the features that provide the most value. We could add new features now, but they would not have much value to the customer. However, delaying the same feature to the next design cycle could have more customer interest with lower costs. In addition, delaying introduction of the feature will save design time now and allow the designer to wait until there are better tools for the design work.

PROBLEMS

3.18 List five attributes of one customer group for fresh fruit and list five needs for fresh fruit. In other words, who is the group and what do they need?

3.19 Some products attempt to satisfy a broad number of market groups. Some products satisfy only a single market group or individual. What are the advantages and disadvantages of each?

3.20 Define five product elements that differentiate product consumers.

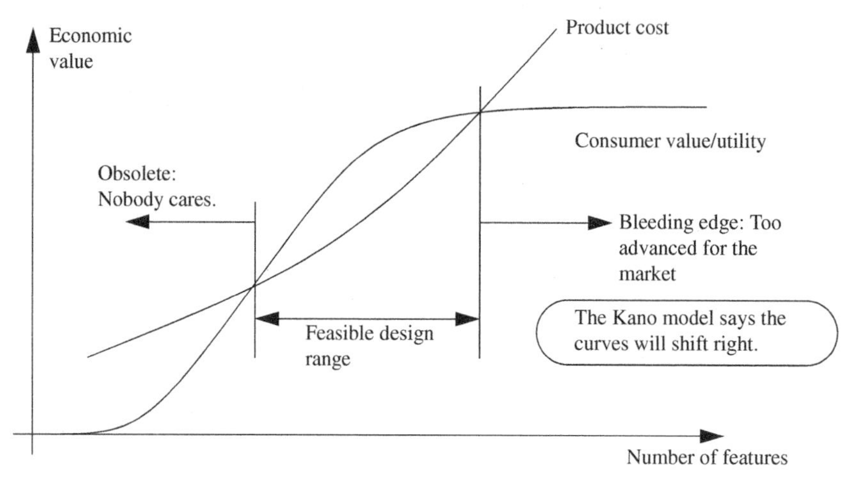

FIGURE 3.10

Selecting an appropriate number of design features.

3.21 What is a legacy feature?

3.22 Are more consumer products revolutionary designs or incremental designs?

3.23 Why are the Kano curves always shifting to the right and down?

3.24 The Kano model works well in rational design space but at extremes it breaks down. Discuss what this means with an infinite number of product functions and excitement.

3.25 Apply the neighbor test to five new computer features. Which of these features pass, fail, or are marginal?

3.26 What is the purpose of the neighbor test?

3.27 Considering the Kano curve, what makes a product obsolete?

3.28 The Kano curve for consumer value/utility plateaus once a large number of features are added. Why?

3.29 A product is developed at the bleeding edge of the market and then kept on the market until it is obsolete. What must happen to the price for it to remain competitive? Use a graph to illustrate the change in value.

3.6 Patents

Patents are a source of ideas.

The patent system was developed to encourage inventors to share ideas with the general public and competitors in exchange for a few years of commercial monopoly. Patents are available for ideas that are new, unique, and nonobvious. Patentable ideas must have some sort of utility, including processes, design functions/components, machines, and business methods. Typical patents that might be found in a laptop computer include a new fan motor for cooling the processor, a circuit for sharing memory between multiple processor cores, an etching process for the integrated circuits, a design for a three-dimensional display, and a one-step owner registration process for new computers. The patents are valid for 20 years after the inventor files the initial disclosure, or 17 years after the patent is awarded. After the patent has expired others are free to use the ideas. It is worth noting that global patent law has been harmonized by the World Trade Organization, so most countries have similar patent structures and procedures.

Patents can be very useful for designers. The patent database becomes an excellent resource for design ideas.

For example, a designer can use it to find alternative designs for items such as power supplies, latch designs, composite material layups, and heart stents. Each patent includes a section outlining the best known implementations that can be used as design guides. If a patent has expired, the idea can be used freely. If a patent is still valid the ideas can be licensed. Licensing patents, or obtaining your own, can be very useful to establish competitive advantage. If a competitor holds a patent, you can design around their protected ideas. A wise designer will use the patent database as an encyclopedia of ideas and not be dissuaded when similar patents are found (see Resource 3.1).

Parts of a patent for compact fluorescent light bulbs are shown in Figs. 3.11 and 3.12. Every patent contains clerical information that includes the title, number, inventor, date of filing, date of award, and abstract. Patents are expected to give references to previous patents that are related. These previous patent numbers can be extremely helpful when searching for similar ideas. It is very likely that other compact fluorescent light patents will also refer to the same patent numbers, and these will appear in a search for those numbers. The background of the invention describes the needs and utility and reference designs showing implementation.

Resource 3.1 Book website: http://www.engineeringdesignprojects/home/content/resources/patents.

The disclosure drawings, description of the drawings, and summary of the invention are shown in Fig. 3.13. Here the inventor describes the best method of implementation, containing the information required to produce a similar design. The figures vary widely between disciplines, including schematics, flowcharts, mechanical drawings, chemical equations, graphs, tables, storyboard drawings, and so on. The text may seem difficult to read at first but the use of words is very specific and will be appreciated after some practice.

The claims provide legal weight to the patent (Fig. 3.13). The claims are written specifically to indicate what is novel in the design, the preferred implementation, and alternative implementations. In the patent, claims 1, 2, and 3 discuss variants of the bulb to broaden the design space covered by the invention. If you suspect that your design may infringe on a patent that is currently valid, read the claims carefully. Sometimes a small novel variation that is not available in the current model will be enough to circumvent or extend the patent. If the technology is protected by an active patent it is reasonable to negotiate a license.

A basic patent search method is outlined in the following procedure. It is wise for designers to do a search when looking for design ideas or as a preliminary step when determining the novelty of a new concept. Although an engineer may do a basic patent search to look for ideas, and possible infringement, he or she will eventually have a patent attorney do a thorough search:

(1) Do a simple search to find patents using some technical keywords.

US 20030223230A1

United States
Patent Application Publication
Li

Pub. No US 2003/0223230 A1
Pub. Date: Dec., 4, 2003

COMPACT FLUORESCENT LAMP

Inventors: Qingsong Li, Irving, TX (US)

Corresponding Address: Munsch, Hardt, Kopf & Harr, P.C. Intellec tual Property Docket Clerk 1445 Ross Avenue, Suite 4000, Dal las, TX 75202-2790 (US)

Appl. No: 10/212,939

Filed: Aug. 6, 2002

Related U.S. Application Data

Continuation-in-part of application No. 29/161,695, filed on May 31, 2002.

Publication Classification
Int. Cl. ... F21V 7/10
U.S. Cl. ... 362/216; 362/260

In accordance with an embodiment of the present invention, a compact fluorescent lamp comprises a spiral compact fluorescent tube comprising a plurality of loops, at least one of the plurality of loops having a different cross-sectional width than a cross-sec tional width of at least another one of the plurality of loops.Cur rent U.S. Classification

International Classification
F21V007/10
Referenced by
Patent Number Filing date Issue date Original Assignee Title
US7045959 Jan 30, 2004May 16, 2006Shanghai Xiang Shan Industry LLCSpiral cold electrode fluorescent lamp
US7053555 Nov 13, 2003May 30, 2006Matsushita Electric Industrial Co., Ltd.Arc tube, discharge lamp, and production method of such arc tube, which enables brighter illuminance
US7264375 Mar 3, 2006 Sep 4, 2007Self-ballasted fluorescent lamp
US7268494 May 9, 2005 Sep 11, 2007 Toshiba Lighting & Technology Corporation Compact fluorescent lamp and luminaire using the same
US7503675 Jan 8, 2007 Mar 17, 2009S.C. Johnson & Son, Inc.Combination light device with insect control ingredient emission
US7862201 Jul 20, 2006 Jan 4, 2011TBT Asset Management International LimitedFluorescent lamp for lighting applications
US7973489 Nov 2, 2007 Jul 5, 2011TBT ASSET Management International LimitedLighting system for illumination using cold cathode fluorescent lamps
US7988323 Sep 29, 2009Aug 2, 2011 S.C. Johnson & Son, Inc.Lighting devices for illumination and ambiance lighting

COMPACT FLUORESCENT LAMP - BACKGROUND OF THE INVENTION

[0001] Many residential lighting products and light fixtures are configured around incandescent bulbs. Homeowners enjoy the warm light, low initial cost, and compact size of incandescent bulbs.

[0002] A different type of lighting product, known as a fluorescent lamp, is also available. A fluorescent lamp comprises a ballast and a glass tube with two electrodes, one at each end. The ballast is used to regulate electric current into the lamp. When switched on, electric current passes through the ballast. Electric current then passes in an arc between the electrodes through an inert gas in the glass tube. Heat from the arc vaporizes tiny drops of mercury in the glass tube, making them produce ultraviolet light, which in turn causes a phosphor coating on the inside surface of the glass tube to glow brightly and radiate in all directions. The most common configuration of a fluorescent lamp glass tube is a straight line. When compared with incandescent bulbs, fluorescent lamps use less electricity and typically last longer. These and other qualities make fluores cent lamps desirable substitutes for incandescent bulbs.

[0003] The general term "compact fluorescent lamp" (CFL) applies to smaller-sized fluorescent lamps, most of which have built-in ballasts and a threaded base that may be installed in a standard incandescent bulb socket. Although the underlying physics is the same, a CFL represents quite a departure from a standard fluorescent lamp. First, the color of light produced by a CFL is nearly identical to that of an incandescent bulb. Also, the threaded bases enable them to fit most standard incandescent bulb sockets. A spiral shaped CFL with a cylindrical profile, such as shown in FIGS. 1A and IB, is currently the most popular CFL. The drawbacks associated with fluorescent lighting products, e.g., coldlooking light, blink ing, awkward sizes and high-pitched noise, have largely disappeared in modern CFLs.

FIGURE 3.11

Compact fluorescent bulb (US patent 20030223230A1).

(2) Read the abstracts to determine relevance. You may find different language to use for the searches.
(3) For patents that are related to your design:
    (a) Read the disclosure for technical design details.
    (b) Read the background for needs (i.e., design motivations).
    (c) If the patent is less than 20 years old, read the claims to see if your work infringes.
(4) Use the patent references to search for older related patents.
(5) Use the patent number to search for related, parallel, or newer patents.

Resource 3.2 For more on patents, see this book's website: www.engineeringdesignprojects/home/content/resources/patents.

SUMMARY OF THE INVENTION

[0004] In accordance with an embodiment of the present invention, a compact fluores cent lamp comprises a spiral compact fluorescent tube comprising a plurality of loops, at least one of the plurality of loops having a different crosssectional width than a cross-sectional width of at least another one of the plurality of loops.

[0005] In accordance with another embodiment of the present invention, a compact flu orescent lamp comprises a spiral compact fluorescent tube comprising of a plural ity of loops of non-uniform cross-sectional width.

[0006] Other aspects and features of the invention will become apparent to those ordi narily skilled in the art upon review of the following description of specific embodiments of the invention in conjunction with the accompanying figures.

BRIEF DESCRIPTION OF THE DRAWINGS

[0007] For a more complete understanding of the present invention, the objects and advantages thereof, reference is now made to the following descriptions taken in connection with the accompanying drawings in which:

[0008] FIG. 1A is a front elevational view of a conventional compact fluorescent lamp and FIG. IB is a plan view showing a distal end of the compact fluorescent lamp of FIG. 1A;

[0009] FIG. 2A is a front elevational view of a conventional lamp reflector and FIG. 2B is a plan view showing a distal end of the lamp reflector of FIG. 2A;

[0010] FIG. 3 is a front elevational view of a compact fluorescent lamp in accordance with an embodiment of the present invention;

[0011] FIG. 4 illustrates light radiation pattern when the compact fluorescent lamp of FIG. 1A is combined with the lamp reflector of FIG. 2A shown in phantom;

[0012] FIG. 5A illustrates light radiation pattern of the compact fluorescent lamp of FIG. 3 when combined with a lamp reflector in accordance with an embodiment of the present invention;

[0013] FIG. 5B is a plan view showing a distal end of the compact fluorescent lamp reflector of FIG. 5A; and

[0014] FIG. 6 is a front elevational view of a compact fluorescent lamp in accordance with an alternative embodiment of the present invention.

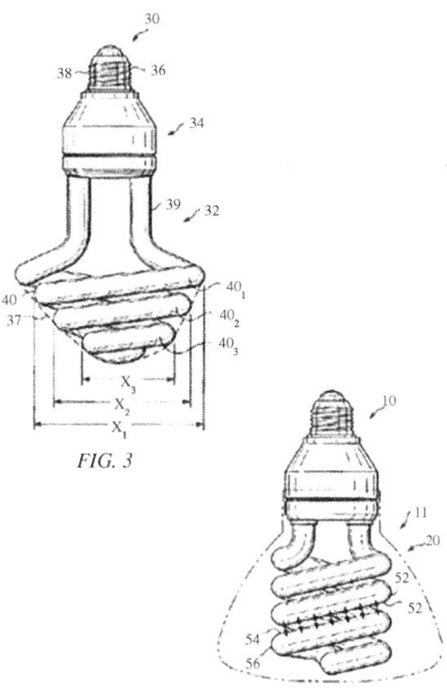

DETAILED DESCRIPTION OF THE DRAWINGS

[0015] The preferred embodiment of the present invention and its advantages are best understood by referring to FIGS. 1 through 6 of the drawings.

[0016] FIG. 1A is a front elevational view of a conventional compact fluorescent lamp 10 and FIG. IB is a plan view showing a distal end of compact fluo rescent lamp 10. Compact fluorescent lamp 10 comprises a compact fluorescent tube 11 and a ballast 12. FIG. 2A is a front elevational view of a con ventional lamp reflector 20 and FIG. 2B is a plan view showing a distal end of lamp reflector 20. Lamp reflector 20 comprises a housing 22 and a cover 24.

[0017] FIG. 3 is a front elevational view of a compact fluorescent lamp 30 in accordance with an embodiment of the present invention. Lamp 30 comprises a compact fluorescent tube 32 coupled to a ballast 34. Ballast 34 may be any ballast now known or later developed. Preferably, ballast 34 comprises a base 36. Preferably, base 36 is adapted for coupling with a conventional electrical light socket (not shown), for example an electrical socket used for incandescent bulbs. In the illustrated embodiment, base 36 has a plurality of threads 38 on the outer surface thereof for coupling lamp 30 with a con ventional electrical socket for incandescent bulbs.

[0018] Tube 32 comprises a proximal portion 39 and a distal portion 40. Preferably, proximal portion 39 couples with ballast 34. Distal portion 40 of tube 32 preferably has a circular spiral configuration and comprises a plurality of loops. Depending on the desired shape or profile, tube 32 may have a more angular spiral configuration, for example triangular, square, rectangular, and/or the like. In the illustrated embodiment, distal portion 40 com prises loops 401, 402 and 403. Preferably, the plurality of loops of tube 32 are of non-uniform cross-sectional width or diameter. The cross-sectional width of loop 401 is X1; the cross-sectional width of loop 402 is X2 and the crosssectional width of loop 403 is X3. As can be seen from FIG. 3, the cross-sectional width of loops 401, 402 and 403 is such that X1>X2>X3. In other words, the loop closest to ballast 34 has the largest cross-sectional width and the cross-sectional width of the loops gradually decrease such that the loop farthest from ballast 34 has the smallest cross-sectional width. If desired, in an alternative embodiment, the cross-sectional width of the loops of tube 32 may be such that the loop closest to ballast 34 has the smallest cross-sectional width and the cross-sectional width of the loops gradually increase such that the loop farthest from ballast 34 has the largest cross-sec tional width.

[0019] As illustrated in FIGS. 1A and IB, each loop of conventional lamp 10 is of the same cross-sectional width. As such, as shown in part by broken lines 13 in FIG. 1A, a profile of tube 11 of fluorescent lamp 10 along its longitudinal axis is substantially cylindrical. On the other hand, as illustrated by bro ken lines 37 in FIG. 3, a profile of tube 32 of lamp 30 along its longitudinal axis is substantially conical. Even if the length of tube 11 of lamp 10 and the length of tube 32 of lamp 30 are the same, the loops of lamp 30 are designed such that the width of the widest loop in lamp 30 is greater than the width of the loops in conventional lamp 10. As such, light from lamp 30 is spread out over a greater area than the light from lamp 10.

[0020] If desired, lamp 30 may comprise a lamp reflector 42 (FIG. 5A). Lamp reflector 42 is preferably coupled to ballast 34 or base 36. Lamp reflector 42 has an inner reflective surface adapted to reflect light from tube 32 to augment light output. When light from tube 32 falls on the inner surface of lamp reflector 42, the light is reflected and directed outwardly away from lamp 30. As illustrated in FIG. 5A, lamp reflector 42 is preferably "funnel-shaped". The shape of an outer surface of lamp reflector 42 is generally concave with respect to a longitudinal axis of lamp 40 with the cross-sectional width of lamp reflector 42 increasing linearly or non-linearly from an end proximal to ballast 34 towards an end distal from ballast 34. Preferably, lamp reflector 42 is narrowest at the proximal end and widest at the distal end. The illustrated shape of lamp reflector 42 enables a higher light output from lamp 30 than a conventional compact fluorescent lamp with a compact fluorescent tube of substantially identical length.

etc.....

FIGURE 3.12

Compact fluorescent bulb (US patent 20030223230A1) (continued).

What is claimed is:

1. A compact fluorescent lamp, comprising: a spiral compact fluorescent tube comprising a plurality of loops, at least one of said plurality of loops having a different cross-sectional width than a cross-sectional width of at least another one of said plurality of loops.
2. The compact fluorescent lamp of claim 1, further comprising a ballast coupled to said compact fluorescent tube.
3. The compact fluorescent lamp of claim 1, further comprising: a ballast coupled to said compact fluorescent tube; and a lamp reflector coupled to said ballast and operable to reflect light emitted by said compact fluorescent tube.
4. The compact fluorescent lamp of claim 3, wherein a shape of an outer surface of said lamp reflector is generally concave with respect to a longitudinal axis of said compact fluorescent lamp with a cross-sectional width of said lamp reflector increasing non-linearly from an end proximal to said ballast to a distal end.
5. The compact fluorescent lamp of claim 3, wherein an outer surface of said lamp reflector is funnel-shaped with a cross-sectional width of said lamp reflector increasing linearly from an end proximal to said ballast to a distal end.
6. The compact fluorescent lamp of claim 3, wherein said lamp reflector comprises a reflector cover coupled to a distal end of said lamp reflector, said reflector cover operable to reduce glare from said spiral compact fluorescent lamp.
7. The compact fluorescent lamp of claim 1, wherein a profile of said spiral compact fluorescent tube along a longitudinal axis of said compact fluorescent lamp is substantially conical.
8. The compact fluorescent lamp of claim 1, wherein a profile of said spiral compact fluorescent tube along a longitudinal axis of said compact fluorescent lamp is noncylindrical.
9. The compact fluorescent lamp of claim 2, wherein a loop of said plurality of loops closest to said ballast has a cross-sectional width larger than a cross-sectional width of any other loop of said plurality of loops.
10. The compact fluorescent lamp of claim 2, wherein a loop of said plurality of loops farthest from said ballast has a cross-sectional width smaller than a cross-sectional width of any other loop of said plurality of loops.
11. The compact fluorescent lamp of claim 2, wherein a cross-sectional width of each of said plurality of loops is staggered with the cross-sectional width of a loop closest to said ballast being the largest and the cross-sectional width of a loop farthest from said ballast being the smallest.
12. The compact fluorescent lamp of claim 1, wherein a configuration of each of said plurality of loops is circular.
13. The compact fluorescent lamp of claim 2, wherein a first loop of said plurality of loops has a cross-sectional width of X1, a second loop of said plurality of loops has a cross-sectional width of X2, and a third loop of said plurality of loops has a cross-sectional width of X3, such that X1>X2>X3.
14. The compact fluorescent lamp of claim 2, wherein said plurality of loops comprise three loops of decreasing cross-sectional widths such that a cross-sectional width of a loop closest to said ballast is greater than a cross-sectional width of the other two loops of said three loops.
15. The compact fluorescent lamp of claim 13, wherein said first loop is closer to said ballast than said second and third loops and said second loop is in between said first and third loops.
16. A compact fluorescent lamp, comprising: a spiral compact fluorescent tube comprising of a plurality of loops of non-uniform cross-sectional width.
17. The compact fluorescent lamp of claim 16, further comprising a ballast coupled to said compact fluorescent tube, wherein said ballast is adapted to couple with an electrical light socket.
18. The compact fluorescent lamp of claim 16, further comprising a ballast coupled to said compact fluorescent tube, wherein a plurality of threads are provided on an outer surface of said ballast to facilitate coupling of said ballast with an electrical light socket.
19. The compact fluorescent lamp of claim 17, further comprising a lamp reflector coupled to said ballast, said lamp reflector operable to reflect light emitted by said compact fluorescent tube.
20. The compact fluorescent lamp of claim 19, wherein an outer surface of said lamp reflector is generally concave with respect to a longitudinal axis of said compact fluorescent lamp with a cross-sectional width of said lamp reflector increasing non-linearly from an end proximal to said ballast to a distal end.
21. The compact fluorescent lamp of claim 19, wherein an outer surface of said lamp reflector is funnel-shaped with a cross-sectional width of said lamp reflector increasing linearly from an end proximal to said ballast to a distal end.
22. The compact fluorescent lamp of claim 19, said lamp reflector being shaped to reflect light emitting from said compact fluorescent tube in a direction outwardly away from said compact fluorescent lamp.
23. The compact fluorescent lamp of claim 17, wherein said plurality of loops comprise three loops of decreasing cross-sectional widths.
24. The compact fluorescent lamp of claim 23, wherein a cross-sectional width of a loop closest to said ballast is greater than a cross-sectional width of the other two loops of said three loops.
25. A compact fluorescent lamp, comprising: a spiral compact fluorescent tube comprising of a plurality of loops, at least two of said plurality of loops having cross-sectional widths different from any other loop of said plurality of loops; a ballast coupled to said spiral compact fluorescent tube, said ballast operable to regulate flow of current into said spiral compact fluorescent tube; and a lamp reflector coupled to said ballast and operable to reflect light emitting from said spiral compact fluorescent tube.
26. The compact fluorescent lamp of claim 25, wherein said compact fluorescent lamp consumes electricity at the rate of thirty watts.
27. The compact fluorescent lamp of claim 25, wherein a light output of said compact fluorescent lamp is sixty-seven lumens per watt.

FIGURE 3.13

Compact fluorescent bulb (US patent 20030223230A1) (continued).

PROBLEMS

3.30 Locate US patent 7264375. Read the abstract and describe how the double spiral makes the patent unique.

3.31 Find a patent for a car-door-hinge design.

3.32 Consider the claims in US patent 20030223230A1. How many of the 27 claims would remain if a patent examiner rejected claim 1?

3.33 Locate a standard that defines the maximum current that can be provided with a USB 3.0 connection.

3.34 For your region, identify the government regulation or law that requires seat belt use in cars.

3.35 Use an Internet search, shopping, or auction site to find a machine that will shake a box with a force of up to 10 G (1 G is the force of gravity). Determine the price and the operating specifications for the machine.

3.36 Locate a nearby company that manufactures products that you use. What other products does the company provide? (As an example, there is a local company that produces floor cleaning systems and I own one of its products. The company produces other floor cleaning products, which are sold in retail stores, as well as commercial floor cleaning products.)

3.37 Find a local interest group that has a professional focus. Good examples include groups such as the IEEE, ASME, IIE, and many others.

3.38 Locate an Australian patent for a swing. Read the claims and determine what the patent covers.

3.39 Review US patent 1836349 for a candy-forming machine and identify the major mechanical components in the figures.

3.40 Use the citations/references in US patent 7564678 to find a link to the original Apple iPod patent. For each patent used, list the patent number.

3.41 Use patent references for compact fluorescent bulbs to trace back to the original patent for the incandescent light bulb. List the patent numbers in the chain you follow.

3.7 Customer specifications

Specifications are the minimum acceptance criteria.

Needs can be vague but can often be used as guidelines to help develop the specifications. However, please be aware that you are promising that your deliverable will meet the specifications, so they must be exact.

Anything less will make your design work more arbitrary and the acceptance of the final design will be arguable. Detailed specifications may seem to slow the design project with minutia that can be set later, but every detail that is skipped will take 10 times longer to add during detailed design, and 100 times longer to add during the build and test phase. Select a set of specifications that you know are feasible, so that you know what you must do and the customer knows what to expect. Attributes of acceptable specifications include the following:

- Define the project performance in detail.
- Include numbers, graphs, diagrams, etc.
- Details are provided first, then text; avoid the vagueness possible with the written word.
- Ensure that specifications allow you to agree how much and what is to be done.
- Avoid feature creep.
- Provide focus.
- Incorporate constraints.
- Define sign-off procedures to close the project.
- Specifications should be defined as minimum requirements, not a wish list.

In some cases the needs will be in the form of specifications. In other cases some of the needs will not be stated. It is important to convert the vague needs into exact specifications, as shown in Fig. 3.14. If things go well, each of the needs will have a corresponding specification. For example, in the figure, specifications 10, 11, and 12 satisfy need C. Need C is addressed by specifications 1, 2, 3, 4, 10, 11, and 12. In this example all of the specifications and needs are related. If any need is not connected, then it is not being addressed, and it should be removed or specifications added. If any specification is not mapped to a need, it should be removed or a corresponding need added.

A complete set of specifications will be exact and testable. At the end of the project they will be used to test the deliverables, as the final acceptance criteria. An example of a working table for specifications is shown in Fig. 3.15. The final technical specifications should be in this form. The specification values in the table will drive many detailed design decisions. If these values change later, detailed design work will need to be redone. In some cases a customer may have essential and optional specification values. Examples include an extra USB port on a computer or a louder volume range for computer speakers. If the "optional" category is not used there will be a tendency to add options as "required" so that they are not forgotten. Each specification should have an undeniable value that can be tested. It is very helpful to specify what tests will be used at the end of the project and what numerical values or ranges are required. For functional elements it is critical to indicate what capabilities are required and how they are to be tested. The needs list may contain customer-suggested specifications and values.

Fig. 3.16 illustrates the effects of detail in specifications. The best option is something quantifiable that can be tested. Qualitative features are acceptable but there is always a question of understanding. For example, a specification may read "brightness adjustment buttons." The customer expects physical buttons on the case, while the designers add two software-only buttons on a computer screen, to save cost and time. Intangible specifications should be avoided because they are very subjective and unverifiable. For example, a specification for "environmentally friendly" may mean recycled paper to the customer, but to the engineers it means minimal waste in production. Again, details are very important. Avoid nonspecific specifications such as "low cost," "nice appearance," "durable," "aesthetically pleasing," or "user friendly." For each specification, ask yourself, "How will I be able to prove this

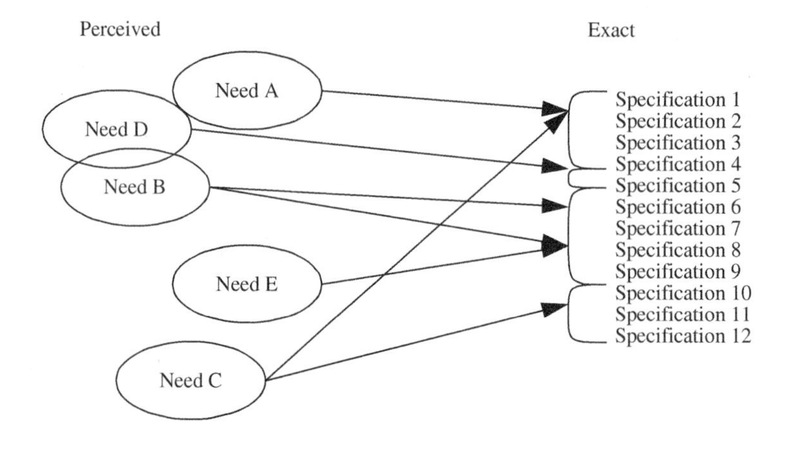

FIGURE 3.14

Mapping needs to customer specifications.

Specification	Required or Optional	Value	Units	Final Acceptance Test or Method	Customer Approved

Note: Consumer product specifications are available on manufacturers and retailer websites. These are not identical to technical specifications and design. Consumer specifications are a mix of technical specifications, embodiment, and design. For example specifications for a computer might include a specific processor number. The technical specification would have referred to the Intel or AMD processor bus protocols, tthe specific processor that was defined as a test case. Likewise the consumer specifications for a canoe paddle focus on the blade shape, weight, color, materials, and total length.The technical specifications for the paddle would include blade break off force, blade drag in water, and mass. The technical specifications define design objectives, the consumer specifications provide the design outcomes. The design specifications tell us what the paddle blade SHOULD DO and the consumer specifications tell us what the blade IS.

FIGURE 3.15

Specification worksheet.

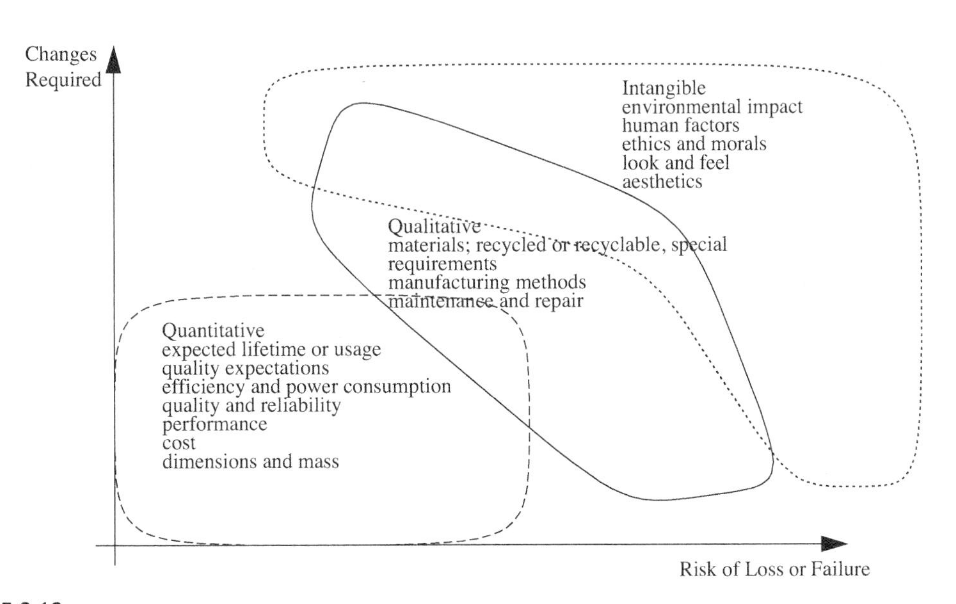

FIGURE 3.16

The spectrum of specifications.

at the end of the project?" A number of good practices for developing specifications are listed as follows:

- Industrial designers get early/preapproval for look and feel, usage model, function, and aesthetics.
- Use benchmark designs when developing needs and specifications.
- If the specifications contain optional requirements, leave these out or connect them to a design bonus.
- Push for evolutionary before revolutionary designs.
- If a customer pushes for open design items to see what is possible or to leave room for change, add these as an option in the specifications. But do not plan to change the specifications without renegotiation.
- Try to talk in general terms that focus on the function instead of solution (e.g., "the automobile should be able to move on ground with a 12-inch variation in height" instead of "the axle clearance should be 12 inches").
- Break requirements into separate measurable or testable values.
- Keep the requirements as simple as possible.
- Avoid vague language; use numbers and technical goals.
- Do not include more specifications than the minimum needed. If extras are added mark them as optional.
- If there are optional specifications, define a point of frustration or use a development bonus.

A reasonable process for developing customer specifications is shown in Fig. 3.17. The process is to (1) approve reasonable and testable specifications, (2) continue looking at untestable or unreasonable specifications, and (3) be prepared to reject specifications. Select specification values so that the process results in a win–win outcome for supplier and customer. Strategically, the supplier wants to have a set of specifications so simple that it could deliver an empty box. Likewise, the customer wants to include as much as possible in the specifications to maximize the benefits. In addition, some project sponsors will push for open items or "loose ends" in projects because they want to see what else is possible, hoping to get extra value. However, these strategic extremes result in a poorer product at the end of the design project.

PROBLEMS

3.42 Why is it important that a customer and supplier sign off, or agree formally, on specifications?

3.43 Write a set of specifications for a ballpoint pen.

3.44 Give an example of a specification for a bucket volume written in quantitative, qualitative, and intangible forms.

3.45 What consumers say they want in a product can be difficult to evaluate. Examples include "ergonomic," "easy to use," "aesthetically pleasing," "responsive," etc. Write down the four examples given and then add another five examples.

3.46 Locate sales material for an automobile that includes specifications. Break the specifications into (a) quantitative, (b) qualitative, and (c) intangible.

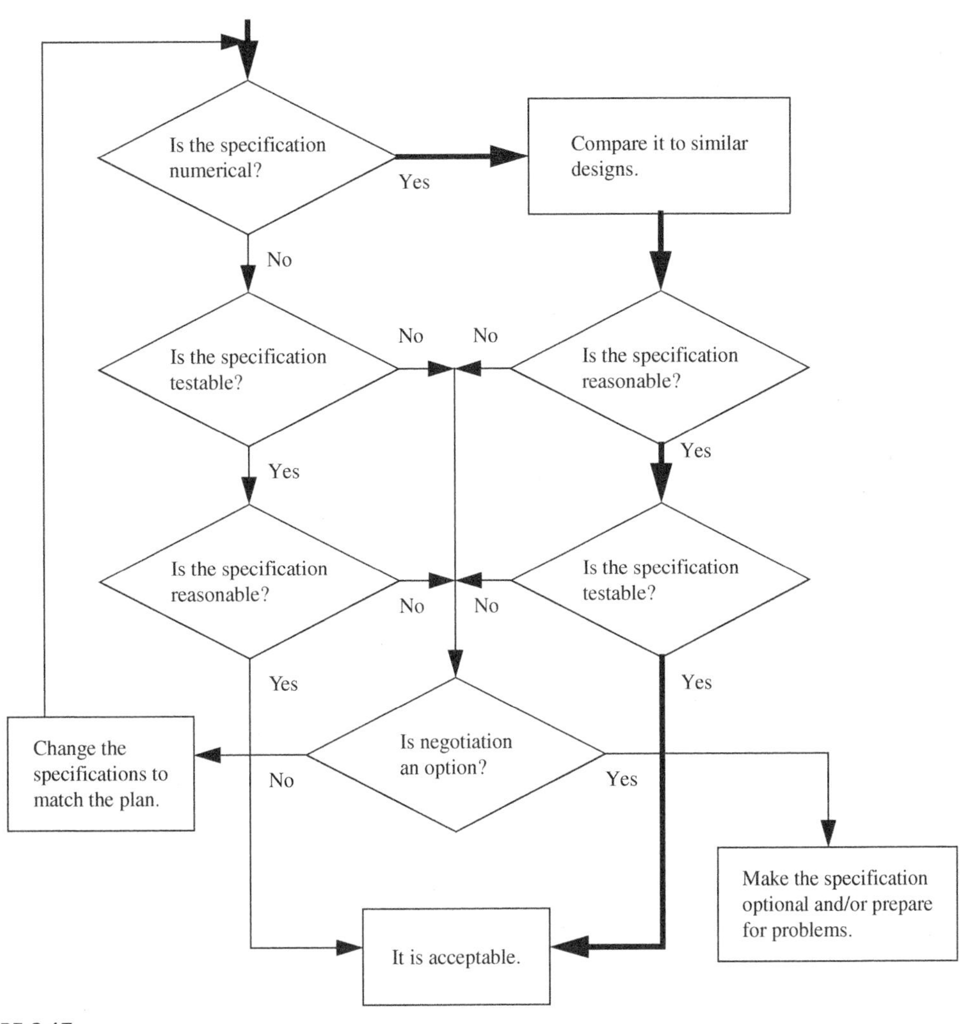

FIGURE 3.17

Screening specifications.

3.47 Complete a needs worksheet for the following customer problem: "I have an issue with seating on a public bus. Each seat is separated with a rail to keep riders from occupying more than a single seat. When the rail is too short riders will put their bags in the next seat. If the rail is elbow height then riders will rest on the rail and lean into the next seat. When the rail is too high the seats start to feel like boxes and people will not sit in them. I need a rail height that will maximize the number of riders sitting in seats."

3.48 Select two similar small passenger automobiles. Create a table that compares the basic specifications of these automobiles. List five other factors that are not listed in the specifications but differentiate the vehicles.

3.49 Mini-case: Aesthetics as a specification

Medical professionals may offer clinics in distant locations. Equipment needed in these locations must be carried and set up at the remote locations. In one case, a piece of equipment was designed to hold a scanner, warm a conductive fluid, hold a laptop, and provide mounts for various probes. The stand was designed to move and position the laptop and probes so that they were close to the patient.

In addition to the technical specifications for the cart, the sponsor asked that the final design be attractive. In response, the design team added "aesthetically pleasing" to the specifications. The sponsors approved the specifications and the design work proceeded. At the conclusion of the detailed design work the sponsor was invited to a formal design review. The frame, mechanisms, and other elements were found to be technically suitable. However, the customer was displeased with the appearance of the cart and did not give approval to release funds for building the design. Given that the specifications for appearance were vague there was no option other than to redesign the appearance of the housings.

In the following weeks the team updated the appearance of the design and consulted with the customer numerous times before receiving approval to move ahead. It was recognized that the aesthetic specification was vague and likely to lead to delays and rejections of designs. Moreover, "aesthetic" should be viewed as a need, not a specification.

How could the process of moving from needs to detailed design be changed to avoid redesign because of vague and/or arbitrary specifications? Is it possible to include artistic appearance as part of the specifications?

3.8 Quality functional deployment

Find the low-hanging fruit.

Quality is a measure of how well the final product meets the customer specifications. It is also a measure of how well the product meets the customer needs. The goal is to select the right specifications to prioritize the design process. A high-quality design will strike the best combination of wants and needs in the customer specifications. For example, a "quality" family car must have a long life, it should have a reasonable cost, but good performance is optional. A "quality" sports car must have high power and excellent handling, but the trade-off is higher cost and shorter life. The specifications set the precision and features in a design. More features and precision mean a higher cost. Customers notice precision below their expectations but higher precision is much less important. When precision increases above customer expectations the customer might not notice or care. For example, a customer expects to spend less for a low-precision component, but it costs less to manufacture. At a high precision level the cost of production is very high, but the customer does not notice or care about the extraordinary precision.

As a supplier, you have limited design resources and many customer needs. The customers can compare your product to those of competitors. You have things you do well and other things you could improve. The QFD method (1) compares your design capabilities with the customer needs, (2) compares the customer needs of your product with those of other competitors, (3) compares the technical aspects of your product with other products, and (4) outlines the effort required for design improvement (Fig. 3.18). The QFD outcome is a ranked list of design features that will produce the greatest benefit for the effort. A QFD, or house of quality, chart is shown in Fig. 3.18. The multistep analysis

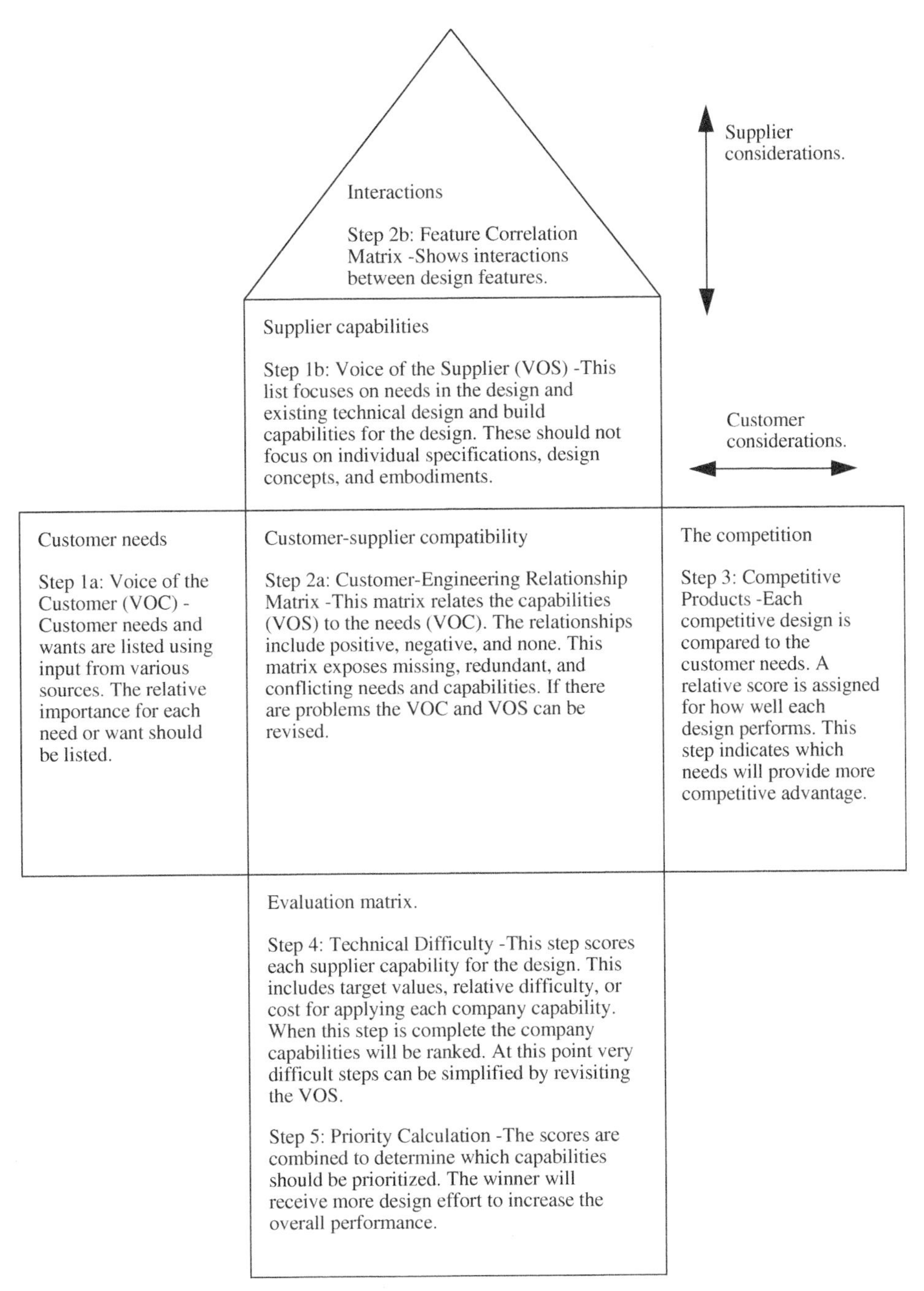

FIGURE 3.18

The house of quality layout and steps.

process begins in step 1a with the VOC, or the customer needs. It is important that these are from the customer perspective, much like consumer specifications. (Note: Quickly skimming ahead a few pages will help when picturing the matrix contents.)

In step 1b, the voice of the supplier (VOS) is expressed as design capabilities and design features that can be controlled during the design. The list should omit obvious or easy functions and focus on the more difficult design aspects or more challenging technical specifications. The customer–supplier compatibility matrix is completed in step 2a with the related VOC and VOS. The score represents how strongly each customer need is supported by the supplier capability. The complication triangle on the top relates each of the supplier capabilities. This can be used to indicate how an improvement on one capability might positively or negatively affect other design functions. Step 3 is a comparison between your product and the competitor's from the customers' perspectives. The scores are relative rankings.

A technical comparison of competitors' products is done in step 4. This reveals what supplier capabilities are actually important to the customer. Finally, in step 5, the scores are combined into numerical totals for each design capability. A capability with the highest score will allow the greatest amount of customer quality improvement, with the least amount of effort.

An example of a needs worksheet for a floor cleaner is shown in Fig. 3.19. The VOC, step 1a, is developed first. If the requirements are not obvious they can be developed using the needs and "whats." These should result in a list of requirements that the customer uses to differentiate products. The customer requirements are related to the final technical specifications but they are not the same. For example, a consumer specification of a quiet coffee grinder might translate into a technical specification of motor precision, housing design, and grinding blades. For the floor cleaner in this example, the customer wants to be able to pick up dirt, move the cleaner easily, and minimize maintenance. Each of the requirements is given a relative importance and the total should add up to 100%. In the example, two different customers are considered. Note that the number of replacement bags is more important to daily users.

VOC Customer Requirements	Uses	What does it mean? (Whats or Hows)	Requirements	Customer value.	
				Daily use.	Monthly use.
	Vacuuming	Dirt removal.	Removes dust particles.	10%	30%
			Removes large debris.	5%	10%
		Handling.	Lower mass.	15%	10%
		Sound.	Quiet.	20%	10%
	Maintenance	Consumables used.	Use fewer bags and filters.	15%	5%
		Bags/filter changes.	Less time changing bag/filter.	10%	5%
		Repairs.	A long life before failure.	15%	10%
	Storage	Small size.	Self contained in a small volume.	10%	20%

FIGURE 3.19

Step 1a, a sample voice of the customer (*VOC*) needs worksheet.

Step 1a: VOC

- This step should include marketing, sales, and other professionals with customer contact.
- Select a target customer group for the design. Avoid the temptation to expand the group to increase the market size.
- Identify the customers' needs, "whats," and requirements.
- Convert the needs to "whats" by asking, "What must be accomplished?"
- Convert the "whats" to requirements by asking, "How will this be accomplished?"
- If customers are able to provide only general needs it may be necessary for the QFD team to develop the "whats" and requirements.
- Requirements should be measurable and differentiate designs. For example, don't use "it can clean."
- For customer value scores ask, "What is your top priority?" Another approach is to ask for an ordered ranking of all of the options.

The VOS should be much easier to develop using knowledge of past projects and technical expertise. As a minimum, the list should include capabilities that will, or will probably, be used in the design work. The VOS capabilities are listed across the top row of Fig. 3.20. These split the work into the

VOS Capabilities and product control characteristics.						
Electric motor.	Impellor.	Agitator.	Air filtration.	Dust collection.	Attachments.	Structure.

Requirements	Dail.	Mon.
Removes dust particles.	10%	30%
Removes large debris.	5%	10%
Lower mass.	15%	10%
Quiet.	20%	10%
Use fewer bags and filters.	15%	5%
Less time changing bag/filter	10%	5%
A long life before failure.	15%	10%
Self contained in a small volume.	10%	20%

Add the VOS capabilities.

FIGURE 3.20

Step 1b, mapping needs to customer specifications. *VOS*, voice of the supplier.

challenging technical design elements, including the motor, air system, and housing. Individually, each of these capabilities requires different skill sets and effort. For example, the structure requires manufacturing and aesthetic expertise, the impellor requires mechanical engineering, and the motor requires electrical and manufacturing expertise. This example is relatively high level, but it could conceivably contain more than 100 features and components. This list could also include a power switch, lights for dark corners, the handle, and much more. The capabilities do not need to be limited to in-house abilities if they can be provided by an outside supplier.

Step 1b: VOS

- The capabilities required for the design work are listed across the top of the chart.
- These should be developed with design, manufacturing, and quality engineering input.
- This should be a list of the challenging functions and components.
- Most businesses are able to identify these tasks by groups and individuals in the company, such as an engine design group.
- The capabilities can include outside suppliers.
- This list does not need to include minor design and manufacturing issues or tasks.
- These are not concepts or specifications, but they are related because they will eventually be used to select specification values and designs.

Once the needs and capabilities are defined, step 2a or step 2b can be completed. Step 2a is shown in Fig. 3.21, with an intersection matrix for the VOC and VOS items. Values are added to the boxes to indicate a strong (9), moderate (3), weak (1), or no (0) relationship. The values can be slightly subjective, but should be based on the current capabilities and designs. An example of a strong positive relationship is the impellor design may have a significant impact on the quietness of the cleaner. An example of no relationship would be the impellor does not affect the time required to change a dust collection bag.

Step 2a: Begin laying out the planning matrix

- The relationship can be determined by asking, "If we change the supplier capability, will it have an impact on the customer requirement?"
- The magnitude of the strong effect is 9 for a very large positive or negative effect.
- The magnitude of the moderate effect is 3 for a smaller positive or negative effect.
- For a score of 1 there is a marginal interaction between design factors.
- These diagrams can also be drawn with circles and triangles as a visual aid, but that is not done here.
- Each capability row is multiplied by the customer value to get an importance score. Higher scores mean that capability has more impact on customer satisfaction.

When the customer–supplier matrix is complete, there should be some strong values in each column and row. Empty rows could mean that capabilities are not needed, or that requirements are not being addressed. The importance rating can be used to highlight supplier capabilities providing the greatest customer value. In this case, the agitator, dust collection, and attachments have notably high scores. The electric motor score is very low, meaning that the customer does not value the motor, although he or she does value what the motor does.

Many of the supplier capabilities interact, as shown in the triangular top of the house in Fig. 3.22. The columns and diagonals relate each design capability to the others. A higher score indicates that the

Scale:
9: Strong positive.
3: Moderate.
1: Weak positive.
0: Little or no effect.

Requirements	Dail.	Mon.	Electric motor.	Impellor.	Agitator.	Air filtration.	Dust collection.	Attachments.	Structure.
Removes dust particles.	10%	30%		9	9	3	9	9	
Removes large debris.	5%	10%		9	9		9	9	
Lower mass.	15%	10%	3		3		3	9	9
Quiet.	20%	10%	9	9	9		3	3	3
Use fewer bags and filters.	15%	5%				9	9		
Less time changing bag/filter.	10%	5%				9	3		9
A long life before failure.	15%	10%	3	3	9			3	9
Self contained in a small volume.	10%	20%			9		9	9	9
Importance Rating Daily user.			270	360	585	255	495	465	510
Importance Rating Monthly user.			150	480	750	180	660	690	435
Average.			210	420	668	218	578	578	473

Importance of the electric motor for daily users:
= [10, 5, 15, 20, 15, 10, 15, 10] * [0, 0, 3, 9, 0, 0, 3, 0]
= 0 + 0 + 45 + 180 + 0 + 0 + 45 + 0 = 270

FIGURE 3.21

Step 2a, the customer–supplier compatibility matrix.

supplier capability interacts with other capabilities. For example, changing the motor would require changes to the impellor and structure of the cleaner. However, the electric motor is entirely independent of the agitator design.

Step 2b: Supplier capability correlation matrix

- The relationship between each of the capabilities is indicated in the upper triangle where the vertical and diagonal lines meet.
- A high score of 2 indicates that changes in one capability will require changes in the other. A score of 1 means minor changes may be needed, while no score indicates that they are unrelated and will not need any changes.

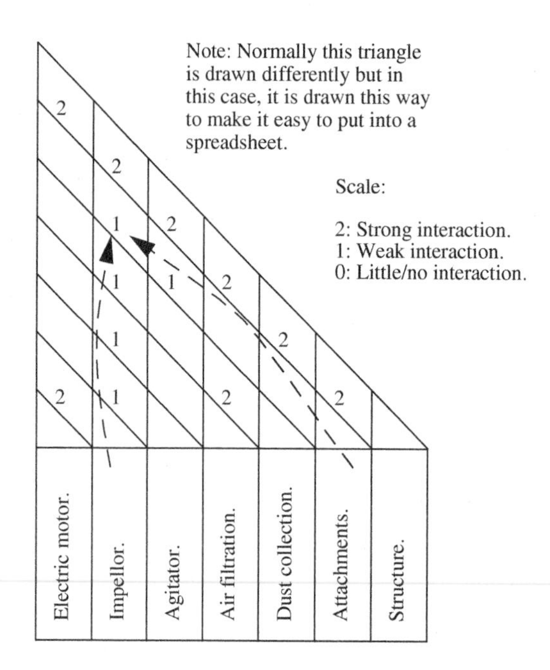

FIGURE 3.22

The supplier capability interaction matrix, step 2b.

- The control characteristics should be reconsidered if, in general, there are more negative than positive effects.
- In this case, the diagram is skewed to the left to make it easy to enter into a spreadsheet. Traditional methods use a centered triangle for visual effect.
- Traditional methods use symbols for scores that are converted to numbers. They may also assign positive values to beneficial interactions or negative values to competing capabilities.

The competition analysis is shown in Fig. 3.23. For each of the customer requirements there is an estimate of how well it is achieved by a design. The same process is repeated for each competitive design. A calculation of a position is made to illustrate how far a supplier capability lags behind the leading competition. In this case, the position for the, motor is −18, meaning that a supplier is ahead of the competition and would not gain any value by improving the capability. However, the score for the air filtration is 33, indicating that the competitors are far ahead in that capability. In this example air filtration is a strong contender for extra design effort.

Step 3: Customer capability importance rating

- This stage should involve engineering, sales, marketing, and any other professional that works closely with the customers.
- Your product and the competitors' are ranked against the consumer requirements.
- The scores range from 1, for no satisfaction, to 5, for absolute elation.
- Some traditional methods will draw graphs for the 1 to 5 scores for each product. This visually illustrates the relative satisfaction of customer needs.
- The customer requirement satisfaction is converted to a supplier capability satisfaction using the consumer importance rating and the requirement−capability matrix.

Scale:
5: Outstanding fulfillment of requirements.
4: Very pleased with performance.
3: Satisfactory response to requirements.
2: The solution partially fulfills the requirement.
1: Does not address the requirement.

Requirements								Current product	Competition A	Competition B
Removes dust particles.		9	9	3	9	9		3	2	4
Removes large debris.		9	9		9	9		2	2	5
Lower mass.	3		3		3	9	9	4	4	2
Quiet.	9	9	9		3	3	3	4	2	2
Use fewer bags and filters.				9	9			2	5	2
Less time changing bag/filter.				9	3		9	4	5	2
A long life before failure.	3	3	9			3	9	4	4	2
Self contained in a small volume.			9		9	9	9	3	3	2
Customer value our product.	60	93	156	63	126	132	147			
Customer value competition A.	42	66	129	96	141	117	150			
Customer value competition B.	30	105	141	48	135	129	78			
Position (Behind the leader)	-18	12	-15	33	15	-3	3			

= [3, 2, 4, 4, 2, 4, 4, 3] * [0, 0, 3, 9, 0, 0, 3, 0]
= 3*0 + 2*0 + 4*3 + 4*9 + 2*0 + 4*0 + 4*3 + 3*0 = 60

FIGURE 3.23

Step 3, customer expectations are compared with competitive products.

- The position is the difference between your product and the strongest competitor. A positive position means that the competition is ahead of the supplier.
- The QFD method requires an existing product and competitors. However, with some creativity it can still be used. If the supplier does not currently have a product, it can project what could be done and how the product would compare. If a supplier does not have competitors, then they can compare existing products in similar markets.

The difficulty of improving a requirement/capability is estimated for the design and manufacturing changes (Fig. 3.24). In this case, a score of 5 indicates that the design or manufacturing effort is trivial. However, a score of 1 indicates that it will take a substantial amount of effort to make a slight

Simplicity scale:
5: Trivial.
4: A routine operation.
3: Careful setup and planning.
2: Complex and multiple steps.
1: Almost impossible.

	Electric motor.	Impellor.	Agitator.	Air filtration.	Dust collection.	Attachments.	Structure.
Importance rating Daily user.	270	360	585	255	495	465	510
Importance rating Monthly user.	150	480	750	180	660	690	435
Average importance.	210	420	668	218	578	578	473
Customer value our product.	60	93	156	63	126	132	147
Customer value competition A.	42	66	129	96	141	117	150
Customer value competition B.	30	105	141	48	135	129	78
Position (Behind the leader)	-18	12	-15	33	15	-3	3
Design simplicity	3	2	2	4	4	4	2
Manufacturing simplicity.	2	4	3	4	4	4	2
Lowest (most difficult).	2	2	2	4	4	4	2
Deployment score.	-7560 6th	10080 3rd	-20025 7th	28710 2nd	34650 1st	-6930 5th	2835 4th

210(−18)2 = −7560

FIGURE 3.24

Steps 4 and 5, technical difficulty and deployment matrix.

improvement. The lower of the two scores is selected to reflect combined difficulty. In this example, the impellor is a very complicated design, but when complete, it should be relatively simple to make by injection mold. Therefore, the difficult design effort makes it a difficult part.

Step 4: Supplier capability simplicity

- Each of the supplier capabilities is considered as it relates to the product.
- If the supplier capability does not require substantial effort it receives a score of 5.
- A score of 1 is used for a supplier capability that requires substantial effort to make a trivial increase in improvement.

- Design and manufacturing are separated to emphasize the different natures.
- The lower of the two scores is used to indicate a difficulty score. A lower score means that the capability will be costly to improve.

The various scores are multiplied to obtain a deployment score, as seen in Fig. 3.24. A higher deployment score makes the supplier capability a better choice for improvement. A higher deployment score indicates that the capability is more important to the customer, lagging behind the competitors, and easy to improve. Negative deployment scores indicate that the supplier is already ahead of the competitors and extra effort would have less importance to the customer. In this example, the air-filtration capability received the highest score and should receive the most design effort to improve. The dust collector and impellor are also excellent candidates for improvement. On the other hand, effort that goes to the electric motor, agitator, and attachments would have little consumer benefit.

Step 5: Supplier capability deployment

- This is a combination of customer value, competitive comparison, and difficulty.
- The values are multiplied to emphasize multiple benefits.
- These scores can be normalized but the relative scores will be the same.
- The highest scores should be the functions chosen for deployment. There is some room for management decisions, but the numbers are strong indicators.
- Deployment means that additional resources will be used to improve the design capability.

The true value of the QFD method is the ability to quantitatively analyze design effort. Although the numbers may be somewhat subjective they provide a greater level of objectivity. One problem the method exposes is a supplier who is so focused on perfecting a design element that they neglect other customer values. In other words, a negative deployment score is a message that it is time to focus elsewhere. The QFD method may be extended to determine specification values by mapping supplier capabilities to technical specifications with deployment matrices. This is very valuable when the supplier capabilities and technical specifications are not similar.

PROBLEMS

3.50 What is the difference between quality and precision?

3.51 Why is QFD called the house of quality?

3.52 What do the top to bottom and left to right directions represent in QFD?

3.53 You are doing a QFD chart for a residential wall-mounted light switch. One of the needs you have heard is that it should "click sharply but quietly."

(a) Convert the need to "hows" and create at least five requirements.

(b) Identify three possible customer groups for testing.

(c) Suggest relative values for each of the customers.

3.54 Your design team is developing the customer specifications for a can opener. You have already established that it will use a hand crank on the side. Define 10 VOS capabilities and control variables.
Note: The can opener will have two handles, a blade to cut the lid, a roller to push the can onto the blade, and a crank to turn the roller. You can change component features to change the feel and performance of the can opener.

3.55 What should be done if the customer–supplier compatibility matrix has an empty column or row?

3.56 What does it mean if any columns or diagonals in the supplier capability interaction matrix are empty?

3.57 How could competitive values be created if there are no competitors?

3.58 Develop a QFD chart for coffee cup features. Use standard cups that are available from local coffee shops and vendors. The objective is to design a new cup with the widest consumer and vendor appeal.

3.59 Describe a process for evaluating if a bicycle is "ergonomic."

3.60 Select a laptop computer manufacturer. Use the memory size in its current product lines to construct a Kano curve.

3.61 Select a company that manufactures computer graphics cards. Develop a table that shows product age, one graphics benchmark, and the product cost. Identify the commodity and niche products in the table.

3.62 Consider standard window glass. You need to describe it using two perspectives: customer specifications and technical specifications. The customer needs should be put in general terms such as solid, insulating, and clear. The technical specifications should be related to physical properties such as thermal resistance, fracture pressure, and transparency. Use a matrix to relate each of the specifications.

3.63 A bathroom fan design has the following requirements. Develop a list of technical specifications that address all of these and use a matrix to verify the coverage.

(a) Clears a regular bathroom in 5 min

(b) Quiet

(c) Normal utility ratings

(d) Fits in a standard hole

(e) Connects to standard ducts

3.64 Develop a set of questions for a customer. The customer currently uses 4-m-high step ladders.

3.65 How are the QFD customer requirements and values obtained?

3.66 Consider a specification for a car that reads "hold five passengers." Give five examples of the different interpretations that may be used in design and testing. For example, is the driver a passenger?

3.67 What are the advantages and disadvantages of numerical specifications?

3.68 Why is testing important for numerical specifications?

3.69 The following vague specifications were provided for a laptop. Rewrite these to be specific and testable.

(a) Laptop will work 10 h.

(b) Screen is viewable in daylight.

(c) Cost will be low.

(d) It will be reliable.

(e) It will be high quality.

(f) It can be upgraded.

(g) It will be aesthetically pleasing.

3.70 Consumer devices must be appealing and it is tempting to add aesthetics to the specifications. Explain how aesthetics could be in the specifications.

3.71 Office chairs have a maximum design weight. Find the technical specifications that define these.

3.72 Briefly describe why each of the following attributes is important when developing specifications:

(a) Detailed

(b) Testable

(c) Clear

(d) Understandable

(e) Unique and not open to interpretation

3.73 The reliability of research resources can vary. Use a search engine, such as Google, to find information about programmable logic controllers. Sort the first 50 information sources into academic, corporate, irrelevant, anecdotal, and unknown.

3.74 What is the difference between searching for information and searching for answers?

3.75 Find the following items for a battery:

(a) technical specifications from a manufacturer's website that include time—voltage curves;

(b) a commercial site that compares batteries from various manufacturers;

(c) a research paper that discusses new materials for increasing battery life;

(d) an application or selector guide that indicates how to select various battery sizes, based on life, space, power, temperature, and more.

3.76 How does learning to research new technologies support lifelong learning?

3.77 Find examples of:

(a) databases of books and standards

(b) retail catalogs with parametric selection tools

(c) industry product guides

3.78 Use Internet auction or sales sites to find new and used prices for a laser cutting machine that can cut stainless steel circuit-board mask materials at least $500 \times 500 \times 0.3$ mm thick.

3.79 Mini-case: Project initiation

The military deals with entrepreneurial project approaches using the Defense Advanced Research Projects Agency (DARPA). The website lists a number of priority technologies as well as requests for general proposals (www.darpa.mil/). See "Opportunities" on the main web page and look for the submission processes for different companies and agencies. An example of a DARPA project could be a new system for nonlethal weaponry. A proposer would complete a proposal for outlining the project details and projected budget and deliverables. DARPA would receive and review the form. If the proposal meets the needs and policies it might be approved and the designer would be expected to deliver as outlined.

The arts community generates many creative ideas but has issues reaching a larger customer base. In response, the Kickstarter website was created to present project proposals (www.kick-starter.com). Site visitors can view the projects and make bids on them. Each bidder contributes a small amount of the minimum needed for the project. If the minimum for a project is met or exceeded it moves forward. The outcomes of the projects vary widely. A common approach is to have different donation levels, and the level of donation is tied to a number of items and additional features. The niche for the website has expanded to include many engineering and technology projects. An example Kickstarter project might be a new type of computer mouse. The project sponsor would post a description of the planned project. Visitors would review the project details and might donate $75 for one mouse at the end of the project, or $200 for three mice. The proposer might require $150,000 before moving forward. What are the common elements in the DARPA and Kickstarter proposal processes?

3.80 Mini-case: Specification drift

Gaming took a massive leap forward in the 1990s as computers became fast enough to expand graphics from two dimensions to three. Two-dimensional games normally had characters that moved on a surface that scrolled as the game progressed. Three-dimensional games allowed a player to move in three dimensions using a perspective view. A few landmark 3D games include Doom (1993), Duke Nukem (1996), and Quake (1996). Customers enjoyed the new generation of 3D games and wanted more like them. The companies that developed Doom and Quake produced sequels that were all commercial successes. Duke Nukem, developed by 3D Realms, was also the subject of a design project called Duke Nukem Forever. The game was announced in 1997 for delivery in 1998. Between 1997 and 1998 game developers shifted from an older graphics processing software library (engine) to a newer graphics engine developed for Quake II. The new graphics library was essential to provide a contemporary appearance. The team developed graphics, wrote software, and modified the library. In 1998, the Unreal engine was released and the team decided to move to it, discarding the work done with the Quake II engine. The next decade was filled with similar technical changes, the addition of new game features, business changes, and missed deadlines. The product has become synonymous with vaporware—promised software that never materializes.

When the game was eventually released in 2011 it was projected to sell 3 million copies but sold only 1.6 million. Compare this to Doom 3 (2004), which sold 3 million copies, and Call of Duty: Modern Warfare 3 (2011), which sold 28 million copies.

In hindsight, the specifications for the project were repeatedly "improved" after substantial design work was complete. Each change resulted in lost time and effort. Ironically, the developers' desire to adapt, to produce a cutting-edge game, resulted in a game that was 13 years late and was criticized for being outdated. It is possible to argue that they should have released the game in 1998 with the out-of-date features and then moved on to a newer version. Investigate the development of Duke Nukem Forever and find 10 events in which the team effectively changed the specifications.

Further reading

Baxter, M., 1995. Product Design; Practical Methods for the Systematic Development of New Products. Chapman and Hall.

Cagle, R.B., 2003. Blueprint for Project Recovery-A Project Management Guide. Amcon Books.

Charvat, J., 2002. Project Management Nation: Tools, Techniques, and Goals for the New and Practicing IT. Project Manager. Wiley.

Davidson, J., 2000. 10 Minute Guide to Project Management. Alpha.

Dhillon, B.S., 1996. Engineering Design; A Modern Approach. Irwin.

Dym, C.L., Little, P., 2009. Engineering Design: A Project Based Introduction, third ed. Wiley.

Gido, J., Clements, J.P., 2003. Successful Project Management, second ed. Thompson South- Western.

Hauser, J.R., Clausing, D., 1988. The House of Quality, May—June. Harvard Business Review, pp. 63—73.

Heldman, K., 2009. PMP: Project Management Professional Exam Study Guide, fifth ed. Sybex.

Hyman, B., 2003. Fundamentals of Engineering Design, second ed. Prentice-Hall.

Kano, N. (Ed.), 1996. Guide to TQM in Service Industries. Asian Productivity Organization, Tokyo.

Kerzner, H., 2000. Applied Project Management: Best Practices on Implementation. John Wiley and Sons.

McCormack, M., 1984. What They Don't Teach You at Harvard Business School. Bantam.

Niku, S.B., 2009. Creative Design of Products and Systems. Wiley.

Rothman, J., 2007. Manage It! Your Guide to Modern Pragmatic Project Management. The Pragmatic Bookshelf.

Ullman, D.G., 1997. The Mechanical Design Process. McGraw-Hill.

Verzuh, E., 2003. The Portable MBA in Project Management. Wiley.

Verzuh, E., 2005. The Fast Forward MBA in Project Management, second ed. John Wiley and Sons.

Wysocki, R.K., 2004. Project Management Process Improvement. Artech House.

CHAPTER

Concepts and technical specifications

4

4.1 Introduction

A Utopian design requires no effort or cost, but does everything.

The needs and specifications create an expectation for what the design should be. The concepts and technical specifications determine how the design can be done. In a strict sense, the customer specifications are used to generate concepts and the concepts are then combined into technical embodiments (Fig. 4.1). In a practical sense, most designers have multiple concepts and embodiments in mind long before the customer specifications are approved and will skip directly to the embodiments. Please note that many designers lump concepts and embodiments under the single label of concepts. After the

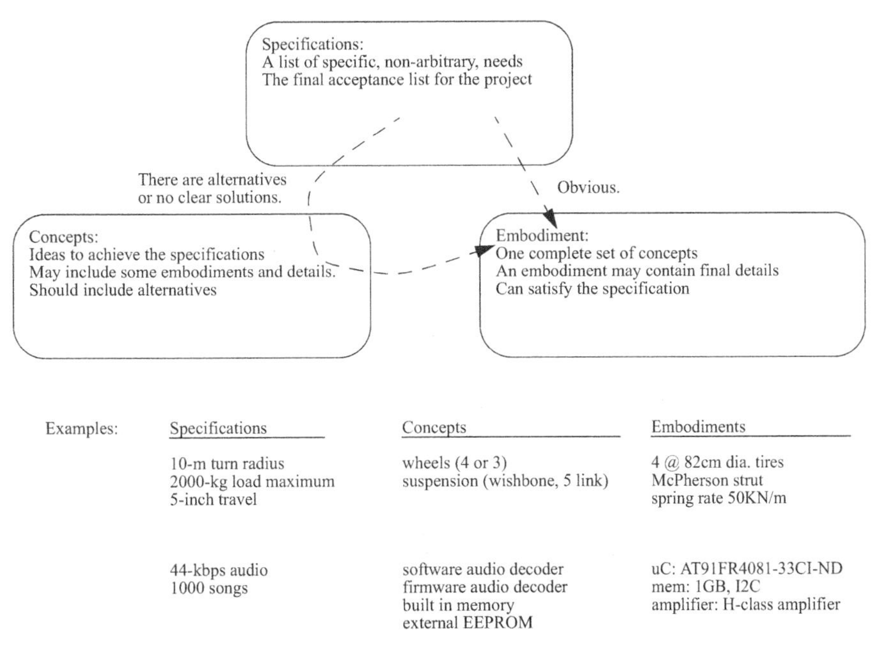

FIGURE 4.1

Specifications, concepts, and embodiments.

Engineering Design, Planning, and Management. https://doi.org/10.1016/B978-0-12-821055-0.00004-9

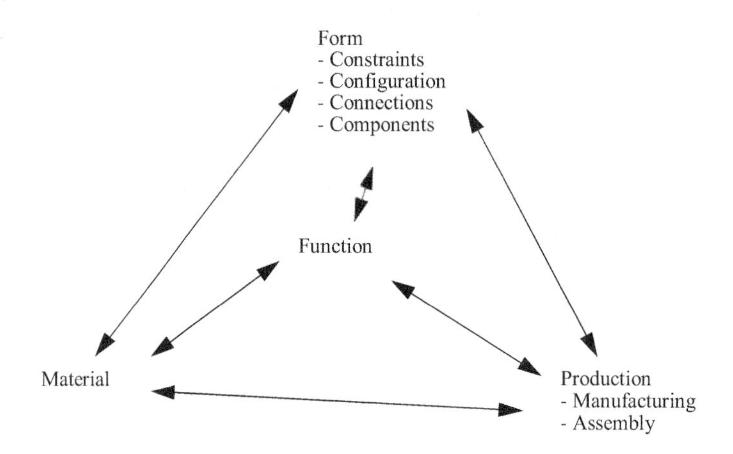

FIGURE 4.2

A domain model of design.

technical details of one embodiment are set, the technical specification will be set. Regardless of the steps taken, the objective is to finalize the technical specification. The technical specifications will be used as the road map for the detailed design. Some examples are provided to illustrate the ever-refining detail that occurs with moving from customer specifications to embodiments. The key is that the embodiments and technical specifications provide enough detail so that the design process becomes more routine and predictable.

A useful cognitive model for describing a design is shown in Fig. 4.2. Many designers separate form and function. In this case, a working design must address all of these elements. As with any model, this one can be expanded as needed for various design types.

PROBLEM

4.1 What is the difference between specifications and embodiments?

4.2 Concepts

Steal the best ideas.

The normal approach to design is to work from the general to the specific. For our purposes, we will split these stages into general conceptual design followed by the detailed design phase (Fig. 4.3). Any design can be described as a number of major and minor parts. Sometimes the parts are obvious, other times the choices for the parts are (1) there are known alternatives or (2) there are no obvious choices. For the parts of the designs that are obvious, the concept generation and selection phases of the design become unnecessary. In any design, some parts will be obvious, while others will involve selection and sometimes concept generation. At the conclusion of the conceptual design approach, it is absolutely required to have concepts that are very likely to satisfy the specifications.

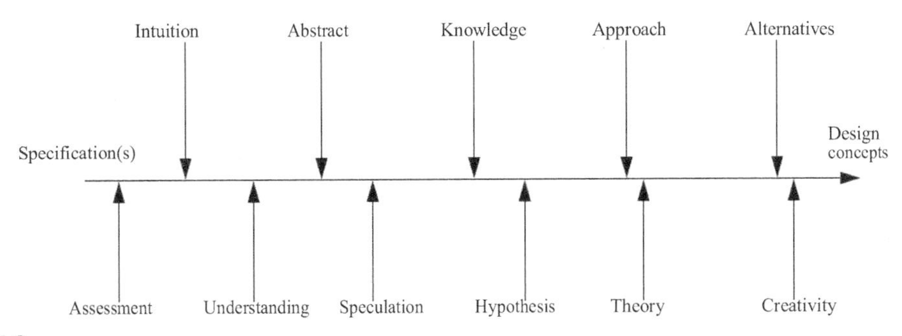

FIGURE 4.3

Factors in concept development.

Concepts could be:

- big-picture ideas or small details
- alternative ways to achieve one or more design functions
- abstract or theoretical ideas
- already widely used and understood
- unknown to the designers
- new parts or technologies
- a design already well known and widely used in the company
- how the system will be used

Examples of concepts include:

- alternative fuel sources (macrolevel)
- business models (macrolevel)
- radio-frequency versus infrared communication (midlevel)
- motor and gear selection, including gear ratios and couplers (midlevel)
- user interface and graphical user interface (GUI) models (midlevel)
- physical layout (midlevel)
- snap fit or threaded fasteners (microlevel)
- class AB or class D power amplifier (microlevel)
- encryption versus obfuscation for security (microlevel)

Each company handles the process of conceptual design differently. The worst is ad hoc, when "it just happens." Companies with specialized departments tend to complete documents for their phase and then "throw them over the wall." The example in Fig. 4.4 shows this structure for a company

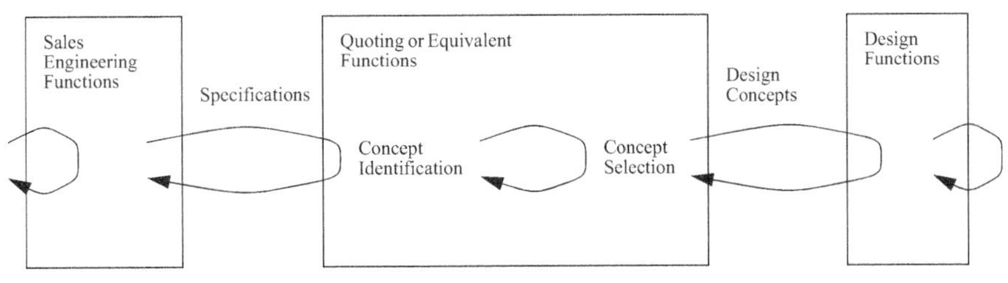

FIGURE 4.4

Traditional functional design (throw it over the wall).

that has a sales engineering division that does the initial work with customers. That division will work with the customer to develop a set of rough, with some detailed, specifications. They then pass these to a quoting department that considers design alternatives. The quoting department selects one that satisfies the customer and fits the business needs. They collect and estimate costs and then compile a bid. If the customer accepts the bid, the details from quoting and sales will be passed to the design department. The design is thrown over the wall, from sales to quoting, to design, and so on. The new owner of the design is in charge but will backtrack to ask about justifications and decisions; however, these are normally set and cannot be changed easily. Note that backtracking occurs when (1) a design choice does not work as expected, (2) there is a change in the definition of the project, (3) critical components change, or (4) manufacturing capabilities change. Backtracking can be time consuming and should be minimized. Sometimes issues will not be resolved by backtracking to help the project move ahead.

A wise designer will prefer designs with obvious solutions or alternatives. Concept generation can be difficult to do well, has higher risk of failure, and should be reserved for those components that require innovation. The highest risk always comes with high-level concept generation. Obviously these designs are exploratory in nature and should be saved for pilot projects. The basic process of concept generation and comparison is shown in Fig. 4.5. In this process, the teams start by examining the specifications and selecting obvious solutions. Concepts are generated for solving more complex problems. The better concepts are developed and compared.

Eventually the best concepts are combined with the obvious solutions in the technical specifications.

The over-the-wall separation of functions allows more focus, and efficiency, but the approach also makes early decisions more costly to change. An alternative is to have groups work concurrently on tasks. In the over-the-wall approach the departments do the functions themselves and then check to see if the results are acceptable. In a concurrent design approach, the process procedures formally involve all of the stakeholder groups. This approach will result in some waste but will reduce the number of poor decisions. For example, if somebody from the design department attends a sales engineering meeting with the customer, he or she may be able to modify a specification value so that an off-the-shelf component can be used, instead of a custom design. This early change could dramatically cut the complexity and cost of a project, but it also takes the designer away from the task of designing. Likewise, if a sales engineer sits in on a design meeting he or she will see the impact of small details in specifications and will be able to establish better specifications in the future.

PROBLEM

4.2 Should all concepts in a design be at a high level only? Explain.

4.3 What is the difference between a high-level macro and a low-level micro concept?

4.4 Is a concept needed if a design choice is obvious? Explain.

4.5 What are two advantages and two disadvantages of over-the-wall designing?

4.6 How can prototypes be used in concept development?

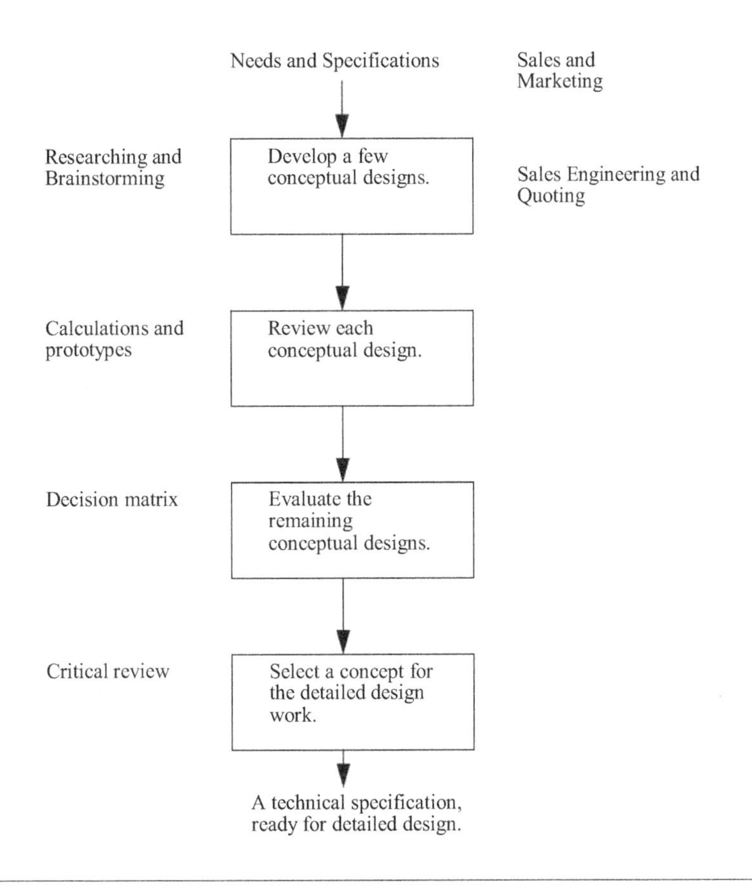

FIGURE 4.5

A simple process to generate a conceptual design.

4.3 Specifications to concepts

Poor designers blindly fit problems to their personal knowledge.

A formal approach to developing and reviewing concepts is given in Fig. 4.6. Basically, the specifications are reviewed and concepts generated to address all of the requirements. After a screening process, the best concepts are chosen and the design moves forward. However, when there are too few alternatives, we try to create alternatives and backups.

The concepts in Fig. 4.7 address various specifications. We can use some combination of these concepts to satisfy the overall specifications. For example, the only concepts that will work are B and D together. But, if we eliminated spec 3, then we could use concepts A and D instead.

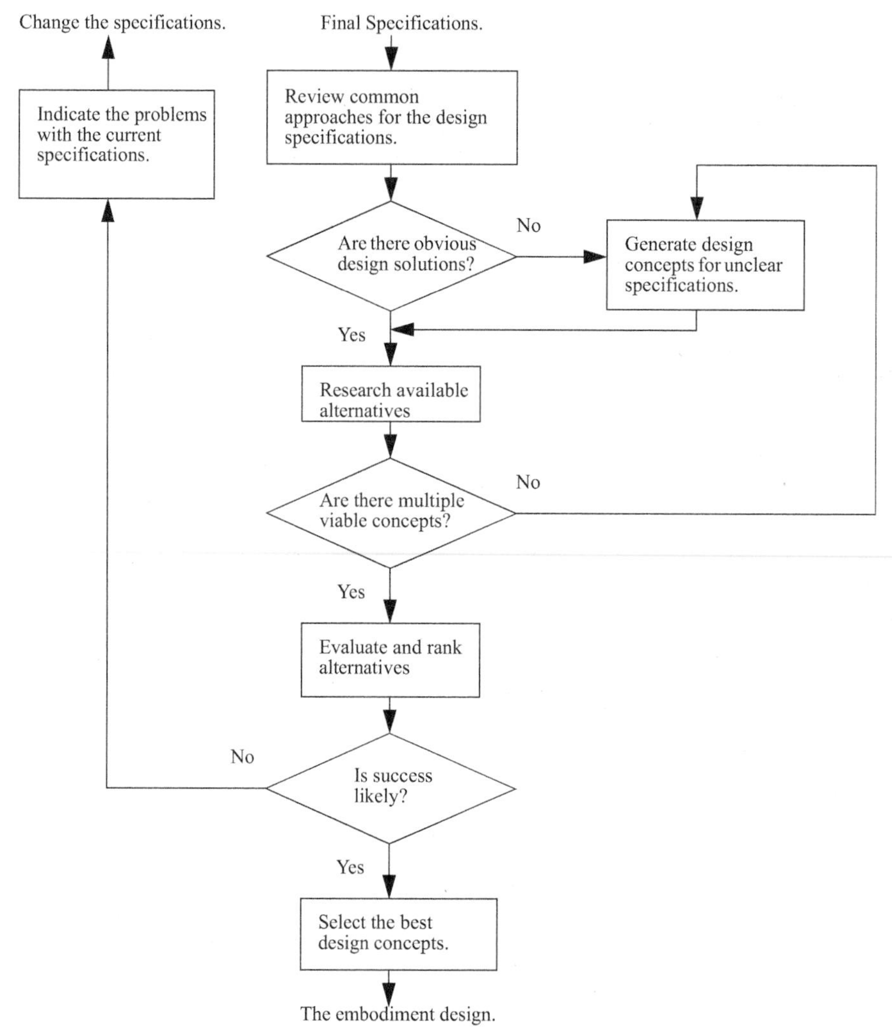

FIGURE 4.6

Approaching design issues.

Specification List	Concept A	Concept B	Concept C	Concept D
Spec 1	y	y	y	
Spec 2	y	y	y	
Spec 3		y		
Spec 4	y	y		
Spec 5			y	y
Spec 6				y

FIGURE 4.7

Specifications to concepts mapping.

PROBLEM

4.7 Why should a design always have alternative concepts?

4.8 Assume Zeke was asked by his supervisor to develop five concepts for a new LED light housing. Zeke found one concept he really liked and then created four more. Two of the five concepts would cost too much to use, and one of the concepts used a technology that did not exist. Is it more unethical for Zeke to present these as five or two design concepts?

4.9 Will a technical specification satisfy all of the customer specifications? Explain.

4.10 Do all concepts need to satisfy the same specifications?

4.11 Is it possible to combine microconcepts into a macroconcept? Explain.

4.4 Representing concepts

A design is a collection of concepts. The most common conceptual design approach is to break a design problem into smaller pieces. The concepts are generated, explored, selected, and then combined into a final technical specification. The concept diagram (CD) in Fig. 4.8 can be useful for visually organizing the concepts and alternatives. This can also be done with a spreadsheet, but will take additional time to review. Some general concepts to remember: (1) include the obvious concepts, (2) show alternatives and determine where there are gaps, and (3) there should be concepts that address all of the specifications. It can be effective to put each of the concepts into the following categories: O, obvious concepts, no other work needed; C, compare the concepts; and G, generate concepts.

Beyond the CD, other design representations can be used to capture design concepts and expose unknown parts of the design. Concepts are normally communicated in a visual or detailed form, to improve the clarity of the overall design. Documentation that would be expected to explain a concept would include the items in the following list, but each discipline has preferred representation techniques. The key is to use the right representation methods. If text is to be used to capture a design

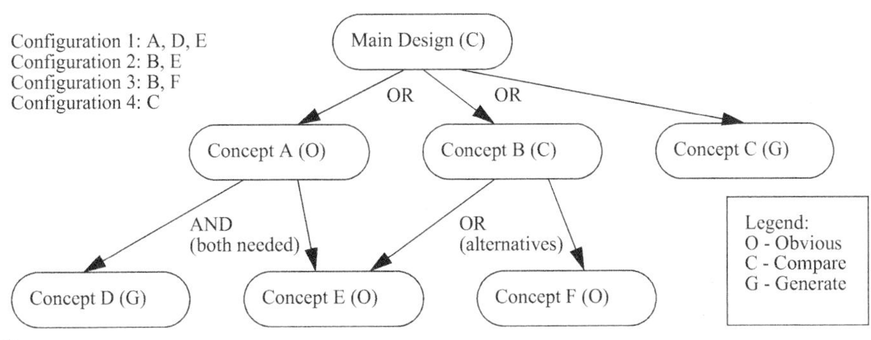

FIGURE 4.8

Concept diagram.

concept, it is important to recognize how easy it is to make the written word vague or arbitrary. When text is used for concept representation, it should be to support the information, not replace it:

- block diagrams of major components
- flowcharts, state diagrams, pseudocode, a list of steps, data flow
- sketches of basic physics
- proof-of-concept calculations to support a design concept
- crude prototypes or bench tests
- existing design implementations
- rough budgets for money, mass, etc.
- key technologies and alternatives, including critical parts
- layout diagrams
- process diagrams
- graphs
- pictorial sketches for look, feel, colors, textures, form, etc.
- GUI sketches
- mechanical sketches using normal drafting practices, including 2D, isometric, and pictorial views

Systems can be viewed as a collection of functions (blocks) that are connected by arrows (Fig. 4.9). These block diagrams show the major parts of a system and the values/materials/other that are passed between them. The advantage of these diagrams is that we can put system parts in "black boxes" to reduce the complexity, while looking at the system on a macrolevel. If the block diagram is sufficient, each of the blocks becomes a stand-alone design problem. (Breaking big parts into smaller ones is a very good strategy.) These diagrams are well used in electrical and computer engineering, but are very useful for other engineering work.

A more advanced version of function-based diagramming is shown in Fig. 4.10.

The method provides a substantial amount of flexibility; however, the rules of thumb are as follows:

- Pick reasonable function boundaries (not too much or too little).
- Conserve energy and material.

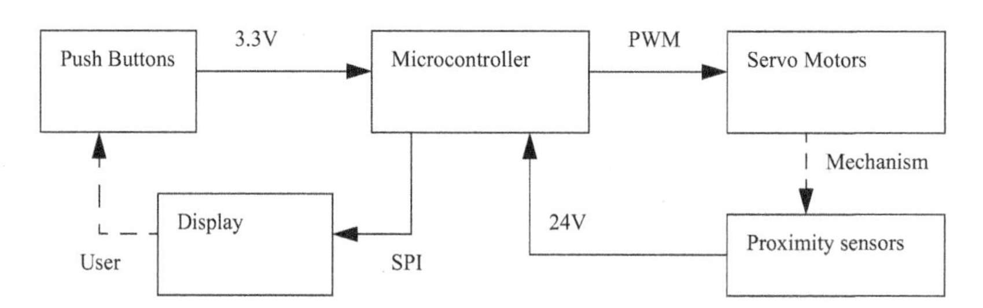

Each of the blocks is a function that accepts inputs (from the left) of materials, information, energy, and more. The function then transforms these to outputs (to the right) in some other form. Functions of the system that depend on each other are linked by the vertical lines. Again the boxes are all considered to be "black boxes" requiring additional design. In some systems, such as IDEF, the functional blocks are broken down multiple times until they reach obvious functions. Again, the approach allows the designer to look at the system from a high level or abstract perspective before attempting more specific conceptual design work.

FIGURE 4.9

Sample block diagram. *IDEF,* Integrated definitions; *PWM,* pulse width modulation; *SPI,* serial peripheral interface.

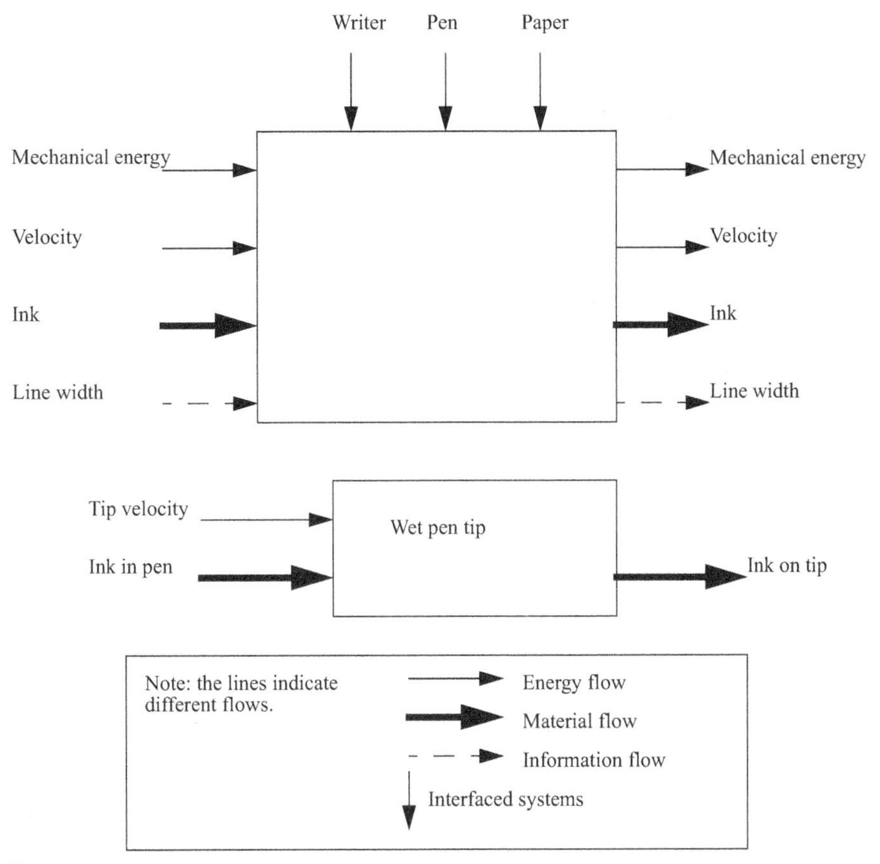

FIGURE 4.10

A sample function diagram for drawing lines on paper.

- Information can be discarded easily, but it takes energy and materials to create.
- Add information flows to determine when things should happen, how they should happen, and how they are related.

Once a functional diagram has been created, follow all of the paths through the system. Consider the following:

- Logical: Follow materials, information, signals, etc., through the diagram and look for logical changes.
- Complete: Does each function receive and output all of the needed values?
- Black box: Can the black box functions be implemented directly? Or is more reduction needed?
- Missing: Is there anything missing that should be there?
- Notation: Use standard notations and units and check for balances.
- Sequence: Follow the sequence of operations.
- Fuzzy: Are there things in the diagram that are unclear?
- Next: Look for open questions that require more work.
- Level: Are all of the functions generally at the same level of complexity?

- Nowhere: Are things used that do not seem to have any effect?
- Unwanted: Are there items that are not needed, or can be used elsewhere for energy or other?

Other technical diagramming techniques include:

- block (or signal flow) diagrams
- flowcharts
- state diagrams
- mind maps
- Ishikawa/fishbone diagrams
- trees and hierarchies
- graphs, with arcs and nodes
- matrices
- networks
- word clouds
- free sketching and doodling
- algorithms and subroutines

PROBLEM

4.12 Consider a portable drink cup that provides cooling. Sketch the concepts, including (a) a flexible pouch, (b) a solid cup with a lid, (c) insulated sides, (d) water-filled ice cube analogs, (e) an electric cooler (Peltier junction), and (f) multiple insulating sleeves.

4.13 Draw a CD for a portable drink cup that provides cooling. The concepts include (a) a flexible pouch, (b) a solid cup with a lid, (c) insulated sides, (d) water-filled ice cube analogs, (e) an electric cooler (Peltier junction), and (f) multiple insulating sleeves.

4.14 Should each technical embodiment include the same number of concepts?

4.15 Why should concepts be represented graphically?

4.16 Draw a block diagram for a coffee maker.

4.17 Draw a function diagram for a bicycle brake.

4.18 Draw a function diagram for a flashlight (a.k.a. torch).

4.19 Draw a block diagram for a wireless computer mouse.

4.20 Use three different visual techniques to show concepts for a remote-control fob for a car.

4.5 Identifying concepts

Do not reinvent the wheel. If you need to solve a problem, consider what has been done already. You may find that you have an excellent solution for your design problem that will let you spend more time and energy on other problems. Even if it has been done before, you will get some clues and new directions. A good way to start this process is to (1) review the specifications and (2) list possible solutions to achieve each of the specifications. These can then be sorted from high-level concepts that address many or all of the specifications to low-level concepts that address only one. At this point, it is fine to have a messy pile of ideas, but eventually it will need to become a CD or something similar.

When identifying concepts the aim is to build an inventory of knowledge to find the low-hanging fruit that is easy to pick. If no good concepts are found, then we will need to make our own. If there are multiple options, we will need to select one; however, the focus is to look and investigate. Basically, all information is good, and if concept generation is needed, it may also prove useful. During this phase, collect good notes and keep an open mind. Strategies that will be useful here include the following:

- Self: Consider what you know about similar issues and available concepts. You will probably know one or more ways to fulfill the specification.
- Previous: Work done during the specification development can be useful here. In some cases a suitable concept may have already been selected. The major risk here is that other good solutions are ignored or overlooked.
- Patents: Patents protect new ideas for 20 years. After that you can use the idea freely. Before a patent expires you can negotiate a license or develop other ideas. Be careful to look for related patents.
- Market: Look to see what is for sale—consumer products, auction items, and industrial products.
- Supplier: Look for supplier-based solutions to your problems. Call them. Check their catalogs and websites.
- Technology: Find technologies that are commonly used.
- Literature: Refer to books, magazines, libraries, the Internet, etc.
- Network: Talk to people you know for ideas and find experts.
- Consult: Pay somebody to help sort the ideas.
- Internal: What are the available capabilities at the company?
- Standards: Are these common components, practices, methods, techniques, etc.?

An ideal outcome at the end of the concept identification phase is to determine what you do, and don't, know.

PROBLEM

4.21 Consider the design of a folding seat that can be carried over long distances.

(a) List five ways the seat could fold.

(b) Locate five different product types and provide pictures.

4.22 How is concept identification different from concept generation?

4.23 As the saying goes, "The best ideas are stolen." What does this mean?

4.6 Concept generation

If there are good and obvious concepts for every part of a design, there is no need to generate new concepts. In practice, each design will need some conceptual work. In innovative design these can be sweeping high-level concepts such as new alternative energy sources. In commodity design, these can be focused on single features or methods such as a new lower-cost lighting element.

Generating concepts carries some risk, so an approach is to generate several viable concepts so that you will have more to choose from and more options if the concept choice fails.

If you have reached the concept generation stage, it is because you could not find enough obvious concepts. Therefore, the way you are trying to satisfy the specifications is not providing useful concepts. To find solutions, you will need to work outside your normal bounds. The process of creativity explores new ways to look at problems and generate alternative solutions. Some common themes in creative design questions include the following:

- Combine or break apart ideas.
- Start with something silly and refine.
- Consider the abstract (e.g., energy, information).
- What approaches are not normally used in a design?
- Most of our ideas will be discarded—but all we need is one good one.
- Look at other things; try not to be too focused.
- Break up the design into smaller, easier to design, pieces.

Concept generation methods are plentiful and each one provides some unique opportunities and values. Some of these are described in the following list. Major variations between these methods are (1) the number of people involved, (2) the level of creativity required, (3) the amount of change expected, and (4) how the creative process is structured. Fig. 4.11 shows the methods in terms of

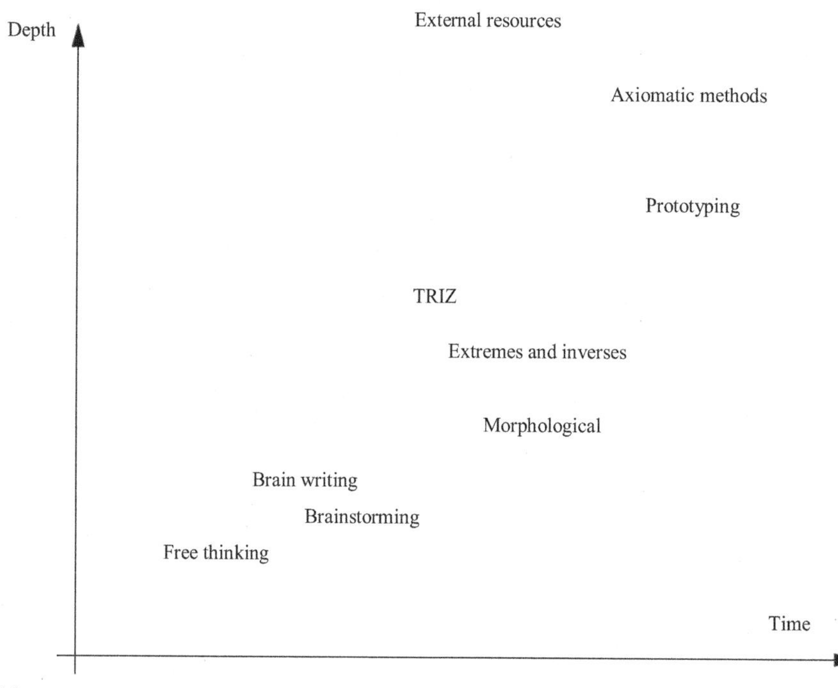

FIGURE 4.11

Concept generation techniques.

the time required. Approaches such as brainstorming do not require much time and forethought. A method such as prototyping requires more time and effort. Each of these methods will have advantages based on problems, individuals, and needs:

- Brainstorming: a technique for generating ideas using free-flowing conversation.
- Brainwriting: a nonverbal approach to brainstorming in which ideas are written down and then passed to others who add more ideas. The 6–3–5 method is a variant whereby six participants create three ideas every 5 min.
- TRIZ: an inventive problem-solving technique.
- Doodling: a free-form method of drawing sketches, including discarding, changing, or combining to find solutions.
- Thought experiments: consider the problem using extreme examples.
- Incremental: combine or modify existing ideas.
- Prototyping: build a model of a process, component, or critical feature and test it.
- Eureka: wait for inspiration.
- Architect: a designated concept creation individual.
- Ishikawa diagrams: a useful technique for capturing cause-and-effect relationships.

PROBLEM

4.24 Are concepts required for all parts of a design?

4.25 Does concept generation involve creativity?

4.26 Does creativity mean that all ideas are new?

4.7 Prototyping

In engineering problem-solving, many assumptions and simplifications are used. The translation from theory to practice sometimes fails when (1) important factors have not been considered, (2) assumptions and simplifications are unrealistic, (3) variables interact in unexpected ways, (4) there are unexpected deviations and events, or (5) there are unintended consequences. These issues are well known for well-established designs such as golf clubs and balls. However, newer designs, or new applications of designs, are at greater risk for failure and warrant prototypes.

The key to prototyping is to identify designs that may not behave as expected. This can occur when designers have little or no experience with a design problem, for example, the feel of a new user interface or tool handle. Some problems are so complex that they defy rigorous analysis, and testing is the only option. Examples of overly complex problems include aerodynamic flow around structures, magnetic flux in iron, and culture growth in a bioreactor.

Scale and approximate modeling are common prototyping techniques. In a scale test, the overall process is reduced in a costly variable such as size, speed, mass, and so on. For example, a large radar antenna array might use 6 receivers instead of 600. These might then be used and tested to identify effects of vibration, humidity, wave reflections, impedance matching, and amplifier noise. The results would then be used to modify the design before the final design is fabricated. An approximate model will focus on some components, but ignore others. These models are often constructed with duct tape, temporary wires, development boards, LEGOs, cardboard, rapid prototype parts, and more.

It is becoming increasingly common to develop digital prototypes. The results of design work are simulated and then displayed. Common examples include finite element methods, surface rendering, circuit simulation, and much more. These prototypes can be developed very quickly, but are subject to the assumptions used in the simulation and software setup.

Examples of prototyping approaches, and tools, follow. Prototypes are created by engineers to test concepts in the early design process. It is worth noting that sometimes a project will be launched and run entirely to produce a prototype before a full-scale design is attempted. It is possible to think of this process as two design projects, or a design project in which the conceptual and detailed design phases overlap:

- Rapid prototyping: There are a number of processes that can quickly convert solid models into parts. The materials are generally plastics or similar materials, useful for testing assembly, aesthetics, and touch/feel. Simpler parts can be made in a few hours.
- Prototype printed circuit boards: Online services and circuit-board mills can be used to produce boards very quickly from computer-aided design (CAD) systems. The typical time frame is less than an hour for a circuit board mill in a design office.
- Development kits: Electronics companies often provide general boards, or reference designs, that can be used to create and test hardware and software before the detailed design work begins.
- Component off the shelf: Modules or major pieces of the system are purchased in a working form. This sidesteps some testing as well as detailed design, but often at a higher cost.
- Sample parts: Most component suppliers will provide free or low-cost sales samples to major designers. Other times standard parts can be purchased, modified, and used.
- Crude: Sometimes a prototype can be constructed in minutes out of cardboard, duct tape, and other common components.
- Hobby parts: The hobby market can be a great source of parts, especially for scaled models. This includes electronics, mechanical, and many other system types (see Resource 4.1).
- Hacked: An existing design can be modified as a proof of concept for a new design. This is a very good approach when developing a new generation of a product.

PROBLEM

4.27 What are three advantages and disadvantages of building prototypes?

4.28 Does a prototype need to prove all of the concepts?

4.29 Does a prototype need to be functional?

4.30 How could a pile of old toys be used to build prototypes?

4.31 Mini-case: Prototyping

A company that makes large sheet metal ducts was looking for a machine to bend flanges. The sheet metal tubes had diameters up to a meter, and 3-cm flanges were added to both ends. The design team had a number of options for bending the ends of the tubes but settled on a method that used a pair of rollers. The team was confident that the design would work and wanted to seek permission to begin detailed design. To reduce the concept risk, the team was asked to develop a prototype test for the process. A crude test was performed on a manual lathe and appeared to produce an acceptable flange. There were some minor issues with a bend radius and cracking near a weld seam. However, the team used the results of the test to argue that the project should advance. Based on their technical assessment they were given clearance to move ahead with detailed design. The design and build process went smoothly and was somewhat ahead of schedule. During testing a number of problems arose, including the roundness of a flange, cracking near welds, material deflection, and pipe slippage. Most of the issues had been observed in prototyping but the assumption was that a larger-scale machine would resolve the issues. To correct the issues the team redesigned, rebuilt, and retested multiple times. Each time the results of the tests were discussed by the team and a new concept was developed to solve new problems. After going through the process multiple times the team was far behind schedule, but eventually the machine met the specifications. The decision to analyze or prototype can be difficult. Often concepts appear to be foolproof, but unexpected problems arise later. However, the time spent in prototyping and analyzing delays the start of the build process. By their nature, prototypes are scaled and approximate models of the final system, and so the results are open to interpretation. Considering these factors, what would be a reasonable process for deciding when to do prototypes, how to interpret the results, and when prototyping should end?

Resource 4.1 *Make Magazine* (http://makezine.com/): an outstanding collection of hobbyist creations that range from crafts, sewing, simple machines, robots, microcontrollers, and more.

4.8 Brainstorming

Basically, brainstorming generates a large number of diverse concepts using a group of individuals. One approach is:

(1) Have a meeting of individuals (6–12 is good) related to the current design tasks.
(2) Make it clear that criticism is not allowed and every idea is good.
(3) Ask everyone to write ideas on separate pieces of paper.
(4) Start going around the room one at a time, and ask for the ideas. (Don't allow criticism or judgment!) After the idea is given, the paper is placed in the center of the table.
(5) This continues until all ideas are exhausted. (Participants should generate new ideas based on what they have heard from others.) Encourage participants to suggest ridiculous ideas.
(6) Go through the ideas in the middle of the table and vote for the best one(s).

During the idea-generation phase it is critical to avoid being negative. Some of the problem phrases and words are:

- That won't work because ...
- I don't agree.
- We already tried that.
- We can't do it because ...
- That is silly.
- We need good ideas, don't waste our time.
- A better idea is ...
- Is that possible?
- You don't know ...

Brainstorming may be less effective if there are voices that tend to dominate the discussion. To overcome this, brainwriting allows ideas to be shared and extended nonverbally. In simple terms, each team member writes some ideas on paper. The piece of paper is passed to another team member. The ideas are read and new ideas are added. This continues until all of the ideas have been passed, or there are a suitable number of responses. After that, the ideas are reviewed for interest. Another round of writing generates a new set of ideas. This can be repeated a number of times before the ideas are reviewed and prioritized critically.

The Delphi technique was developed to work with separated experts. The process begins with a generalized questionnaire sent to the individuals, often via email. The responses are reviewed, and a new, more specific, questionnaire is created. Multiple rounds of reviews and questionnaires may occur before the process ends. The result will tend to be a group consensus that includes all voices.

PROBLEM

4.32 Brainstorming does not allow criticism of ideas until the end of the process. Why?

4.33 Assume you are in a brainwriting session. You are asked to generate concepts for new methods of cleaning dishes after a meal. Write a list of 20 different ideas. The ideas can be serious or silly, but they must be related to the problem.

4.34 Write a possible first- and second-round Delphi questionnaire for chair design issues.

4.9 Morphological matrix methods

A morphological attribute list can be used to break a problem into parts. Fig. 4.12 shows an example of a morphological matrix for a new gardening tool. The problem statement is used to identify three terms: gardens, plants, and tools. For the word "garden" a few different forms are listed. For example, a garden could be hanging or a lawn. These lists can be generated with free thinking, a thesaurus, a web search, and so on. Similar lists are generated for the remaining terms. The terms form a morphological matrix.

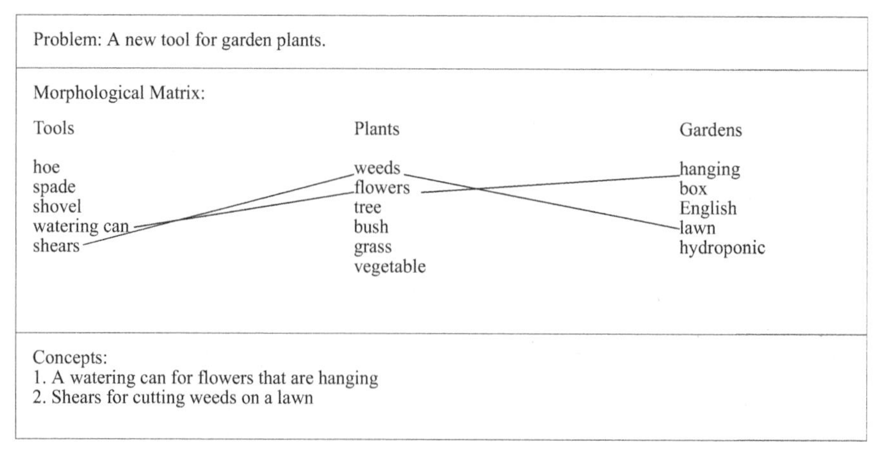

FIGURE 4.12

Morphological menu matrix.

To generate concepts, a word is selected from each column to form an abstract sentence. The first combination would be "hanging weeds hoe." The question is, "How could this be made?" A design team might reply, "With an extended hand tool that reaches into a hanging basket to remove weeds." If done exhaustively this matrix would yield $5 \times 6 \times 5 = 150$ possible combinations.

PROBLEM

4.35 Develop a morphological matrix for "a cooler for a hot car." Use at least two columns with five words to generate 10 concepts.

4.10 Free thinking

Sometimes a concept solution is not obvious because we are looking at the problem, solution, or system with a limited perception.

There are a variety of approaches used to suggest new mechanisms or perspectives for satisfying specifications. In simple terms, if you find that you are stuck, look at the terms and ideas in this section to conceive new directions for thought.

If we assume that we know the solutions, but just can't see them, we can use the list of mental barriers provided by Niku (2009):

- false assumptions and nonexistent limitations
- typical solutions
- being overwhelmed and making things more difficult than they are
- incomplete or partial information
- information and sensory saturation
- associative thinking
- misunderstanding
- inability to communicate properly
- emotion-, culture-, and environment-related barriers
- falling in love with an idea
- improper methods of solution
- overabundance of resources

If you get stuck generating concepts, consider alternative problem-solving strategies. In these cases you can use the list in Fig. 4.13 to prompt thoughts.

PROBLEM

4.36 Discuss five abstract prompts that can be used for planning a human mission to Mars based on experiences with moon missions.

4.37 List one example for each of Niku's mental barriers.

4.38 Give five examples of technologies that can be adapted to keep shoes clean on muddy days.

Divide problems - if a problem seems too hard to solve then divide it into smaller, more approachable, parts.
Combine - join two or more parts to get a new function.
Question assumptions - review all constraints and goals and make sure they are sensible.
Information completeness - assess the available information for i) too little, ii) too much, and iii) conflicting.
Analogies - look for similar problems.
Detach - a few times step back and question what you are doing.
New perspective - find another way to look at the problem.
Replace - find a component to use in place of a traditional.
Adapt - use existing solutions in a new way.
Simplify - reduce complexity, eliminate competing ideas, prioritize.
Theory - consider the problem in abstract terms.
Question everything - ask questions about everything, especially the obvious.
Keywords - list keywords and then search for other uses - consider an Internet search or thesaurus.
Extremes - What are the limits for problem solutions? How far can the solution be pushed before it stops working?
Inverses - Flip parts of the problem, including inside-outside, up-down, rotation-translation, voltage-current, flow-pressure.
Reduction - Break a large problem into smaller parts that are easier to solve.
Approximation - Close may be good enough.
Arbitrary - If there is no obvious choice then pick something.
Negligible - It may not be significant.
Simplify - Details can camouflage the obvious.
Analogs - Put a problem in more familiar terms.
Reversible - Some systems only work one way, others can be used backwards.
Cause and Effect - Effects don't change causes unless the system is reversible.
Scalable - A smaller design may work at a larger scale, or sometimes not.
Zero-sum - In some form everything that goes in must come out.
Entropy - Things always settle down unless they are stirred.
Unintended consequences - Small changes sometimes have bigger effects.
Shortcuts - Always take longer.
Focus - The most important issue comes first.
Assumed - Normal things are not always needed.
Coupled - Things may, or may not, be independent.
Complete - Is is complete and does it have to be?
Characterize - Find a simpler way to describe it.
Redirect - If the solution isn't working then change the problem.
Failure - A negative outcome provides information.

FIGURE 4.13

Abstract prompts.

4.11 Deconstruction

Conventional thinking allows great progress by building on the knowledge of the past. For example, a chair should have a seat, a back, and four legs. This then allows room for incremental improvement, including adjustability, padding, wheels for rolling, flexibility, arm rests, and appearance. A chair designer simply assumes the chair geometry and moves forward to new materials and adjustments. Convention supports evolutionary improvements to designs. Revolutionary designs need to establish new conventions. Deconstruction is a process of undoing assumptions to create design freedom.

Deconstruct a design by listing the essential features. Each of these is then questioned. At first the method will feel awkward, until some interesting outcomes arise. Some of the questions to ask include: "Could this be eliminated?" "What other ways could it be done?" "Do the other steps really need this

Using a disposable coffee cup:

i) Hot coffee is poured into the insulated cup and a lid is put on to keep the coffee warm.

ii) The lid has an opening so that the coffee can be sipped.

iii) The cup has a shape that is easy to carry and sit on a table.

iv) The cup is made from low cost materials that can be recycled and do not contaminate the coffee.

Contrary questions:	Contrary answers:
Why does the:	
1. coffee need to be hot?	1. The coffee can be heated as the drinker sips.
2. coffee need to be poured?	2. The coffee only needs to be poured if it is made in another pot.
3. cup need to be insulated?	3. Insulation is not needed if the coffee and room are the same temperature.
4. cup need a lid?	4. No lid is needed for a sealed container.
5. lid need to be put on?	5. Don't put a lid on.
6. coffee need to stay warm?	6. It could stay at room temperature until it is sipped.
7. lid have an opening?	7. The coffee could seep through a porous surface.
8. coffee need to be sipped?	8. Use a straw.
9. cup need to be carried?	9. The cup can be put in a pocket.
10. cup need to sit on the table?	10. The cup can be sticky and cling to any surface.
11. cup need to be low cost?	11. The cup can cost more if each use costs less.
12. cup need to be recycled?	12. Make the cup reusable.
13. cup not contaminate?	13. The cup could add sugar, creamer, or flavor.

Useful ideas:

Keep the coffee at room temperature until sipped and then heat.
Make the coffee in the cup.
Put it in a sealed contained to eliminate a lid.
The coffee is sipped through a porous surface or a straw.
The cup could add sweeteners, creamers, and flavors.
Make the cup fit into a pocket or stick on surfaces.

The concept:

The coffee is sold in low cost, sealed, room temperature bags that a user connects to heating/flavor unit. As they sip the coffee the unit heats and flavors the quantity consumed. The heat/flavor unit has a clip so that it may be clipped to a belt, pocket, strap, etc.

FIGURE 4.14

Deconstruction of a disposable cup.

step?" A simple example for a disposable coffee cup is shown in Fig. 4.14. In this example the assumption that the coffee "must be hot" is questioned; the response is that the only time heat is important is "when drinking." The idea that is gleaned is that the coffee can be heated just before drinking. The result is a component in the cup that actively heats the coffee.

The process of deconstruction will be unique for every attempt, but it always generates new ideas. Of these, some will be ridiculous or impractical. However, the method will produce a few ideas that can be used as alternatives to conventional thinking.

PROBLEM

4.39 Deconstruct a standard traffic stop sign.

4.40 Deconstruct a sheet of writing paper.

4.12 TRIZ

The theory of inventive problem solving, TRIZ, is a systematic method for solving problems using an inventory of ideas and problem solutions developed from thousands of Russian patents (Altshuller, 1988). The method's steps define a specific problem, restate it in abstract terms, find generic solution approaches, and apply the generic solution to the specific problem (Fig. 4.15). The basic TRIZ method provides 39 contradiction parameters, listed in the figure. The concrete problem needs to be converted to one parameter to maintain, and another to increase (or decrease). Once encoded, the TRIZ matrix is used to suggest possible solutions (see Resource 4.2). The solution numbers relate to 1 of the 40 TRIZ principles. The final step is to convert the principles back into concrete solutions.

Resource 4.2 Book website TRIZ spreadsheet: http://www.engineeringdesignprojects.com/home/content/concepts.

PROBLEM

4.41 Use TRIZ to find four approaches to decrease the weight of a bag used to carry books.

4.13 Back-of-the-envelope calculations and functional prototypes

Many great ideas were first written on a handy scrap of paper such as an envelope, napkin, or menu. Likewise, many ideas were tested with simple models constructed with an ERECTOR set (Meccano), random toy pieces, cardboard, and other discarded materials. It is part of a quick progression of an idea to a simple test. Good designers often do "back-of-the-envelope" (BOE) calculations to determine if an idea has any merit, or if it is a waste of time. Of course, it is not necessary to find an envelope to do the work, but the crude and fast approximations are key. These methods determine solution viability and parameters within an order of magnitude.

To reemphasize the point, when solving problems, quick notes, sketches, and calculations are incredibly valuable. Some of the reasons that a designer will sketch or calculate include:

- to determine if a concept is possible
- to get the key ideas on paper and then play with them
- to try some what-ifs
- to capture ideas for later
- to communicate with somebody else

The 39 Contradiction parameters:

1. Weight of moving object.
2. Weight of stationary object.
3. Length of moving object.
4. Length of stationary object.
5. Area of moving object.
6. Area of stationary object.
7. Volume of moving object.
8. Volume of stationary object.
9. Speed.
10. Force.
11. Stress or pressure.
12. Shape.
13. Stability of the object composition.
14. Strength.
15. Duration of action by a moving object.
16. Duration of action by a stationary object.
17. Temperature.
18. Illumination intensity.
19. Use of energy by a moving object.
20. Use of energy by a stationary object.
21. Power.
22. Loss of energy.
23. Loss of substance.
24. Loss of information.
25. Loss of time.
26. Mater or substance quantity.
27. Reliability.
28. Measurement accuracy.
29. Manufacturing Precision.
30. External harm affect object.
31. Object causes harm.
32. Ease of manufacture.
33. Ease of operation.
34. Ease of repair.
35. Adaptive or versatile.
36. Device complexity.
37. Detecting and measuring is difficult.
38. Extent of automation.
39. Productivity.

The 40 Solution Principles:

1. Segmentation.
2. Extraction/removal.
3. Local quality.
4. Asymmetry.
5. Combining.
6. Universality.
7. Nesting.
8. Counterweight.
9. Prior counter action.
10 Prior action.
11. Cushion in advance.
12. Equipotentiality.
13. Inversion.
14. Spheroidality.
15. Dynamicity.
16. Partial or excessive action.
17. Moving to a new dimension.
18. Mechanical vibration.
19. Periodic action.
20. Continuity of a useful action.
21. Rushing through.
22. Convert harm to benefit.
23. Feedback.
24. Mediator.
25. Self-service.
26. Copying.
27. Inexpensive vs. expensive.
28. Replacement of a mechanical system.
29. Pneumatic/hydraulic construction.
30. Flexible membrane or thin film.
31. Use of porous materials.
32. Changing the colors.
33. Homogeneity.
34. Rejecting and regenerating parts.
35. Physical and chemical transformation.
36. Phase transformation.
37. Thermal expansion.
38. Use strong oxidizers.
39. Inert environment.
40. Composite materials.

General contradiction:

Increase: 13. Stability of the object composition.
Maintain: 11. Stress or pressure.

General problem solutions from the TRIZ matrix:
2. Extraction/removal.
33. Homogeneity.
35. Physical or chemical transformation.
40. Composite materials.

Abstract

Concrete

A specific design problem or contradiction.

Problem: Increase car tire life. This means similar traction for a longer travel distance.

Specific design solutions:

2. As the tire wears allow rubber to fall out maintaining the tread depth.
33. (Nothing obvious).
35. As the tire wears the rubber could harden.
40. The tire rubber could be mixed with fibers to change the wear patterns.

FIGURE 4.15

The TRIZ process.

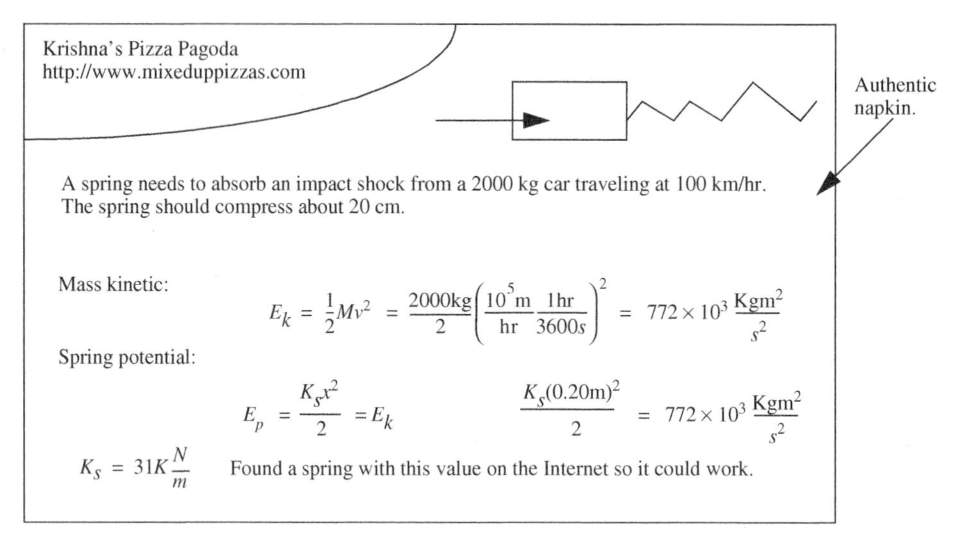

Krishna's Pizza Pagoda
http://www.mixeduppizzas.com
Authentic napkin.
A spring needs to absorb an impact shock from a 2000 kg car traveling at 100 km/hr.
The spring should compress about 20 cm.
Mass kinetic:
$E_k = \frac{1}{2}Mv^2 = \frac{2000\text{kg}}{2}\left(\frac{10^5\text{m}}{\text{hr}}\frac{1\text{hr}}{3600s}\right)^2 = 772 \times 10^3 \frac{\text{Kgm}^2}{s^2}$
Spring potential:
$E_p = \frac{K_s x^2}{2} = E_k \quad \frac{K_s(0.20\text{m})^2}{2} = 772 \times 10^3 \frac{\text{Kgm}^2}{s^2}$
$K_s = 31K\frac{N}{m}$
Found a spring with this value on the Internet so it could work.

FIGURE 4.16

Back-of-the-envelope spring-coefficient calculation.

The key to these approaches is to oversimplify the problem. Some BOE strategies to help the process are given in the following list. An example of a quick calculation of a spring value is shown in Fig. 4.16.

- Question: What is the important question about the idea or problem?
- Parse: Reduce or eliminate the unimportant or less important details.
- Approximate: Sacrifice accuracy for fast results. Reduce complex parts to simple equivalents—a complex cross section of a car becomes a sphere, an I-beam cross section becomes a rectangle.
- Sketch: Draw stick figures, free body diagrams, doodles, cartoons, mechanism positions, isometric parts, layouts, etc.
- Energy: Calculate the energy added or removed from the system using the internal kinetic and potential energy.
- Extremes: Consider the extremes of motion or operation to determine feasibility and needs.
- Order of magnitude: Accept that the results will be too high or low but should be in a range from 1/10 to 10 times the actual value.
- Flow: Draw lines that show how force, flow, heat, etc., flow through components and systems.

Some examples of BOE calculations are shown in Figs. 4.16–4.18. The calculation in Fig. 4.16 simplifies a complex problem by simply approximating the kinetic and potential in a mass spring impact. The spring coefficient calculated will not be the final value, but it shows that the design could work and provides a rough value. If this did not work, the napkin and the idea go into the garbage. The wheelchair lifter in Fig. 4.17 began with some sketches of wheelchair positions. Adding some more detail makes a possible solution obvious. It often takes a few envelopes before the right mechanisms are found. The initial design for a tension member (Fig. 4.18) starts with a sketch of a couple of geometries. Looking at the force flow lines highlights where failure might happen, and the tensions in these areas are used to guess the necessary thickness. Of course, in the lower case the shear modulus should have been used, but this is a BOE calculation so we know it will need to be around 1 cm. If the

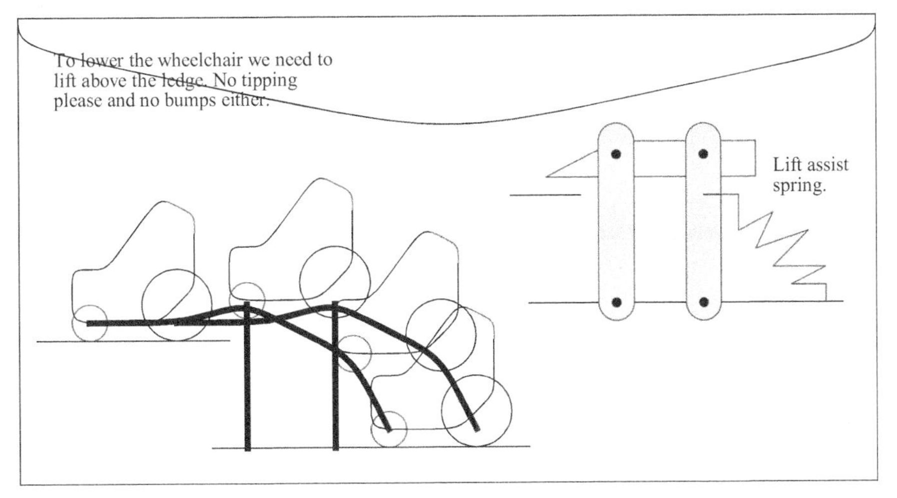

FIGURE 4.17

Back-of-the-envelope wheelchair-lowering mechanism.

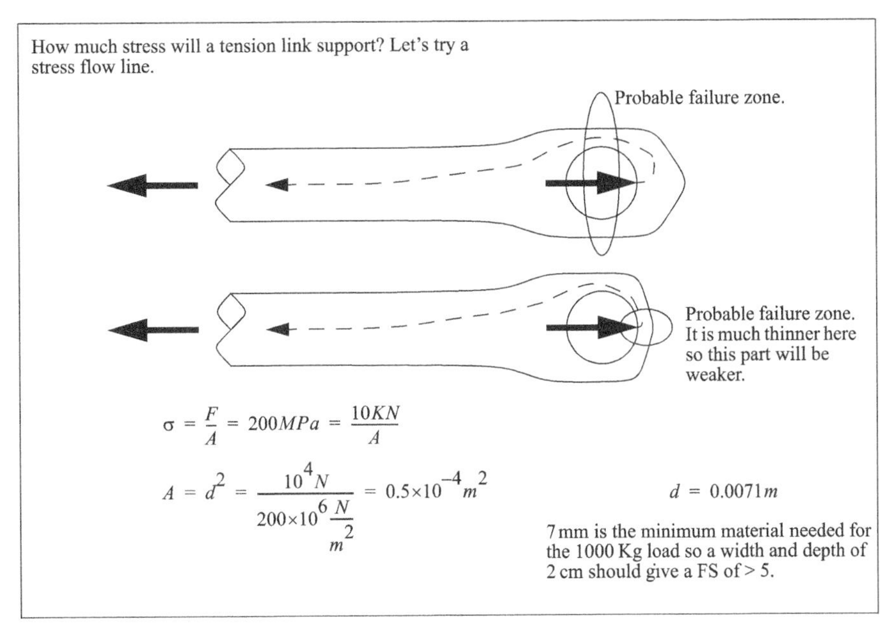

FIGURE 4.18

Back-of-the-envelope dimension estimate. *FS*, factor of safety.

thickness were 2 cm, then the design would probably have a factor of safety (FS) of more than 2. For each of these three examples detailed engineering calculations must come later.

BOE calculations are often followed with the construction of simple prototypes. The two main reasons for building a prototype are to see how it feels and to get an idea of how/if it works. A well-chosen prototype will help make the decision to pursue an idea or to rule it out. For mechanical designs, typical prototyping materials and methods include:

- building toys: ERECTOR set, Meccano, LEGO;
- craft materials: straws, popsicle sticks, toothpicks, tape, glue, cardboard (see Resource 4.3);
- structural duct tape;
- woodworking: basic cut wood pieces and common household hardware;
- hacking: similar designs are cut, joined, or modified for the new design;
- CAD: digital models can quickly provide an analysis;
- rapid prototyping: CAD models can be produced quickly in plastic-like materials for reduced function testing;
- computer numerical control: computer-controlled machines can be used to quickly mill or laser cut parts;
- bench tests: standard laboratory equipment is used to verify basic process inputs, parameters, and outputs.

Like the BOE calculations, prototypes are needed to answer very specific questions. As a result, the key is to build the least amount possible but still increase confidence in a design idea. The sketches in Fig. 4.19 show the progression of thought from crude sketches to a functional prototype.

Resource 4.3 *Make Magazine* (www.makezine.com), a hobbyist community dedicated to designing and building interesting devices, is a good resource.

PROBLEM

4.42 Do a BOE calculation for a mass of 100 kg dropped from 10 m, attached to a steel cable, to estimate the cable size required for the dynamic forces.

4.43 Use BOE calculations for the total cost of a bridge with 100 metric tonnes of steel.

4.44 Use BOE calculations to estimate the number of people required to pass 10-kg sand buckets over a mile using a firefighter's brigade approach.

4.45 Design a kinematic mechanism to remove a slice of toast from a toaster and put it on a plate.

4.46 Consider a new three-legged kitchen chair with a back. Develop a prototype by (a) sketching and (b) building with paper and/or wood.

4.47 Design an L-shaped bracket to hold a shelf to a wall. There are two screw holes on the wall and the shelf sides of the bracket. Estimate the points of maximum tension or compression.

4.14 Factor of safety

Assumptions allow us to simplify problems enough to avoid deadlock. Examples include idealized equations for stress, standard material properties, manufacturing tolerances, work methods, and how a design will be used. It is up to the engineer to keep track of the assumptions and provide a

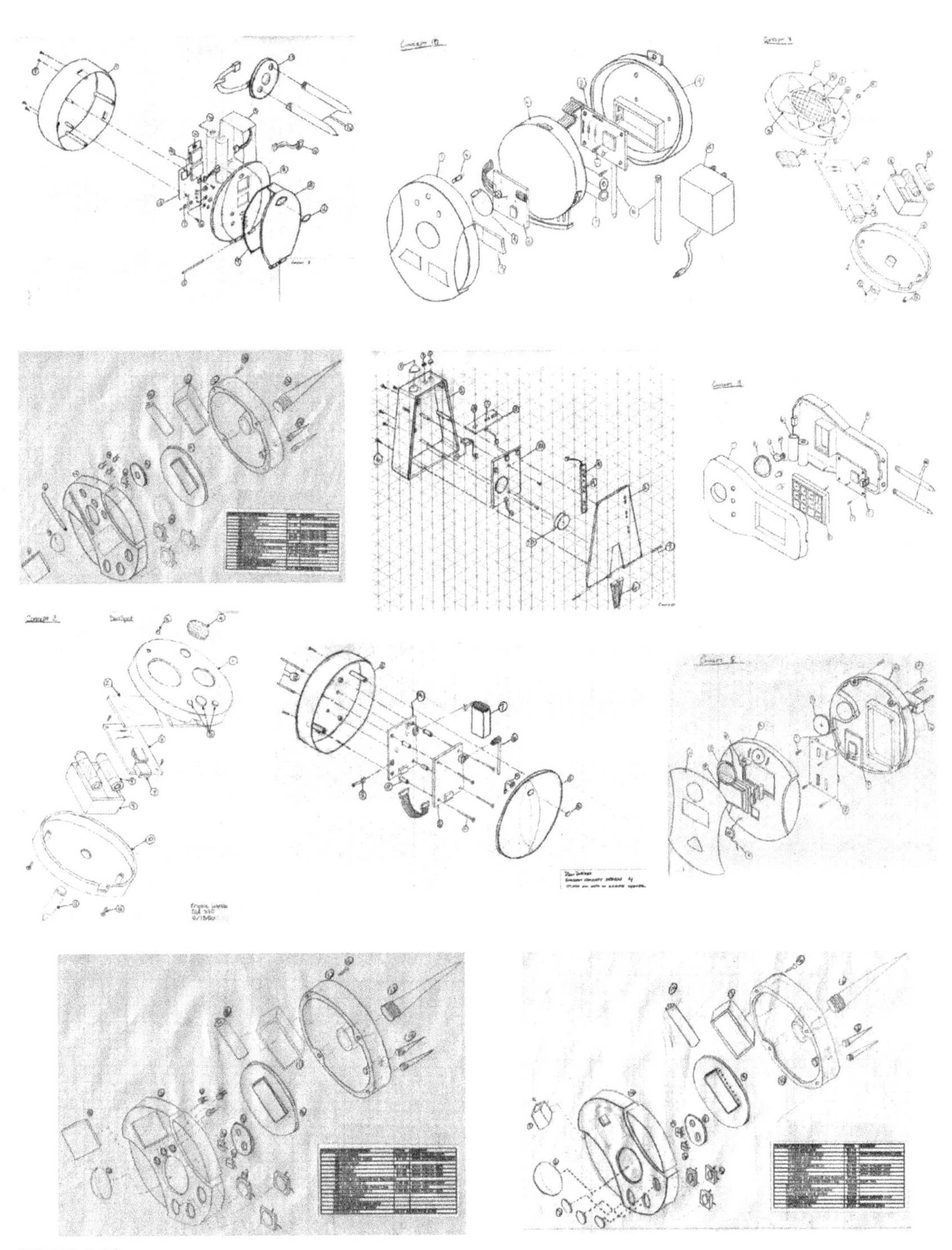

FIGURE 4.19

Photos of crude sketches, preliminary designs, and functional prototypes.

Courtesy of D. Godfrey, E. Ligeski, S. Palasek, D. Parker.

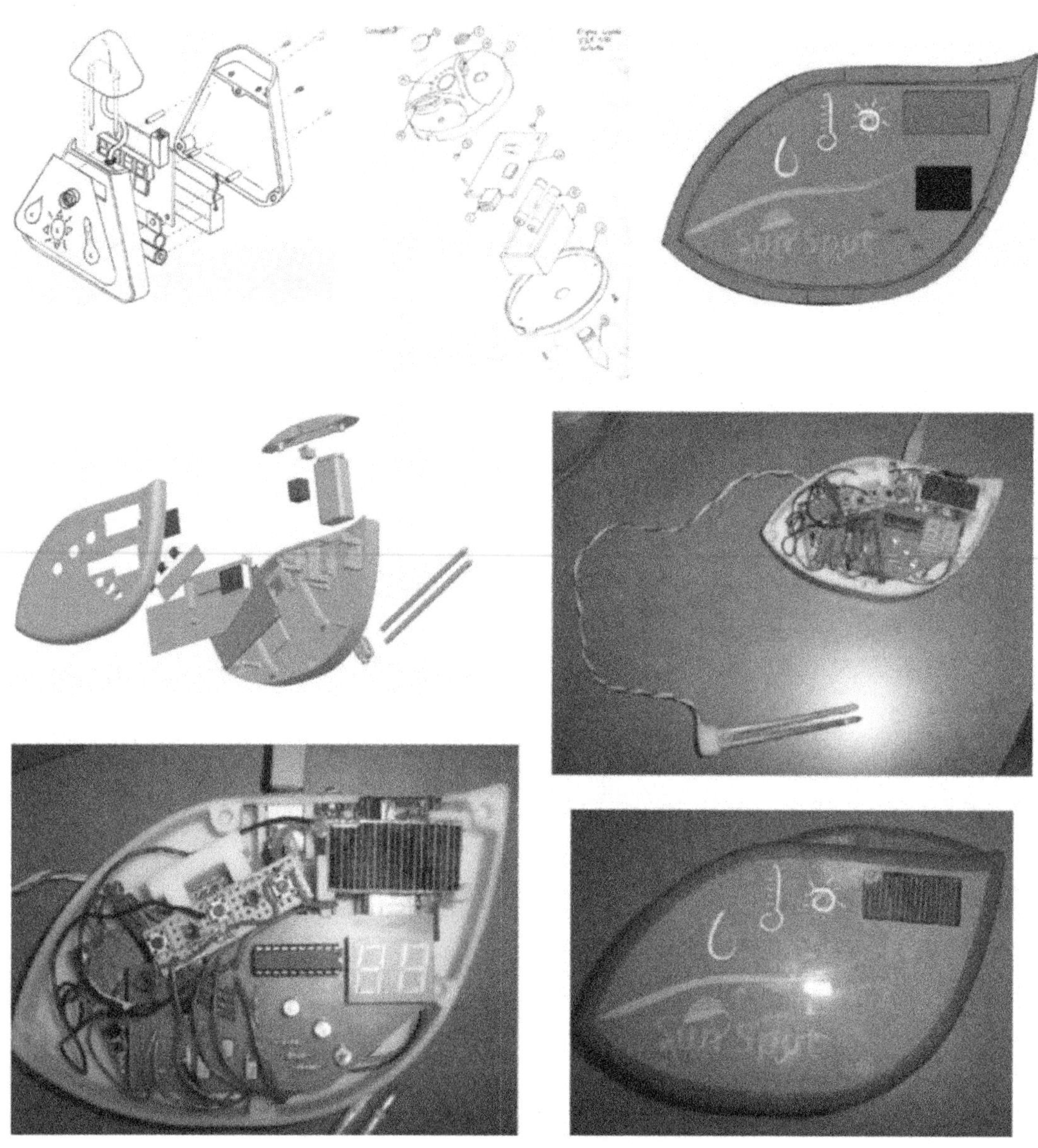

FIGURE 4.19, cont'd

margin for error. The FS is the preferred method. If the FS is 1.0, then the design is at the usage limits. If the FS is 3.0, then the design can take up to three times the design specifications. If the rated passenger load is 300 kg with an FS of 2.0, then the car will fail if 601 kg of passengers are applied.

In mechanical design, a higher FS often results in more material, mass, and cost. For mature designs the FS can be lower, sometimes even 1.0. Consider the aerospace industry, which places a premium on weight; a lower FS such as 1.1 is very important. The other extreme would be a playground equipment maker that would select a higher FS to emphasize safety over mass.

A wise designer will adjust the factor of safety upward in the observance of uncertainty, variability, contingency, and consequences of failure. Some of the things an experienced designer will watch for are as follows:

- Unfamiliar: The theory, design, or other aspects are new or poorly understood. This concern decreases with experienced input.
- Static versus dynamic: Static loads are generally well behaved, but dynamic loads may lead to short-duration overloads.
- Disasters: The potential for major unforeseen events such as earthquakes, tsunamis, plane impacts, etc.
- Catastrophic failure: When a system will fail suddenly with no warning. The better case is a system that fails gradually in a way that is noticed or provides warning.
- Specifications for materials and components: Values from suppliers and third parties are more suspect.
- Manufacturing variation: The manufacturing controls for tolerances and work methods are poorly maintained.
- Parallel systems: When three bolts attach a support, it is better to assume that only two of the bolts are carrying the load.

A simple method for estimating a factor of safety is given in Fig. 4.20. First and foremost, knowledge, K, is key. If you don't understand it, get knowledge and/or overdesign. Building prototypes and performing experiments can be a very fast way to increase knowledge. Knowing the worst-case variation in the process, S, will gauge how much must be added for the worst case. Finally, more dangerous designs require a greater margin for those we protect. The aerospace industry has worked diligently for the past century to drive all of the unknowns and variations to negligible values, and is able to design safely with a very low FS.

PROBLEM

4.48 Calculate and estimate an FS for a steel cable carrying a bouncing mass.

4.15 Concept selection

An amateur will have an idea and then work to eliminate alternatives.

When presented with multiple alternative concepts, we want to use the opportunity to pick the concept that will give us the best outcome for the lowest cost. This point is the most critical to establishing the remaining project work. It is an opportunity to stop, assess the options, and pick a good path.

$$FS = \left(\frac{100\%}{K}+1\right)\left(\frac{S}{100\%}+1\right)\left(\frac{H}{100\%}+1\right) \quad \text{eqn. 4.1}$$

FS = Factor of Safety
K = Knowledge (100 is all, 0 is none) e.g. mechanical knowledge but new application is 50
S = Statistical variations of factors. e.g. materials +/-10, user overloads up to 150
H = Hazards of failure (100 is catastrophic, 0 is unimportant)

For example, consider a bridge design
K = 50 - We know the theory but we have not built a bridge before.
S = 10 - The manufacturing process is good and the materials very consistent.
H = 100 - If the bridge falls it will cause loss of life.

$$FS = \left(\frac{100\%}{K}\right)\left(\frac{S+100\%}{100\%}\right)\left(\frac{H+100\%}{100\%}\right)$$

$$FS = \left(\frac{100\%}{50\%}\right)\left(\frac{10\%+100\%}{100\%}\right)\left(\frac{100\%+100\%}{100\%}\right) = 2(1.1)2 = 4.4$$

FIGURE 4.20

A simple estimation of the factor of safety.

All too often, people rush past this stage so that the design work begins sooner. (Note: Rushing to detailed design is an all-too-common problem.) While this does give the appearance of apparent progress, it hides all of the extra work that will need to be redone and all of the future problems. It is better to pause here and make a very pragmatic decision about the options for executing the project. The problems that happen very often are that people pick a solution because it is their idea, the other ideas seem like more work, somebody else recommended it, or people don't want to spend the time to look into it. Again, do not rush through the concept selection step.

In this step it is important to (1) eliminate the unsuitable concepts, (2) compare the remaining concepts objectively, and (3) select a good concept. During the concept generation stage you will have generated a number of infeasible ideas. In that stage, these can be valuable to help generate new ideas, but at this point they are just distractions. For example, there is no point in spending time evaluating an infeasible concept such as nuclear reactions to power a car. Some simple filters for vetting these concepts are as follows:

- Some concepts are clearly out (e.g., nuclear)
- Is it possible?
- BOE calculations—energy, clock speed, stress, time, power
- Do the basic parts exist?
- Are they patented?
- Prototype
- Sketch
- Similar designs in use
- Risk
- Complex or simple?
- Am I choosing this because it is my idea?

Many design concepts can be ruled out or shown to be plausible using BOE calculations. These are often crude but effective oversimplifications. For example, a simple calculation of energy loads compared with the available power can expose power shortages or indicate that more than enough power will be available.

- Basic stress/strain/buckling: simple loading cases, possibly including stress concentrations
- Power balance: input versus output
- Rate of change: required velocities, slew rate
- Scalability: What happens when it gets larger or smaller?
- Speed: clock speeds, frequency
- Economy of scale: consolidate complexity and cost

A procedure for screening concepts is shown in Fig. 4.21. In general terms, poor concepts are discarded, good concepts are kept, and questionable concepts are explored. This process should be applied to each concept before ranking and choosing.

We can improve the assessment process if we separate numerical measures from minimum/maximum measures, as shown in Fig. 4.22. For example, if we are looking for a system that will allow a color touch screen, then the concept can be ruled in, or out, quickly in step 1. After this, the other numerical specifications can be assessed with a weighting approach. In step 2, the target value for each specification is compared with each concept. Each of the concepts can be clearly ruled in or out. After this, the remaining concepts are viable and worth consideration. Examples of target values and ranges are as follows:

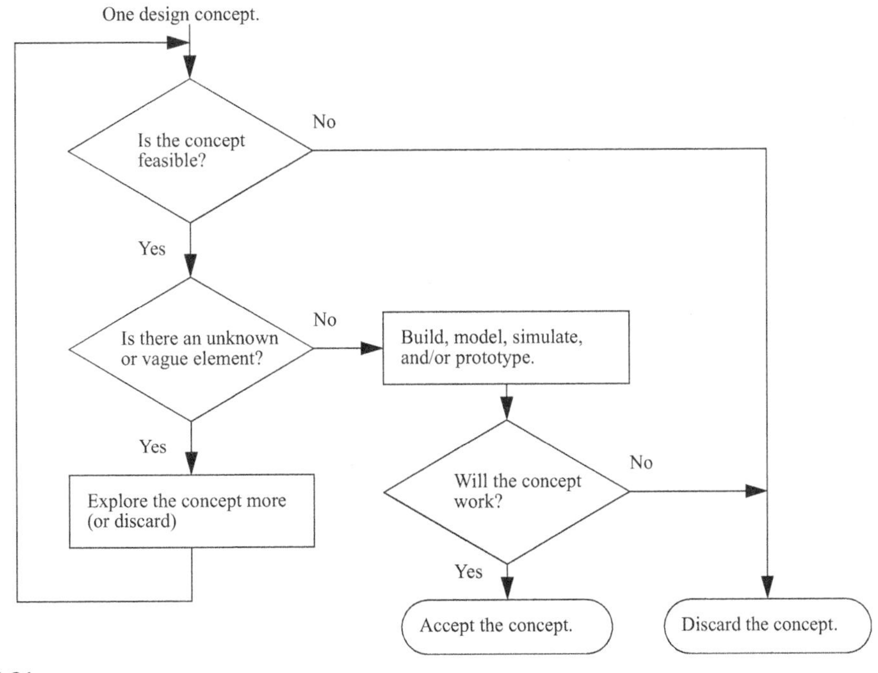

FIGURE 4.21

Reviewing design concepts.

Step 1: Requirement Matrix — this is for things that are clearly yes or no

		Concept 1	Concept 2	Concept 3
Required	Specification 1	y / n	y / n	y / n
	Specification 2	y / n	y / n	y / n
	Specification 3	y / n	y / n	y / n
	Specification 4	y / n	y / n	y / n
	Specification 5	y / n	y / n	y / n
Optional	Specification 1	y / n	y / n	y / n
	Specification 2	y / n	y / n	y / n
	Specification 3	y / n	y / n	y / n
	Specification 4	y / n	y / n	y / n
	Specification 5	y / n	y / n	y / n

Step 2:Performance Matrix — do once for each design — this mustuse numerical values only

		Target Value/ Range	Concept 1	Concept 2	Concept 3
Required	Specification 1	e.g. 23–64	value (in/out)	value (in/out)	value (in/out)
	Specification 2		value (in/out)	value (in/out)	value (in/out)
	Specification 3		value (in/out)	value (in/out)	value (in/out)
	Specification 4		value (in/out)	value (in/out)	value (in/out)
	Specification 5		value (in/out)	value (in/out)	value (in/out)
Optional	Specification 1		value (in/out)	value (in/out)	value (in/out)
	Specification 2		value (in/out)	value (in/out)	value (in/out)
	Specification 3		value (in/out)	value (in/out)	value (in/out)
	Specification 4		value (in/out)	value (in/out)	value (in/out)
	Specification 5		value (in/out)	value (in/out)	value (in/out)

FIGURE 4.22

Specification-oriented decision matrices.

- Lower cost is better. Below $54 is essential.
- The target load is 50 N. Below 45 N is unacceptable, above 60 N is not important.
- Manufacturing processes must use machining that can be done in-house.
- Expected in-service lifetime is 10 years or more.

Designs must meet all of the specifications to be acceptable. In a good design, there will be multiple options that meet the required specifications. The optional specifications can be used when a choice is not obvious. Or, a judgment about how well a specification can be met is another useful metric.

PROBLEM

4.49 What problems might occur if a designer takes personal ownership of an idea?

4.50 Should the number of concepts increase before concept selection? Explain.

4.51 What are reasonable options if a concept would probably work with substantial effort?

4.52 What should be done if a concept is not clear enough to accept or reject?

4.53 Assume that a portable drink cup has the requirements (a) hold 1 L, (b) maintain a cool temperature, (c) allow small sips/drinks, and (d) be portable. Construct a requirement matrix for a portable drink cup that provides cooling. Sketch the concepts, including (a) a flexible pouch, (b) a solid cup with a lid, (c) insulated sides, (d) water-filled ice cube analogs, (e) an electric cooler (Peltier junction), and (f) multiple insulating sleeves.

4.16 Decision matrices

A simple decision matrix is used to relatively compare alternative concepts against one reference concept, Fig. 4.23. The process begins with the selection of criteria for comparison. One concept is chosen as the baseline reference and given scores of 0 for each criterion. The other concepts are compared as better, "+," or poorer, "−." The concept columns are summed to provide a relative score for each concept. The steps for the method are:

(1) Select one design to be a good medium. This does not need to be the final design.
(2) List the conceptual designs as columns, put the reference design in column 1.
(3) List the selection criteria as rows. These should be closely based on the specifications.
(4) For each of the criteria, compare each concept to the reference. If it is clearly better enter a plus "+," if it is clearly worse use a negative sign "−."
(5) Total the "+" and "−" values to get a total for each concept.
(6) The design with the highest score is often judged the best candidate for detailed design (although other designs may be chosen).

Complex relationships can be compared using a weighted decision matrix, Fig. 4.24. For this method, each criterion is assigned a relative importance by selecting a weight. In this example, specification 3 is eight times more important than specification 2. Each concept is scored on a scale for each specification. So concept 2 is a good choice for specification 1 but a poor choice for specification 2.

Evaluation Criteria	Concept 1	Concept 2	Concept 3	Concept 4
Specification 1	0	+	+	+
Specification 2	0	-	+	0
Specification 3	0	+	-	+
Weighted Total	0	+1	+1	+2

Highest ranked ↗ (Concept 4)

FIGURE 4.23

Decision matrix.

Evaluation Criteria	Weight	Concept 1	Concept 2	Concept 3	Concept 4
Specification 1	0.5	1	5	3	2
Specification 2	0.1	2	1	5	2
Specification 3	0.8	2	3	1	3
Weighted total.		0	2	1	1.0+0.2+2.4 =3.6

Winning concept.

FIGURE 4.24

Weighted decision matrix.

The scores in each of the columns are multiplied by the specification weight and then summed. The column with the highest value is the better design. The steps in the method are:

(1) A column is created for each conceptual design.
(2) A row is created for each specification or need.
(3) A numerical weight is given to each specification, or need, to indicate the relative importance.
(4) The matrix is completed using scores for how well each concept satisfies each specification. The ranking is often done relative to one of the design concepts, with the middle of the scale being the first concept. A scale of -3 to $+3$ is reasonable.
(5) Using the specification weights, the column values are multiplied and added to get a score for each concept.
(6) The concept with the highest score is judged the best candidate for detailed design, although other designs may be chosen.

This method can be used quickly, but is prone to a few problems, including (1) the linear weighting scale does not capture the nonlinear utility of design features, (2) a number of unrealistic designs can be included to give the impression of a real comparison, and (3) the matrices can be misused to justify bad designs. If you are using this method and find yourself adjusting the weights, then you are probably trying to get the design you want instead of the design you need. And, if there are designs that do not meet specifications, they should not be considered in the first place.

PROBLEM

4.54 Why are weighted decision matrices used instead of regular decision matrices?

4.55 Why does a decision matrix use a baseline, or median, design?

4.56 Should the highest matrix score always be the concept chosen?

4.57 Do the decision matrices need to use all of the specifications and concepts?

4.58 A night-light must run for 2 h, provide 1 W light output, and mount on a 2-cm-thick board.

(a) Create concepts for the night-light.

(b) Create a decision matrix.

(c) Create a weighted decision matrix.

4.17 Embodiment design alternatives for a technical specification

Concepts are ways that a design could be done. Embodiments are the way the design will be done. Basically each embodiment is a separate technical specification. A good embodiment design includes enough detail so that components and parts can be developed independently. Key attributes of embodiments include:

- configuration
- format
- framework
- hierarchy
- composition
- form

Typical components in an embodiment include:

- software code using stubs and/or an application programming interface
- state diagrams for sequential systems
- timing diagrams
- mechanical sketches using normal drafting practices, including 2D, isometric, and pictorial views
- calculations and equations to support the design concept
- test data
- electrical schematics and data sheets for major components
- block diagrams for electrical, software, data flow, and control systems
- general budgets for money, mass, heat, power, etc.
- flowcharts
- physical prototypes
- selected component sources, costs, data sheets, specifications, etc.
- universal markup language for software design
- formal process diagrams, piping and instrumentation diagrams, or similar
- algorithms and pseudocode
- GUI sketches
- product component breakdown diagrams
- details presented as lists, tables, or numbered steps
- aesthetics: sketches, solid models, color schemes, surface finishes, and mock-ups
- a parts list, or bill of materials, for major or strategic parts
- suppliers for critical parts, including availability and cost

PROBLEM

4.59 Discuss the statement "An embodiment design is similar to a customer specification."

4.60 Should an embodiment design include all major components? Explain.

4.61 Mini-case: Things change

Electric motors can be used in wheel hubs to directly drive a vehicle. A student team was redesigning an electric vehicle to move from a single rear-drive axle to direct-drive systems on each wheel. The geometry of the wheels was not standard and it took some time to locate a motor supplier that could supply the four drive motors. The student design team ordered one motor to build a prototype of one wheel. Satisfied with the testing result, the team moved ahead with the design and received approval to purchase parts. Three additional motors were purchased, for a total of four drive motors. While waiting for the other three motors to arrive, the team made the other drive motor parts. When the motors arrived the team began assembly of the custom-made parts and the motors. It did not take long to discover that the hole patterns in the three new motors did not match the original motor. In addition, some of the geometry was not the same as the first motor. The team investigated and found that the motors were the same model but some design changes were made that resulted in the differences. The result was that the team needed to spend additional time modifying the drawings, modifying some parts, and remaking a few others. Was it a good idea for the team to order one motor for prototype testing? Should they have ordered four at the same time? Should the team have waited until the motors arrived before building the other parts? Should the experience be considered a normal variation or does the team need to change the way they test and plan?

4.18 Intellectual property

During rapid change patents may only protect obsolete ideas.

The process for managing intellectual property is shown in Fig. 4.25. Copyrights, industrial designs, and trademarks provide decades-long protection for creative works, including writing, images, shapes, logos, and product forms. Patents provide 20 years of protection for a useful thing or process. Patents are used to prevent competitors from copying new designs and guarantee revenues for an inventor. Designers must be aware of the patent system when doing concept generation. If a patent has expired, the idea can be used freely. A new concept can be patented. An idea covered by a current patent should be avoided unless it can be licensed. In cases where a competitor has used a patent to block competition, companies will develop new technologies or circumvent patents. Circumvention is a process of finding loopholes in poorly written patents. Other forms of legal protections include the following:

- Trademarks: These are small symbols, names, or designs (marks) that are used to distinguish one product from others. A registered trademark uses the symbol ™ or ®. This prohibits the unauthorized use anyplace else in the country of registration.
- Copyright: A copyright is used to restrict the right to copy or perform certain creative works. Copyrights generally exist until 70 years after the author's death, in most cases. Copyrights can be registered (an optional step), but when the copyright is to be assigned or licensed to another party, it should be registered.

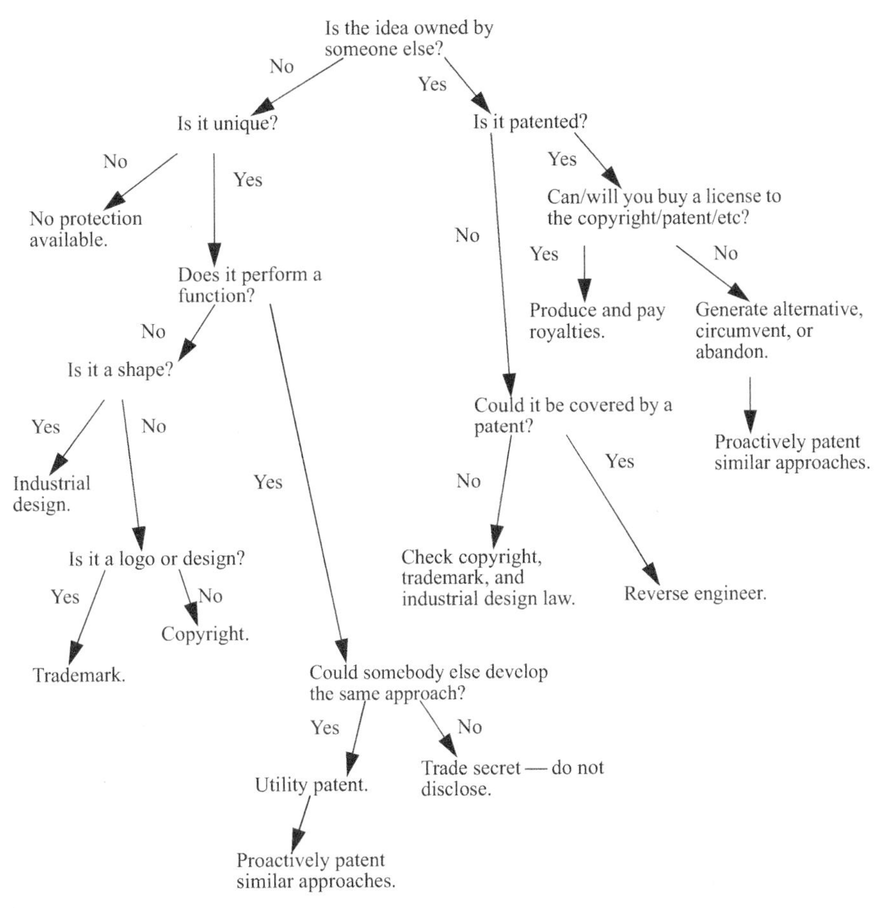

FIGURE 4.25

General selection of legal protection.

- Industrial design: This is a protected design that is novel and original and generally refers to a sculpture, shape, configuration, or pattern that is aesthetic. The functional components cannot be considered. This can be registered for 5 years and then 5 more.
- Trade secrets: This is similar to a patent except there is no public disclosure and it may include information or other nonpatentable things. Generally a trade secret permits a business advantage over the competition ("industrial know-how").

The patent application process is outlined in Fig. 4.26. Patentable ideas must be unique and nonobvious. For example, changing a seat fabric would not be patentable but changing the method for mounting the fabric might. Unique means that the idea was not patented or published before. The uniqueness of a patent is verified by a search of the patent database and other public resources. If patentable, a patent lawyer will write and submit the patent applications. The patent office reviews the patents for validity and may return the patent application for modification. If accepted, the patent is listed in public patent databases.

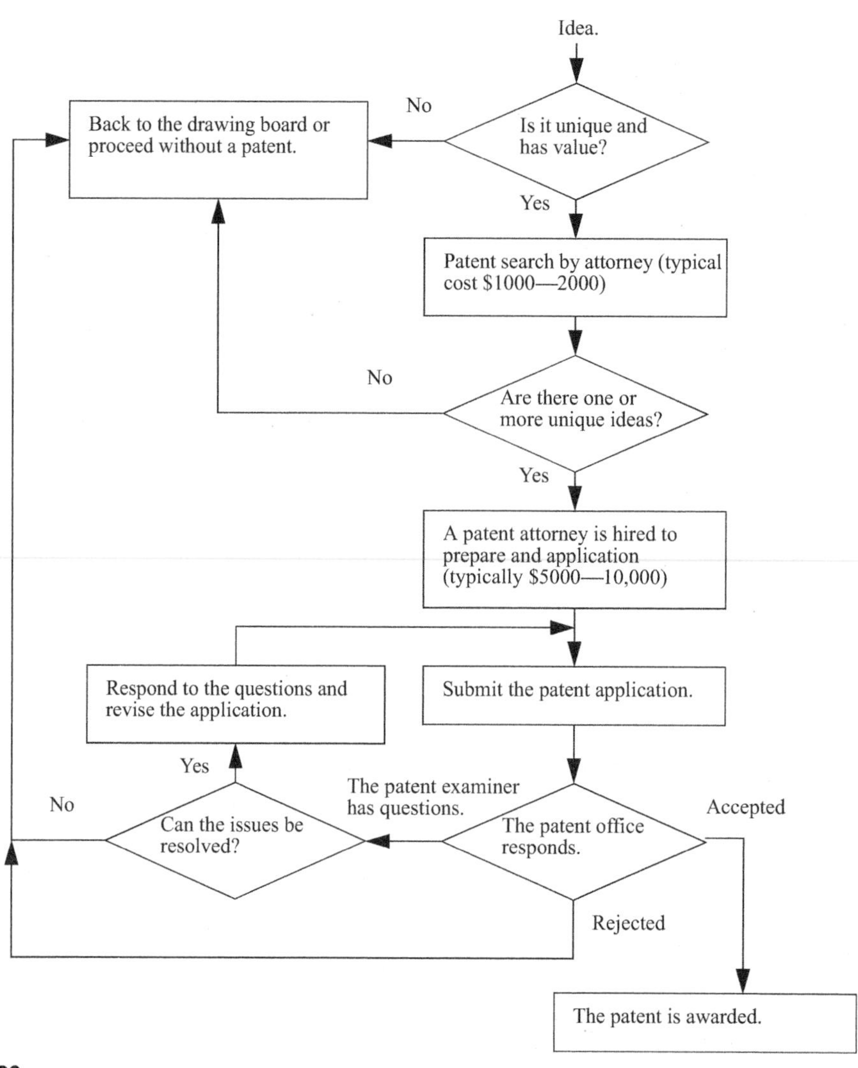

FIGURE 4.26

The patent process.

A patent is a recognition of legal rights. The patent holder is responsible for detecting infringement and enforcing legal rights. For example, an inventor might hold a patent for a display integrated into a car windshield wiper. He or she may find the product being sold without permission. The inventor could hire a lawyer to get the company to pay royalties or take them to court. Some other key points about patents include the following:

- Patent rights can be assigned or transferred legally.
- Patents are normally registered by country, with some exceptions such as the European Union.

- Disclosure of the idea before the patent application is submitted can invalidate the patent, although some countries provide a grace period of 1 year.
- You must enforce the patent yourself in civil court.
- Employees don't own patent rights.

The first major step taken in the patent application process is a patent search. The existing patent databases are searched to find patents for solving similar problems or with similar claims. The outcome of the process should be a list of related patents and an understanding about patentability. The related patents, and their language, will be used when writing the final patent application. The following steps outline the rudimentary process. Designers may conduct patent searches to (1) see if a new concept has already been patented, (2) determine if a new idea can be patented, and (3) look for solutions to problems that are already patented. A sample approach is given in the following:

(1) Go to a patent search website (e.g., www.uspto.gov or www.google.com/patents), and search using some relevant terms.
(2) Look for similar, competitive, or older patents.
(3) Read the patents and note the following items:
  (a) specific phrases and words
  (b) patent numbers
  (c) people and companies
  (d) the claims
(4) Use the most relevant details to search again:
  (e) Each patent refers to similar patents. Searching for the patent numbers is a fast way to find similar patents.
  (f) Key terms can be used for a more targeted search. This is necessary because the language used in patents does not necessarily agree with regular technical terms.
(5) Continue until all similar patents are found.
(6) Read the patents very carefully. The most important section is the claims.

Engineering designs must be protected before patents are awarded. Generally, this means that the designs are kept secret. Secrecy and patent priority are determined through careful systems of documentation and record keeping. The most standard legal tool is the nondisclosure agreement (NDA). The agreement is signed before intellectual property secrets are shared. The terms of the agreement define what is considered secret, how long it will be secret, and legal liability if disclosed. It is very common to have NDAs signed by new employees, company visitors, sales representatives, suppliers, customers, and so on. When new employees are hired, they are often asked to sign a noncompetition agreement that bars them from working for competitors for a certain period after they leave the company.

PROBLEM

4.62 What cannot be patented?

4.63 What is an NDA used for?

4.64 What is the difference between a trade secret and a patent?

4.65 For a ballpoint pen:

(a) Identify and suggest a design concept to improve one unappealing feature.

(b) Develop five concepts for a similar competitive product with evolutionary improvements.

(c) Develop three revolutionary redesigns.

4.66 Use brainstorming to develop concepts for putting on shoes.

4.67 Determine suitable intellectual property protections or other strategies for the following items:

(a) A manufacturing process for a new shape of pasta

(b) A recipe for a carbonated drink

(c) A coffee cup with a new retail logo

(d) Software that uses a new algorithm for weather prediction

4.68 Describe the difference between design objectives and constraints.

4.69 Bouncy, Inc., designs exercise equipment and is planning to develop a new device to be used while watching television or using a computer. The general concept chosen is a modular weight system that can be attached to wrists and ankles. The weight can be adjusted to suit the user. The specifications include a mass range from 1 to 10 kg in 1-kg increments. The users' wrist diameter will range from 20 to 60 cm. The users' ankle diameter will range from 40 to 80 cm.

(a) Develop a set of concepts to implement the system. There must be at least five concepts that would work. Omit concepts that are not practical.

(b) Rank the concepts and recommend a first choice and a backup concept.

(c) Develop a conceptual design. Create a rough budget and materials list. Each major component should be listed.

4.70 Farmers in Nicaragua grow coffee beans in large enough quantities to sell. Unfortunately they do not produce enough to attract international buyers. A local buyer purchases the beans below market value from many farmers and collects them into a larger batch that can be sold to international buyers. The farmers would be able to make more money if they were able to sell directly to international buyers, retailers, or customers. Develop concepts that would achieve this objective. They may be a combination of business and technology approaches.

4.71 Design a coffee maker that uses green beans, ready for roasting. Green coffee beans must be processed in a number of steps before they become a cup of coffee. The core steps of the process are (a) pick the berry, (b) remove the beans, (c) clean and polish, (d) sort into grades, (e) age, (f) roast, (g) grind, (h) add hot water, and (i) remove the grounds from the coffee. Some consumers choose to grind the beans themselves. It is rare for a consumer to roast his or her own coffee beans. Develop concepts for a coffee maker that will accept whole green beans and then roast, grind, and brew the coffee in a single machine.

4.72 List five things people may say to sabotage brainstorming.

References

Altshuller, G., 1988. Creativity as an Exact Science, Translated by Anthony Williams. Gordon & Breach, NY.
Niku, S.B., 2009. Creative Design of Products and Systems. Wiley.

Further reading

Baxter, M., 1995. Product Design; Practical Methods for the Systematic Development of New Products. Chapman and Hall.
Dym, C.L., Little, P., 2009. Engineering Design: A Project Based Introduction, third ed. Wiley.
Dhillon, B.S., 1996. Engineering Design; A Modern Approach. Irwin.
Heldman, K., 2009. PMP: Project Management Professional Exam Study Guide, fifth ed. Sybex.
Hyman, B., 2003. Fundamentals of Engineering Design, second ed. Prentice-Hall.
Ullman, D.G., 1997. The Mechanical Design Process. McGraw-Hill.

CHAPTER

People and teams

5

5.1 Introduction

People are the greatest resource in any business. Reasonably motivated employees will do your work in reasonable quantities and qualities. Empowered and motivated employees will take ownership of the work and do an outstanding job. The key is to hire the right people, give them what they need, and let them do their jobs. It is a matter of mutual benefit (Fig. 5.1).

5.2 Individuals

Models can be used to simplify complex systems, analyze situations, and make decisions. Like any tool, models depend on assumptions and simplifications that reduce accuracy, but in the absence of alternative methods for understanding they provide value. The model shown in Fig. 5.2 was developed by Maslow as an explanation of the relationship between personal needs and state of mind. In simple terms, when a person is worried about survival he or she will focus on it as the highest priority. If the needs at the bottom of the pyramid are fulfilled, the focus changes to the higher-level needs. As an employer you expect productivity, and to get it you must make sure that the first three levels are solid. If you want innovation you will need to address the first four levels. Related models exist that are often used to explain people in the workplace, such as the following:

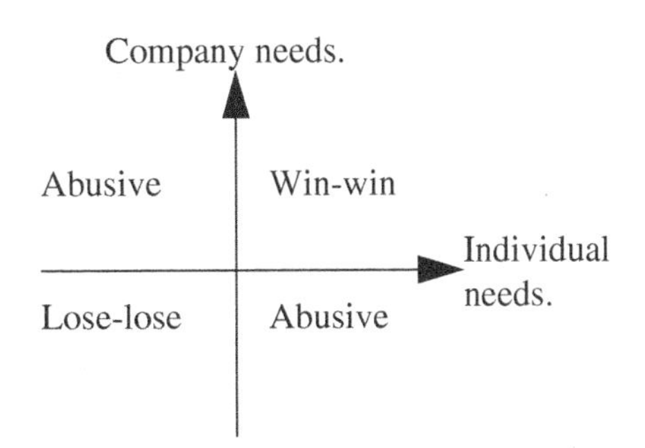

FIGURE 5.1

Satisfying goals.

Engineering Design, Planning, and Management. https://doi.org/10.1016/B978-0-12-821055-0.00005-0

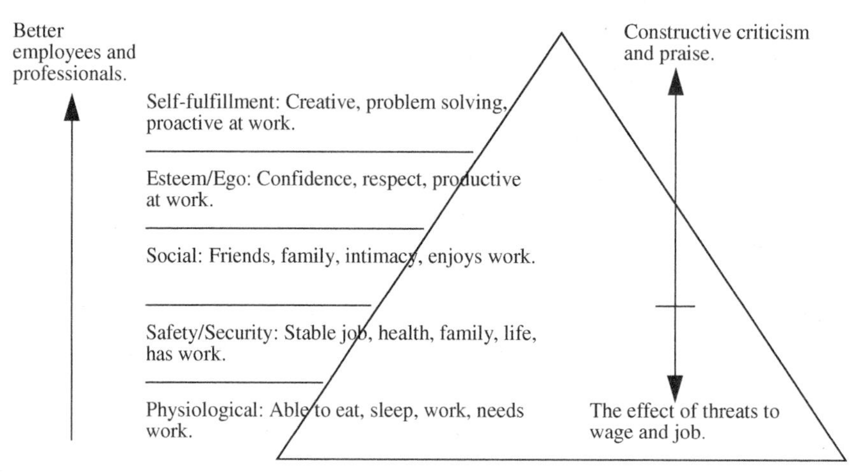

FIGURE 5.2

Abraham Maslow's hierarchy of needs.

- Hygiene theory: Preventing dissatisfaction is a function of creating a good work environment. This is often achieved with recognition, pay, benefits, and working conditions.
- Achievement theory: Motivation can be encouraged by recognizing past achievements, promoting professional affiliations, and granting authority.
- Expectancy theory: Motivation comes from an expectation of positive outcomes.

Employees will be productive when they are safe and part of a social group. Employees will be creative when they feel appreciated and empowered. Likewise, withdrawing any of Maslow's lower needs will probably undermine any of the higher needs. For example, a company that announces lay-offs are coming should not expect high levels of productivity and creativity.

In the presence of change, things become uncertain. When things can change, people tend to focus on the worst case and act accordingly. You should always consider change as another source of trouble. The unknown brings uncertainty and possibly fear. Even if it is not obvious, it is probably still there subconsciously. Naturally, this undercuts productivity. In an effective organization people are not worried unnecessarily.

There are a few concepts, shown in the following list, that are useful when considering people. In their own ways, they want to become great and you can either fight them or help them. If you are lucky, they will help you, too:

- More: Everybody has unfulfilled hopes and dreams and things they want to do. They also have regrets. To assume that the way somebody is now is who they want to be and they will never change is an amateur mistake. Find out what people want to be and help them get there.
- Flaws: It can be difficult to admit personal flaws to yourself and even harder to admit these to others. Allowing others to recognize your personal flaws will make them more understanding of you, and make them more comfortable asking for help.
- Cognitive dissonance: People can fully accept two or more facts but ignore the disagreement between them. In simple terms, strong feelings will overrule logic. Logical structures are constructed to explain the irrational, often accompanied with "I don't know why, I just know it is wrong." We all do this.

- Psychic: Just because you believe or feel something doesn't mean others know it. Telepathic communication is generally ineffective.
- Offensive and defensive: A human reaction to feeling vulnerable is to be defensive. If you push a little harder they will become offensive.
- Perspective: It is easier to force ideas into what you know. It takes more effort to expand what you know.

PROBLEMS

5.1 Models such as Maslow's are easily criticized for being oversimplified and lacking subtlety. Can the same be said for lumped models in free body diagrams (FBDs) and schematics? Explain.

5.2 Where could the following people be placed on Maslow's hierarchy?
- **(a)** A new parent
- **(b)** An employee who just received a large raise for performance
- **(c)** An employee who has been moved to a new department with no friends
- **(d)** An engineer who has been earning many patents

5.3 How could you use hygiene theory to improve the productivity of somebody who is avoiding work?

5.4 What role does praise have in achievement theory?

5.5 Use expectancy theory to describe the effects of negative comments from the customer.

5.6 Use Maslow's hierarchy to create a new procedure for mentoring employees.

5.7 Fit the concepts (i.e., more, flaws, cognitive dissonance, psychic, offensive and defensive, and perspective) to the following statements.
- **(a)** She should know what the problem is.
- **(b)** I know it will never fly but I want to try anyway.
- **(c)** You should be happy that you have a job.
- **(d)** If I were him I would not do that.
- **(e)** I tried but the problem was not defined properly.
- **(f)** She made that mistake last year and I am sure nothing changed.

5.2.1 Personal growth

Show interest to get interest.

Employers and employees are expected to engage in continuous improvement. To improve, it is essential to identify personal areas of weakness and goals for improvement. Needless to say, it is hard to have an objective perspective. A few basic prompts are given in the following list to help find self-improvement opportunities:

- Strengths: Use your strengths effectively.
- Mistakes: Be willing to try new things, make some mistakes, and learn.
- Assess: Look at what you do to find opportunities for improvement.

- Continuous: Be looking for ways to improve all the time.
- Learning: Stay up to date in your profession.
- Waste: Business time and resources should all contribute to some goal.
- Confidence: Being overconfident will reinforce problems. Being underconfident will make you unable to make decisions.
- Criticism: Every critic will tell you something you can use, but it is not always in what they say.
- Listen: Pay attention to advice and signals from other people. They can tell you things you would not recognize about yourself.

A simple process for continuous improvement is shown in Fig. 5.3. Before starting any work, plan your approach, process, and expected outcomes. During and after the work, you should be reviewing your progress and asking what could be improved. When done, you should honestly review how the work went and plan to adjust and adapt next time.

Understand your strengths and weaknesses.

The biggest trap for professionals is to become "firefighters." These people make progress by solving problems. Their ability to solve problems is impressive, but it often hides the fact that they did not prevent the problems in the first place. Being able to solve problems is not a negative skill. However, it is bad when it is the normal approach to projects. The most effective professionals will plan and execute smoothly and rarely need to deal with major problems. The difference is that a firefighter is very noticeable, while a planner sits quietly, avoiding problems. An analogy is a doctor who successfully detects heart problems or a doctor who helps heart attack recovery. Fig. 5.4 shows the steady, proactive work style and the firefighting work style. A planner is proactive and works ahead. The firefighter ignores issues until they become problems. These are solved by brief periods of intense activity. It is important to recognize the difference between apparent and effective progress and recognize that the work style that firefighters use will cost more.

The following are some excellent strategies to help change your methods and approach:

- Visualize: Close your eyes and imagine yourself doing the task and succeeding. This technique is used in sports often.

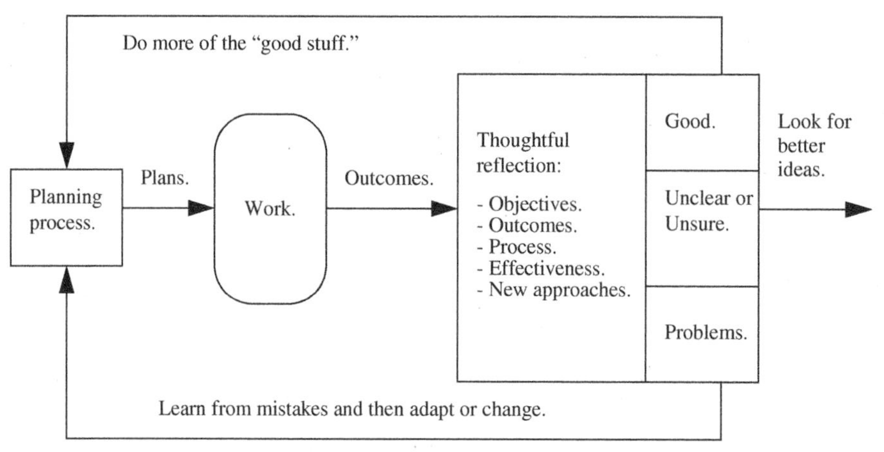

FIGURE 5.3

Assessment for improvement.

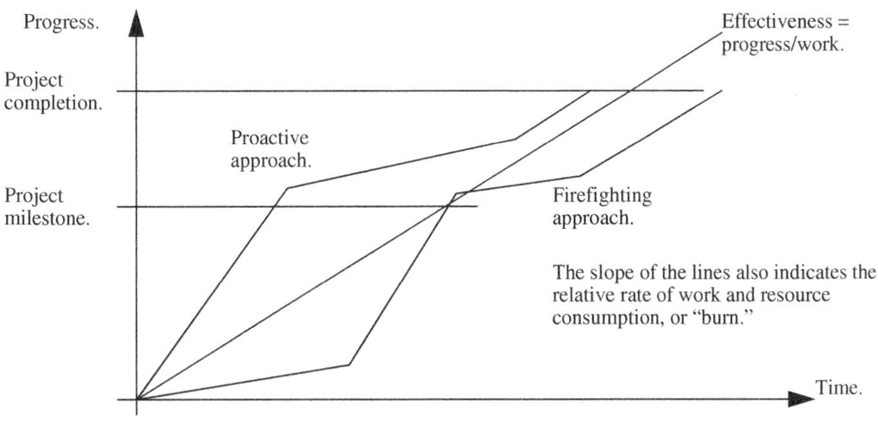

FIGURE 5.4

Capacity utilization.

- Goals: Set small goals to pursue. Once you reach them, create another attainable goal. Trying to change too much at once will often lead to frustration.
- Positioning: Put yourself in roles where you are more likely to succeed. Find people or activities that have already succeeded.
- Learn: If you make mistakes, learn; don't quit.
- Proactive: Work ahead and leave time for mistakes and reflection.

PROBLEMS

5.8 What does constructive failure mean?

5.9 Compare proactive and reactive work approaches.

5.10 List five benefits of self-assessment.

5.11 Explain the saying "Use your strengths and compensate for your weaknesses."

5.12 What are five methods you can use for self-assessment?

5.13 What are five methods you can use for self-improvement?

5.14 A very common professional approach is to begin each workday by reviewing progress from previous days and planning work for the day. How could assessment be integrated into this process? Explain your approach using numbered steps.

5.2.2 Learning

Understand what you know and what you don't know.

We are always learning and teaching others, becoming effective engineers, and helping others do the same. Understanding the learning process will help when you are delegating tasks, training new employees, acquiring new skills, or assessing professional progress.

The model of learning shown in Fig. 5.5 starts with simple exposure to knowledge. The value of this model is the ability to assess learning and understanding. The model can also be used to develop learning strategies. For example, new employees can have a narrow focus, beginning with observing

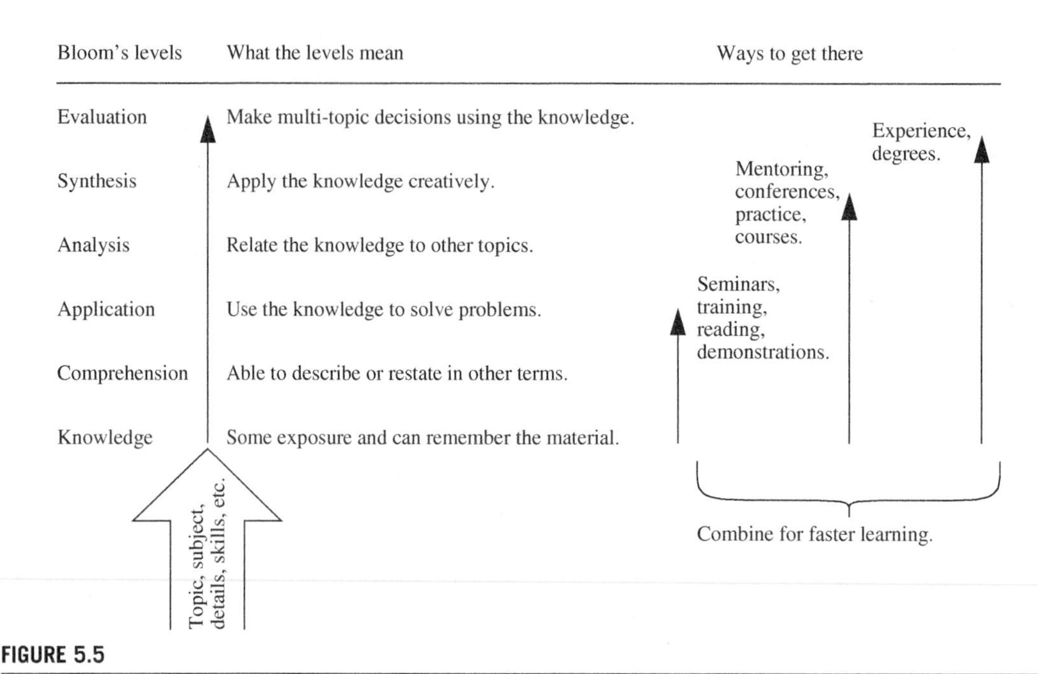

FIGURE 5.5

Bloom's taxonomy of learning.

other engineers work (knowledge), having them help the engineers (comprehension), giving them tasks to apply their knowledge (application), having them critique the work of others in design reviews (analysis), having them lead their own projects (synthesis), and having them review project proposals (evaluation). At any point in time, we are at multiple levels in this hierarchy for different knowledge. The model can also be used to identify shortcomings in individuals and plans. For example, if somebody says, "My work does not reflect what I know," then there is a good chance that he or she is at the "comprehension" level but has been asked to work at or above the "application" level.

An effective company will understand what knowledge it needs and at what level of capability. It will then match or nurture employees for those roles. This process often involves years of effort and is a problem for managers using short-term objectives. Fig. 5.6 shows the body of knowledge for professionals over their careers. They enter the workforce with a wealth of general knowledge built in the education system. Over time, exposure to specific knowledge widens their knowledge base and experience makes them able to work at higher knowledge levels.

When developing education plans for yourself or others consider the following factors:

- Interest: Some topics are more entertaining and easier to digest. Others require more time and patience.
- Style: People have different learning styles. Hands-on people are said to be concrete learners, while people who are more comfortable with written and symbolic knowledge are called abstract learners.
- Individual: Learning alone is difficult because there is nobody else to talk to about the material.
- Discipline: Prolonged time with difficult topics requires substantial motivation.

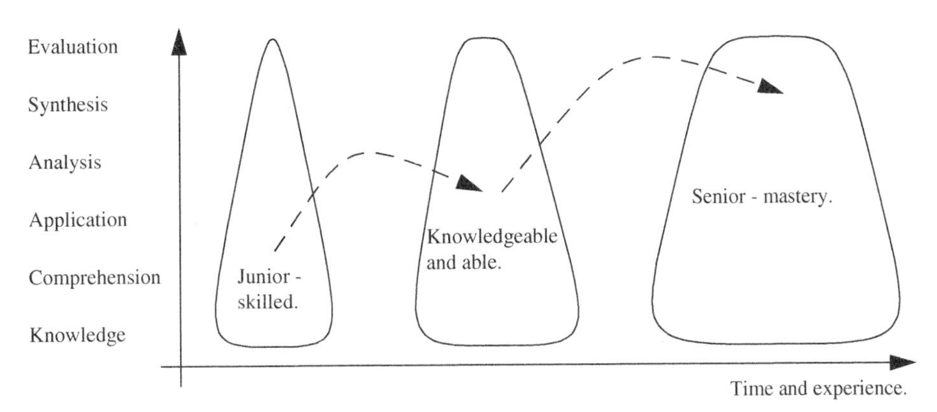

FIGURE 5.6

Professional knowledge.

- Reluctant: Learning is much easier when people are positively motivated, curious, and interested. Being forced to learn something you don't like is rarely effective.

  There are a number of strategies that will enhance learning, such as:

- Teaching: An excellent way to learn is to teach somebody else.
- Context: Provide a reason or application for knowledge that is being taught.
- Repeat: Repeat the details to the learners a different way, or have them paraphrase what you told them.

PROBLEMS

5.15 Bloom's taxonomy of learning is widely used in education to describe a current knowledge level and plans for advancement. How can this be applied to learning to ride a bicycle?

5.16 Assume an engineer has application knowledge of the wave-soldering process. What methods could be used to increase his or her knowledge to the level of synthesis?

5.17 Bloom's taxonomy is applied separately to each subject and topic. At any time, a person will have many topics he or she recognizes. Should a senior engineer have evaluation-level knowledge of many topics?

5.18 What is the difference between concrete and abstract learners?

5.2.3 Attention and focus

Don't work hard, work smart.

Manufacturing equipment adds value only while it is running. Many companies consider a machine running 80% of the available time to be excellent. Much of the "downtime" comes from factors such as maintenance, shift changes, changing tooling, setting up new jobs, and so on. The same lesson applies to people as they take breaks, switch between tasks, go home for rest, and so on. Understanding the relationship between tasks and how people work makes it possible to be more effective.

The first major factor to consider is exhaustion. Fig. 5.7 shows some hypothetical relationships between the time spent at work and the productive time. A highly productive employee may be at work for 40 h in a week, but has effectively worked for only 35 of those hours, while an average employee may have done the equivalent of 25 productive hours. Toward the end of a normal workday or week exhaustion begins to have an impact on productivity. For example, a good employee might be 80% effective on Monday, Tuesday, and Wednesday, but this decreases to 70% effective on Thursday and Friday. One week he or she was paid overtime to work on Saturday and the productivity went down to 60%. Simply put, as people work longer periods of time their productivity goes down and will produce lower productive returns for the wages.

The next factor in productivity is task switching. Tasks can be crudely categorized using (1) time to completion and (2) complexity. When a task is complete, there is a sense of perception and achievement, a reward. For short-duration tasks there can be a frequent series of rewards. Some people thrive in this environment. When doing longer tasks, the rewards are much larger but come less often. Simple tasks, such as responding to email, can be done quickly, while larger tasks require more time to focus. The time to focus on the complex task can require minutes or more, and once interrupted the focus is lost. People working on complex tasks become easily frustrated when small interruptions break their focus and require time to refocus. An example of losing focus for a complex task is shown in Fig. 5.8. The obvious lesson here is that interrupting complex work will have a larger impact on productivity. This is why many knowledge workers, such as computer programmers, prefer to work at odd times of the day when other people have gone home. Productive professionals will often book times without interruptions when they need to do large tasks.

The number and mixture of task types also have an impact on effectiveness. Fig. 5.9 shows the diminishing effectiveness as the variation and number of tasks increase. Consider an engineer who is working with suppliers to find parts for a new project. If all of the work is with a single supplier for similar parts, he or she can be relatively effective. If the job is changed slightly to work with multiple suppliers for the same parts, the work then involves substantially more detail and organization.

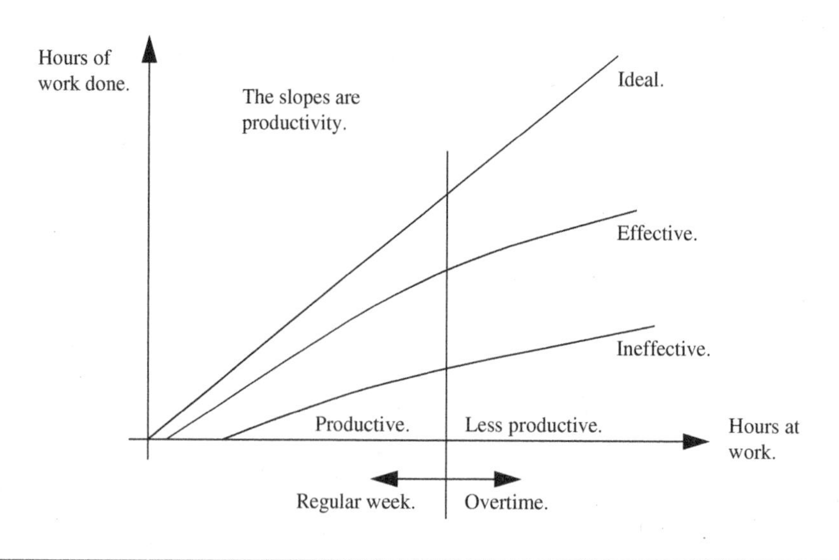

FIGURE 5.7

Diminishing return with overtime.

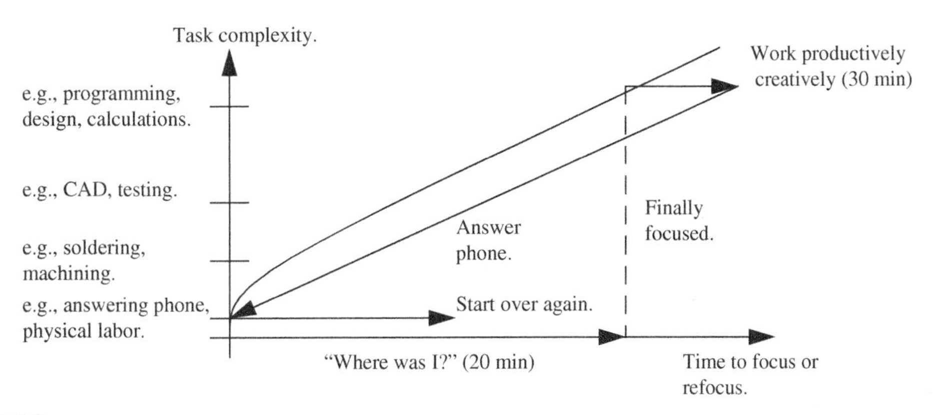

FIGURE 5.8

Interruptions and focus; an example for 60% efficiency.

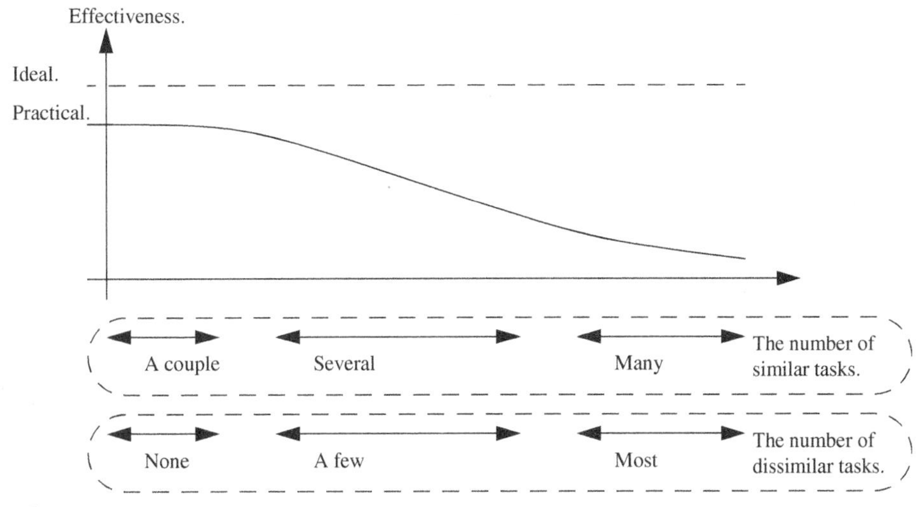

FIGURE 5.9

Multitasking.

This has an obvious impact on effectiveness. Another example is the types of tasks. If an engineer who was working on purchasing for 40 h per week is reassigned to 20 h on purchasing and 20 h supporting sales, he or she will be less efficient at both. In general, switching tasks requires time to wrap up the current tasks, change your frame of mind for the new task, and reorganize your desk with the new tools. An example of this effect is somebody who is working on email, phone, calendars, making notes, and writing budgets. Each time the employee shifts his or her gaze on the screen there is a dwell time. And, he or she will tend toward the most satisfying task, so the email is done first and the budget is done last.

Some general rules for making the most effective use of time are:

- Separate: For short and long tasks, separate the activities in time or by employee.
- Group: Simple tasks should be grouped together during the day or week so that they don't interrupt complex tasks.
- Unavailable: Book blocks of uninterrupted time for large and complex tasks.

- Responsive: Book regular blocks of time when you are available.
- Recognize: Other people face the same dilemmas; recognize this when you feel inclined to interrupt.

PROBLEMS

5.19 What is productivity?

5.20 Can actual productivity ever reach 100%?

5.21 How does refocusing attention reduce productive time?

5.22 How can a professional increase his or her effectiveness when he or she has a mixture of large and small tasks?

5.23 A reasonable limit for a work week is between 40 and 50 h. After that time, nonsalaried employees often receive a higher hourly pay rate, but their productivity drops. Assume that employee productivity, *P*, is described by the following function. The pay rate, *C*, is also indicated. Develop a new function for the cost per ideal work hour, *V*.

$$C_{regular} = 20.0t$$

$$C_{overtime} = 40.0t$$

$$P = 1 - \rho^{0.30t}$$

$$V = \frac{C}{P}$$

5.3 Organizations

Companies are people working together with similar objectives. In a perfect world, they would selflessly agree to work in perfect harmony toward a completely understood goal. In practice, there are many individuals with distinct personalities. They are not the same, they do not have identical goals, they approach work and life in different ways, and they have different motivations. Nothing in this section should be a foreign concept, but as a professional, a more deliberate approach is required when working with others and this begins with the ability to describe and understand ourselves and others. In other words, the purpose of the following section is not to teach some new aspects of human nature but to provide concepts and terms that we can use to describe, understand, and solve problems.

5.3.1 Motivation

Compensate for every negative action by doing 10 positive actions.

Motivation comes from a variety of positive and negative sources. These are different for everybody but, if you listen, people will tell you what is needed for motivation. On the other hand, Maslow's hierarchy can be an excellent way to predict negative and demotivating factors. Motivation is always a risk in times of change, as shown in Fig. 5.10. Demotivation often happens lower in the organization where there is less knowledge about change and ability to participate in decision-making. The important thing to remember is that your employees will always feel less secure than you do and as a result will be less motivated. However, you can motivate people by informing and empowering them. In other words, the more you give, the more you get.

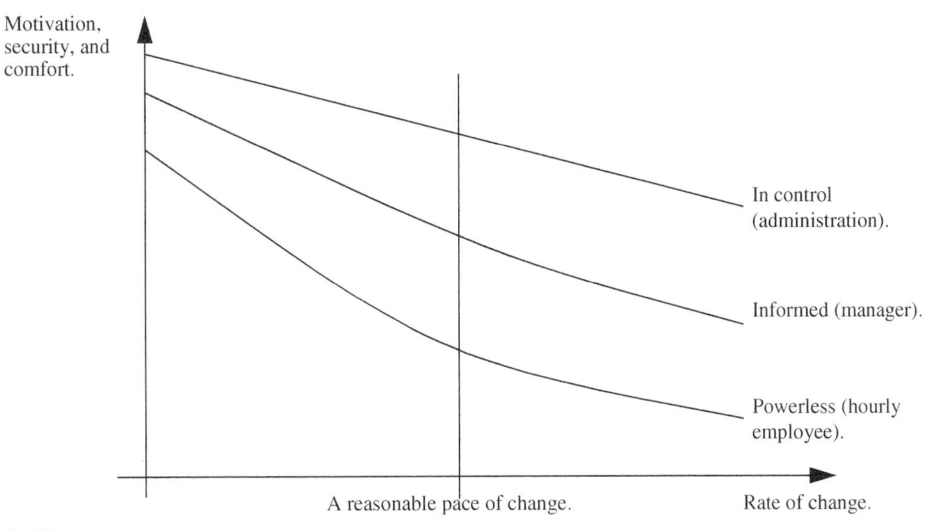

FIGURE 5.10

How people respond to change.

Motivation can be a combination of factors, including those shown in Fig. 5.11. It is unreasonable to expect all the positive factors on the right, but if you strive to minimize the negatives, on the left, and maximize the positives, then everybody will benefit. This applies to managers and employees alike. For example, if there is a boss who micromanages, and employees who resent him or her, then everybody loses.

When people are demotivated there are a number of signs that say it may be time for action:

- Passive: People respond passively when active responses would be expected.
- Resentment: People feel that others are being treated better.

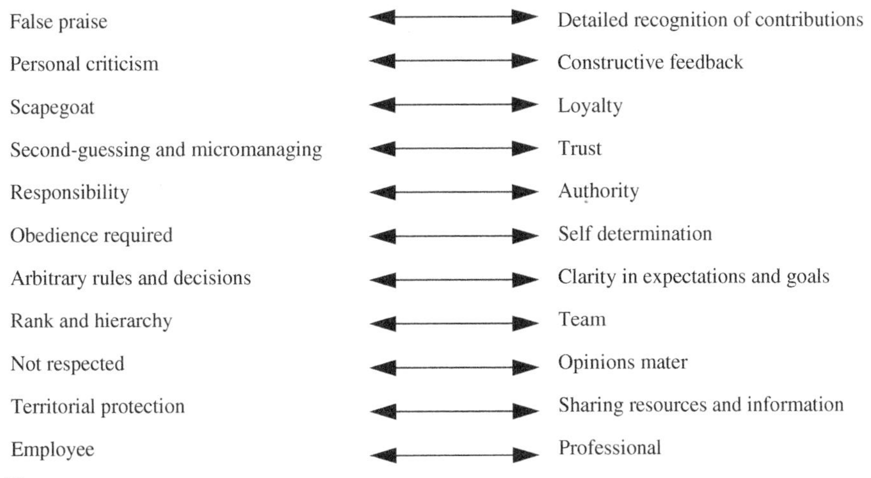

FIGURE 5.11

Factors that influence motivation.

- Anger: Anger is a common response to a threat.
- Groups: People cluster into closed social groups.
- Absent: There are a large number of vacation, sick, or interview days.
- Quiet: Communication is forced and brief. Information is not given until requested.
- Perception: Your instincts are useful here, too.

PROBLEMS

5.24 How is motivation connected to control and power?
5.25 What would widespread cynicism indicate?
5.26 List 10 factors that would decrease employee motivation.

5.3.2 Politics

The term "politics" carries a significant amount of negative baggage, but it just describes how large groups interact. The goal of politics is to unite people for the better good. Of course, everyone has a different view of "uniting" and "good." In the corporate world the political objective is to get the entire organization (and beyond) to work toward a common goal: the success of the company.

The organization chart for a company dictates the general political structure from the top, with the board of directors and chief executive officer. Legitimate power (authority) is distributed down through the organization from manager to employee, to support the mission, vision, objectives, and goals. When there are peers, legitimate power is replaced with referent power (respect) based on merit, seniority, reputation, and similar factors. The ability to make things happen (power) is then a combination of authority and respect. When decisions need to be made, the people look to those with the appropriate type of power. An example of the power spectrum is shown in Fig. 5.12.

In the workplace political problems normally occur when people do not unite (willingly agree to work together) or don't understand the company goals (the good). The greatest political problems arise

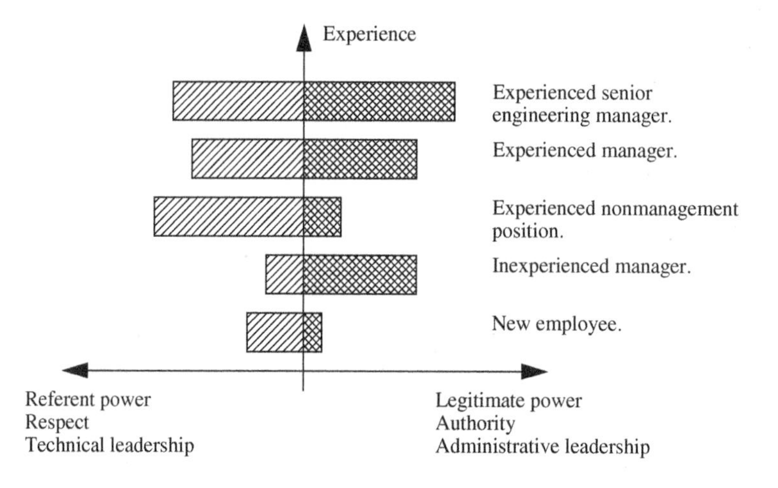

FIGURE 5.12

Types of power.

when there are conflicts between peers with referent power, or between referent and legitimate authority. Political warning signs include the following:

- Ideology and absolutism: rejection of others' points of view
- End-justifies-the-means philosophies
- Social camps and groups: a social separation forms with subgroups

Decision-making is the point at which all of the political forces come together. If a decision is to be made, there will be people directly and indirectly interested in the alternatives. Political effort will be required for decisions that (1) require significant effort, (2) use significant resources, (3) affect individuals, or (4) increase the rate of change. Even if you have the authority to make the decisions unilaterally it is better to assume that you are facilitating decision-making. There are some political strategies that are typically employed to enforce decisions. Anything other than win-win will have a long-term effect on morale, efficiency, loyalty, trust, respect, and authority. In other words, do not force authority on anybody unless there are no alternatives. And, if you must make negative decisions, make them rarely:

- Positive
  - Win-win: Finding a mutually beneficial outcome is the best approach. This may be achieved through compromise.
- Negative
  - Manipulation: This is a poor short-term strategy—if you mislead and use people it will always lead to bad feelings and problems.
  - Tit-for-tat: This is also known as an-eye-for-an-eye. In an adversarial situation, this will go through a phase of repeated damage. Eventually it should lead to some level of truce. At the extreme level this is war.
  - Domination: This requires a long, constant application of energy to force an agenda. This can make a result happen, but at a tremendous cost. Eventually this will fail.
  - Isolation: This is withdrawal from other groups. In the short term, this will reduce demands but will eventually result in mutual loss.
- Necessary
  - Decision-makers: Getting these people on your side will help you make and execute decisions.
  - Consensus building: Involving people in shaping decisions will make them more likely to support the outcome.
  - Friendly: Being friendly can make people less skeptical and more trusting.
  - Informed: Tell people what is happening to remove the fear of the unknown.
  - Favors: Build favors by helping others.

The term "personal agenda" carries a negative connotation, but it is better to accept these as things that are important to people and not to impose value judgments. Always remember that to the other person his or her agenda is important and right. Some examples of personal agendas are listed. Find out what personal agendas other people have and respond thoughtfully:

- Wanting a raise or promotion
- Personal recognition
- Union strength and protection
- Company benefit
- Moral issues
- Friendship

PROBLEMS

5.27 What is the difference between referent power and authority?

5.28 Some trivial event that nobody can remember has risen to the level of animosity between two engineers, Peter and Paul. Yesterday Peter threw away a coffee that Paul had left on a central table. Two days ago Paul had used one of Peter's drawings as a tablecloth for lunch. Peter and Paul have been doing these things for weeks. What is the name for this type of negative behavior?

5.29 What is the office equivalent of exile?

5.30 Is a personal agenda personal or selfless?

5.31 What is a good political outcome?

5.3.3 Loyalty and trust

Webster's Dictionary definitions:

Faithful: (3) firm in adherence to promises or in observance of duty: conscientious; (4) given with strong assurance: binding <a faithful promise>; (5) true to the facts, to a standard, or to an original <a faithful copy>.

Loyal: (1) unswerving in allegiance: as a: faithful in allegiance to one's lawful sovereign or government, b: faithful to a private person to whom fidelity is due, c: faithful to a cause, ideal, custom, institution, or product.

Trust: (1) a: assured reliance on the character, ability, strength, or truth of someone or something; b: one in which confidence is placed.

Loyalty is a measure of how well people adhere to our needs, as shown in Fig. 5.13. To expect absolute unconditional loyalty is unreasonable at best, but people will try their best. To expect loyalty you must give it at the same level, or higher. Faithfulness is the degree to which somebody will opt to be loyal.

Loyalty is based on trust. It is easy to confuse trust with loyalty. When you trust somebody, you know what they might do, even if it is not what you want them to do. And, when you give them authority, your trust determines what you expect. Given that trust is based on an understanding of a person, it is built over time, as shown in Fig. 5.14. If you have a new employee, you assign authority and assess the results. Over time you develop a better understanding of what that person will do in different circumstances. Trust will tell you what somebody will do given a set of circumstances. If you trust them to do what you need, you will have natural loyalty. If you want an outcome that does not match what you would normally expect you must "spend from the bank of loyalty." Or if somebody is faithful, you will not need to negotiate for the loyalty. Confidence and trust are related.

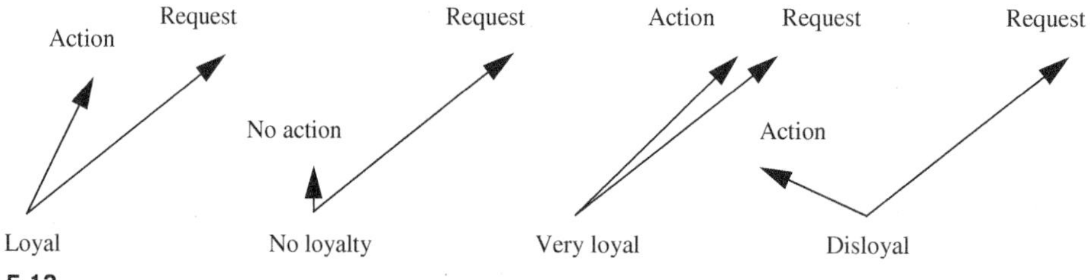

FIGURE 5.13

Different levels of loyalty.

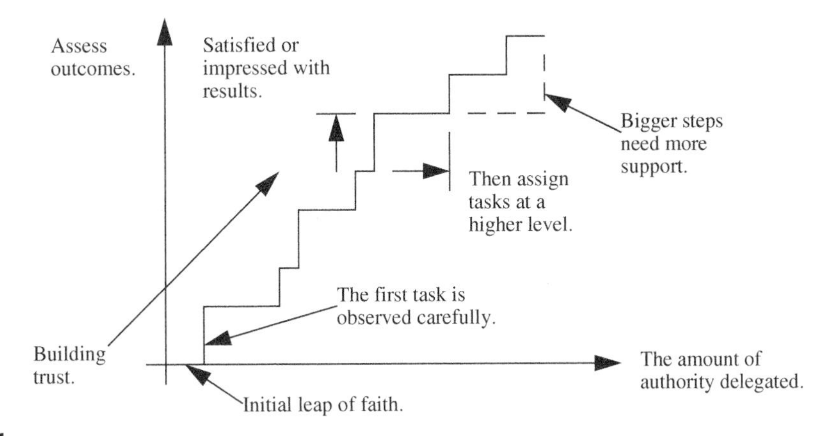

FIGURE 5.14

Building trust.

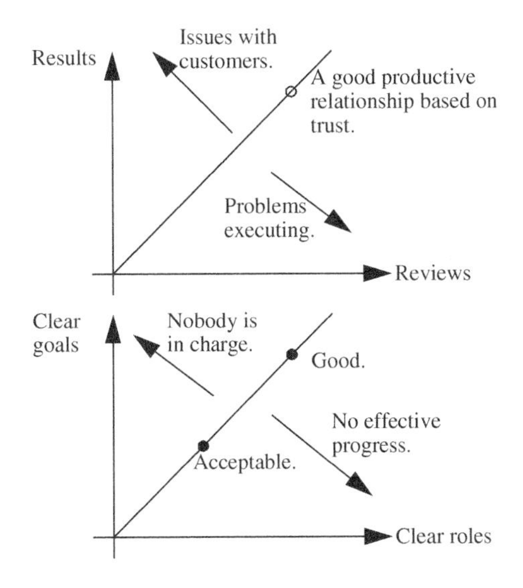

FIGURE 5.15

Trust and customers.

Trust is also an issue when dealing with customers, as shown in Fig. 5.15. Throughout the life of a design project, the customer will be watching activities and outcomes to see if they match his or her expectations. Even if the results are not exactly as expected but they are consistent, there is the groundwork for trust. If the customer can see how the results satisfy his or her needs then he or she will be able to understand the outcomes, building trust. If the results do not match, most customers will not try to understand the differences and trust is never built. This means that if you want customers to trust you, you must (1) understand what they need, (2) give them what they need, and (3) when you can't, explain why it does not match their needs.

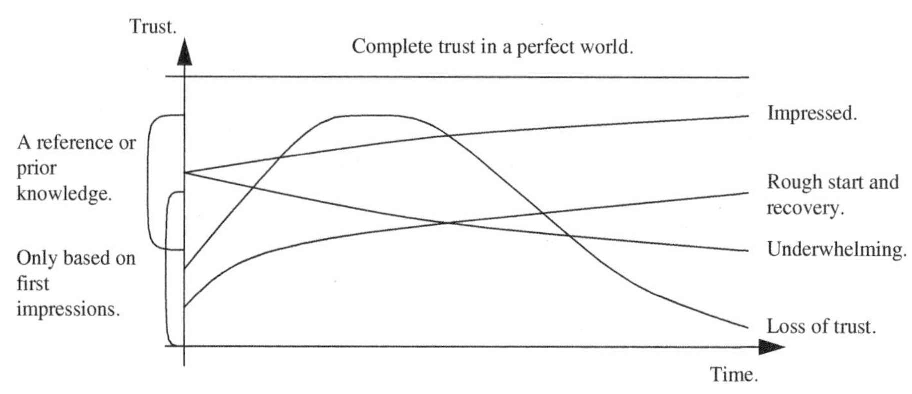

FIGURE 5.16

Trust over time.

Fig. 5.16 shows how business trust tends to evolve over the life of a design project. As always, referrals and first impressions can have an impact on trust. The time that has the most impact on trust is the beginning of the project. Use the first few interactions to increase trust, and the level of authority will be greater, and work will advance faster. Please note that this applies to all aspects of business relationships, including major corporations, suppliers, customers, and employees.

Factors in building trust include the following:

- Ethics: An important factor in trust, it makes you more predictable.
- Experience: The best predictor of future outcomes is the past.
- Approach and justification: Describing what you do will help build understanding and trust.
- Honesty and transparency: These help build understanding.
- Realistic: Recognize the positives and negatives.
- Platitudes: Using extreme descriptions will indicate you cannot see subtleties.
- Confidence: Faith in others breeds faith in you.

PROBLEMS

5.32 Are loyalty and obedience the same thing?
5.33 How do first impressions influence trust?
5.34 Is it reasonable to expect trust in a new relationship?
5.35 Is it reasonable to expect loyalty in a new relationship?
5.36 A useful strategy with new customers is to select a small project to earn trust before moving to larger projects. What is the advantage and disadvantage of this approach?
5.37 You know that an engineer will always leave at noon on Friday. You ask the engineer to stay until 4:00 p.m. on Friday but he still leaves at noon. Is this a trust or loyalty problem?
5.38 Is it easier to trust realistic people, even when you do not agree with them? Explain.

5.3.4 Responsibility and authority

Hire good people, give them what they need, and then stand back and let them do a good job.

Authority is the ability to make decisions and act on them without intermediate review. Accountability is the eventual review of the results. Responsibility is the combination of the two, as shown in

Fig. 5.17. In a healthy environment people are held accountable for the authority they have, and while developing, they can be given more authority than accountability to encourage growth with some safety. Of course, when employees have no authority, accountability and responsibility are the same thing. This lack of control undermines safety, morale, and productivity.

Some approaches used in delegation are shown in Fig. 5.18. When accountability exceeds authority people have less control and have a greater risk of negative personal consequences. If there is benevolence, a manager may shift positive accountability to those responsible and keep negative accountability. In a malevolent situation, negative accountability will be shifted to those responsible and positive accountability kept. A manager with little trust will assign and review work in smaller increments; this is known as micromanaging. In a micromanaging situation the manager never feels at ease and the employees are aware that they have no trust or authority.

Assigning authority is relatively simple: let everybody know who is allowed to do what. Then step back and do not interfere or undermine unless the person asks for help. The key elements are to make sure that the decisions that can be made are (1) giving clearly defined authority to the employee, (2) recognizing the authority to others by clear communications, and (3) escalation guidelines for your involvement. For example, a project engineer might be given a budget of $10,000 for purchasing parts. The budget is approved (giving authority) and the purchasing department is told to process any purchases until $9000 (recognizing authority), when the manager should be informed (escalation).

Praise the good work and provide constructive strategies for the poor.

Accountability can be more difficult. As a manager you will be able to review work and look for positive contributions, even if they are not what you expected. The key is to (1) recognize the positive

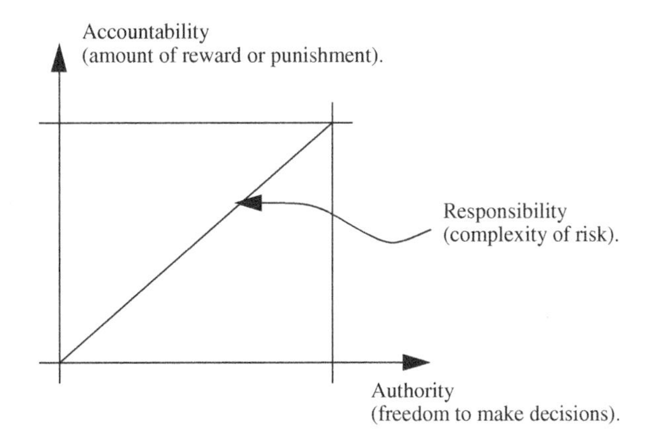

balanced: authority = accountability
faith: authority > accountability
irrational: authority < accountability
responsibility = sqrt(authority^2 + accountability^2)

FIGURE 5.17

Authority, accountability, and responsibility.

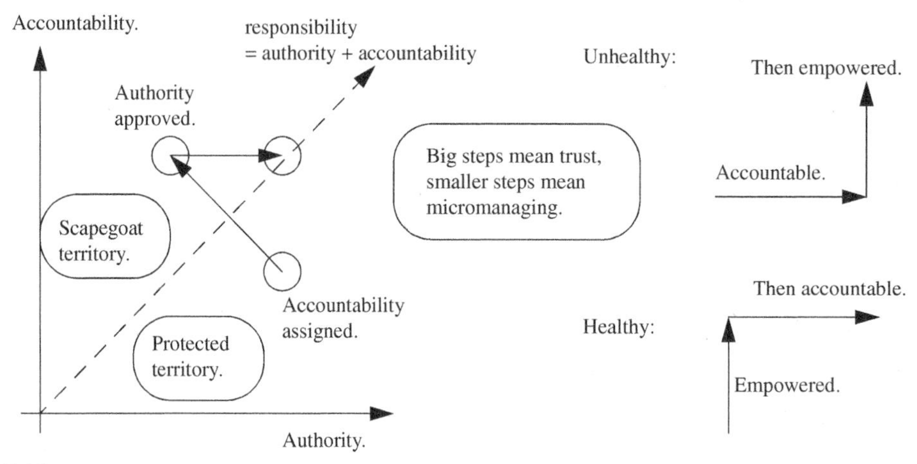

FIGURE 5.18

Recognizing delegation patterns.

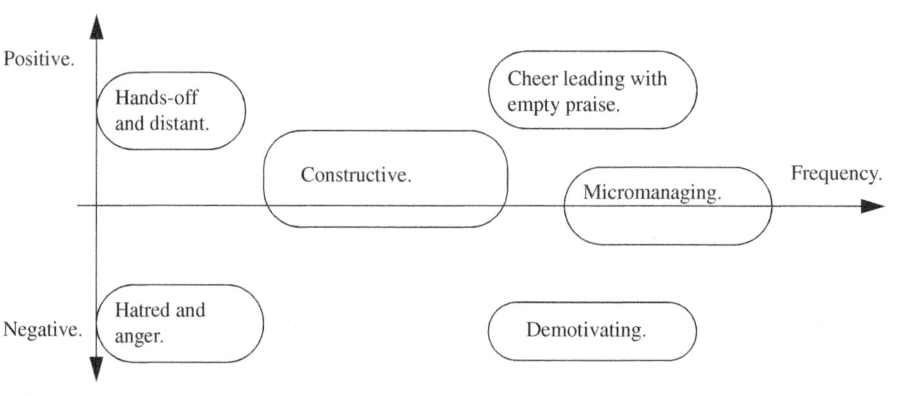

FIGURE 5.19

Proactive feedback.

outcomes as beneficial, (2) provide constructive feedback about items that do not meet the needs, and (3) discuss a win-win approach to increased responsibility. The other issue is how frequently feedback is offered. Depending on the level of authority this may be daily, weekly, or monthly. Most organizations do an annual review, which is meant to be comprehensive, to adjust salaries. Fig. 5.19 shows a spectrum of feedback content and frequency. The ideal strategy is regular feedback, but not intervention, with positive feedback and constructive suggestions. The worst case is infrequent and inconsistent feedback that is punishment oriented.

PROBLEMS

5.39 What is the difference between accountability and authority?
5.40 What is a healthy relationship between accountability and authority?

5.41 Describe micromanaging with respect to accountability and authority.

5.42 Consider Maslow's hierarchy.

- **(a)** Does accountability move up or down the pyramid?
- **(b)** Does authority move up or down the pyramid?

5.4 Managing individuals in organizations

The basic tasks a manager is expected to do include (1) leading, (2) directing, (3) planning, (4) coordinating, and (5) assessing. Being a senior person and directing others with more topical knowledge and experience than you can seem intimidating at first, but it does get easier with time and experience. If you can show people that you are competent you will get respect. And of course, try not to take out your anxiety on others, or demand compliance, or be closed to criticism. Look to others for suggestions and feedback. Be adaptive.

By accepting that all of us have strengths and weaknesses we can better understand ourselves and look to others to help complement our abilities. A number of required situations that a manager can handle are shown in Fig. 5.20; these should be considered to be minimums. However, we should develop strengths and work toward preferred habits. For example, there may be a time when a customer does not know exactly what he or she wants from a design project. But the manager should gently guide the process toward detail and certainty.

Management by objectives is a popular approach to employee empowerment. The employees are asked to propose their work goals in support of the organization's needs. Management approval makes them responsible for the objectives with both authority and accountability. The ability to influence work assignments also increases employee self-control and motivation. One approach is outlined as follows:

(1) Make a prioritized list of what you need them to do and what you want to do (optional).
(2) Have them develop a list, also prioritized, of what they would like to do and what they want from you.
(3) Sit together and review the lists, preferably as professional peers.
(4) Find those items that are wants and needs for each.
(5) Then, go through the lists by priority and pick some goals and eliminate others.
(6) Leave some items off the lists. Do not try to do everything.

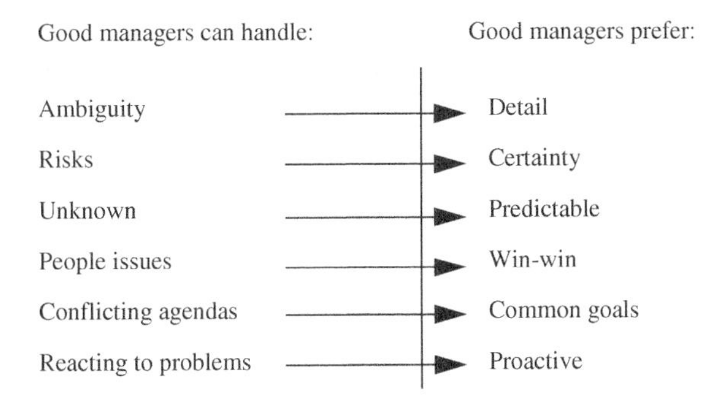

FIGURE 5.20

Good managing habits.

PROBLEMS

5.43 How could management by objective increase employee morale?

5.44 A manager is known for her ability to gamble on risks and win. Is this a healthy perspective?

5.4.1 Leadership habits

Leading and commanding are not the same.

Leadership is not just a task for managers. Leaders make great team members, and a great team is made up of leaders who work well together. There are a number of good practices that managers can or should do for their people:

- Understand people
  - Understand individual motivations (intellectual, contributing, working with people, etc.).
  - Know what people need to thrive.
  - Be aware of issues in the group.
  - Understand motives for themselves and others.
  - Listen.
  - Understand that everyone wants to contribute in his or her own way.
  - Help people want to follow.
- Motivate
  - Provide recognition for progress and work done well.
  - Encourage and allow people to apply their strengths.
  - Act as a motivator: enthusiastic, committed.
  - Avoid nonproductive busy work.
  - Allow people some freedom to choose what they do.
  - Focus on the successes and try to get more.
  - Be considerate of negative and opposing opinions.
- Nurture
  - Give people tasks that they can succeed at and the needed tools.
  - Delegate responsibility (chance for reward), meaning accountability and authority (ability to produce).
  - Encourage intellectual and professional growth.
  - State concerns and offer suggestions.
  - Nurture motivation and creativity.
  - Expect everybody to learn from mistakes.
  - Do things that are not efficient, but for the benefit of others.
  - Develop people (to advance); involve people in your activities or let them replace you.
  - Provide feedback along with suggestions.
  - Buffer people but do not isolate.
  - Provide free work time to advance.
  - Focus on personal improvement based on what they have done before, trying to set new personal records.

- Support
  - Be flexible and adapt; things change.
  - Make decisions with the people they affect.
  - Deal with the little problems now, but focus on long-term goals.
  - Be flexible and adaptive.
  - Enhance project image.
  - Consider what to do at points of frustration to (1) move to backups, (2) backtrack, (3) stop.
- Be knowledgeable
  - Understand the objectives of the project.
  - Use clearly stated evaluation tools tied to the organization.
  - Structure processes to manage complexity.
  - Look for "orphan tasks" that need people or resources.
  - Listen and respond thoughtfully.
- Be trustworthy
  - Be honest and transparent.
  - Keep people aware of reasons for decisions and actions.
  - Build individual and team trust.
  - Do not micromanage.
  - Admit lack of knowledge or mistakes.
  - Be selfless.
  - Do not be pompous.
  - Be ethical.
  - Remain calm under pressure.
  - Be proactive.
  - Negotiate sponsor trust.
- Remember
  - Uncertainty comes from lack of control and knowledge. Change causes uncertainty and possibly fear. New directions bring uncertainty.
  - Just because something is irrational doesn't mean it is not real to somebody.
  - It is easier to find motivation than to try to create it. Put people in a position where they can contribute and get value.
  - Everybody should get value from their jobs. Equating money with human value is an amateur mistake.
  - Over time the priorities and outlook of people change.
  - Recognize where people are with personal knowledge.
  - When people under you are growing (with new authority) keep the accountability.
  - You cannot lead people you do not respect.

We all have bad habits. But part of the continuous improvement process requires identifying these and trying to lessen and eventually replace these habits. Identify some of the habits in the following list that you need to improve on:

- Dehumanize and marginalize
  - Assume that a paycheck makes an employee do whatever is asked; relate pay with motivation.
  - Make decisions alone.
  - Ignore issues and worry until they become problems.
  - Let people take care of themselves.

- Rely on self-preservation.
- Compare an individual's effort to the strongest performers.

- Bully
  - Use threats to get things to happen.
  - Demand military discipline and rules.
  - Use fear to promote productivity and creativity.
  - Punish all mistakes so that they will not occur again.
  - Discipline in public to teach others a lesson.
  - Discuss the virtues of being able to fight and then move forward.
  - Dictate actions.
  - Associate failure with punishment.
- Pompous and arrogant
  - Do not consult people before making decisions.
  - Micromanage and second guess every team decision.
  - Make decisions to take to the group.
  - Make decisions that affect other people, and then tell them.
  - Use stricter rules for those who "need it."
  - Micromanage.
  - Criticize without benefit.
- Selfish behavior
  - Make yourself indispensable.
  - Delegate accountability down but keep authority: "My head is on the block so you need to do what I say."

The book *How to Win Friends and Influence People* by Dale Carnegie (1981) has stood the test of time. It uses a very positive approach to working with people and grooming relationships. The principles from the original book, listed in Fig. 5.21, suggest that you be polite and friendly to make people comfortable and show interest to have them engage you on a personal level. This list should not be a surprise, but it does provide a way to self-evaluate skills and plan for improvement.

By its nature, leadership is based on respect to make decisions and lead the team. It is natural that you will encounter negative and differing opinions. Regardless of how or what is said, try to find the value. Even the most critical person has some level of interest in any topic. If you are competent and capable you will accept that there are multiple views and they all have value. It is incredibly beneficial to have somebody who will give you blunt criticism. Even if you still disagree after discussions, there will be a level of respect for both opinions. Consider the following points during those potential conflicts and avoid the temptation to react:

- Irrelevant: Is the difference of opinion important? All too often it is not.
- Impartial: It may be your idea, but don't be afraid to let it go.
- Ask: Find out why a person disagrees.
- Listen: The other person may be right.
- Explain: Your viewpoint may not be obvious.
- Concede: Allow differences in opinions, they deserve the chance to try it their way, even if it may lead to some failure.
- Agree: Be prepared to accept their opinion in place of yours.
- Respect: Always accept the value of their opinion, even if you can't agree.
- Unified: If necessary there should be an agreement on action, or recognized differences, to allow the team to move forward.

Fundamental Techniques in Handling People
1. Don't criticize, condemn, or complain.
2. Give honest and sincere appreciation.
3. Arouse in the other person an eager want.

Six Ways to Make People Like You
1. Become genuinely interested in other people.
2. Smile.
3. Remember that a person's name is to him or her the sweetest and most important sound in any language.
4. Be a good listener. Encourage others to talk about themselves.
5. Talk in the terms of the other person's interest.
6. Make the other person feel important and do it sincerely.

Twelve Ways to Win People to Your Way of Thinking
1. The only way to get the best of an argument is to avoid it.
2. Show respect for the other person's opinions. Never say, "You're wrong."
3. If you're wrong, admit it quickly and emphatically.
4. Begin in a friendly way.
5. Get the other person saying "yes, yes" immediately.
6. Let the other person do a great deal of the talking.
7. Let the other person feel the idea is his or hers.
8. Try honestly to see things from the other person's point of view.
9. Be sympathetic with the other person's ideas and desires.
10. Appeal to the nobler motives.
11. Dramatize your ideas.
12. Throw down a challenge.

Be a Leader: How to Change People Without Giving Offense or Arousing Resentment
1. Begin with praise and honest appreciation.
2. Call attention to other people's mistakes indirectly.
3. Talk about your own mistakes before criticizing the other person.
4. Ask questions instead of giving direct orders.
5. Let the other person save face.
6. Praise the slightest improvement and praise every improvement. Be "hearty in your approbation and lavish in your praise.
7. Give the other person a fine reputation to live up to.
8. Use encouragement. Make the fault seem easy to correct.
9. Make the other person happy about doing the thing you suggest.

FIGURE 5.21

The key principles in Dale Carnegie's *How To Win Friends and Influence People.*

PROBLEMS

5.45 Is it possible for somebody to be a nurturing bully? Explain.

5.46 What are three things you could do when you are wrong or have made a mistake?

5.47 What are five strategies to use when somebody says something incorrect?

5.48 When does Carnegie say you should talk most?

5.49 Assume a team has been working on the design of a new communications tower, but is far behind schedule. You have been appointed to replace the previous team leader. Without any prior knowledge of the problem, what are 10 strategies you might consider before meeting the team?

5.50 Review the principles from Dale Carnegie. Select three that you do well and write a summary of methods you use. Select three principles that you do not do well and suggest a method of improvement.

5.51 List the Carnegie principles that deal with criticism.

5.4.2 Delegation

When responsibility is assigned you need agree, disagree, or say what you will do instead.

Helping people grow is a simple cycle: assign a new task, the work is performed, the outcomes are assessed, and plans are made for the next delegation cycle. Eventually the delegated task becomes a permanent responsibility. A reasonable approach is outlined in Fig. 5.22. The ideal process is to proactively develop employee abilities when you have the flexibility to deal with the process of growth (i.e., things will differ from what you expect). If you react to an overloaded work schedule by delegating, you often rush and do not have enough time to mentor and support, and the result will be less beneficial. However, in either case, the underlying goal should be to foster employee growth using task assignment. Please recall that delegation means that responsibility is assigned by giving authority to the employee, along with accountability.

Some important points to consider when delegating are given in the following list. Essentially, don't use delegation as a method to clone yourself. Use delegation to develop people with complementary approaches, to make you more effective:

- Use delegation for training.
- Clear expectations are critical.

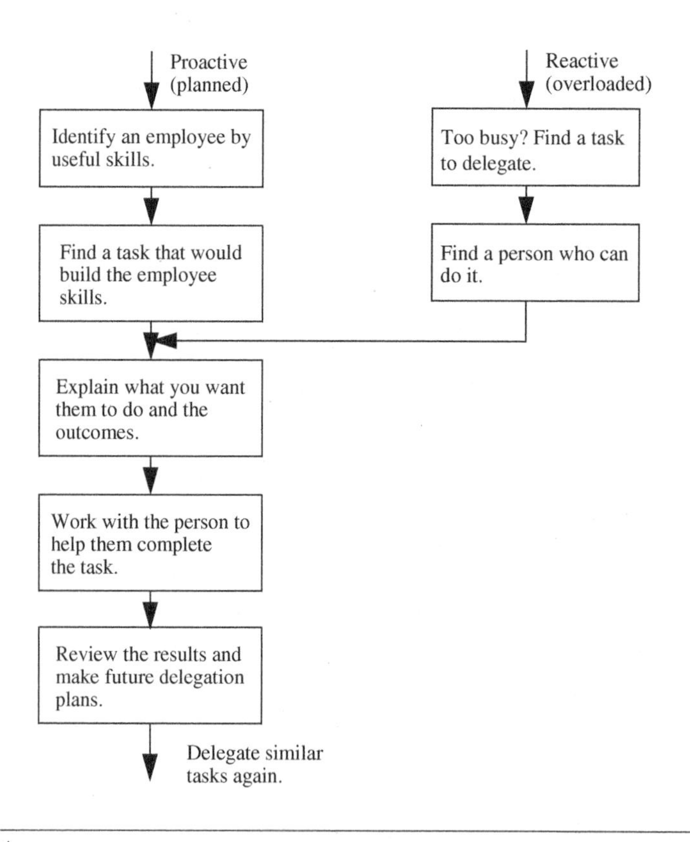

FIGURE 5.22

Delegation flowchart.

- Provide resources and access during the learning phase.
- Recognize that time spent early will be paid back many times in the future.
- When delegating, you can learn a few things that you simply assumed.
- The key to delegation is to assign authority ahead of responsibility.
- Welcome mistakes as long as they come with learning and growth.
- Don't delegate a job you wouldn't do yourself.
- Consider delegating to consultants, subcontractors, or suppliers.
- Delegation helps you move up in a company, but it becomes harder as you are promoted.
- Adjusting your work expectations can expand the ability to delegate. Even if the results are different from your personal standards, they might be as good or better.

PROBLEMS

5.52 What are 10 benefits of delegating?
5.53 What are 5 costs of delegating?

5.4.3 Making inclusive decisions

If you are always behind you will never catch up.

If employees have no control they will not willingly accept responsibility. The best way to alienate employees and peers is to make decisions and then dictate the results to the group. A good practice is to seek input on every decision and accept feedback constructively to achieve group approval. If a decision is unanimous people will be faithful. If somebody disagrees, but has had a chance to sway the decision, then he or she is much more likely to be loyal to the results. Fig. 5.23 illustrates a method for making decisions that will have an impact on others. It is assumed that a decision must be made, and that all options are consistent with the organization's policies and philosophies. It is worth noting that there are some decisions that do not need group approval because (1) they do not affect people on the team, (2) the team has already agreed to guiding principles for the decisions, or (3) there are factors that preclude the team from participating. When decisions have been made that affect others it is essential that they be told and have a chance to respond. This process can be very intimidating for insecure and/or junior managers because it appears to undermine their authority, but being open and honest will increase loyalty and, it is hoped, faithfulness.

PROBLEMS

5.54 Describe decisions you have seen, or made, that led to each of the three problem-solving outcomes. In particular (a) a great decision, (b) an unpopular decision, and (c) a bad decision.
5.55 Does silence mean agreement? Explain the other possible meanings.
5.56 More control leads to more security and productivity. Explain the cost of excluding other people from decisions.
5.57 What are the factors in a bad decision?

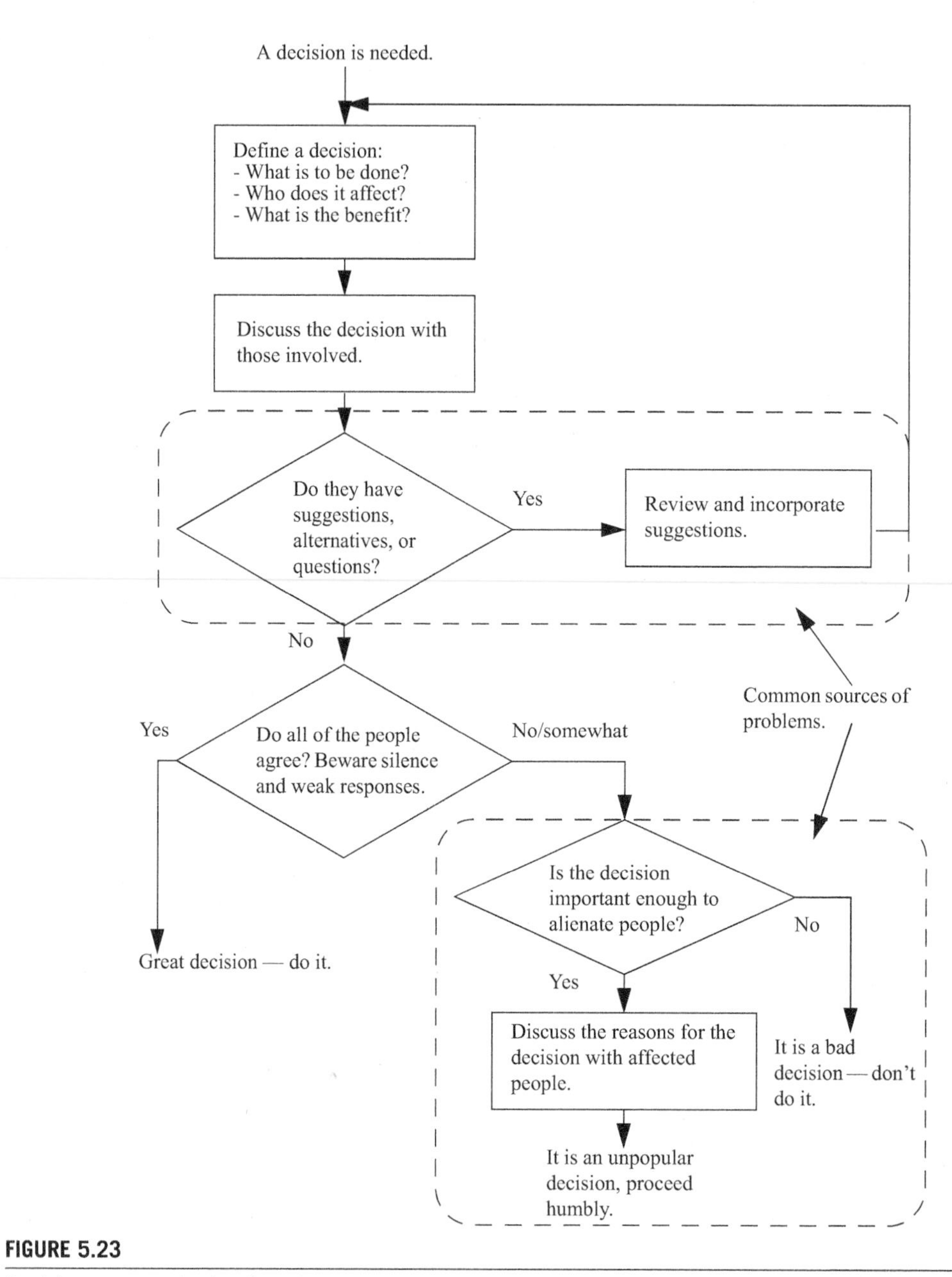

FIGURE 5.23

Decision communication flowchart.

5.4.4 Wellness and productivity

Everybody has problems at some point in their life. A callous response is to discard problem employees, but other employees will recognize the lack of loyalty and become demoralized. The right thing to do is to help them overcome the problems, improve their life, and become productive and fulfilled. One approach is to use Maslow's model to identify the problems and work with them to

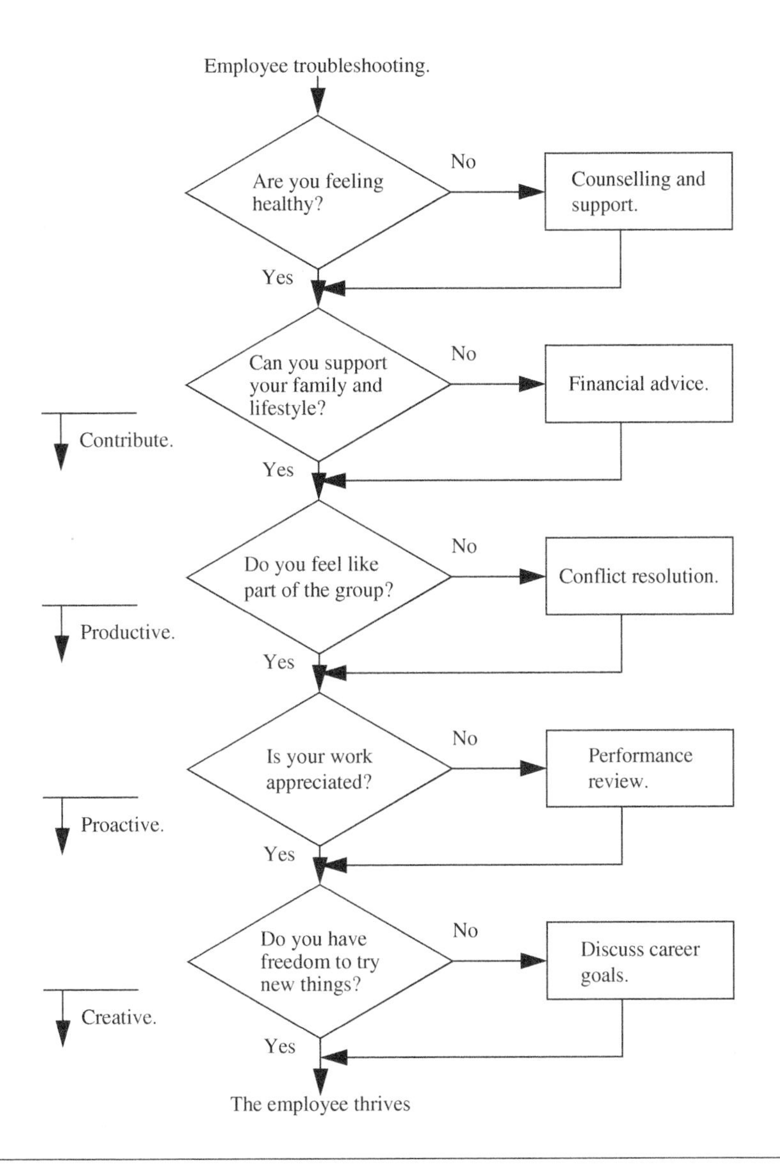

FIGURE 5.24

Solving problems with Maslow.

develop a strategy to overcome the issues (Fig. 5.24). The model has obvious limitations, but it can be a useful starting point.

PROBLEMS

5.58 Assume you are supervising a group of employees. There are productivity problems and you assume that they are because the employees are "having issues." Develop two examples of problems for each level of Maslow's hierarchies.

5.59 You have an employee who is productive but rarely creates new work to do by herself. How could you use Maslow's hierarchy to explain the current situation? What strategy can be used to make the employee proactive and eventually creative?

5.4.5 Conflicts and intervention

When problems arise they come in different variations, including (1) all of the team with one person, (2) some of the team with one person, (3) some of the team with some others on the team, and (4) one person with one person. If more people are involved it becomes harder to solve the problem. There are often specific issues that started the problem, and there may have been an escalation over time. Common sources of conflicts are outlined in the following list. It is better to identify these and focus on nonpersonal issues during the conflict resolution process to avoid the risk of criticizing an individual:

- Nonpersonal
  - Staffing
  - Equipment and facilities
  - Responsibilities
  - Schedules
  - Capital spending
  - Costs
  - Technical opinions
  - Administration
  - Priorities
- Behavioral
  - Excessive—exceeding goals or expectations, expecting more from others
  - Undermotivated—not interested
  - Untalented—not able
  - Antisocial—will not interact productively
  - Distracted—has interest but not focus
  - "Us versus them"—philosophical rift
  - Anger and emotion
  - Roles—there is confusion about work authority and accountability
  - Power—inappropriate use of power
  - Style—personal styles don't match
  - Oversensitive—misinterprets others' actions
  - History—people are not ready to let go of issues
  - Threat—an outstanding threat creates tension
  - Harassment—based on personal feelings but crossing the line of acceptable behavior
  - Personality

There are a number of conflict resolution mechanisms indicated by the Project Management Institute, listed next. It is best when the individuals and groups resolve the issues themselves, but when they cannot be overcome it may be necessary to move up the administrative chain. Once this happens the major tools start to include salary- and employment-based options. A general problem-solving approach is shown in Fig. 5.25.

This method identifies a reasonable approach to conflict in a team:

- Forcing
- Smoothing or accommodating
- Compromise
- Confrontation or problem solving
- Collaborating
- Withdrawal or avoidance

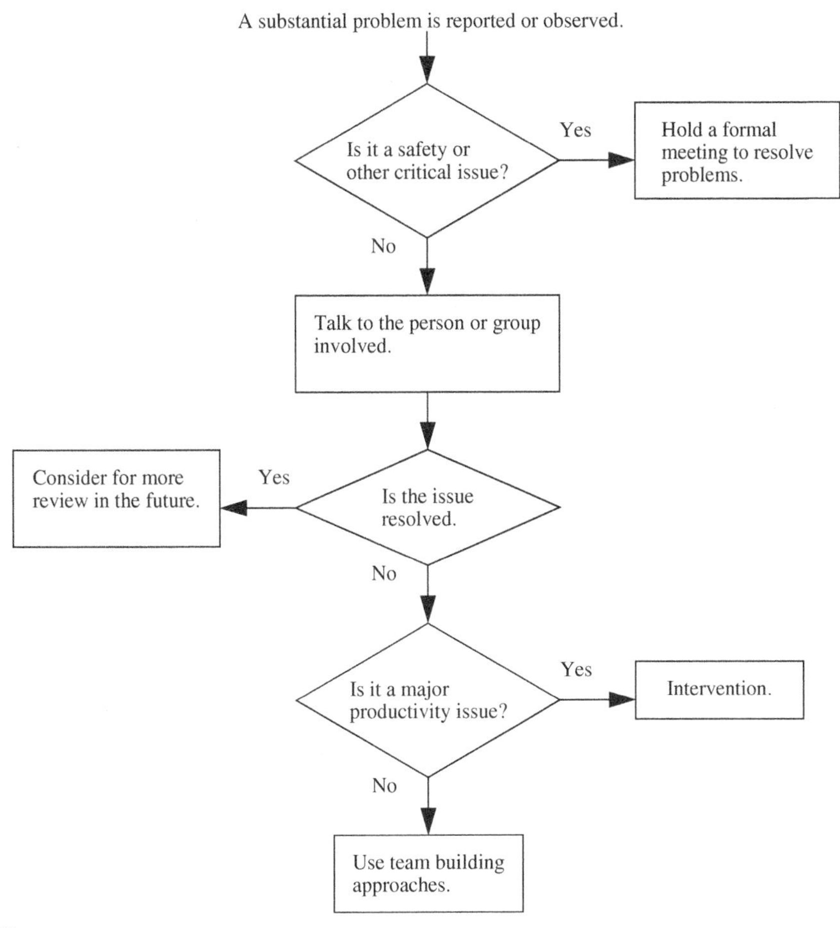

FIGURE 5.25

Conflict resolution.

PROBLEMS

5.60 As a manager you are faced with a situation in which one employee punched another. Describe three different outcomes and the process to get there.

5.61 What factors will lead to administrative intervention?

5.62 One possible outcome of intervention is that a team is disbanded. List three sources of conflict that would be a reason for disbanding a team.

5.4.6 Hiring and promotion

When hiring employees, consultants, suppliers, or customers you need to consider (1) technical abilities, (2) potential for trust, (3) potential for loyalty and faith, and (4) potential for productivity. The same is true for promotions, where you are effectively hiring a person for a new higher-level position. These factors must be assessed quickly during an interview for a new employee. Some of the items to consider are shown in the following list and can be used for planning interviews, or if you are being interviewed. The hiring and promotion decision is based on the probability of long-term productivity and mutual benefit. Avoid hiring, or taking a job, if the fit is poor:

- Intangibles
  - Reputation
  - First impressions
  - Eagerness and enthusiasm
  - Trustworthy independence in work
  - Personality
  - Self-awareness and vision
- Business
  - Financial expectations
  - Gives you something you need
  - Wants some things you have
- Strategic
  - Previous work
  - Fits a clear need
  - A mutually beneficial relationship
  - Is innovation needed?
  - Is the work a commodity in nature?
  - Corporate and individual philosophy

PROBLEMS

Problem 5.63 What are 4 major factors to consider in hiring?

Problem 5.64 Create a list of 10 interview questions that would test the 4 major areas to consider in hiring.

5.5 Teams

A team can extend your work capacity and expand your abilities.

There are physical limits to our productive work hours, and our knowledge and work methods both enable and limit our effectiveness. Teams provide more available work hours and a broader coverage of work methods and knowledge. A great team would match the technical and personal skills needed for a project. Together the members would be much more effective than any of them could be alone. It would be nice to say that there is a unique way to form teams, but the process is imperfect:

- Monocultures: People are the most comfortable with similar people and will choose them if given a choice.
- Available: There are never enough people available for all of the work.
- Unique: Each person is different and there is no easy way to quantify personalities.
- Predict: You cannot always predict what is coming or what will change in the job or the people.

Individuals have specialties and work methods that will make them effective at some tasks, but not others.

We create teams to do more than is possible for a single person and to combine knowledge and differing approaches to be more productive. The key concept is that a team should be more than just a duplication of members, but instead it should be composed of complementary skills and approaches. To be successful, teams must embrace the concept of diversity of personalities, approaches, and skills. Typical factors that are considered when developing teams are (1) technical knowledge and skills, (2) personality (introvert, extrovert, detailed, conceptual, etc.), and (3) working styles (methodical, binge working, strategic, etc.).

- Everybody must recognize differences in styles.
- Use the differences to make the team stronger.
- Compensate for your individual weaknesses.
- Exploit your strengths.
- Recognize that a lack of diversity is a problem.

A reasonable method for composing teams is seen in Fig. 5.26. It is an iterative process in which you try to match skills and then look for problems in personality and availability. It would be unusual to assemble the perfect team, but at some point you must accept the team as suitable and move ahead. When you are on a team, you need to accept that there will always be some sort of mismatch.

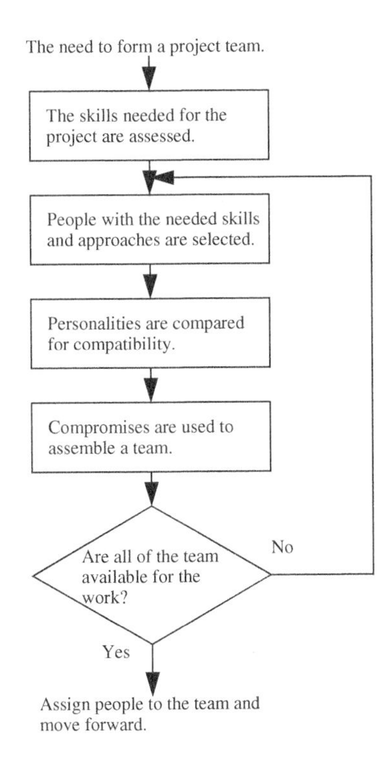

FIGURE 5.26

Methodical team planning.

PROBLEMS

5.65 The saying goes, "You go to war with the army you have, not the army you want." How does this apply to team work?

5.66 Should you try to assemble a team with matching opinions? Why?

5.5.1 Skills matrix

To assign a team you need to have an understanding of routine and unique issues in the design project. To begin, develop a list of challenges that are expected during the project. Focus on those problems that (1) are unusual, (2) are not commonly done, (3) require a special skill, or (4) must be done precisely. You may be able to develop this list yourself but others will also provide valuable input. An example of listing and scoring these issues is shown in Fig. 5.27.

Score	Sample items
1–5	Customer is hard to work with. Customer has special needs. Requires special technical knowledge. Requires special business knowledge. Requires innovation. Multidisciplinary aspects. Regulatory requirements.

FIGURE 5.27

Project challenges.

Required/critical project skills.	Score.	People available for the project.				

FIGURE 5.28

Skills matrix.

Once the project challenges are understood, it is time to begin fitting people to the team. The skills matrix in Fig. 5.28 can be useful for ensuring coverage of project needs. This matrix will be used in the process of team selection in Section 5.5.2. In Section 5.5.3, we will look at assessing people's approaches. As we pick people for the team we compare their abilities to the project needs. After this first pass we will know where we have duplication and where there are holes.

PROBLEM

5.67 Create a skills matrix for house-painting projects. The team of three will need to find customers, prepare the spaces, paint walls, paint corners, paint special features, inspect the final results, show the customer, and handle payment. Each of the three individuals cannot have more that half of the skills.

5.5.2 Profiling

A personal history tells us the most about how someone works and his or her personal approach. If it is your first time working with him or her you will need to count on information that is (1) self-reported, (2) heard secondhand, or (3) provided by references. When your knowledge is sparse it is always good to watch closely and see how people approach the smaller problems. At times this may involve a couple of small or noncritical tasks. With experience you will become better at understanding how people work and will find it easy to construct a profile like that in Fig. 5.29.

A crude but useful starting point for understanding people is to put names on productive behaviors. Some of the common terms are itemized in Fig. 5.30. The names and definitions of these traits are secondary; you can use or create your own, if you would prefer. However, the key is to be able to describe your employees and tasks in common terms.

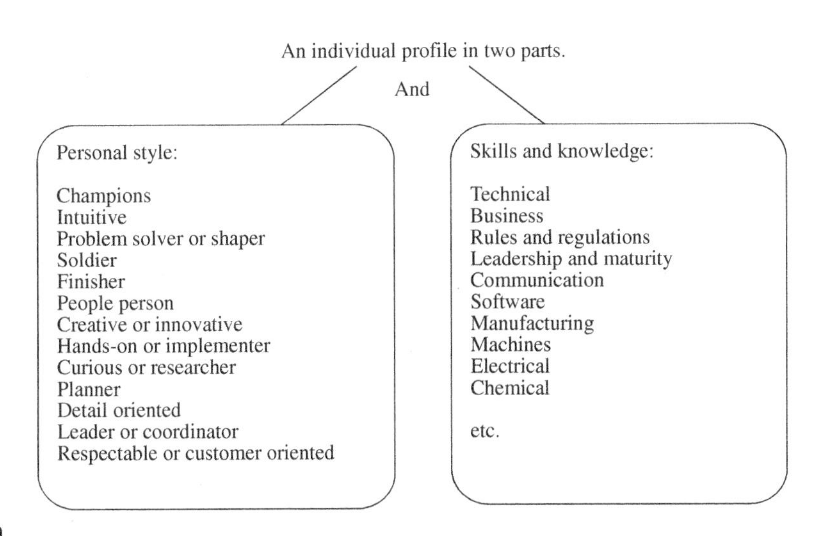

An individual profile in two parts.
And
Personal style:
Champions
Intuitive
Problem solver or shaper
Soldier
Finisher
People person
Creative or innovative
Hands-on or implementer
Curious or researcher
Planner
Detail oriented
Leader or coordinator
Respectable or customer oriented
Skills and knowledge:
Technical
Business
Rules and regulations
Leadership and maturity
Communication
Software
Manufacturing
Machines
Electrical
Chemical
etc.

FIGURE 5.29

A model of team members.

Champions:
- A champion is somebody who is personally driven to see a project from beginning to end.
- They are a source of energy (and leadership) for others.
- They tend to take and do what needs to be done.
- If there is a strong champion care must be taken to ensure the supporters do not disengage.
- Identify these people and engage them strategically. They will willingly become overcommitted.

Intuitive:
- Often able to tell when "things don't feel right."
- In the absence of a solid direction is able to make a reasonable choice.

Problem Solver or Shaper:
- Uses a methodical approach to solve problems.
- Can reduce complex problems to manageable pieces.

Soldier:
- A solid worker who performs the tasks as needed.
- Is not slowed by adversity.

Finisher:
- Toward the end of the project the energy level goes up.
- Focused on satisfying the project goals.
- Takes care of all of the small details to close.

People Person:
- Great at working with others.
- A social leader that promotes.

Creative or Innovative:
- Will generate new ideas an approaches easily.
- Able to go against convention.

Hands-on or Implementer:
- Produces physical systems easily.
- Has a good understanding of practical issues.
- Not afraid to prototype and test.

Curious or Researcher:
- Enjoys looking at new knowledge.
- Is able to search for knowledge over long periods of time.
- Can ask questions that expose misunderstandings.

Planner:
- Like clarity about the future.
- Feels best when things are occurring normally.

Detail Oriented:
- Very good at noticing small details. Often catches careless mistakes.
- Important for involved tasks.

Leader or Coordinator:
- A person with respect from the team. Leads by example.
- Naturally supports people in trouble.
- Can unite people to work together.
- Provides a rational voice when the team is getting off track.

Customer Oriented or "Comes Across Well":
- Has an approach that conveys confidence and respect.
- Provides professionalism when working with outside groups.
- A good communicator.

FIGURE 5.30

Project professional attributes often used for peer description.

PROBLEM

5.68 Identify five project professional attributes that describe you.

There have been numerous approaches used to formally categorize personalities. The most common approaches use a few rating criteria ranges such as (1) introverts/extroverts, (2) concept/detail-oriented, (3) subjective/objective, and (4) decisive/flexible. People are placed on the scales using self-reporting, tests, group review, and other methods. Once complete the person will have a set of scores on the scales that define his or her personal style (see Resource 5.1). One popular approach was developed by Myers and Briggs using the categories (1) introverted to extroverted, (2) sensing to intuitive, (3) thinking to feeling, and (4) judging to perceiving. Their approach requires the person to complete a questionnaire. The answers are then scored to determine a numerical score between two extremes. The numbers do not imply any judgment; they are just indicators. It is very common for engineers to rank as mostly introverted, with a bias to thinking and judging. The strategic value of these scores is a relative comparison of team members.

- Introverted and extroverted: Extroverted will work with other people for energy and introverted is self-driven.
- Sensing and intuitive: Sensing people prefer concrete evidence; intuitive people will accept implied and theoretical evidence.
- Thinking and feeling: Thinkers work better with logical relationships, while feelers use associations and analogies.
- Judging and perceiving: Judging people prefer true/false situations, while perceivers accept "gray areas."

Resource 5.1 Textbook website: http://www.engineeringdesignprojects.com/home/content/humans.

Personality traits will change over time, especially when amplified by stress and exhaustion. Fig. 5.31 shows a set of scales for personalities. For example, when things are calm and fresh a person might be seen as enthusiastic, but this could degrade to offensive under pressure. If an individual is very sensitive to the effects of stress you should plan for the modified personality traits, minimize stressful deadlines, and minimize individual exhaustion. If problems arise, remind team members that the personality shifts are short lived.

Technical skills and knowledge are much easier to categorize. Technical abilities are a function of knowledge and experience. For employees who are new to the organization, have new assignments, or are recently promoted, it is useful to review what they know and what they don't know. Most companies watch these people carefully and can assess their knowledge using the outcomes from projects.

PROBLEMS

5.69 Can a "people person" be "detail-oriented"?

5.70 Locate a Myers–Briggs or similar survey on the Internet. Complete the form and submit your scores. Which do you agree with?

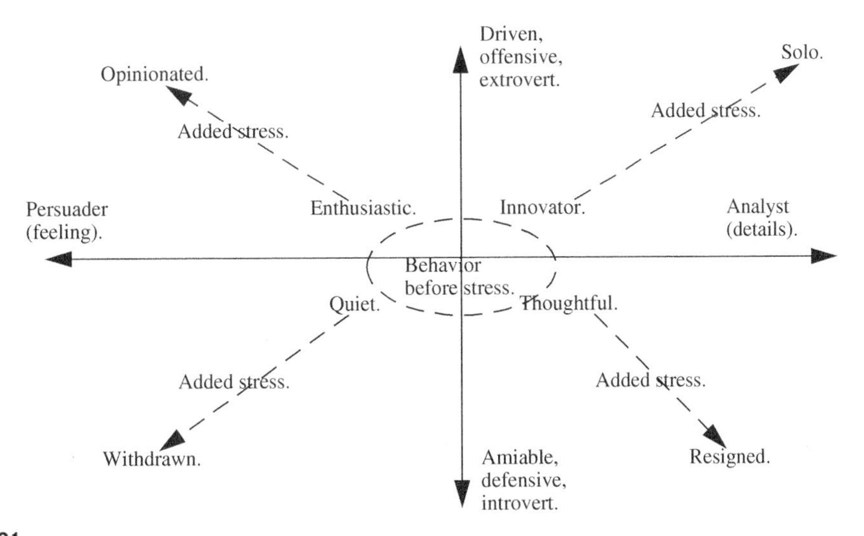

FIGURE 5.31

An example of working styles and the effects of stress.

5.71 Why do we use personality indicators even if they are very crude estimators?
5.72 Martin Luther King, Jr., said, "The ultimate measure of a man is not where he stands in moments of comfort and convenience, but where he stands at times of challenge and controversy." How does this apply to people in stress?
5.73 What Myers–Briggs indicators would be well suited to checking tolerances on engineering drawings?
5.74 Should personality indicator scores be used in hiring? Explain.
5.75 Peter is very opinionated when he experiences stress at work. What is he probably like without stress?
5.76 Why is it important to consider pre- and poststress personalities when creating teams?
5.77 Propose an alternate method to using personality typing.

5.5.3 Personality matching

Matching technical knowledge and skills is much easier than picking personalities that will be productive. In general you need some sort of natural leadership to move through difficult decisions and tasks. Some projects will need substantial interactions with clients, detailed design work, practical implementation, creative design, and more. In addition, making teams socially appealing makes them more effective.

Fig. 5.32 shows a spectrum of personalities that should be considered in team formation. The three regions inside dashed lines show groups that do not mix well. Some managers will attempt to mix personality extremes and deal with personality issues as they arise. For example, mixing a group of high-energy and quiet individuals will result in a split environment. Other managers will group employees with similar personalities to minimize conflicts. For example, all voices will be heard in a team of normally quiet individuals.

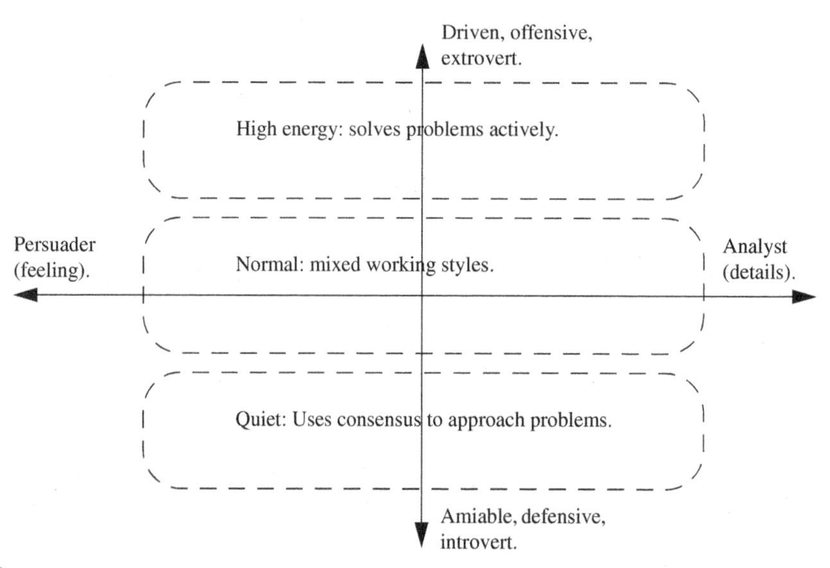

FIGURE 5.32

Schemes for building successful teams.

- Blending (in the absence of strong personalities): Combine the teams considering primarily skills, schedules, obligations, social skills, etc.
- Grouping (with extreme offensive/defensive personality types): A team with strong offensive/extrovert members should not have defensive/introvert members. The same is true for strong defensive/introvert team members.

As shown in Fig. 5.33, a similar division occurs for teams with a split between emotional individuals and those who are technically driven. In these situations the detail-oriented people ignore the persuaders and do what they think is right. The persuaders become frustrated because they have no influence. As indicated before, these methods are crude but can be used to identify and manage potential problems.

Each team member brings multiple skills. These should be balanced across the team. Team roles are identified in many sources, including Ullman (1997):

- Organizer/coordinator: mature and able to guide decisions and structure;
- Creator: generates concepts and solutions for difficult problems;
- Gatherer or resource integrator: able to find key resources to solve problems;
- Motivator or shaper: provides motivation to the team to overcome problems;
- Evaluator: able to assess results and make judgments;
- Team worker: prefers and encourages group work;
- Solver: implements plans without wavering;
- Completer/finisher or pusher: a pragmatic team member that focuses on the end results.

PROBLEMS

5.78 List three advantages of blending and three advantages of grouping.

5.79 Will a driven persuader be a bully? Explain.

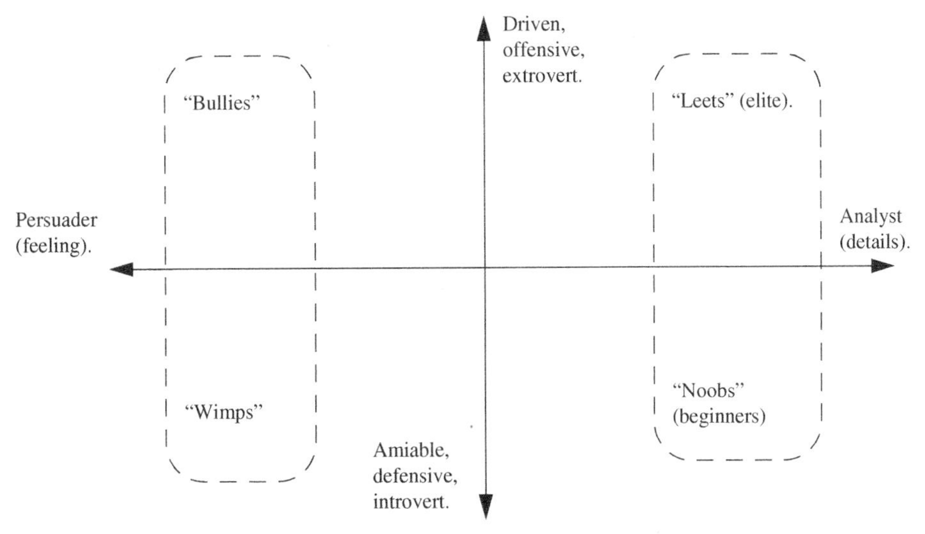

FIGURE 5.33

Schemes for building problem teams.

5.80 In a meritocracy, you are only as good as your work. In a seniority environment you are defined by your experience. What are three disadvantages and three advantages of each?
5.81 Give three examples of extreme personalities that may cause team problems.
5.82 Consider the categories in the table. Write one thing you do for each category.

Subjective.	Objective.
Decisive.	Flexible.

5.5.4 Managing teams

Communicate.

If you understand the composition of your team and the role of each team member, you will be able to communicate the details, help people understand their contributions, understand your relationship to other team members, and make them more able to be productive. It also helps a manager assign responsibility. Some guidelines for a successful team are listed as follows. The key in all of these principles is clarity, authority, and reward:

- Productivity: A well-defined and realistic set of goals, purpose, and methods will lead to an enthusiastic team.
- Composition: Combine technically complementary and agreeable personalities.
- Methods: Lay out the problem-solving styles and techniques. Also indicate what, if any, deviation is permitted.
- Frequent gratification: Devise goals and objectives to develop frequent feelings of accomplishment.
- Bonding: Team members should spend time together to develop personal relationships.
- Responsibility: People must understand what they need to do and what authority they have over resources, people, and decisions. This must be clear to each individual.
- Goals and objectives: It should be clear what is expected, how it should be done, and when it is due.
- Escalation: Indicate when a problem should require intervention, in other words, when they should ask for help.
- Approvals: Specify what approvals are required and who will be making the decision.

The roles for team members, and other stakeholders, can be formally listed in a linear responsibility chart (LRC) or "responsibilities list," shown in Fig. 5.34. The LRC is then used for activities such as labor estimation, budgeting, assigning tasks, generating schedules, and job costing. When compiling an LRC you will want to consider the following factors:

WBS Item (task)	Assigned Person	Deliverable	Authority	Approval
CAD work	Mary	Mechanical Drawings	CAD operator 1 day	Joe
CNC machining	Joe	Parts	6 hours of CNC	Customer
PCB assembly	Vlad	Tested Circuit Board	2 days of assem/test	

FIGURE 5.34

Linear responsibility chart. *CAD*, computer-aided design; *CNC*, computer numerical control; *PCB*, printed circuit board; *WBS*, work breakdown structure.

- Experience: Teams that have worked together before will understand responsibilities and may not require as much management.
- New: Inexperienced teams need to have responsibilities spelled out to avoid overlap and ignored tasks.
- Responsibility: Delegators must remember to keep authority and accountability together.

When responsibilities have not been clearly assigned and communicated there are a variety of problems that arise. Issues also occur because of personality types. Sometimes these issues can be detected and fixed before they become a problem, but it is better to plan ahead and never have the problems. The following list can be reviewed and compared with previous team experiences:

- Orphan: An important task is not assigned to anybody and everybody assumes somebody else is doing it.
- Frustrated: Somebody is asked to work on a task but does not have the authority or resources to complete the work.
- Personalities: People play political games with resources needed by a perceived enemy.
- Scapegoats: A manager keeps authority but passes accountability and/or blame to subordinates.
- Confidence: A team does not have confidence in the project, one another, or the managers.
- Communication: The team does not communicate within itself, with customers, or with managers.
- Personality approaches
  - When introverted individuals are encountered
    - (i) Encourage them to speak out.
    - (ii) Make team members aware of the introvert's message.
    - (iii) Use introverts to "check" extroverts.
  - Extroverted individuals
    - (i) Somebody is needed to slow them down and allow others time for thought.
    - (ii) Make extroverts listen to others, and repeat.
    - (iii) Ensure they don't dominate the group.
  - Fact-oriented people
    - (i) Need to be encouraged to think wildly.
    - (ii) Need to plan/set goals before starting.
  - Concept-oriented people
    - (i) Need to be grounded in detail and should be asked to explain in detail.
    - (ii) Need to be kept on track.

- Encourage team members
  - Assess—Know personal motives and resolve with the team.
  - Goals—Understand objectives and goals and compare personal actions to team goals.
  - Communicate—These individuals communicate often and feel free to speak their mind.
  - Together—The team picks goals, objectives, and outcomes and then acts together.
- Discourage negative behaviors
  - Selfish—They do not care about the team or other team members. They put their needs first.
  - Vague—They are not clear about what they are doing and why they are doing it.
  - Resentful—They have negative feelings about other team members.
  - Anger—Personal anger is used as a regular tool in interactions.

Once a team is formed and starts to work together for the first time, there are a number of generally accepted stages. In the forming, storming, and norming stages, there is a period of adjustment and learning about the others. Eventually the team begins truly performing. As the team is redistributed at the end of the process there may be some feelings of loss. These stages can be used for predicting some of the social dynamics in a newly assembled team:

- Forming: Bring the team together.
- Storming: The people on the team interact sometimes unpleasantly.
- Norming: A social structure emerges.
- Performing: The team is productive.
- Adjourning: The realization of the end of the project.

PROBLEMS

5.83 Describe 5 approaches a manager might use to improve the effectiveness of a design team.
5.84 What are three advantages of an LRC?
5.85 You notice that in a team of 10 engineers, two talk 90% of the time and four never talk. What is probably happening?
5.86 Consider the forming-to-adjourning team model. What can an engineering manager do to help a team through the first three stages of forming, storming, and norming?

5.6 Ethics

Morals are personal beliefs, quite often related to social and religious norms. Ethics are a set of professional rules that preserve the trust for a profession and maintain legal boundaries. For example, you may consider it immoral to drink alcohol. However, if you work for a company that manufactures alcohol, your ethical duties are not to discourage consumption but instead to make the product safe for consumption. Another example is a patient who has done something the doctor considers morally repugnant; the patient can expect a high level of care from the doctor because of the ethical obligations. As professionals grow they will develop a clear distinction between their personal morals and professional ethical decisions, as illustrated in Fig. 5.35. By definition every profession must follow a code of ethics. When a professional has a conflict between morals and ethics he or she is obliged to choose ethics. In cases in which these differences cannot be resolved, engineers will often change employers.

The basic principles of ethics are a remarkably simple priority of obligations. For most professions these are (1) humanity, (2) state (law and government), (3) profession, (4) employer, and (5) you. Yes, your personal

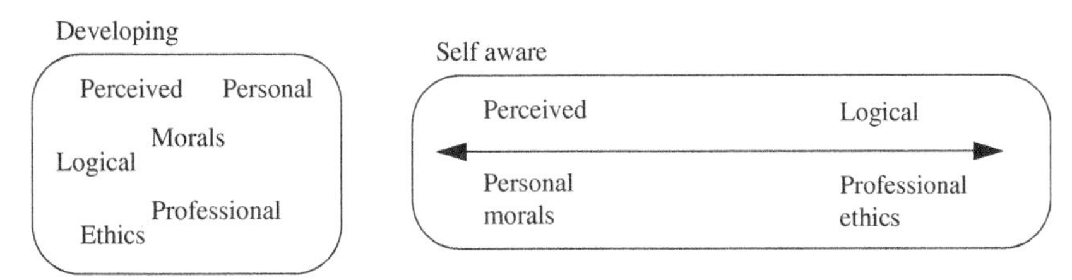

FIGURE 5.35

Impartial decision-making.

needs morals come last. We need to follow the ethical codes because we will do work that is not easily understood. People trust us to "do no harm." As professional engineers, we need to encourage that trust for our work. When we are trusted we can do so much more. Some of the basic ethical principles are listed as follows:

- Be fair, impartial, and loyal.
- Put public safety ahead of all other responsibilities.
- Honor integrity, courtesy, and good faith.
- Disclose other concerns that might influence your decisions at work.
- Only do technical work and express technical opinions that are well supported.
- Build trust.
- Observe and protect secrets.
- Be aware of the hierarchy of duty.
- Do not ask others to do things you would not do for ethical reasons.
- Cooperate with other professionals when working on common projects.
- People are individuals not defined by culture, gender, race, disability, history, or personal preferences.
- Be accountable for your work.

Codes of ethics are often voluntary for professionals; however, in extreme circumstances violating the terms can result in loss of license, loss of employment, criminal charges, and civil lawsuits. In practice, many ethical issues are hard to identify, subjective in interpretation, and difficult to resolve. Some common types of ethical lapses are:

- negligence;
- failure to consider the safety and well-being of the ultimate user or consumer;
- failure to correct a situation dangerous to the public;
- failure to follow guidelines, codes, and standards;
- certifying work without verifying the content;
- failure to point out potential problems if directions are not followed;
- doing work you are not qualified to do;
- not disclosing conflicts of interest when you might gain personal profits from professional decisions;
- not acting in a "respectable" way;
- not disclosing information when requested officially.

Ethical cases in which the problem is open to interpretation are called ethical dilemmas. The following lists some common issues. For each of these cases it is up to the professional to assess the possible outcomes of actions or inaction:

- The worst of two evils: For example, a factory is polluting the local village, but to close the facility would eliminate 50% of the regional employment.
- Ethical or moral: For example, would you be able to select one member of your family to send away so that the others would be allowed to remain?
- Suspected hazard: Some problems are suspected but not certain. Is the suspicion enough to act on?
- Whistleblowing: An employer is violating public trust and you must observe a higher-priority loyalty to the public.
- Conflict of interest: Impartial decisions are tainted by personal interest. In these cases you have something to gain personally when you are trusted to make an impartial decision.
- Serving two masters: When you are acting as an intermediary between two parties in a dispute. Care must not be taken to lean toward the party that hired you.

A conflict of interest is probably the most common ethical issue. In these cases you may have some personal gain when making decisions that should be impartial. A simple case is that a sales representative often buys you lunch after a sale. It is possible that you may be purchasing from that supplier because of the lunch. Now imagine a trip to a "training session" at a tropical resort. In these cases you should refrain from making those decisions or disclose fully all of the potential benefits you may gain. Note that even if you do not benefit, the appearance of conflict of interest is unethical, too.

Whistleblowing is a particularly good example. Engineers are often privy to confidential technical information about products, processes, and so on, and, from time to time, will become aware of dangers to an unsuspecting public. Sometimes these are clear, other times not. In these circumstances the individuals are obliged to act as an advocate and attempt to remedy the situation. If attempts to remedy the problem fail, then they are required to report these problems (whistle blow), regardless of personal considerations. Simply leaking information without taking appropriate steps first is not ethical. These cases do carry great personal risk for professionals, who will often lose their employment. However, in these cases the whistleblower is often able to recover personal losses with a lawsuit. The basic steps to be followed in the whistleblowing process are:

(1) detection of a problem;
(2) investigation of the problem to form sound technical opinions;
(3) attempt to remedy the problem with employer/client;
(4) if a "standoff" occurs, contact of the appropriate government body to begin the whistleblowing procedure.

Commercial transactions are not bound to the same ethical rules as professionals but there are a number of fundamental expectations. At times these can be modified contractually, but in those cases the issues are clearly written and require mutual agreement. When considering the following list, think of it from the perspective of a customer, supplier, employee, or manager:

- expectation of good estimates of cost, time, and results;
- frequent, honest, complete communication;
- important details and agreements in writing;
- disclosure of problems and issues as soon as possible;

- fair value for fair pay;
- confidentiality;
- deliverables that meet agreed specifications;
- value of the work and time of others;
- timely payments and billing;
- respect and trust that others will do their work;
- listening to others for input and concerns.

Codes of ethics are common in companies. The motivation for these codes can be the result of proactive management decisions in response to negative incidents, legal requirements, customer requirements, supplier requirements, or public requests. Regardless of the motivation, they are useful guidelines to preserve the integrity of the company. Common elements of corporate ethics policies are:

- worker/workplace health and safety;
- safety of the consumers and general public;
- conflicts of interest, including gifts and ownership of company stock;
- bribing officials;
- insider information;
- workforce diversity;
- equal opportunity;
- privacy and confidentiality;
- bias and harassment based on gender, sexual orientation, lifestyle, race, origin, military service, etc.;
- conflict resolution;
- impartial and independent review;
- environmental protection;
- exploitation of workers by international suppliers.

Note that engineers are ethically bound to observe the laws of other countries in business; e.g., bribes outside the country can lead to criminal charges.

PROBLEMS

5.87 Develop a list of five conflict-of-interest situations and indicate the level of severity.

5.88 Is the death penalty for murder an ethical or moral issue? Explain.

5.89 Ethics define a hierarchy of responsibility. For engineering, the general public comes first. Who comes first in medical ethics?

5.90 Given that nuclear weapons can be used to hurt people, is it ethical for an engineer to participate in design work?

5.91 What ethical problem is associated with accepting lunch from a sales representative?

5.92 In whistleblowing, the safety of the public overrides the obligation to the employer. When should an employer expect confidentiality?

5.93 Should you tell your employer if you are doing consulting work in your free time? Explain.

5.7 Professionalism

Do it right the first time and then move forward.

Engineering is a profession with a mission of protecting the public trust. This means that engineers must be competent, effective, and ethical. Competence is based on a good education in engineering, followed with practical experience. Effectiveness is determined by a set of skills that help us get things done correctly. Ethics help the general public trust our decisions by knowing that we put their safety ahead of everything else.

5.7.1 Time management

Plan your time as if you are planning for someone else.

There is always more work than time. You will never be able to do everything. The same is true for people you manage. The key to time management is to pick the most important tasks and do them first. Skill is required to objectively assess the relative and absolute importance of tasks. A good place to start is to consider the tasks before you accept and start them. It is always easier to say no and never start the task, than to accept it and do a poor job. If you accept a task make sure that you can do it successfully without a negative impact on other work. Ask the following questions for new work:

- Who will benefit from the task?
- What is the real outcome of the task?
- Where will I find the resources and time to do it?
- When does it need to be done?
- Why is it needed?
- How long will it take to complete?

The best way to manage your time is to create a rational plan to reach your goals and then follow it (Fig. 5.36). However, even with the best plans things change and unexpected events occur. Less-detailed plans will allow more time for task selection. A reasonable approach to task selection is shown in Fig. 5.37. An effective professional will set up a work environment that favors proactively working ahead. On the other hand, Fig. 5.38 shows a problematic approach to task selection that relies on reactive firefighting.

Time planning needs to be adapted to personal work styles. People are generally aware of their approach to doing work, but by continually assessing and adjusting your expectations and approaches you can become more effective. During project work, a log can be used to track your time. Please note that in many cases these logs are required for tracking job costs when employees work on multiple projects. At regular intervals it is useful to stop and review work, asking questions to reveal good and bad practices:

- What am I spending too much time on?
- What am I not spending enough time on?
- What could I do better?
- Are there other ways I can improve?
- Am I doing unnecessary tasks?

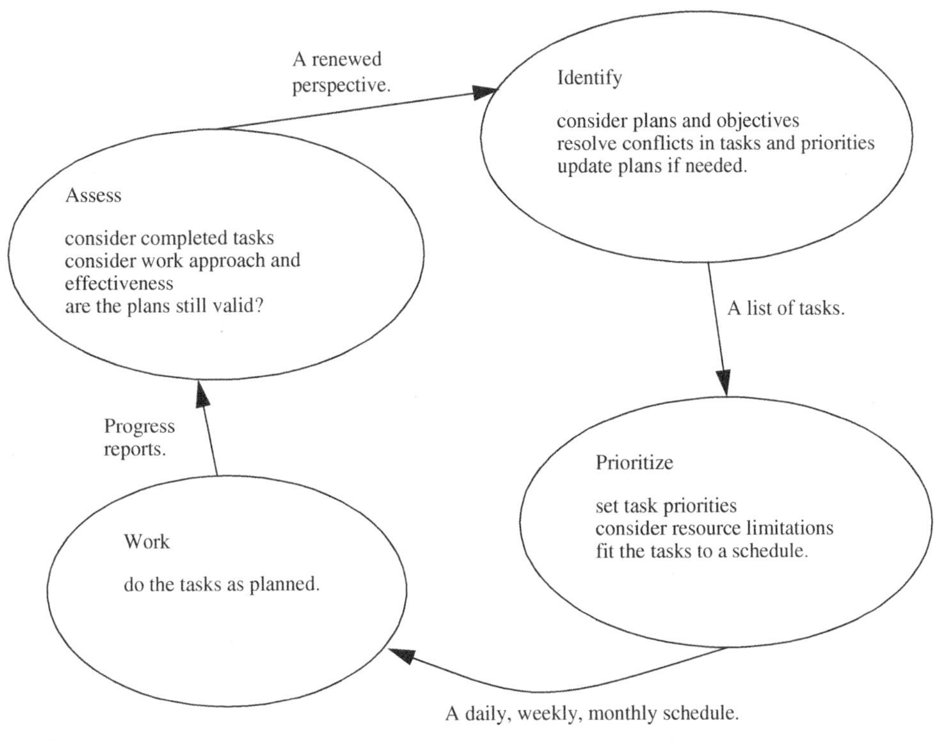

FIGURE 5.36

The time management cycle.

- Are there any tasks I could delegate?
- Am I wasting anybody else's time?
- Are there tools or methods that would help me?

When assessing your progress it is also helpful to consider how you are feeling. Obviously feeling good probably means that you don't have major time planning issues. Uncomfortable or bad feelings can often be indicators of problems. When using personal feelings for assessment, consider the following:

- Frustrated: the processes don't match your capabilities or expectations. This often occurs because:
  - items take longer than expected (poor time planning);
  - unexpected items come up (did not recognize risk);
  - results fail or are thrown out (the motivation or specifications for the work were not clear);
  - debugging isn't completed (small details and steps that are skipped or rushed become big problems later);
  - emotions are in the way (create emotional distance between yourself and the work decisions);
  - the to-do list is too long (consider eliminating, changing, or delegating).
- Lost: not sure what to do next or about the current situation:
  - not sure why you are doing things or what the value is (poorly understood goals and objectives);
  - too many things to choose between (poor planning and prioritization);

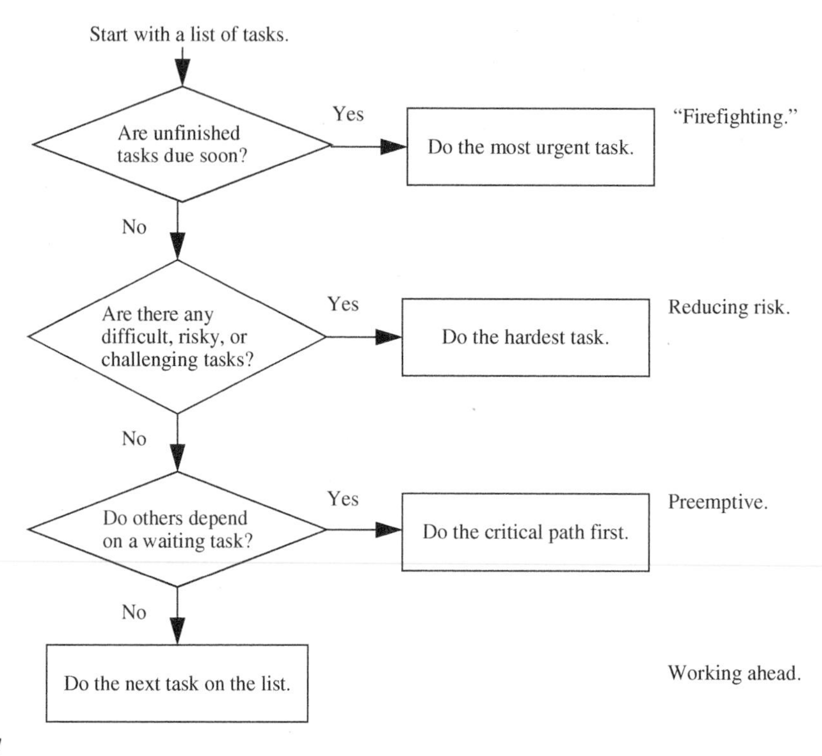
Start with a list of tasks.
Are unfinished tasks due soon?
Yes
Do the most urgent task.
"Firefighting."
No
Are there any difficult, risky, or challenging tasks?
Yes
Do the hardest task.
Reducing risk.
No
Do others depend on a waiting task?
Yes
Do the critical path first.
Preemptive.
No
Do the next task on the list.
Working ahead.

FIGURE 5.37

Prioritizing: "What could be done next?"—the good version.

    - overcommitted (taking tasks for wrong reasons without cost–benefit analysis);
    - overwhelmed by details (use plans and systems to organize tasks and details);
    - indecisive (review motivations for a decision);
    - worried about making mistakes (follow the plan).
- Rushed: always feeling like you are behind:
    - spending too much time on simple things (poor planning makes little problems into big ones);
    - last-minute work (procrastination, or delaying difficult tasks);
    - deadlines (working to deadlines instead of being proactive);
    - firefighting (waiting until problems need to be fixed, and then getting rewarded);
    - switching tasks often (switching tasks costs time; reducing switching will be more efficient);
    - squeaky wheel (do tasks in sequence and avoid pressure to change priorities);
    - weekends and nights are spent thinking of work (create some distance to refresh yourself and be more effective).
- Disappointed: things are not done the way you expect:
    - delegation (adjust expectations and work to get the best results—and trust).
- Unproductive:
    - too much socializing (schedule it with things such as breaks, lunches, and outings);
    - odd jobs (recognize tasks with no value);
    - interrupted often (set times to accept and block email, people, phone, etc.);

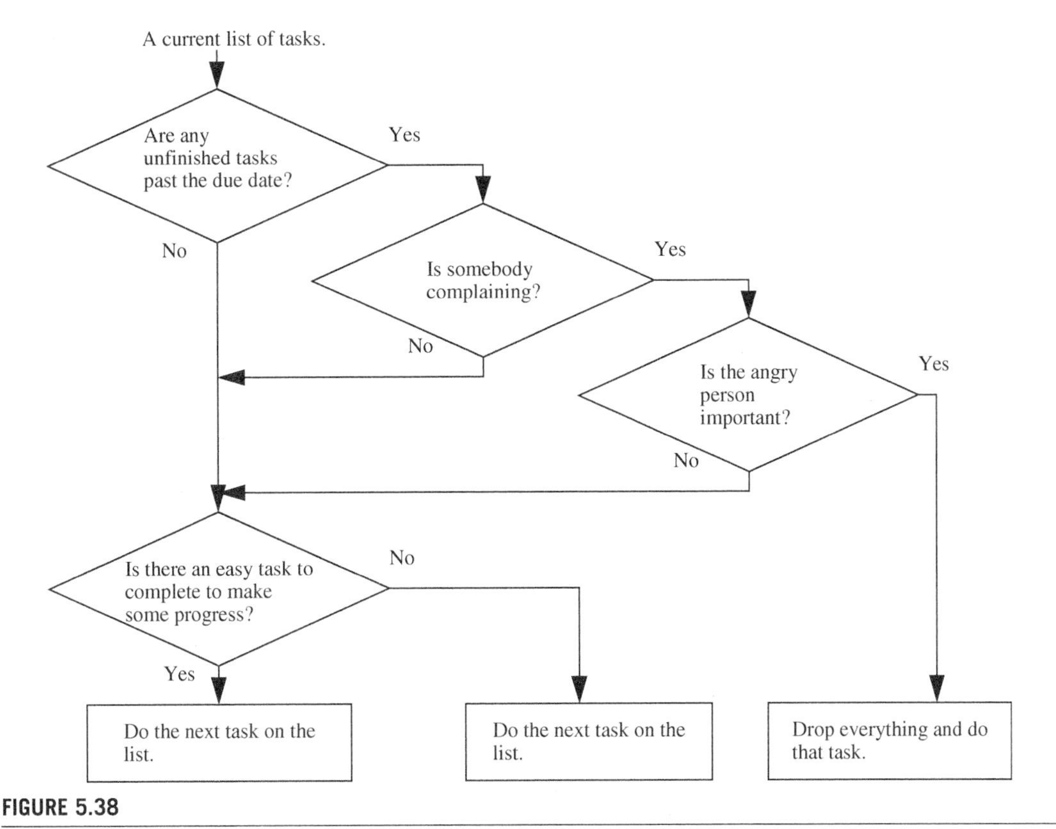
A current list of tasks.
Are any unfinished tasks past the due date?
Yes
No
Is somebody complaining?
Yes
No
Is the angry person important?
Yes
No
Is there an easy task to complete to make some progress?
No
Yes
Do the next task on the list.
Do the next task on the list.
Drop everything and do that task.

FIGURE 5.38

Prioritizing: "What must be done next?"—the bad version.

- overinformed (filter information);
- avoiding tasks (accept the value of the task, do not plan around it);
- saying "yes" (you agree to do more tasks than you can do well).

- Focused: focus can be very good or very bad:
  - no end in sight (review the overall progress and create logical subgoals);
  - blinders on (frequently check the big picture plans);
  - distracted (consider the value of the task, it may also be a lack of rest).

There are a few simple axioms that can be used to focus your scheduling efforts. If you can improve a few of these it will have a noticeable impact on your effectiveness and reduce the negative feelings about work:

- Focus
  - Learn to say "no."
  - Have clear motivations, objectives, and goals.
  - Eliminate uncertainty.
  - Paying attention to detail saves time later.

- Avoid waste
  - Focus on results instead of activity.
  - Train and trust others to reduce oversight.
  - Create blocks of uninterrupted time for working.
  - Question all meeting and communication times for value. Plan for separate social time.
- Prioritize
  - Develop a method to prioritize tasks.
  - Triage information and requests. Create a hidden pile for paper and email that is "not quite garbage."
  - Decide to do the hard things first.
- Plan instead of reacting
  - Plan in detail.
  - Pick a plan and stick to it (see "Say 'no'").
  - Avoid making decisions "on the fly."
  - Expect the unexpected, and keep some extra time as a contingency reserve.
  - Prioritize planning over firefighting.
  - Create systems and methods to organize messy tasks and information.
  - Delegate tasks to train others or for expertise.
- Be objective
  - Make good practices normal habits.
  - Learn from mistakes such as time misestimates.
  - Make yourself dispensable, then others can do your work easily too.
  - Accept that other people do things well also, just differently.
  - Develop a sense for when you are lost or locked in.
  - Question your decisions for rational versus emotional motivation.
  - Be prepared to give up on your ideas and decisions. It is not personal.
- Say "no"
  - Many people have great ideas that will require your time.
  - How often do you have a great idea?
  - Even if you like the idea, screen it.
  - If you decide to do something new, decide what you will give up.
  - Is it "more important" than other work you have right now?
  - How does it help other people who are depending on you?

PROBLEMS

5.94 How does the saying "put first things first" apply to time management?
5.95 What are three advantages and disadvantages of saying "no" to an optional request?
5.96 What is firefighting in task management?
5.97 List 10 strategies you could use if you are missing too many deadlines.
5.98 Another saying is "The squeaky wheel gets the grease." How could this apply to your work priorities?
5.99 List 3 factors critical in setting task priorities.

5.7.2 Being organized

Do the hard things first.

There is a huge quantity of information to process and it can easily consume a majority of your day. It comes through a number of pathways, including (1) email, (2) phone messages, (3) written requests, and (4) verbal requests. Some engineers and managers report hundreds of emails per day. When you consider that somebody can send an email to hundreds of people with little thought, the need to act on many emails is suspect. When handling email, and other messages, there are some tricks that are very helpful. The best is to use email filters. These can take incoming email and put them in folders or forward them by looking at the sender, topic, and content. Use an email folder for "spam" to reduce the volume of bulk commercial and unsolicited email. Other email can be sorted into other folders for specific jobs, urgent tasks, and others as appropriate. The key is to sort the email so that you can deal with one project at a time and can minimize task and job switching. A general approach to email triage is shown in Fig. 5.39. The key skills are to identify information that is not important and can be deleted or filed. When forwarding email, or acting on it yourself, consider that it will take time from other work. When asked to do tasks, don't be afraid to say "no" if it does not add value. Sort the other email so that you can deal with it at the right time.

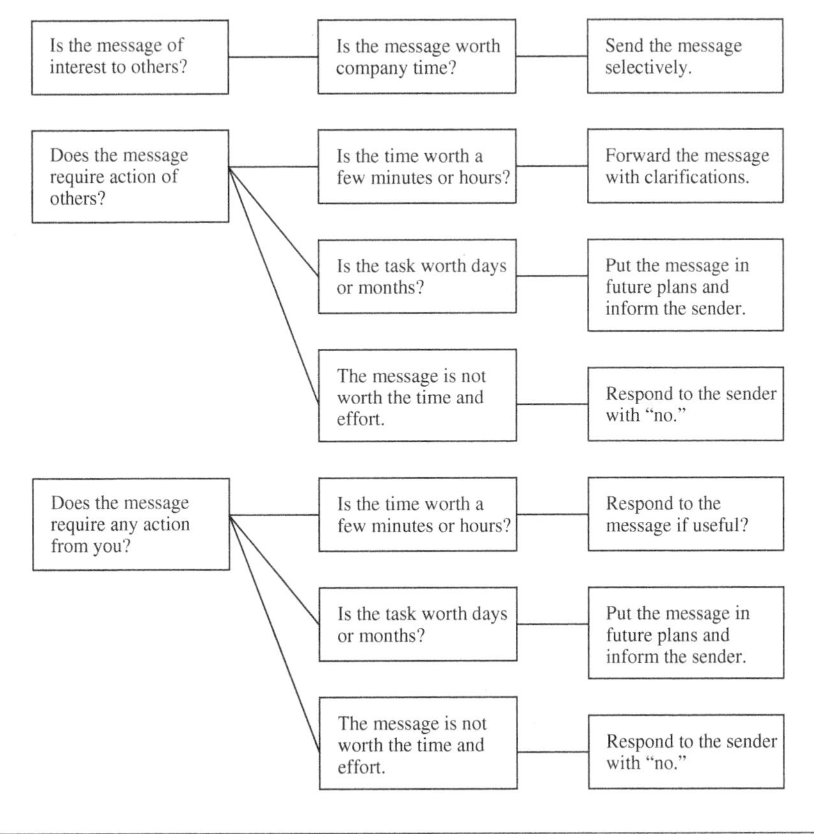

FIGURE 5.39

Sorting personal email, mail, and memos.

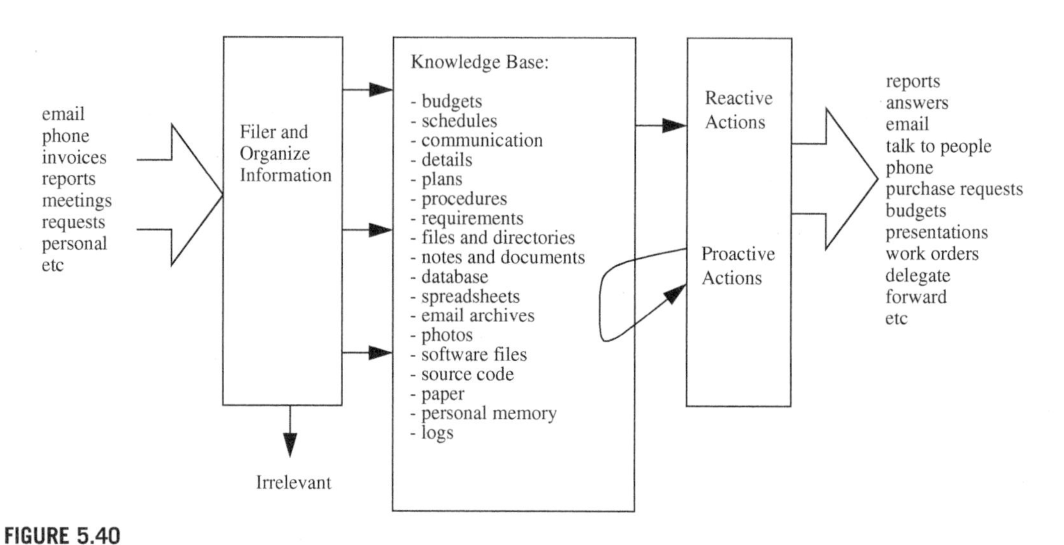

FIGURE 5.40

The information view of a project manager.

Each professional has a knowledge base that is a combination of technical, business, personal, project, customer, experience, and much more information (Fig. 5.40). As new information is processed it is compared with the knowledge base and decisions are made. Of course, the knowledge must lead to action. Actions can be proactively planned or be reactions to change and unexpected events. Some of this knowledge is written or stored in a computer. Other elements are known or understood by the professional. Consider an email from a customer who asked for a progress update and includes a new feature request. You read the email and split it into the two issues in your mind. For the progress request you look at the Gantt chart and see that the project stands at 78% and is 5 days ahead of schedule. Within 5 min, the response is written along with a note that says that you will discuss the feature change with the team. The requested change is added to the spreadsheet for team meetings. At this point it does not require any more thought until it comes up at the meeting. The item is raised at the team meeting and the group agrees that it would add 3 weeks and $20,000 to the budget. The response is made to the original email and the email is archived. In total the two items in the email were processed separately and in total took a few minutes of the manager's time. To continue the example, 2 weeks later a progress report is written for the customer, as dictated by the project plan, a proactive action.

Each professional will develop his or her own preferred set of tools for organizing. Some of the major tools are shown in the following list with some hints. Larger companies will often use groupware that offers a number of functions for sharing calendars, computer directories, contacts, email, and more. A forward-looking software system allows real-time sharing of document, spreadsheet, and presentation files. Spreadsheets have become a fantastic tool for organizing information. Project management software is available from a large number of suppliers and as an Internet service. These simplify the standard project management tasks:

- Email
  - Use filters and labels to sort email as it arrives.
  - Spam filters can be effective when trained.
  - Archive old email for records.

- Phone
  - Use texts for fast communication, even in meetings.
  - Voicemail keeps personal messages.
  - An Internet-connected phone can allow Internet access while traveling.
- Calendar
  - Use to store past events.
  - Useful to remind of future events and prevent conflicts.
  - Coordinate with others using shared calendars.
- To-do list
  - A list of regular tasks helps proactive approaches.
  - Useful to identify urgent/reactive tasks.
- Contacts
  - Enter contact details as people are met; enter all business cards.
  - Log information about a person.
  - Use contact logs (when, why, outcome).
  - Create contact lists, e.g., personal, internal, customers.
- Notes
  - Keep track of informal and formal decisions.
  - Provide a reminder of what was said and by whom.
  - Keep track of useful details you have found.
- Project tracking sheets
  - Track job expenses.
  - Track employee hours by job.
  - Track changes, major events, decisions, and test results.
- Writing space
  - Sticky notes.
  - White boards.
  - Magnets.

A pause to reflect and review the work can help reorganize tasks. One approach is to end each evening by concluding all tasks, even if it means putting them back on a to-do list. The first task every morning is to set the tasks for the day. Some questions to prompt the process are shown in the following list. Details can be found by reviewing the proactive and reactive needs for the day, the week, and the month as listed on to-do lists, on schedules, in messages/email, on calendars, and on other tracking tools. A reasonable to-do list is created for the day. Once the list is done, the plan is set and there should be great resistance to changing the list. A number of professionals will read email only once a day to avoid frequent work changes. As the day proceeds, the tasks are removed from the list. If done correctly this can seem like a little reward to give extra energy to reach the next goal. This approach is very well used by distance runners. Tasks that last longer periods can be divided into shorter tasks:

- What should I be doing today?
- What are my priorities?
- What decisions do I need to make?
- What should be done at the end of the day?
- Who should I work with today?
- Do I need to report results to somebody?
- What do I do if I run into trouble?
- How has my work been received?

When managing employees it is necessary to monitor and plan their work on a regular basis. We hope that they could organize their time as well as or better than us. When employees have difficulty organizing their work it can be useful to help them prioritize. If you find that you have to organize everybody, then you are micromanaging (not a good thing). While teaching junior employees how to manage their time, contact may be needed a couple of times per day.

PROBLEMS

5.100 Examine your email program or web service. Identify the method for sorting and archiving email. As a minimum, create groups for personal, school/work, and project emails. Tag or move emails in the inbox to these groups. Demonstrate the results with screen prints or personal demonstrations.

5.101 In your email program/service find the filter or autosort rules. Create rules that will sort emails with a keyword in the title, such as "project," and automatically move them to a project folder or archive. Send an email from another computer or person with the tag in the title. Print a screen capture or demonstrate the automatic sorting function.

5.102 Examine your email spam, inbox, and trash folders for the past week. Sort your emails into (a) irrelevant, (b) information, (c) brief task, and (d) long task.

5.7.3 Diversity

It is a simple truth that on a daily basis engineers will work with people different from themselves. These diverse groups of people vary in many ways, including culture, ethnicity, gender, philosophy, disability, desires, expectations, and much more. When we work best we assume that everybody is an individual and distinct. At our worst we take one or more of these traits to stereotype and guide how we interact with other people. A mild example is assuming that somebody who lives near the Louvre loves art, and then beginning every discussion with Impressionist painting. Another example would be to assume that all women are caring and want to mentor younger women. Yet another is to assume that older white males are naturally biased or that Asian women are not. In many cases people use these stereotypes with the best intentions but it often leads to problems. We are all different. The best way to work with others is to assume that their worldviews are the most important, and be flexible with yours. The worst way to get along is to assume your worldview is the most important and people should be flexible with you. Of course, the win-win approach will produce the best results. A very limited list of these differences is shown as follows, but can be easily expanded:

- Religion: an endless source of conflict, even within the same religion
- Politics: allegiances to political groups or philosophies
- Country: nationalism and patriotic opinions
- Culture: a distinct group
- Class: an inherited or entitled attribute
- Financial status: how much money you make
- Age: perceptions of ability and contributions
- Gender: gender differences and relationships

The paramount issue in diversity is stereotyping. This means that we make some observations or assumptions about a group and then assume that they apply to everybody from that group. This becomes a problem when we act on those assumptions. In a similar line of thought, everybody will respond differently to stereotyping. The whole thing becomes messy quickly. The safest strategy is to wait for a person to define his or her identity.

The following list shows a number of triggers that lead to problems. If you are going to a particular location in the world it can be valuable to purchase a guide to local customs and beliefs. However, if you are not sure, look for cues and learn quickly:

Behavior

- Forcefulness
- Gestures—hand signs and motions, pointing
- Looking—eye contact
- Physical—touching, personal space
- Food—there are many issues here
- Common sense—practices that have developed and are accepted as normal

Social

- Class—being aware of class such as upper class versus servants
- Dominant versus subservient
- Financial status
- Familiarity—use of names and titles
- History—previous events and understandings

Philosophy

- Religious beliefs (this is a very big one to the point of life threatening)
- Sexual beliefs
- Political—freedom, democracy, religion, socialism, legal systems
- Superstition—numbers, rituals, omens
- Property of others
- Morals and ethics

Physiology

- Age
- Gender
- Hygiene and cleanliness—shoes, washing, cough/sneeze/sick, etc.
- Physical differences in appearance or abilities

A particular note of value concerns humor. Jokes work because people find them reasonably abnormal.

Depending on the culture, things may seem completely normal (unfunny) or too unusual (absurd or offensive). So keep in mind that any type of joke or humor may not be received the way it was intended. It is sad to say, but the best approach is to avoid humor, but when somebody says something funny try to find the humor.

The following list gives a few examples of cultural issues for education purposes. You can try to guess the culture:

- Showing the bottom of your shoes, or touching them, is offensive.
- If your nose is running you should sniffle; using a tissue is unclean.
- The host must eat first.
- To eat all of your food says the cook did not make enough.
- If you do not eat all of the food it says that it was not good.
- You are expected to eat (messy food) with your hands.
- Using a person's given name is demeaning or overly personal.
- You must drink from a shared cup.
- You are expected to wear shoes in a house.
- You should agree even if you don't understand.

The English language has adopted content from many others. In turn, it has been adopted and adapted by many cultures. The major differences are accents, spelling, phrases, and colloquial terms. Formal English has two major variations, US and International, that are supported by spelling- and grammar-checking software. A few examples of language deviations include the following:

Spelling differences between US and International

- aluminum/aluminium
- center/centre
- color/colour
- pleaded/pled

Regionally offensive words

- Shag, bum, fanny, randy

PROBLEMS

5.103 What does "polite" mean?

5.104 The company has just hired a new employee, Matt, who has tattoos from the tip of his toes to his hairline. Some of the images include religious icons and nudity. A deeply religious engineer, Ezekiel, is offended and refuses to work with the new employee, citing the nudity. The new employee explained that the tattoos are part of his ethnic background. Provide three approaches the company could use.

5.105 Is bias caused by diversity? Explain.

5.106 A male employee has been the subject of many "blond" jokes and eventually yells back, "I am not defined by the color of my hair." What does he mean?

5.7.4 Entrepreneurship

A crisis means a change that creates both threat and opportunity.

After a few years of work experience some engineers decide to develop a new product, process, method, business idea, and/or company. Engineering knowledge and design project skills provide a good basis for becoming an entrepreneur. Those who follow this path often mention the reward of building a business, creating something unique, freedom to do what you like, and eventually financial rewards. This section will outline some of the business knowledge and anecdotes for first-time entrepreneurs.

The key to starting a business is opportunity. You will need to be able to do something that somebody else is willing to purchase. It is often better to focus on industrial or specialized needs, as they tend to be easier to develop, build, and sell at a reasonable profit. Consumer products can be very difficult to develop because they add the complexity of marketing, distribution, sales, service, and manufacturing. The best business ideas often come from problems that engineers observe at work. Moreover, many engineers launch businesses doing something that their former employer needs. Increase the chances of success by using what you know well to solve problems you know exist. Start with incremental solutions instead of trying to revolutionize markets.

After identifying an opportunity it is time to develop a business plan. As a minimum these plans outline the money required to complete the project, a timetable for the work, and payments. Better plans also include details of needed facilities, people, and equipment. Great plans will also consider risks and alternatives. For an engineer these can be seen as extended project plans with details added for business considerations. Business plans are used to find business partners and secure funding from some of the groups listed as follows. A business plan also helps the business maintain clarity through the early phases:

- bank loans less than $100K to start
- mortgages less than $100K
- angel funding less than $1M
- venture capital less than $20M
- small jobs

Once the business is running, it requires personal discipline to stay focused and do the work. There is no employer to encourage you, money will be in short supply, and you will need to do many jobs. However, following the business/project plan will be the best path. Fig. 5.41 shows some of the problems that cause entrepreneurs to stray from their plans and lead to failure. You may lose customers if

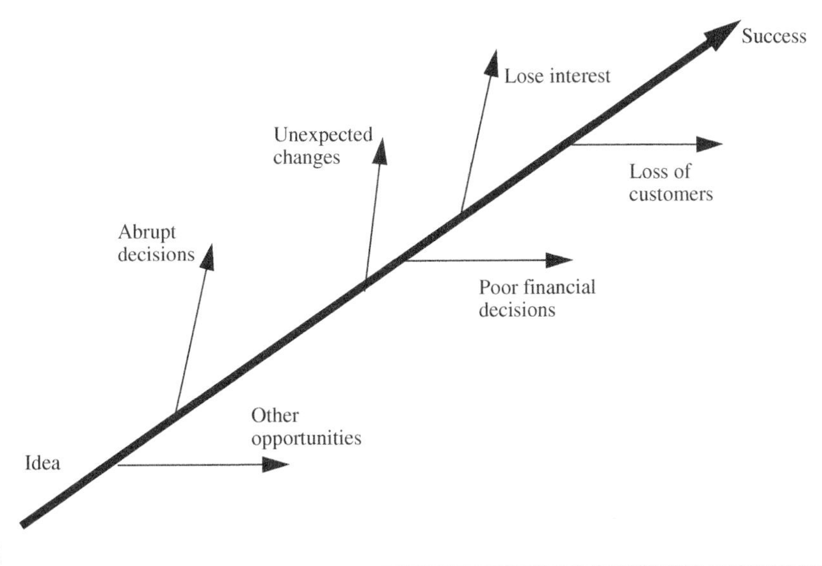

FIGURE 5.41

Entrepreneurial distractions.

deadlines are sparse or the work is below standards. Unexpected issues do arise, but a good plan should accommodate these variations. Making decisions using instinct leads to many problems and should always be avoided. Finally, other attractive opportunities will arise during the startup. Deciding to change the business direction may seem like a good idea but it puts the project back to the beginning. During this phase it is important to stick to the plan.

After the first few projects companies will begin to grow and the approach to business will change. There is a delicate balance involved in adding costs to increase the business capacity. Successful entrepreneurs will tell stories about buying used equipment, cleaning toilets themselves, getting friends to deliver products, having family members answer phones, and so on. Adding people and resources at a conservative rate, so that they can be trained and deployed effectively, will help build a solid foundation for future growth.

As companies grow, the management approach must also grow. Quite often the entrepreneurial skills begin to impede the company. Entrepreneurs love the daily operation of the business they built. It can be hard to distribute control and trust to other employees. Knowing when to let go is the last major step needed by entrepreneurs. Professionals who move beyond entrepreneurship often say that they learned to "hire talented people, give them what they need, and stand back and let them do their jobs."

Entrepreneurs should always be introspective to keep themselves on track. Some questions and thoughts are provided to prompt the progress assessment process:

- Creativity is fun but invention is 1% inspiration and 99% perspiration.
- Am I treating each day like a regular job?
- What can go wrong, and what can I do to prepare?
- Is the plan working?
- Am I on track for success?
- Do I really need to spend money? Can I buy something used? Can I do without? Can I improvise?
- Can I get similar results but spend less money?
- Am I confusing speculation (maybe) with opportunity (is)?
- What mistakes have I made, and what did I learn?
- Do I have a clear picture of where I am and what I need to get to my goal?
- Do I know I have a customer who will buy my product?
- Am I spending too much?
- Am I trying to be all things to all people?

One area of failure for a number of engineers is sales. It doesn't matter how great your product is if nobody knows it or wants to buy it. Successful entrepreneurs will find ways to mix with potential customers, suppliers, and allies. They will listen to find out what others are doing. They will let others know what they are doing.

When possible they build new personal and business relationships. Networking is a name for making new contacts through people you already know. Being able to build a supportive network is incredibly important in business. This is well beyond the scope of this book, but a good place to begin exploring these relationships is in the Carnegie books. Customers found through networking will have some referential trust and are much more likely to lead to sales. Sometimes your network will lead customers to you. The alternatives to networking are advertising and cold calls. Advertising exposes a wider audience to your products, with substantial expenses for media print, trade shows, television, radio, and so on. Cold calls are the worst case, in which you phone or visit a customer who does not know you. Regardless of the path to the customer you must provide solutions they want or need at a cost that satisfies everybody. If you can keep your customers satisfied they will return and may even refer other customers.

When you succeed it will be very rewarding. When you fail, learn from your mistakes: fail constructively.

The following lists a few of the items to consider. Knowledge and support for creating small businesses are available in many forms. Consider free government consulting, paid services, business associations (e.g., chambers of commerce), consultants, professional societies, and so on.

Motivation and failure

- Most new companies fail.
- Success sometimes takes a few times.
- Very few patents lead to products.
- Things change.
- Persistence!
- You will be poor for a while.
- It often takes years.
- Save your money until it is needed.
- Be poor before you start to make money.
- Feast and famine.

Factors in success

- Persistence and enthusiasm
- First things first: value-added only
- Do what you do well, and the hard things first
- Pragmatic
- Being frugal
- Productive mistakes: reasonable risk
- Manage your employee: you
- Firm objectives: flexible goals
- A clear customer

Factors in failure

- Spending freely on secondary needs
- Following unrelated opportunities
- Spending time on the unessential
- Hiring staff too soon
- Not recognizing weaknesses
- Becoming paralyzed by fear
- Valuing "firefighting"
- Failure to plan
- Trying to be everything to everybody
- Adopt the not-invented-here philosophy
- All inspiration, no perspiration

Items to remember

- Permits and licenses (e.g., business)
- Regulation—OSHA, FAA, FDA
- Zoning
- Employee laws
- Taxes
- Accounting
- Health insurance

PROBLEMS

5.107 Define entrepreneur.
5.108 List five places to find opportunities.
5.109 How much value do brilliant ideas have? Explain.
5.110 Consider the saying "You need to spend money to earn money." Does this mean that you will earn more if you spend an extra $10,000,000? Explain.
5.111 What will probably happen to an entrepreneur without focus?
5.112 What should an entrepreneur do if his or her business fails?
5.113 Given all of the details and risks, why would anybody want to be an entrepreneur?

5.7.5 A professional image

Focus on what you do well, compensate for the rest.

Simply put, a professional is someone who makes a contribution and helps others do the same. Each workplace is different. Understanding the variations can be critical in first impressions and long-term success. Some general pointers can be of particular help:

- Dress: Dressing appropriately can be key. Some places are extremely casual, others very formal. And, expectations will change for special events and customer visits. Many professionals will keep an extra set of dress clothes at work, just in case.
- Attendance: There are a variety of options here. In all of them you need to let people know where you are and what you are doing. The most relaxed environment is off-site work, from home or on the road. Many companies allow flexible schedules as long as the required work is performed. Most companies have strict on-site work hours.
- Reporting: Know who you should talk to and how often. A good situation will have you talk to your supervisor and colleagues on a daily basis as need requires. The extremes are hands-off (watch out for problems) or overbearing micromanaging (confidence issues).
- Seniority: The higher your position in an organization the more you are expected to set an example.
- Deportment: Be polite. Avoid inappropriate topics.

Some, sadly true, anecdotes are given in the following list. Please do not attempt any of these at work. And, if you see somebody else on this track, try to help them break the pattern and succeed:

- Sleeping in: always being a few minutes late for work gets noticed quickly.
- Asleep at work: one case found sleeping under a desk; another, they got pictures.

- Don't show up: staying home sick without calling in.

Some additional notes of value:

- Rushing past simple design details and taking shortcuts often leads to time-consuming problems.
- Debugging always takes longer than expected.
- Things that are "left for later" never get easier or go away by themselves.
- Extra time spent on design saves even more on debugging.
- If you want to do something, the time and cost estimates will be optimistically low. If you don't want to do something, they will be pessimistically high.
- For everything you decide to do, there is something else you must stop doing.
- Always prefer written agreements; they take priority over verbal agreements and are easier to prove.

PROBLEMS

5.114 What are three professional risks when working from home?

5.115 Professionals often cite "learning to say no" as one of the most professional lessons. Why is this important?

5.116 A professional image should inspire trust. Provide three examples of professional behavior that could decrease trust.

5.117 If an employee is having marital problems, what is the highest level of performance that should be expected? Use Maslow's model to explain.

5.118 What are the differences between ethics and morals?

5.119 Engineers hold many company secrets. When engineers change employers they take that knowledge with them. Are they allowed to use it?

5.120 What is the likely outcome if you praise an engineer in a meeting? What is the likely outcome if you criticize an engineer in a meeting? Use Maslow's hierarchy to justify your answer.

5.121 If a company fires an engineer, does the employee have to observe a nondisclosure agreement (NDA)?

5.122 What are the advantages and disadvantages of critical feedback?

5.123 Responsibility is a combination of authority and accountability. Explain what happens when somebody has only a) accountability or b) authority.

5.124 If an employee tells you that he or she feels like he or she is micromanaged, what is the likely problem?

5.125 What are the key elements to mention in a personal introduction?

5.126 As a manager you need to introduce customers to your employees, peers, and managers. Develop a single-sentence introduction that describes your engineer Joe to the customer if:

(a) Joe will be helping you;

(b) Joe will be making decisions about the client;

(c) Joe will not be working with the client.

5.127 Self-assessment:

(a) Pick a partner.

(b) Assess your partner using the categories listed. Each category should receive a score from 1 to 4. The total score must not exceed 10. Do not show your partner the score while you are developing the numbers.

(c) Write down similar numbers for yourself.

(d) Talk to your partner and compare the rating numbers.

For each skill rate your knowledge using 1 = nothing, 2 = some, 3 = proficient, 4 = knowledgeable, 5 = expert

Practical Skills
design
manufacturing and mechanical
electrical, electronics, and software
laboratory work
general computer use
writing
others ________

Theoretical Skills
design
manufacturing and mechanical
electrical, electronics, and software
mathematical problem solving
materials and chemical
others ________

People and Business Skills
teamwork
leadership
finance and accounting
administrative knowledge and office procedures
legal knowledge and skills
human resources
communications and presentations
others ___________

5.128 Personality typing:

(a) Find an Internet-based personality rating tool. There are many free examples available.

(b) Have each team member take the test separately.

(c) Compile the numerical scores for the team.

(d) As a group decide if all of the scores are reasonable.

References

Carnegie, D., 1981. How to Win Friends and Influence People. Simon-Schuster.
Ullman, D.G., 1997. The Mechanical Design Process. McGraw-Hill.

Further reading

Bryan, W.J. (Ed.), 1996. Management Skills Handbook; A Practical, Comprehensive Guide to Management. American Society of Mechanical Engineers, Region V.
Cagle, R.B., 2003. Blueprint for Project Recovery—A Project Management Guide. AMACON Books.
Carnegie, D., 1981. How to Win Friends and Influence People. Simon-Schuster.
Charvat, J., 2002. Project Management Nation: Tools, Techniques, and Goals for the New and Practicing IT Project Manager. Wiley.
Davidson, J., 2000. 10 Minute Guide to Project Management. Alpha.
Fioravanti, F., 2006. Skills for Managing Rapidly Changing IT Projects. IRM Press.
Gido, J., Clements, J.P., 2003. Successful Project Management, second ed. Thompson South-Western.
Goldratt, E.M., Cox, J., 1984. The Goal: A Process of Ongoing Improvement. North River Press.
Heldman, K., 2009. PMP: Project Management Professional Exam Study Guide, second ed. Sybex.
Heerkens, G.R., 2002. Project Management. McGraw-Hill.
Kerzner, H., 2000. Applied Project Management: Best Practices on Implementation. John Wiley and Sons.
Lewis, J.P., 1995. Fundamentals of Project Management. AMACON.
Marchewka, J.T., 2002. Information Technology Project Management: Providing Measurable Organizational Value. Wiley.
McCormack, M., 1984. What They Don't Teach You at Harvard Business School. Bantam.
Nelson, B., Economy, P., 2005. The Management Bible. Wiley.
Portney, S.E., 2007. Project Management for Dummies, second ed. Wiley.
Professional Engineers Ontario Code of Ethics, Section 77 of the O. Reg. 941.
Rothman, J., 2007. Manage it! Your Guide to Modern Pragmatic Project Management. The Pragmatic Bookshelf.
Salliers, R.D., Leidner, D.E., 2003. Strategic Information Management: Challenges and Strategies in Managing Information Systems. Butterworth-Heinemann.
Tinnirello, P.C., 2002. New Directions in Project Management. CRC Press.
Verzuh, E., 2003. The Portable MBA in Project Management. Wiley.
Verzuh, E., 2005. The Fast Forward MBA in Project Management, second ed. John Wiley and Sons.
Wysocki, R.K., 2004. Project Management Process Improvement. Artech House.

CHAPTER

Decision-making

6

6.1 Introduction

Habit: Set priorities to deal with high risk first.

Decision-making skills are important in the following situations:

- You want somebody else to make a decision.
- You need to make a decision.
- You are supporting a decision-making process.
- You are creating a decision-making process.
- It is possible to argue that everything we do involves making decisions.

The process of making decisions involves (1) establishing the decision context and motivation, (2) generating and identifying good alternatives, (3) comparing objectively, and (4) making a strategic decision. To make a decision you must begin with the following listed elements to set the context and motivation for the choice. Fig. 6.1 illustrates the basic approach whereby infeasible solutions are eliminated and the remaining options are ranked:

(1) A model of success (mission and vision)
(2) Plans to achieve it (goals and objectives)
(3) A clear business decision to make (problem identification)
(4) Background information for the decision (critical analysis)
(5) Motivation to make the decision

PROBLEM

6.1 Can a decision be made when there is only one option?

Engineering Design, Planning, and Management. https://doi.org/10.1016/B978-0-12-821055-0.00006-2

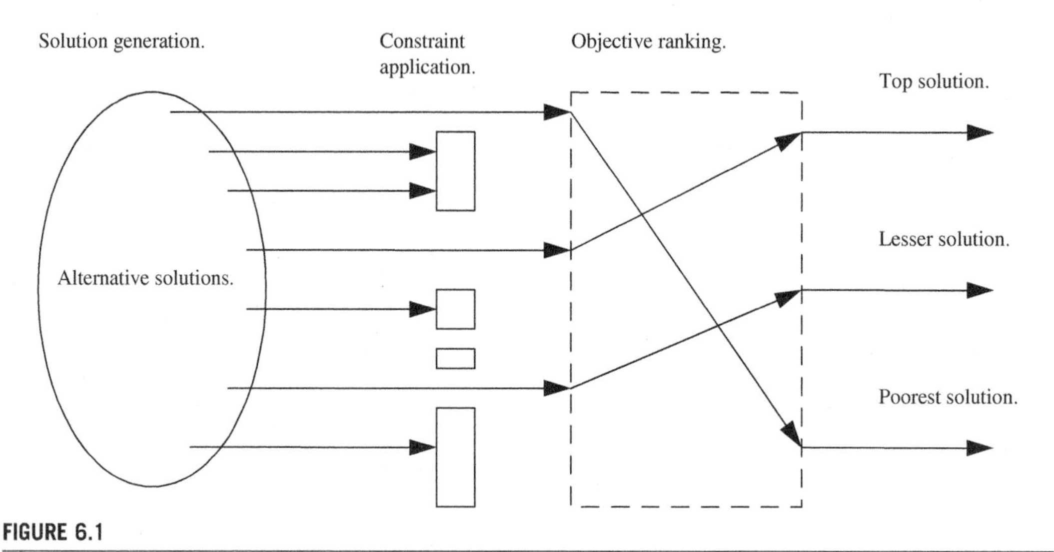

FIGURE 6.1

Making decisions.

6.2 Critical thinking

Fig. 6.2 shows a reasonable way to process a decision. When things do not follow this script, you can adapt, but try to answer all of the questions. Be aware when you are making assumptions. If things feel unclear, ask questions. If you don't know enough, you can always decline to decide.

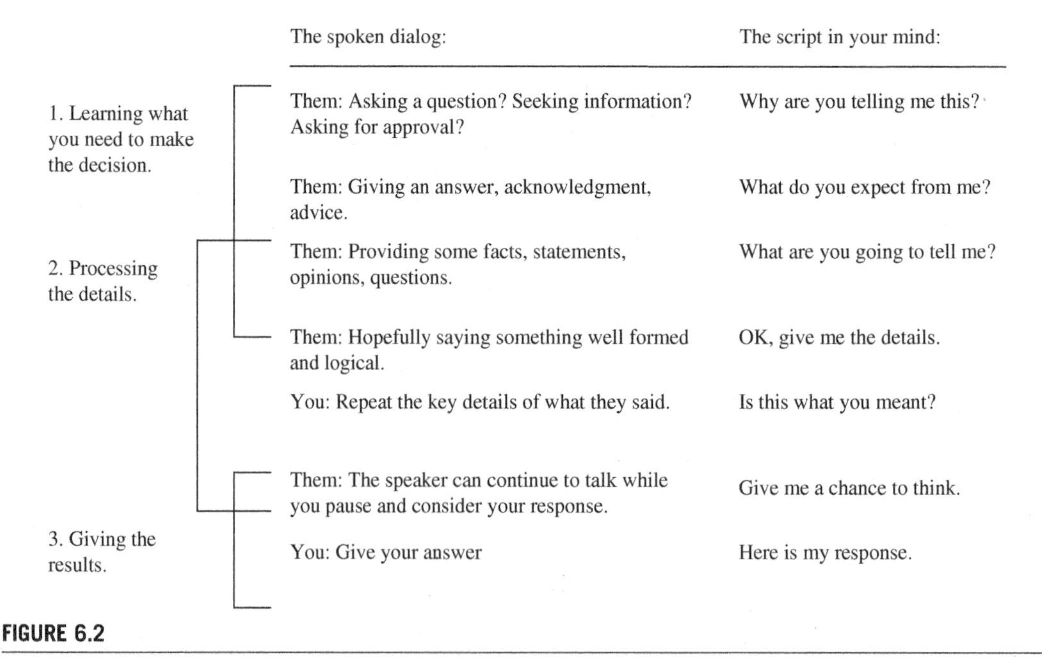

FIGURE 6.2

Listening to make decisions.

PROBLEM

6.2 Could decision-making be summarized as listen, think, then decide? Explain.

6.3 Critical analysis

Look for the low-hanging fruit.

Being "critical" has a negative connotation. But, critical thinking is a process that is essential for processing information. Simply accepting everything as correct and complete is naive. Practically, even the best knowledge has shortcomings; this does not imply any judgment. A critical thinker will consider knowledge and recognize the strengths and weaknesses. He or she will then be able to act effectively. A critical thinker will attempt to fit pieces of knowledge into categories. It can be hard to make decisions when things are unknown or uncertain, but a failure to make decisions may harm a project. When the information required for making project decisions is incomplete or less than perfect, you must understand the issues. Some of the classical categories used when classifying are listed in Fig. 6.3. Knowledge can be crudely categorized as shown in the figure. When listening to information it is worth sorting it into levels of completeness. General models are useful for parsing and understanding information. Information is many small facts that we connect with a framework called knowledge. The relationships are used when applying the knowledge to other problems. The relationships help match the information to the model and then allow you to use it to fill in the missing pieces.

You should do all analysis impartially and avoid a sense of personal ownership of ideas. One approach is to assume that the decisions, and justifications, were developed by another person. Regardless of the source, you must consider alternatives and the possible outcomes. The choices in the following list should be used for each of the alternatives before selection:

- Yes: It is acceptable in the present form.
- Maybe: Additional modification or justification is needed.
- Conditional: It is acceptable with limits or constraints.
- Alternate: Replace it with an alternative.
- No: It is not acceptable.
- Defer: You cannot make a decision.
- Partial: Part, but not all, of the solution is rejected.

To do an analysis, you need to absorb information, organize it into some sort of logical structure, and then compare it with what you understand. Fig. 6.4 shows the general cycle.

When doing analysis, you should understand enough to categorize the information and be prepared to make decisions using it. If you do not understand it (positively or negatively) you are not ready to make a decision. The simple sequence in Fig. 6.5 shows a possible logical sequence for analyzing technical recommendations. In basic terms, look for a logical progression, from understanding the problem to the approach and the final solution. To use this model, you listen or read, and then fit what you know to the sequence. At the end you review to make sure that each of the pieces is accurate and supported by previous steps. If things are not clear, ask questions.

Reject it if you cannot understand it or you disagree with key steps. Accept it if you understand it and agree with all of the steps. Occasionally you will disagree with a recommendation but decide to

Completeness:
(Best) General Model - A collection of laws and theories that can be used to develop new knowledge (extrapolation).
Specific Model - A collection of laws and theories that explain limited knowledge (interpolation).
Theories - A way of explaining the observations and anecdotes.
Observations and Anecdotes - A collection of stories and ideas (data points).
Speculation - Crude ideas of what might be true.
(poorest) Unknown - The issues and implications are not clear.

Rigor:
Calculated - The calculations use exact techniques subject to approximations.
Conclusion - An outcome of a logical process.
Measured - Experimental or other data with quantifiable errors.
Assumption - A educated guess that may not be completely correct.
Arbitrary - Like a coin toss to pick something.

Accuracy:
Veracity - The information is dependable and has been tested over time, comes from reliable sources.
Factual - Details.
Dependable - The source is trusted and is aware of information quality, peer reviewed.
Second-hand - As information is retold it loses accuracy.
Omissions - The information is incomplete because of lack of knowledge or intentional bias.
Untested or Unknown - Interpretations, unverified sources, nonexpert opinions, the result of interpretations or transcriptions.

Awareness:
Reliable or Unreliable - The level of certainty about the truth.
Quantitative or Qualitative - A recognition of numbers versus attribute.
Objective or Subjective - Impartial or perceived interpretations or measurements.

Objectivity:
Circumstantial or Anecdotal - Specific information is presented as a general case.
Biased - Strong emotional content, values (good and bad) are over-represented.
Suspect - Unfounded opinions, not aligned with conventional wisdom, the source seems unsure or non committal.
Speculation - Ideas that have not been tested or screened thoroughly.

Predictable:
Functional - Pieces that interact, resulting in a more complex function.
Transformative - A set of processes alter the basic form.
Procedural or Scripted - A series of steps.
Chronological - Events that happen over time (e.g., seasons).
Association - Events are related.
Analogy - A model of something known is used to draw conclusions.

Common Sense:
Logical - Everything has a clear dependence on other items.
Spatial - Physical presence.
Dependent - One thing leads to another, but it is not reversible. Also known as cause-and-effect.
Zero-Sum or Closed System - All that goes into something, must come out, even if in a new form.

FIGURE 6.3

Seven filters for critical thinking.

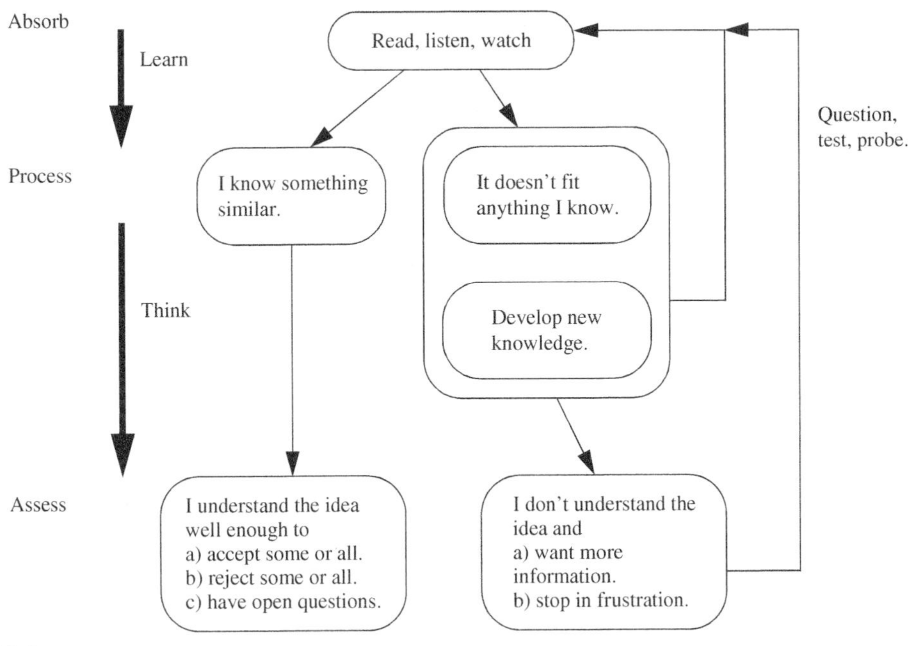

FIGURE 6.4

Knowledge assessment in critical thinking.

support it; do it knowledgeably. Presenters who are new to the process or who have had questions will normally start with the pessimistic process. As trust is built the reviewers will tend to review optimistically.

Analysis is an inexact process and requires that you quantify your results. The following categories provide some general benchmarks for each element analyzed. Some analysis steps will be clear, and obvious, for the decision-making process. Information that is incomplete should be used with suspicion. If the analysis is incomplete you should recognize the lack of certainty and consider that while making decisions:

- Not important
  - The problem does not add anything to the final project outcome.
  - There are better and clearer alternatives.
- Incomplete
  - The problem goals are phrased in words having vague terminology.
  - It is not clear what the benefit will be.
  - Rule of thumb: "You cannot visualize the solution."
- Clear
  - The problem can be described clearly with numbers, figures, lists, and testable results.
  - The success of the solution is testable.
  - Acceptability criteria are clearly defined.
  - There is nothing "left for later."

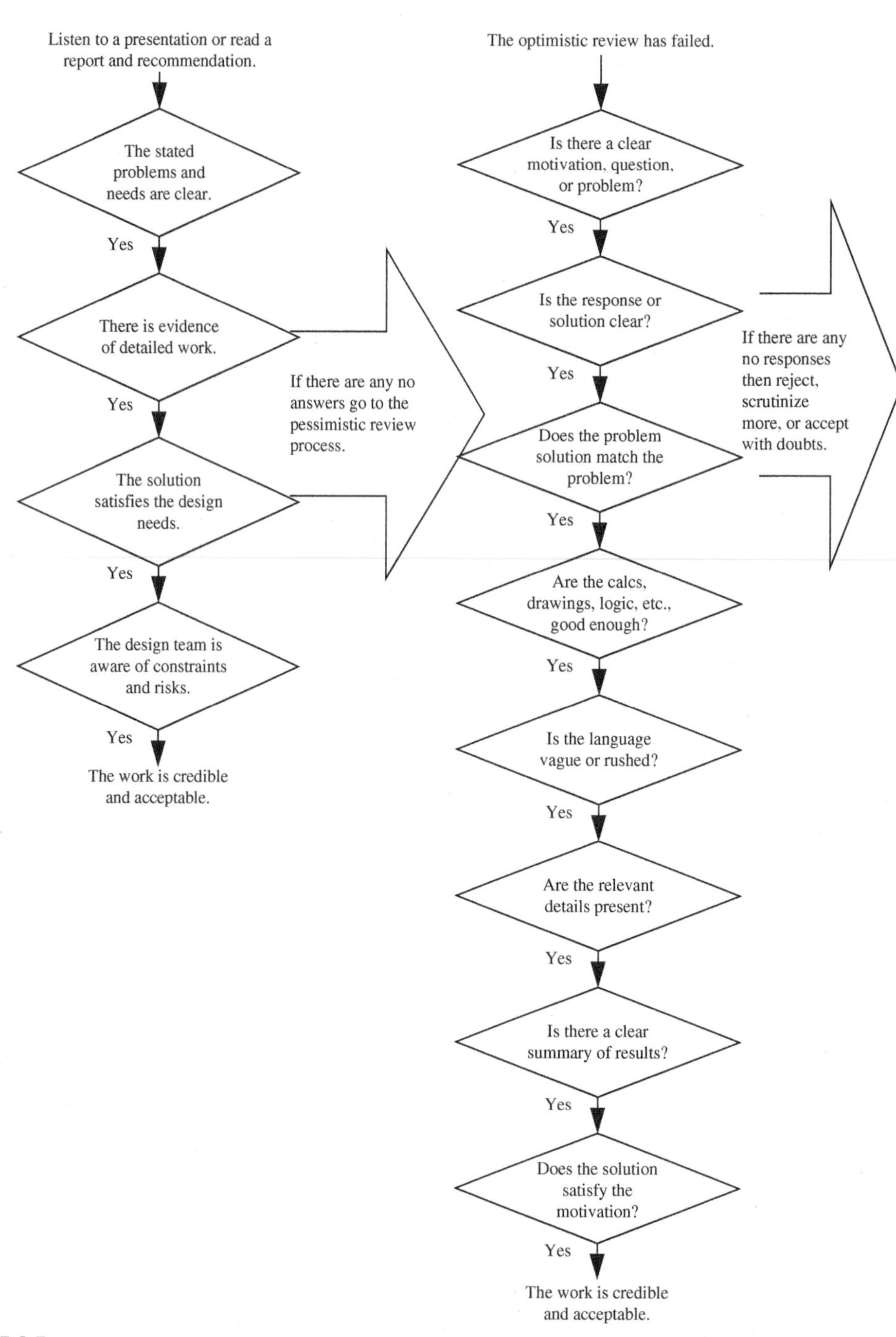

FIGURE 6.5

An optimistic first, and pessimistic second, thinking process for assessing technical solutions.

PROBLEMS

6.3 What does it mean to be critical?

6.4 Select an important critical filter to use when:

- **(a)** listening to a company advertisement
- **(b)** reviewing design calculations
- **(c)** reviewing test data
- **(d)** somebody is proposing a risky project
- **(e)** an engineer is explaining why he or she needs a budget increase of $20,000

6.5 If you do not like any of the decision options, can you create your own?

6.6 If you do not understand the decision options, what should you do?

6.7 If your friend presents a project plan, are you likely to use an optimistic or pessimistic approach?

6.8 Assume you are presenting a solution to a technical problem. What are the advantages of preparing for an optimistic versus a pessimistic review?

6.9 Which type of review is more likely to expose technical design problems?

6.10 Junior engineers often feel uncertain about their knowledge. People asking for decisions do not always have a logical reason or rational recommendation. What should you do if the decision is not clear?

6.4 Selecting between alternatives

There are very few optimal solutions—but there are good ones.

A simple decision-making process is outlined in Fig. 6.6. After the analysis is done you should check for overall clarity, and then decide if it is doable and beneficial. When choosing between alternatives you will find that each has a benefit and cost. In other words profit = benefit − cost. The opportunity cost is the profit of a decision not made. Always remember that valid decision options include saying no or not deciding.

Just because you can do something, doesn't mean you should.

When decisions have clear benefits there is often little need to consider additional factors. But, when decisions are not clear, or are somewhat arbitrary, some of the approaches in the following list can be used to positively or negatively sway decisions:

- Predictability: Sometimes "mundane" is good. Please note, this does not mean that work must be boring or uninteresting. It is a terrible idea to take unneeded chances for personal interest but at professional peril. Challenge can be stimulating, but it brings risk.
- Quantity versus quality: There is a classic trade-off between how much you can do and how well you can do it. Without more resources and costs, an increase in one means a decrease in the other.

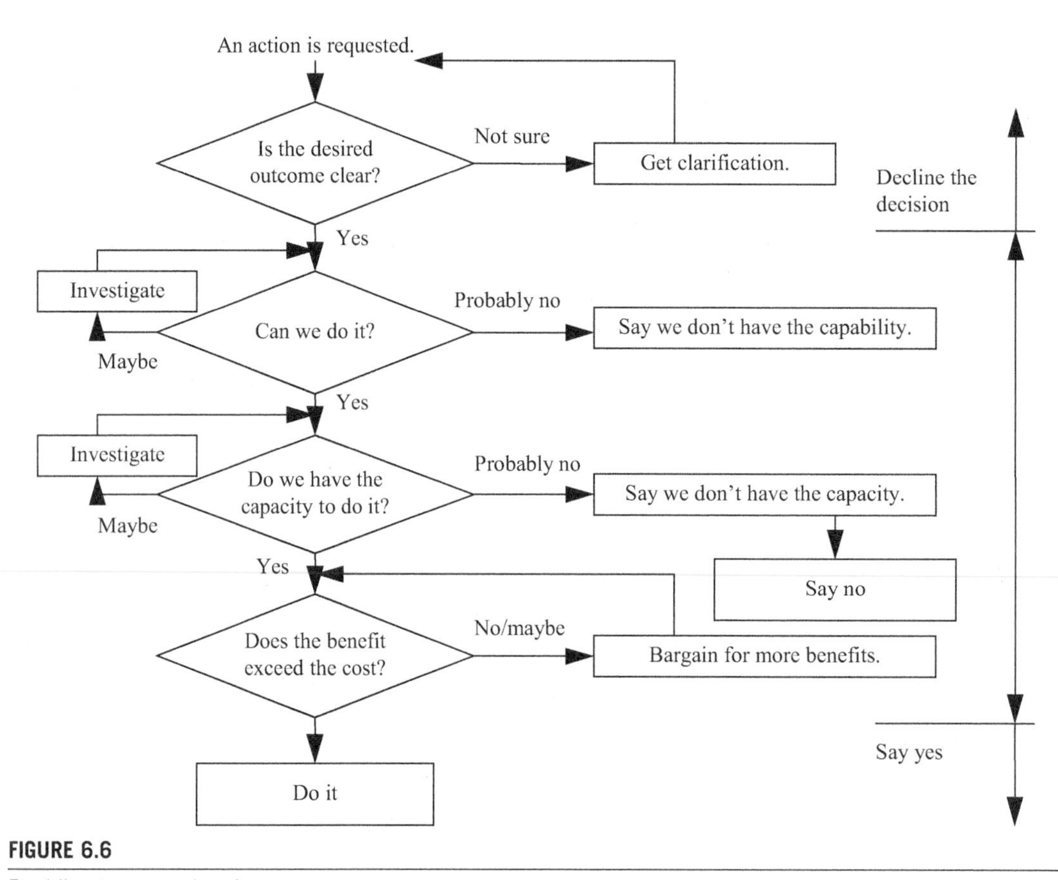

FIGURE 6.6

Deciding to move ahead.

- Switching modes: Over time our decision-making approach might change, including tolerance for risk, available resources, and business priorities. For example, a project might begin as research to develop a prototype, but when a competitor begins to produce a similar product the project may need to be accelerated with success-oriented goals.
- Spending options: Options could include reserve time in a schedule, unused resources, or unscheduled employee time. Consider options as a valuable resource. Do not spend them foolishly; make sure that giving up an option gets other value back.
- Opportunity cost: Whenever you make a decision to do something, you will eliminate the chance to do something else.
- Frustration: From time to time things will not work and you need to make a decision to stop and change directions.
- Cause and effect: Causes will lead to an effect, but effects will not lead to a cause. From experience we know that causes and effects happen together, but it is not always clear which ones are the causes.

Good results often come from a less ambitious task done well, but rarely from an overly ambitious task done adequately.

There are a number of mistakes made by decision-makers. In some cases the mistakes will appear to conflict, but this indicates the complexity of making good decisions:

- Inaction: delaying decisions that result in a delay of progress
- Reckless: rushing into decisions that result in higher costs, higher risks, or fewer benefits, i.e., skipping analysis
- Aversion: delegating critical decisions to others
- Magic: assuming that computers will do things that you don't understand
- Rush: selecting the first viable option that comes along to make progress without waiting for good alternatives
- Overwhelmed: too many options, resulting in excessive analysis

PROBLEMS

6.11 If somebody is known for making rapid decisions, what mistakes could they make?

6.12 A customer wants to add an expensive project specification, without increasing the quoted price. Describe three decision options.

6.13 Mini-case: Quick

Sadao was shopping for a used car and came across a model with a reduced price, having all of the features he wanted, but the paint color was wrong. Another customer was also looking at the car, clearly interested in buying. Sadao weighed the price and features against the color and decided he wanted the car and did not have time to delay. He found a saleswoman, Rin, and bought the car. On the way back to his new (used) car he noticed a similar car in the color he wanted. The price and features were the same as the car he had purchased. He pleaded with Rin, who said his purchase was final. Sadao left the car lot feeling a little less happy with his purchase. How does the concept of opportunity cost apply to Sadao's decision?

6.5 Triage

Decisions come in a variety of forms that can easily distract a professional from the highest priorities. As you need to make decisions it is wise to sort these so that you use your time for the most benefit. Simple decision sorting criteria are listed:

- Small/medium/large: How long will the decision take?
- Question: Is it a decision with a simple answer?
- Unclear decision: This is a poorly formed request.
- Project scope change: This is a casual request with a major impact on the project.
- Costless decisions: These are issues that are easy to process with no costs, and probably no benefits.
- Just because: Some decisions are mass mailed to a larger group.

A more systematic approach to triage is to develop a priority scale such as the one given below. Essentially, deal with brief and urgent project decisions first. Leave less critical decisions until later. For decisions that require substantial time and are unimportant, do these later or decline the decision:

5—Immediate action (hours)
Cases
- Will result in the total failure of the project
- Change in corporate priorities
- Loss of the customer

Responses
- Communicate the issue with the stakeholders and develop an action plan
- Validate the information, vet the options, and decide

4—As soon as possible (days)
Cases
- There is a good chance of failure
- Will result in major time and cost savings
- Customer is in financial trouble

Responses
- If it is a small contractual task, respond now

3—During project review (weeks)
Cases
- Will have an impact on part of the project
- There are possible cost and time savings
- The customer requests changes with no financial backing

Responses
- If a large contractual change is required, put it on the schedule
- Adjust schedule

2—Can wait (months)
Cases
- Will result in inconvenience
- Customer suggestions

Responses
- Respond with no, or request for change, or request funds

1—Put on the maybe list (during downtime or never)
Cases
- Casual suggestions
- It is information for other people

Responses
- Put in a "later" pile
- Forward to somebody (not one of your people)

0—Ignore

One important but undesirable decision-making tool is improvising. This is necessary when the plan has failed. We must make fast decisions when there is no time to replan properly. Basically, it is project planning at a faster pace but with more uncertainty. This is a very good skill to have, but it is bad when you need to use it. For example, you are at work and the sole of your shoe separates. You improvise and use tape to secure it until you can fix or replace the shoe. But you do not want to use tape as a regular tool when all you needed to do was plan ahead and buy new shoes. A good improviser will make decisions that observe the current plan. Poor improvisers will make decisions that go against the plan and will complicate things later.

PROBLEM

6.14 Assign the following items to a triage category. State assumptions if needed:

(a) An invitation to a party at a customer's house

(b) A request for a copy of a budget spreadsheet for the accounting department

(c) A 10-page progress report to put in the project binder

(d) An email opportunity to help an investment banker who needs to move money out of a wartorn country

(e) An important equipment shipment that is delayed 2 days

(f) A critical project employee quits

(g) An engineer says he or she needs an extra month for a task

(h) A device prototype has failed

6.6 Project decisions

When managing a plan there are a few critical steps and outcomes. The process shown in Fig. 6.7 could be done formally or informally on a daily, weekly, or monthly basis. A wise design team will also compare all new information and decisions through this process. The outcome of this process should be a plan that is in control and on track.

An example of a project review decision is shown in Fig. 6.8. Basically the project should continue if there are substantial benefits that the project is likely to provide.

PROBLEMS

6.15 When is it necessary to revise a project plan?

6.16 When is it necessary to stop a project?

6.17 What is escalation?

6.7 Solving formal problems

Formal problem solutions use theoretically rigorous approaches to reach answers. Real problems are often very messy but it is necessary to extract the essence of the issue, typically a simplified model. By contrast, when in school most students routinely solve well-formed problems with just enough information to reach a unique solution. In practice, real problems have a wealth of irrelevant details, arbitrary outcomes, variability, and subtle interactions.

When solving problems, the flowchart in Fig. 6.9 can be useful. This can avoid a few of the obvious problems encountered by young professionals. While these decisions are apparently easy, they can be hard to make when in the situation.

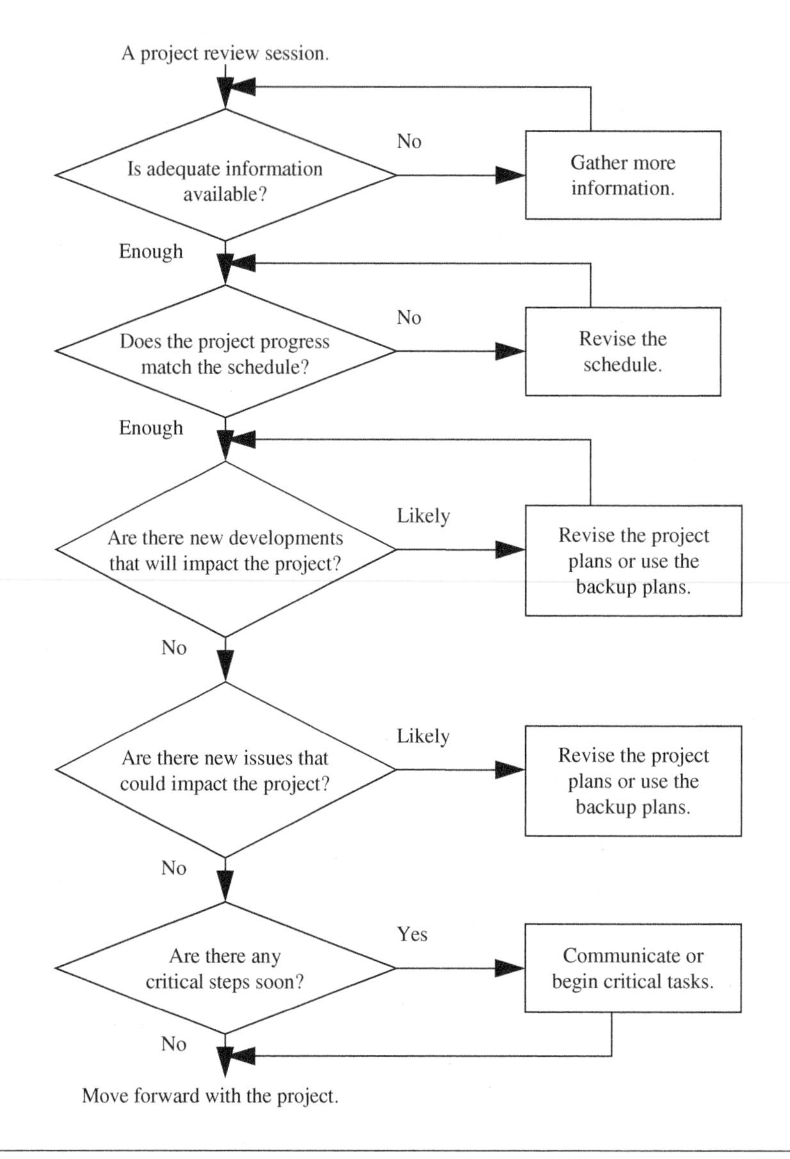

FIGURE 6.7

Critical review of project progress.

Typical problem-solving strategies are as follows:

- Reduce a large problem by breaking it into smaller problems that are easier to solve.
- Negate small factors that have little influence on the solution but add complexity.
- Extract the core problem model that matches some engineering model (e.g., circuits, statics, algorithms, equations).

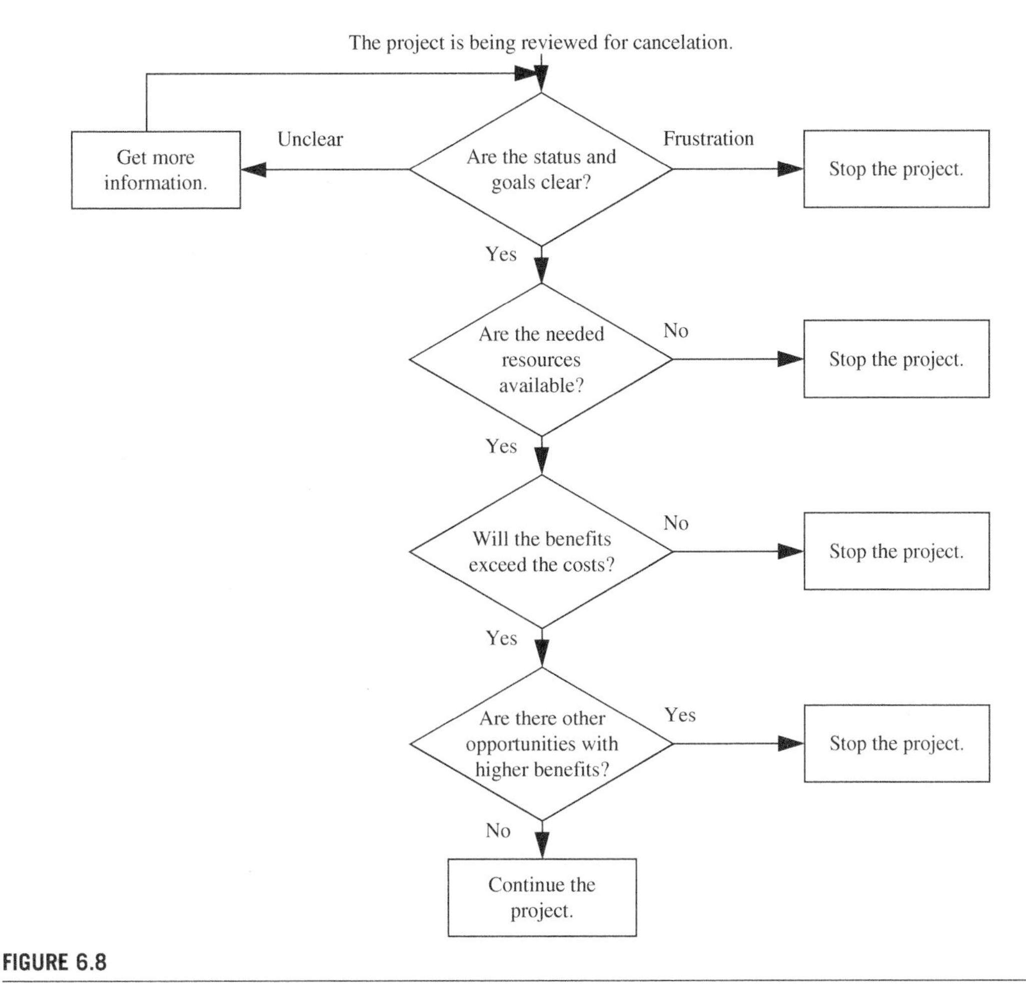

FIGURE 6.8

Deciding to stop a project.

- Simplify factors that are excessively detailed.
- Eliminate factors that do not have any bearing on the problem of interest.
- Approximate values that cannot be measured.
- Recognize unknowns.
- Decouple things that are connected to make the problem easier to solve.
- Borrow: Find parts of the problem that have been solved before to simplify the overall problem.

Developing a problem statement is an important first step. A few questions that can help this process are:

- What do I know about the problem?
- What do I know about this family of problems?

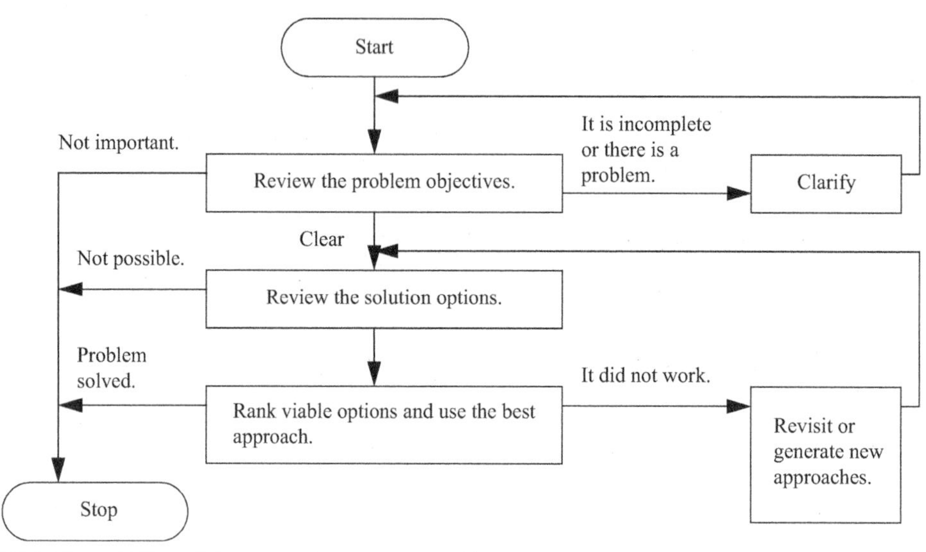

FIGURE 6.9

Methodical problem solving.

- Why do I want to solve the problem?
- What steps are required to solve the problem?
- What am I missing?
- Are there ways I can simplify the problem?

Problem-solving approaches include the following:

- Scientific: Is the theory true or false?
- Adequacy: Is it acceptable (e.g., Will it handle the power applied)?
- Parametric: What is the best value (e.g., What size should the component be)?
- Range: How much can it vary (e.g., What is the largest and smallest size allowed)?
- Characterize: How will it behave (e.g., What is the response to an impulse/impact)?

Fig. 6.10 shows the problem-solving method as a procedure. Fig. 6.11 shows how the problem-solving process is modeled functionally. Although both are modeling the same activity, the models give a markedly different view. If both are applied to assess a problem-solving process they will identify more issues. The key to using the model is to ensure that each of the steps is addressed and that they build progressively and logically from the previous step.

It is nice when a problem is simple and requires only a single-step solution. However, as the design projects become more complex the problem must be reduced to smaller steps. Although the approach shown in Fig. 6.12 seems obvious, it is surprising how many professionals will try to dive into a problem in the hopes a solution will arise in some moment of intellectual transcendence. It is always better to plan and work effectively. In other words, work smart (plan) and don't work hard (skip planning).

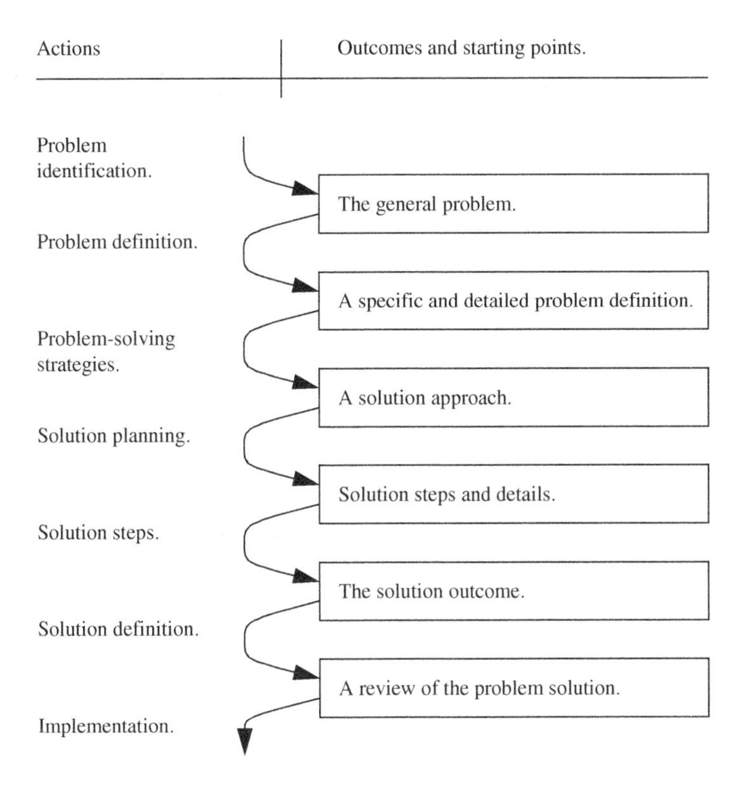

FIGURE 6.10

A seven-step procedural model of problem solving.

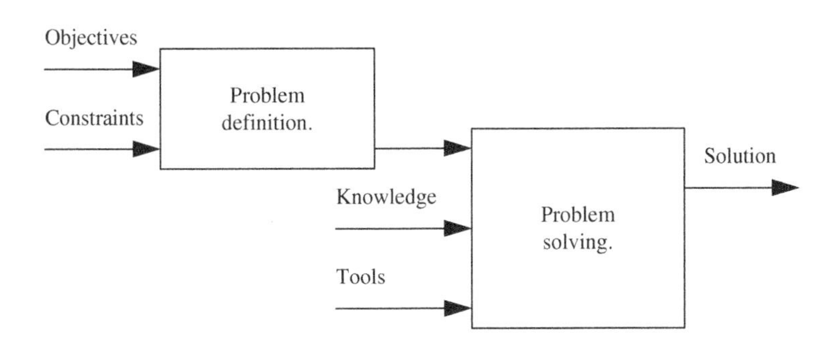

FIGURE 6.11

A functional model of the problem solving process.

PROBLEMS

6.18 Design a method to fire a mass 10 m upward with a compressed spring. Use a seven-step problem-solving method. Hint: Use energy.

6.19 How are free body diagrams or circuit diagrams used to reduce engineering problems?

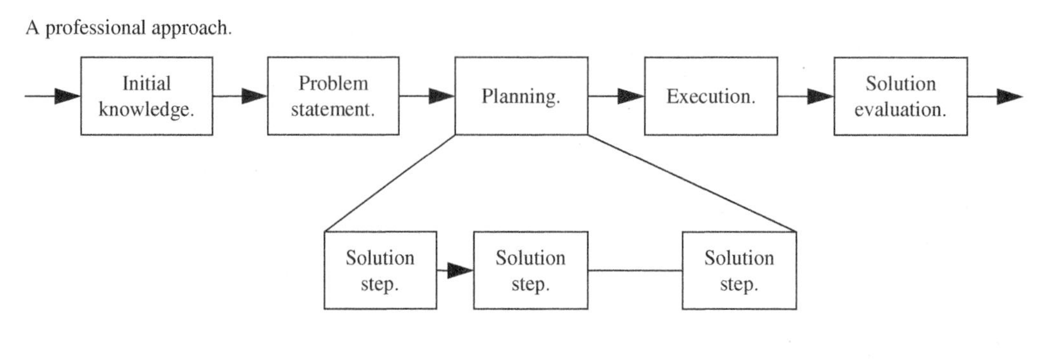

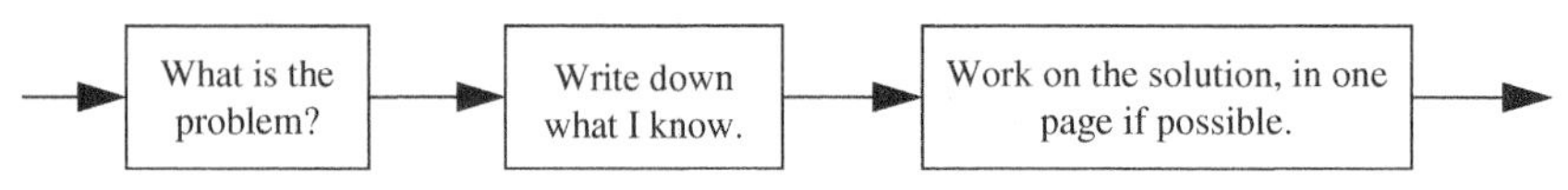

FIGURE 6.12

Problem reduction.

6.8 Risk

Unappreciated risks will lead to failure.

Wise business people will identify unsatisfied opportunities that have benefit and, of course, some risk. They will then work to increase the benefit and decrease the risk. Opportunities that are put on a good trajectory create resources to assume new risks and reap new benefits. An important factor to consider is that as risk goes up, so should the profit in the project. So when a high-risk project fails it is paid for by high-risk projects that succeed; low-risk projects can return lower profits. Projects without profits are good for hobbies but not business.

The approach to handling risks is itemized in the following list. The greatest risk tends to dictate the overall risk for the system, so address the greatest risks first and the overall risk goes down faster. If you cannot find a way to eliminate a high risk, develop backup or contingency plans to reduce the impact of the risk:

(1) Identify risks.
(2) Assess the danger of risks.
(3) Address the greatest risks first to reduce the overall risk.

Anecdotally, the concept of risk management has been the most important factor in design project success. With student project teams I have seen risky design decisions pushed back in favor of other, more comfortable, decisions and work. By the time the decision has been made the damage has been done and there is no time to recover. The best teams look at the work to be done and dive into those items that carry the most risk first. The worst teams do the easy work first to get it out of the way to leave more time for the hard tasks later. The issue with junior designers is that there is no experience with recognizing the risks and then estimating the likelihood and severity. Often optimism is used to justify moving ahead prematurely.

As an example, consider two different teams working on the same project for a new manufacturing process. Team A develops a set of ideas and picks one so that they can move forward to detailed design. Team B also generates concepts and then spends extra time building prototypes, researching, and doing calculations. Team B starts detailed design weeks later and the detailed design takes roughly the same time. The difference is that even though team A begins building sooner, they require a longer time for debugging and design changes. Team B spends less time debugging, and when things go wrong they have a decent fallback strategy. At first team A looks more active because they start design sooner and get to the problems sooner.

Most engineers have been educated in the statistical tools for analyzing the probability of failure. These are very valuable when dealing with complex projects, or when very detailed risk scenarios and numbers are available. What is here can be considered a "back-of-the-envelope" approach. Early in the design process the simple scoring system in this chapter is used to identify issues and overall impact. The best design teams will use this, or something similar, to direct their work, to reduce risk and increase the chance of success.

There is a probability that (1) something we want will fail to happen or (2) something bad will happen. The likelihood, severity, and probability combined constitute the risk. Examples of failures could include wasted time, extra costs, lost business, loss of morale, loss of confidence, serious damage, loss of project, or loss of business. Fig. 6.13 shows categories that can be used for describing the severity of a risk.

The remainder of this section and of Section 6.8 describes a system for identifying and scoring risks for different categories using a 0 to 5 scale. A score of 0 indicates there is negligible risk. A score of 5 indicates something that is likely to be a problem with, and have major negative impacts on, the project. At the beginning of any project the risk will be highest and should decrease as the work proceeds. In a corporate design the overall risk might start near a 2 on this scale, while an entrepreneurial design may start at 4 or higher. A wise designer will try to drive the design down to a risk level of a score of 1–2 before committing substantial resources to the project. If there are any critical risks, the highest score dominates. So, if a design has multiple parts, the worst score for any part may be the overall score for the project. The risk can be reduced by having design alternatives, even if the backup has a higher risk factor. Risk comes from uncontrollable events. These vary based on the projects, industries, people, and more. As a company gains experience it can develop an inventory of problems and statistical measures of occurrence and impact.

Risk = Likelihood * Severity

Likelihood	Severity
Always - It will occur one or more times during the project Probable - Is likely to happen during the project Possible - Happens during a project Rarely - Does not happen often Never - Very unlikely to occur	Catastrophic - Project is not completed - Does not meet major specifications Critical - Reduced functionality - Impending failure Performance - Does not meet minor specifications - Excessive costs - Misses deadlines Graceful degradation - The product goes out of specification sooner than expected Negligible - The effect goes unnoticed or does not require any effort

FIGURE 6.13

The risk equation.

Aside: This section provides a starting point for identifying risks and scores based on historical events. In this case the events are from a collection of incidents covering hundreds of projects conducted for industrial sponsors by undergraduate, graduate, and part-time students. The majority of the projects were for production equipment, test equipment, and new product designs. The success of these projects is gauged by the successful delivery of the projects and eventual use by the sponsors. The methods outlined in this section, and the book as a whole, typically result in a success rate over 95%, and an eventual use rate of more than 3/4.

The scale used in this section is somewhat arbitrary in the numerical range, but it is significant in magnitude. The numerical values are not intended to be precise, but used for quickly categorizing issues. Categorical descriptions are provided to help assign a risk score. Naturally the worst score dominates:

0—Safe; occurrence, rare; impact, insignificant
1—Good; occurrence, sometimes, OR impact, minor
2—Acceptable; occurrence, sometimes, AND impact, minor
3—Risky; occurrence, often, OR impact, major
4—Hazardous; occurrence, often, AND impact, major
5—Dangerous; occurrence, always; impact, catastrophic

Rules of thumb for managing the risks are listed as follows. Simply put, quickly deal with the high scores first:

- A score of 4 or 5 for any part of the project is a major issue and should be addressed quickly as a high priority.
- As the project progresses, the scores should be reduced to 0 or 1 by the delivery time.
- Constantly review and adjust the scores as higher risks are addressed.
- If a company has a standard system for risk assessment and mitigation, use it.

Typical project risks are given in the following list and described in greater detail in the following sections. Recognizing and managing these risks is critical to project success:

- Market: The project is overly ambitious or unrealistic.
- Technical: The design work is technically challenging.
- Purchasing: Critical parts may not be available as needed.
- Cost and schedule: It will be difficult to control project work and resources.
- Staffing: Disruption of the project caused by employee changes.
- Management: A mismatch of the project with corporate objectives.
- External: Factors that can be hard to predict or control.

PROBLEM

6.20 How could we mitigate the effect of a flaming garbage bin?

Once risks are identified, the most significant can be reduced or eliminated. In some cases the approaches will be obvious, such as finding supplies for critical components. Others may require more elaborate approaches. Examples include the following:

- Avoid: Pick design solutions that do not involve that problem.
- Assume: Decide to accept the risk and move ahead anyway.
- Prevent: Take actions to reduce the chance of occurrence.
- Mitigate: Reduce or eliminate the negative impact if the problem occurs.
- Contingency reserve: Have backup resources (e.g., money) if the problem occurs.
- Management reserve: Have backup resources for unforeseen problems across multiple corporate projects.
- Transfer cost risks to a third-party insurer: The insurance becomes a budget item.
- Transfer cost risks to a subcontractor.
- Reduce design risks by employing a specialized subcontractor.

PROBLEMS

6.21 What are the benefits of risk?

6.22 How are risks managed?

6.23 Asteroids strike the earth on a regular basis. Some are large enough to create damage on a very large scale. In 1908 a smaller asteroid exploded over Tunguska, Russia, destroying a forest area over 10 by 10 km. The same event over a major city would be catastrophic. Much larger asteroids pass the earth on a regular basis. A larger asteroid could end life on Earth. Estimate the likelihood and severity of a major asteroid strike using a scale of 0 = none to 5 = absolute. How great is the risk?

6.24 Is it acceptable to have a risk of 4 or 5 at the beginning of a project?

6.25 Mini-case: The pressure chamber

A student design team worked for 7 months to design and build a pressure- and temperature-controlled test chamber. The combined mechanical and electrical design team had invested over 1000 h in the design and construction of the test systems, including the use of some innovative methods for rapidly heating and cooling the chamber. The control system was based around commercial hardware and software made by National Instruments. The control team had also done excellent work and had invested over 600 h in the design and testing of the LabVIEW software. Progress was excellent, and the team had met all of the specifications a month ahead of schedule. As the project neared conclusion the sponsor was asked to visit to see the machine in operation and sign off on the final specifications.

The evening before the sponsor's visit was very busy as the students finished troubleshooting software, hardware, and the cooling system. When they went home the system was ready for testing. The next day they arrived a couple of hours early to verify the operation and do a trial run of the tests. The control computer would not start. The error messages made it clear that the disk had failed. The team reported the issue to the faculty manager. The manager had asked the students to make backups often and he asked them, "How much of the program was lost?" The team reported that they had not made backups and had lost all of the work. Luckily the team was ahead of schedule and was able to re-create the LabVIEW program. With the experience of developing the program before, it required only 200 h to re-create. The final customer sign-off was a month late, but the design easily passed.

What are three lessons from this experience?

6.26 Companies will pay for insurance, to transfer risk. Companies without insurance must pay for failures themselves, potentially threatening the financial health of the business. When should a business assume risk or purchase insurance to transfer risk?

6.27 How can a company insure itself?

6.9 Market

Two design projects may have the same technical needs, but have much different positions in terms of customer need. Fig. 6.14 shows a spectrum of design types. The two good design types are to use existing designs in new markets or to introduce new designs to existing markets. These two strategies make the project purely technical or marketing. Doing a common design for an existing market is easier, but harder to do well. The worst case is to have a new technical design for a new market. These design projects appear very attractive because they might create very profitable monopolies. However, they rarely do. So what does this mean? Learn to work comfortably in the moderate risk range. When you are doing a low-risk project, strive for design excellence. When faced with a high-risk project be very apprehensive.

A simple scoring system for design projects is shown in the following list. Use this when considering new design projects. Understand what you are getting into before you accept the challenge. Assuming that you have reviewed the designs and decided to move ahead, then you can use the following strategies for acceptance:

5—Disruptive design. This creates a new market or kills an old one; great for entrepreneurs and inventors, not companies.

- Contracts should allow total project failure as acceptable.
- Define a point of frustration to end the project.
- Use extensive prototyping and market studies to reduce the risk.

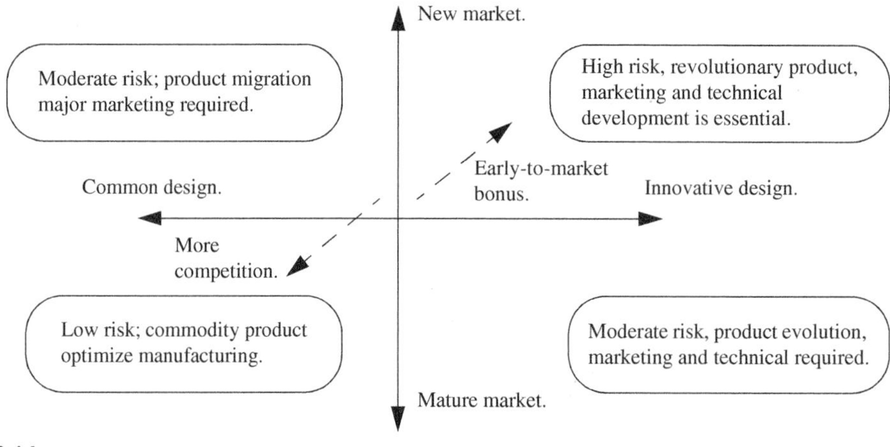

FIGURE 6.14

Market placement of new products.

- Have backup design options.
- Be prepared for significant losses and very high initial product costs.

4—Innovative design for existing market. This requires substantial work to produce.

- Contracts should allow technical failure as acceptable.
- Define a minimum set of deliverables for the project that are noninnovative.
- Tie innovative design aspects to bonuses.
- Prototypes, testing, and early design iterations will decrease the risk.
- Have backup design options.
- Be prepared for possible losses and high initial product costs.

4—New markets for existing design. If you have to develop a market, it is a substantial problem.

- Separate technical and marketing success.
- Use cleanly defined specifications for project acceptance.
- Use the marketing department for testing and incorporate the feedback.
- Be prepared for possible losses and high initial product costs.

2—Common design and evolving market. This is predictable; a good place to patent to lock out competitors.

2—Mature markets and evolving designs. These are already developed, easier to sell.

- Define basic and new features.
- Define cost and time targets.
- If using a newer design type, keep the older one as a backup.

1—Commodity. These have large markets and many competitors. Innovators cut costs in manufacturing and distribution.

1—Established design. The design is a standard type, very similar to another done before.

- Emphasize cost reduction for the final product.
- Use existing resources and methods for design modification and fabrication.

0—Established technology. Components can be purchased and have been available for years from well-established suppliers.

- Purchase and then resell the design.

If the project team or customer does not have enough understanding of these concepts they are likely to accept riskier designs with higher objectives and be more likely to fail. If you are willing to accept the higher risk, do so knowledgeably.

PROBLEMS

6.28 What is the risk score for convincing people to use a new energy source?

6.29 What is the risk score for a new cup holder in an automobile?

6.30 What is the risk score for an automobile with an extra row of seating?

6.31 List 10 examples of commodity products.

6.32 List 10 examples of evolutionary and revolutionary products.

6.10 Technical

The design risks are a higher-level view of a design project. When we are able to focus on the technical aspects we can consider separate components, methods, and procedures. Consider the curves in Fig. 6.15 for cost and risk. Early in the process, we are in the research and development phases, hoping to develop demonstrations and prototypes that support the design ideas. These can be intellectually stimulating and are fine for speculation and early design projects, but they are far from being salable and profit generating. To pay for these exploratory projects we need to focus on the standard designs, or designs using newer technologies with lower risks and costs. In common language, the left side of this graph is the bleeding edge. No design will go from research to consumer product in a single cycle; instead it will occur over many iterations. And, over the life of the design the volume produced will increase, so the cost per design will drop quickly. This means the early customers will pay a premium for high development costs and low production volumes.

The stages of development for new designs are seen in Fig. 6.16. As progress is made and setbacks occur, many design projects are canceled. In fact, many of these stages occur at different companies and institutions. For example, a new surface coating may be developed at a university on a microscopic scale. A corporate research lab develops a demonstration that covers many square centimeters, and it is patented. Support products are developed and then sold to a component maker that uses it in a new machine design that is sold to consumers. The coating technology has passed through three sets of hands. Of course, for the one concept that started at the research stage there may be hundreds that were abandoned. (Note: From a purely commercial perspective, research is a risky process with high costs and little return; hence universities receive public funds to begin the process.)

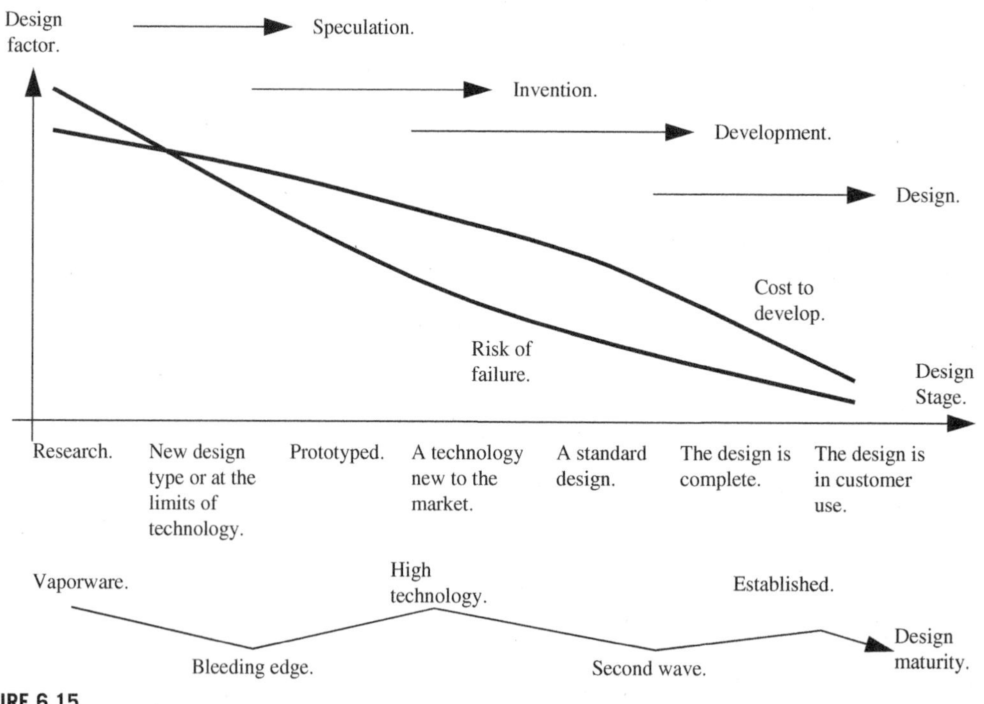

FIGURE 6.15

Design risk issues.

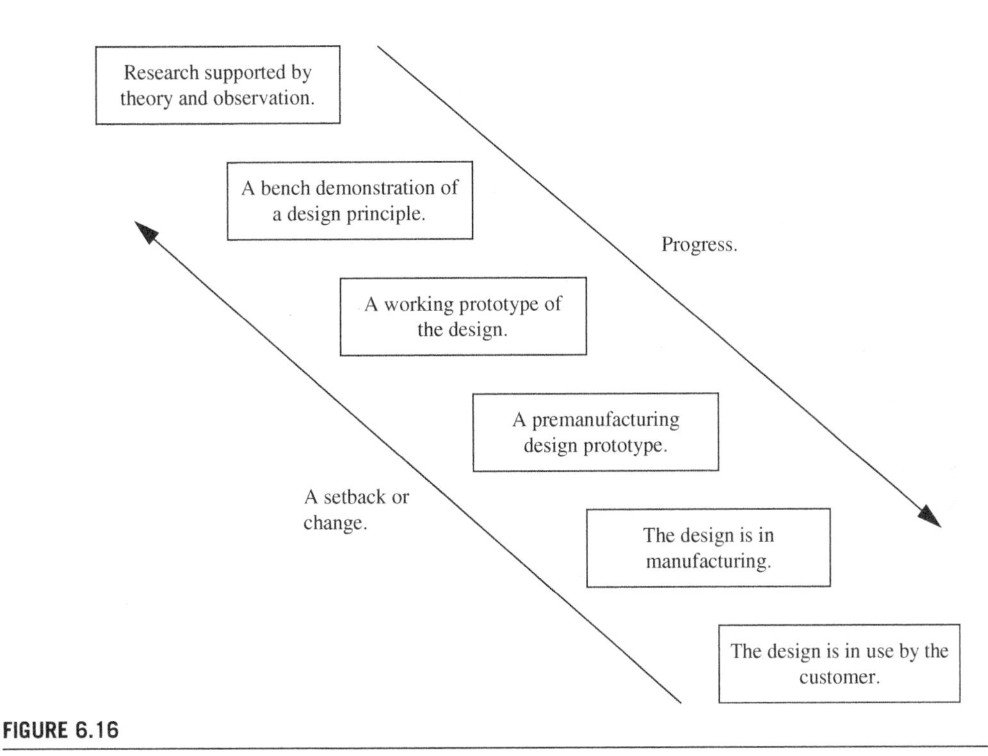

FIGURE 6.16

Moving from research to design.

Normally, during specification and concept development a number of technical design tasks are identified. For each of these we need to estimate risks. The following criteria are useful for identifying the general risk factors. As before, the higher scores should be addressed first. Some possible strategies to reduce the risks are also provided:

5—Research. The design depends on major questions or knowledge that has to be researched.
- Delay the project until the concept is more developed.
- Eliminate from the design options.

4—New design. The design includes theoretical physics, new algorithms, etc.

4—Technology limited. The design can be accomplished with only one approach.
- Create a working prototype as quickly as possible. More prototypes are better.
- Perform basic tests for suitability.
- Consider alternative implementations.

3—Prototype only. The system has been proven with a limited prototype, but has not been fully implemented.
- Test it and try to break it.
- Incorporate more design features into a second-generation prototype.

2—New technology. Components have been available for less than 2 years or from a single source or unknown suppliers.
- Use sample parts, demonstrations, applications, etc., to assess the benefits and costs.

1—Standard design. This is a standard system or component design that has been designed and documented before, but not by the team.

1—Established design. The design is a standard type, very similar to another done before.

- Emphasize cost reduction for final product.
- Use existing resources and methods for design modification and fabrication.

0—Established technology. Components can be purchased and have been available for years from well-established suppliers.

- Purchase and then sell the design "as is."

Beyond basic components, project risks resulting from complexity are hard to quantify, but effectively the relationship is exponential, as shown in Fig. 6.17. This means that doubling the parts count of a design more than doubles the complexity. Experienced design teams will find ways to cut the risks by breaking the design project into separate and discrete pieces.

When approaching a technical design, set specifications, select design concepts, build prototypes, and make decisions to drive the risk to a score of 2 before moving to detailed design. When there are high scores, use alternatives to reduce the overall risk. As a simple example, consider two design alternatives with a 50% chance of failure each. Begin with one alternative, and if failure occurs the backup is used; the result is the chance of failure drops to 25%. If we have three similar alternatives, the risk of failure drops to 12.5%.

Some other concepts that can help when considering technical design issues are given in the following list. In general, the key problem with technical challenges is that familiarity is used as a main reason to accept or reject concepts:

- Unproven technologies often hold promise, but can be very difficult to use.
- The "not invented here" syndrome leads to extra work.
- The simplest design that meets the objectives is the best—if it works.
- Sometimes components do not do everything the sales literature claims.
- Things are normally more difficult than they look.
- Student ownership of ideas can blind them to reason.

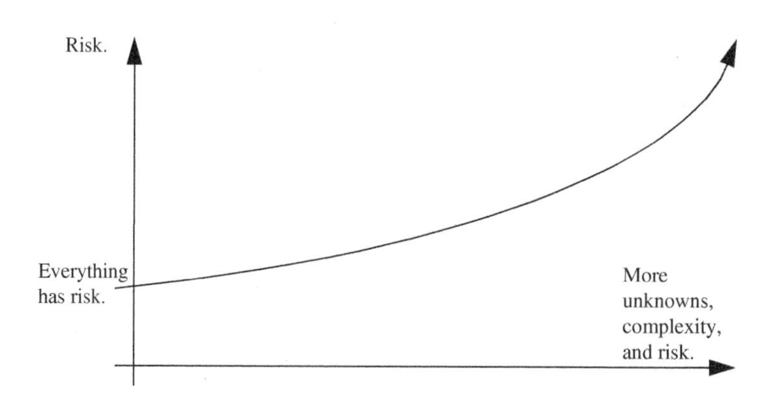

FIGURE 6.17

Relating risk to project scope.

PROBLEMS

6.33 List 5 examples of new technologies that were recalled for safety reasons.

6.34 Consider a successful demonstration of a new 5 × 5 cm flexible computer display. What supplemental steps are required before it will be available as a product?

6.35 People who buy new technologies accept higher risks; the more conservative consumers avoid the first generation of any new technology. How does this influence market size and development costs?

6.36 Why are prototypes valuable? Are there different levels of prototypes?

6.37 Mini-case: Material changes

A customer required a machine to apply adhesives to rubber moldings for automotive parts. The long rubber parts would be held in place while the adhesive was wiped along the length. The adhesive would begin as a liquid. After it had been applied as a liquid, a solvent would dissolve and leave a gummy glue. The part would then be removed from the machine and assembled in a separate operation.

The project was accepted and work began. The work was progressing well, and the design reviews were passed without any major issues. The progress of the team was ahead of schedule when the project entered the build-and-test phase. During testing the team found that the liquid adhesive did not wipe evenly along the surface. The team postulated that the issue could be corrected with a different viscosity. The customer was contacted, and the issue was discussed. The team and the customer decided to move forward by seeking an alternative adhesive. The customer talked to his material supplier and managed to find a reasonable alternative. After a few days the project team received a couple of unlabeled bottles of the new adhesive. Testing began again and the team decided to work late to compensate for the time lost. The new material appeared to be working well and the project team ended the day before midnight.

On the way home, one of the project engineers began to feel ill. He stopped his car and his nausea terminated orally. This event was noted by a passing police officer. After some concerns about intoxication the engineer was free to go on his way. The next day the team members found that they had all experienced similar health issues and they told the manager. The manager told the team to stop using the new adhesive, until it could be investigated.

The manager called the customer to discuss the issues resulting from the new material. The customer then checked with the supplier who revealed that the new material used a different solvent. This new knowledge was used by the team to redress safety issues. The result was that the machine was moved to an area with better ventilation, and when working close to the adhesives the team members wore masks.

Consider the issues, and suggest three different ways this problem could have been avoided or minimized.

6.11 Procurement and purchasing

Purchasing parts, services, and materials brings some definite benefits, including (1) using external expertise, (2) reduced in-house workload, and (3) potential cost savings. The main motivation is that you want to focus on what you do well to make money, and take advantage of the expertise of others. This can also reduce the overall design complexity and dependence on internal resources. However, any part purchased outside or from another division of a company carries some element of risk such as variations in cost, time, and technical specifications. The ranking system given in the following list provides a quick way to apply a score to each part in the design. Obviously the scoring should focus on special parts in the design. To save time and effort, it is wise to ignore commodity parts such as surface-mount discrete components, nuts/bolts/washers, generic raw materials, and so on:

5—No known supplier.
- Do it yourself if you can.
- Eliminate from the design if possible.

4—Special item: requires supplier design.
- Work with suppliers to ensure design compatibility.
- If critical, consider multiple design suppliers. Avoid single-supplier situations.
- Order early.

4—Limited supply: the item or material is not readily available.
- Try to find a more common alternative.
- Order ahead of time in bulk.

3—Standard item: special order.
- Order quickly and add some reserve time for late delivery.

2—Commodity item: back-ordered.
- Build in delays and look for alternative suppliers with inventory.
- Consider a design that will allow substitutions.

1—Commodity item: available now.
- Look for alternative suppliers and lower costs.

0—You have the thing.
- Ensure that it is available for you.

This list of risks is far from complete but has been developed using successes and horror stories involving part identification and purchasing. Aside from the basic factors, there are other variables (Fig. 6.18). For example, if a design is complex, there is more chance that a supplier will encounter problems that could change the cost, delivery date, or deliverable specifications.

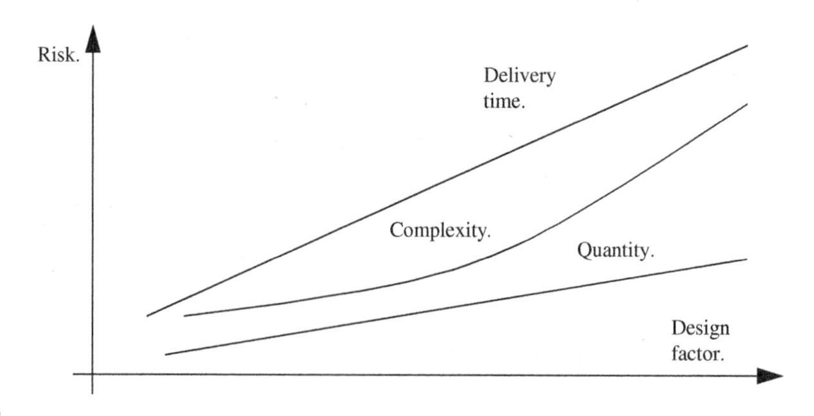

FIGURE 6.18

Supplier risks.

Junior designers will often think of suppliers idealistically. Seasoned designers know that suppliers try their best but make mistakes and have problems too. If you count on a supplier to deliver parts on schedule, as ordered, you will be disappointed when you can least afford it. Some of the unexpected issues to anticipate when selecting purchased parts are:

- A standard item, such as an integrated circuit, motor, or gear set, is out of stock or back-ordered.
- Items are lost or damaged in shipping.
- The items shipped do not match the specifications.
- Rare items are promised but then removed from stock and production.

PROBLEMS

6.38 Assume your company uses a current-generation part that has reasonable costs and average performance. A new component would offer excellent performance at a lower cost. What risks should you anticipate with the newer component?

6.39 Using a single supplier provides the benefits of simplified ordering. What are the risks of using a single supplier?

6.40 Mini-case: Under pressure

A student design team was chosen to construct a pressure vessel that would need to withstand a complex load case and deflect in a very specific way. The vessel was designed to deflect appropriately while meeting the standard engineering codes. As a pressure vessel it had to be constructed by a licensed welder, so it was sent to a local fabricator. The fabricator constructed the vessel but had to substitute a thicker end cap, a normal and safe practice. The change was not visible, but it was noted on the welding certification documents. The fabricator delivered the vessel and documents to the team.

The team had been proactive, and the pressure vessel, was received 4 weeks ahead of schedule. The team accepted the delivery and signed off, and the fabricator was paid. The team focused on other courses and waited for the other components to arrive. By the time all of the parts had arrived, the project was back to the original schedule. The students completed the build phase of the project and began testing.

The tests did not go well because the pressure vessel was not deflecting enough. Debugging stretched well beyond the allotted time, and the project that was weeks ahead of schedule was now weeks behind. Eventually, in frustration, they began to suspect the pressure vessel. They called the welding fabricator and eventually determined that the end cap was thicker than the one they had specified. The fabricator refused to correct the work on the pressure vessel, because it was accepted with the certification documents showing the changes. The students checked the paperwork and did see the substitution. Having exhausted the funds, the project was completed unsatisfactorily.

The project provided a number of valuable lessons including (1) verify parts when they arrive, (2) check the paperwork, (3) when ahead of schedule do not slow down, (4) do not assume anything is correct, and (5) special parts carry more risk. Moreover, it verified that spending time on detailed review is much more efficient than debugging.

Outline a part acceptance procedure the team could have used to detect the problem.

6.12 Cost and schedule

Small decisions can have a huge impact on final design cost. As the design progresses these values drop. In many ways these correlate to the design risk, but are more relaxed. In very simple terms the cost of the project can vary when the design does; risk is a good indicator of how much the design may change:

5—Rough concept. The design has been imagined but not fully detailed.
- Look for designs with a remote similarity to get a ballpark cost estimate.
- Push for more specific designs, including prototypes.
- Be prepared for likely losses, if the design fails.

4—Revolutionary. The design is trying a new technical approach to an old problem.
- Use the existing designs as a starting point.
- Develop costs for the major components using estimates and purchased components.
- Be prepared for possible losses if the design fails.

3—New. The design requires elements that have not been executed before.
- Develop a project plan and budget including detailed variability estimates.

2—Common. A similar design has been done before by others, and details are available.
- Use the expertise of others to estimate costs and project needs.

1—Standard. The design is common and well understood.
- Use standard design rates.

0—Complete. The design work is already complete for a similar design done by the team.

A high-cost risk might result in extra expenses that decrease the project return on investment (ROI). A prolonged schedule would have a similar risk. Therefore, two strategies for a high-cost/prolonged-schedule risk are to increase the ROI or reduce the risks. Fig. 6.19 shows the reduction of cost risk as a project advances. The time to consider these issues is before detailed design. The best strategy to reduce these risks is to push for details and certainty. Any loose ends will result in unexpected costs and extra work. But then again, about 1% of the time things will magically cost less and take less time.

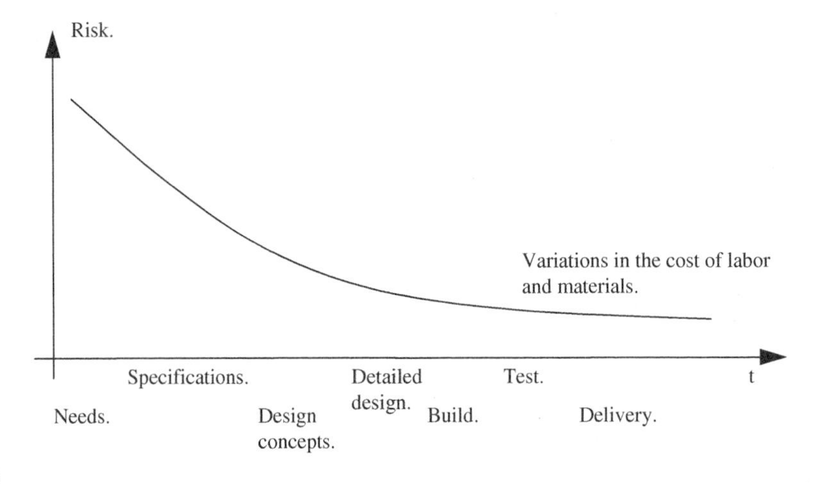

FIGURE 6.19

Cost variations over the life of a project.

PROBLEMS

6.41 Is it reasonable to go from a rough concept to a sellable product in one step? How and/or why?

6.42 The cost and schedule risks should decrease as a project progresses. The two factors are the amount of work remaining and the number of decisions. What happens when something unexpected occurs?

6.43 A project manager must be aware of the risk in a project. What should the manager do if the risk is increasing?

6.13 Staffing and management

You go to work with the team you have, not the team you want.

In an ideal world there are an unlimited number of people who are able and willing to do all of the project work. In reality there is always a shortage of technical skill and an abundance of work. Some of these factors are shown with crude curves in Fig. 6.20. During the course of a design project, employees will be needed on other projects, be hired, be fired, go on vacation, have health problems, be promoted, and more. If a project is short-lived, these are often negligible, but as a project duration approaches months and years, these situations become very real. A poor situation is to have a single employee who can do a critical job, is overworked, is underpaid, and is unhappy. A good situation is to have multiple employees happy, free, and able to do all project tasks well. In practice, there is a balance.

For knowledge-based employees there are a number of project phases that should be considered, as shown in Fig. 6.21. When considering risk, a project that covers more phases over a longer time will require more people with more approaches, over that time. A good strategy for dealing with these phases is to break a large project into smaller projects.

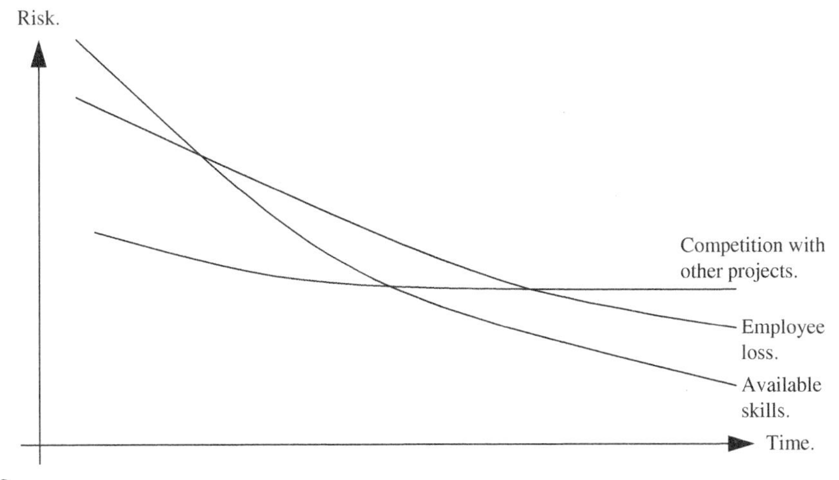

FIGURE 6.20

Variations in project labor risks.

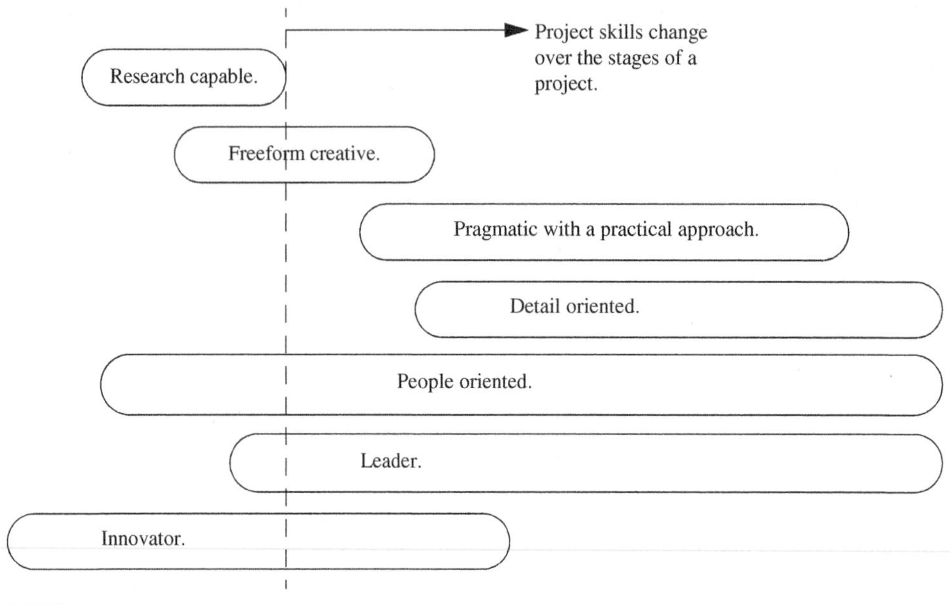

FIGURE 6.21

Recognizing individual employee contributions.

Some risks for staffing are shown in the following list. It is expected that multiple risk factors will be identified for each person on the project. As always, deal with the highest risks first and, when possible, use backups to reduce overall risk. Sometimes this will mean hiring or training proactively for critical staff skills:

5—Nobody can do the job.
5—The workplace is dangerous.
5—Key personnel have an impending personal issue.
5—There is open animosity and disagreement.
5—The project is very remote or dangerous.
4—Only one person can do a job and is overbooked.
4—The job can be done only with training or education.
4—A union is about to go on strike.
4—The team is larger; for example, 20 reporting to one person.
4—People are overcommitted.
3—The workplace is hazardous or remote.
3—Only one person can do a job.
3—All of the people are working at full capacity.
3—There is a union with many work issue grievances.
3—Key personnel have ongoing family issues.
3—There are occasional disagreements and unresolved issues.
3—There are labor issues: unions, morale, turnover.
3—It is the first time the team has worked together.
3—The project is conducted partially off-site.

2—The task is routine but complex or time consuming.
2—There is a good working relationship between all.
2—People have been assigned reasonable workloads.
1—The task is routine and can be done by anybody.
1—The team has worked successfully on a similar project before.
1—There are very good relations with the union and workers.

PROBLEMS

6.44 Can a leader be a detail-oriented innovator?

6.45 Assume a large company does an audit of the design department and finds that 19 primarily identify themselves as leaders, but only one identifies himself as people oriented. What would you suggest as hiring priorities?

6.46 List five items that might change the work a team is doing. Student examples are fine here.

6.47 List five ways to reduce staffing risks for a project.

6.48 Some industries have gone through phases of outsourcing for various reasons. Use the reduction in risk numbers to explain why:

(a) After a strike a company closes a division and moves the work to a nonunion supplier.

(b) A consulting firm is hired at twice the labor rate to do a critical design task.

6.14 Organization

Politics, personalities, people, and priorities are all factors that can have an impact on a project. Even in the greatest design company these will be issues. A good situation is a stable and healthy organization where everybody works as a team with common goals. A poor situation is a company where senior management uses worry about job stability to promote evolving competition built on turf and personality wars. In practice, an organization is a collection of people and there is a mix of all of these issues. Go through the list and consider issues on a regular basis. If there are some higher risks, deal with those before they become problems:

5—A change in ownership or senior management team.
5—There are no resources or space for the project.
5—People are unaware of the objectives and do not have any authority.
5—The project type is completely new to the organization.
5—"Turf wars" stop many projects.
5—Upper-level managers have no vision or control.
4—There are many conflicting personal agendas.
4—The project goals keep shifting.
4—There are complex standards and regulations.
4—Upper-level managers change goals often.
3—Micromanaging is occurring.

3—The project goals are unclear and poorly understood.
3—There are aggressive competitors.
3—There is substantial competition for project space and resources.
3—"Turf wars" lead to conflicts.
3—There is no direct access to upper-level managers.
2—People are familiar with the design work.
2—There are people or small tasks that are new to the team.
2—There is unclear or conflicting communication (e.g., artificial deadlines and forgotten work items).
2—The customer representative does not fully understand the project.
1—People are empowered and informed.
1—The manager delegates freely and effectively.
1—Upper-level managers use a careful and deliberate approach to project work.

PROBLEM

6.49 What are the risk numbers for the following cases?

(a) A manager attends a meeting with a project team and demands a change.

(b) Employees meet an hour a day with the boss to receive work instructions.

(c) The entire management of a company division is fired.

(d) A supervisor books meetings in the afternoon because another supervisor cannot attend.

(e) An employee fills in for a project manager with a family emergency.

6.15 External

Outside of the company there are a number of issues that you may not be able to resolve, including those shown in the following list. However, recognizing these provides a basis for planning. For example, weather issues can be handled with disaster plans; for new legal issues, consultants can be hired; for infrastructure systems, backup generators and water towers can be used:

5—Weather: hurricanes, tsunamis
5—Major earthquakes
5—Legal: the project outcome may be outlawed or restricted
5—Customer: does not understand the project outcomes or project process
4—Weather: tornadoes, floods
4—Legal requirements for the project are changing
3—Weather: blizzards, sandstorms
3—Minor earthquakes
3—Legal: the laws and regulations for the project are extensive
3—Customer: aware of the project goals and process but not involved in the process
2—Infrastructure: frequent power and water losses
1—Customer: is knowledgeable and actively engaged in the project

PROBLEMS

6.50 The European Union harmonized standards across many countries. How would this affect the external risk score?

6.51 Mini-case: Cascading failures

Japan is considered a very stable country that boasts many advanced manufacturing facilities and produces almost a tenth of the world's wealth. However, when the risks of Japan are rated objectively, they have a number of concerns. These risks were profoundly realized in 2011, when a magnitude 9.8 earthquake, which produced a tsunami, caused a major nuclear disaster and major electricity shortages. Many people were injured or killed and many industries suffered devastating setbacks. A lapse in the global supply chain left a number of industries starved for components and products as Japan recovered. IHS Automotive was one of the companies affected. One of its products, an airflow sensor, was used in 60% of the global automotive production. When the factory was closed for repairs, the world's automotive industry slowed dramatically.

The global supply-chain management community took note. The event was a reminder that even very stable companies have risks. And, it served as a reminder to spread the risk over multiple suppliers, in multiple countries.

Could IHS Automotive decrease its supply-chain risk by adding earthquake and power loss protection? Are there other strategies they could have used?

6.16 Risk analysis

Recognize risks so that you can avoid them, or add them into the plans.

Risk factors can be considered alone in simple cases, but when risks are associated through a complex design the factors can be combined statistically if numbers are available. One common industrial tool for numerical analysis is failure modes and effects analysis. The method presented in Section 6.8 assumes that detailed numbers or the time needed for analysis is not available. The method builds on the previous risk scores and adds amplifying factors for frequency and severity. By multiplying the risk, frequency, and severity the critical risks will become evident, as shown in Table 6.1. In this case the

Table 6.1 Risk assessment table.

Project risk item	Risk score	Frequency score	Severity score	Total assessment score
Union issues	3	2	3	$3 \times 2 \times 3 = 18$
New design type	4	3	3	$4 \times 3 \times 3 = 36$
A rare power supply is needed	4	3	2	$4 \times 3 \times 2 = 24$

priority list would be (1) new design type, (2) power supply, (3) union issues. A good manager will address more than one of these at a time, but given a choice the highest priority comes first:

Frequency
5—Will happen many times during the project (>95)
4—Occurs often (>80)
3—Happens half the time (50)
2—Sometimes happens (<20)
1—Rarely happens (<5)
0—Will not happen
Severity
5—Catastrophic: loss of life, major equipment damage, project terminated
4—Serious: injury to personnel, loss of significant equipment, or the project may fail
3—Major: a large change in the plan, schedule, or budget
2—Minor: some loss of time or minor expenses
1—Negligible: inconvenient
0—No negative effects

PROBLEMS

6.52 Are there any risk factors that should not appear in a risk assessment table?

6.53 Prepare a safety risk assessment score for boiling water in a microwave oven.

6.54 How can a total assessment score be decreased?

6.55 Construct a business risk assessment table for an inventor with a working prototype of a toilet seat lifter but who has not found a manufacturer or retailer willing to buy the product. Make assumptions as necessary.

6.17 Design alternatives

Design involves a collection of coupled high- and low-level decisions. Fig. 6.22 shows a decision tree for a conceptual design, in this case using component selection, and processes for decisions. In this example, we will choose to start with component A, but use component B if it fails. Other than that, there are no other alternatives presented. So the next question will be: How risky is this design? The following steps can guide the process of creating these diagrams. The diagrams can also be represented in tables; spreadsheets are very useful for automatically recalculating risk values:

(1) Identify critical choices in the design. These are normally high-level strategic (left side) or low-level (right side) decisions that are technology limited.
(2) Begin to add design choices to the diagram/table. If there are low-level choices (technology limited), they should eventually link to the upper-level choices.
(3) Look at the diagram to find places without alternatives; if they are high risk, consider alternatives.
(4) Look for low-level choices that are too vague and require more detail. Are the lowest-level choices something that can be purchased or is already made in-house?

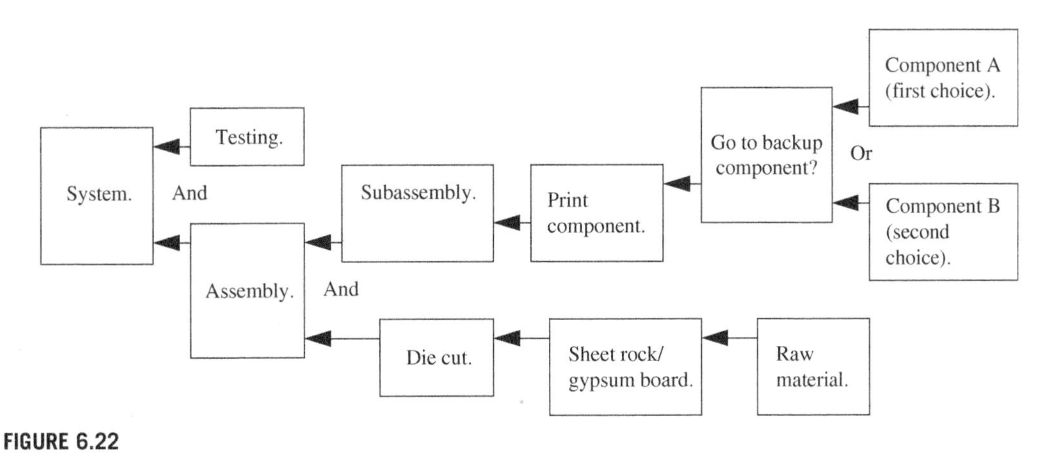

FIGURE 6.22

Design decision tree for dependencies and risk.

The basic 0 to 5 scores discussed before can be used to assign risks to the lowest levels of the design. These can then be combined from the lowest levels to the highest to calculate overall design scores. The first step is shown in Fig. 6.23 in which the risks for each block are written. Please note that at this point the process becomes abstract and the details of the design can be temporarily ignored during analysis, simplifying the process.

The next step is to combine the risk scores as shown in Fig. 6.24. Where there are "Or" alternatives we select the lowest risk, a simplistic approach. For sequential risks we simply add these. For parallel "And" risks, the values are added. As already stated, it is possible to do a statistically valid analysis, but this method will expose opportunities to eliminate risk quickly. And, as the design work progresses, the overall design score will be reduced quickly.

The value of this tool is to drive the decision-making process. Some of the key strategies are outlined in the following list. It is worth considering that during the conceptual stage of the design it is acceptable (but not desirable) to have scores of 4 or 5, but the objective of the design work is to reduce risk to a 3 before moving to detailed design. In detailed design the options should be reduced to a 2, and eventually a 1 or 0:

- Develop alternatives: When risks are high (say a 3, 4, or 5), alternatives should be developed. If two alternative design options have a 50% chance of failure by themselves, they will have a 25% chance of failure when used as primary and backup.

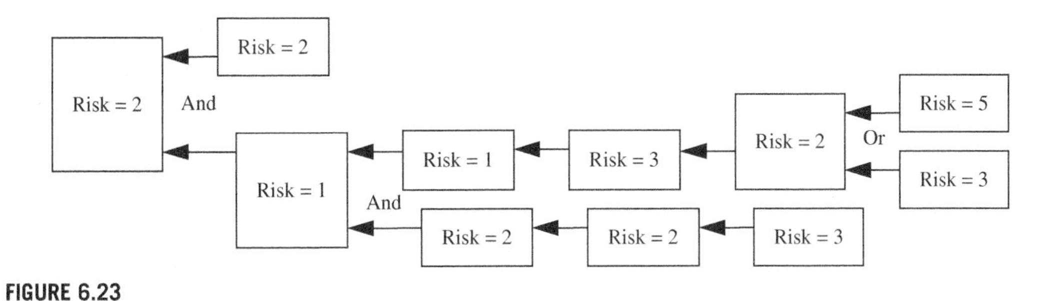

FIGURE 6.23

A risk is assigned to each of the main design parts.

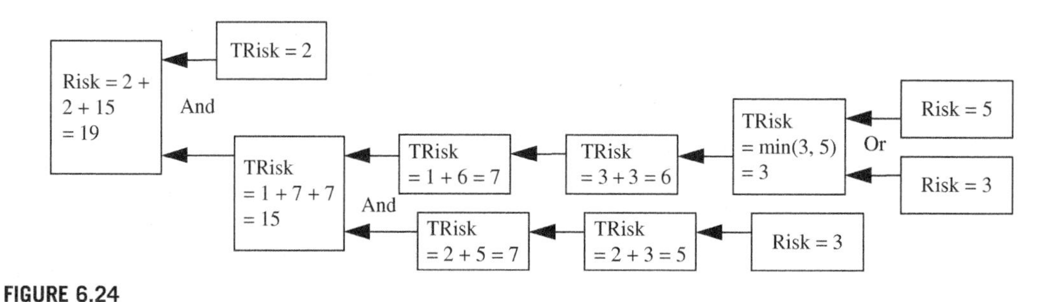

FIGURE 6.24

A risk is assigned to each of the main design parts.

- Develop prototypes: A design risk of 4 or 5 can be reduced to a 3 if a prototype is produced. Refining the prototype can reduce the risk to a 2.
- Purchase development kits or samples: When a purchasing risk is high, or a design risk is high for purchased components, it can be helpful to purchase development kits, obtain samples for testing, obtain design notes, or study existing applications.
- Benchmarking: Higher cost and design risks can be reduced by benchmarking with similar designs. For example, if a similar design can be purchased for $50, the final production cost can be crudely estimated. Taking the design apart and doing analysis can provide a basis for estimating design issues and solutions.
- Consulting: Unknown factors are part of overall risk. The strategic use of consultants and research can reveal unknown factors and reduce the risk for design, purchasing, and cost.
- Focus on the highest risks first: Well-run design projects should not have surprises. A directed strategy of dealing with the highest risk factors first is critical. Unresolved risks can lead to failures later when substantial time and resources have been committed. It is better to identify bad decisions early in the design process.
- Design and build in pieces: If one piece is built and tested the failure risk may be 50%. If two pieces are built before testing, an individual failure risk of 50% increases to 75%. And, now that there are two pieces, there is twice as much work to debug. So, build and test the smallest pieces possible.

PROBLEMS

6.56 If design plan A has a 75% chance of working and the backup plan has a 75% chance of working, what is the combined chance of a successful project?

6.57 A design relies on two concepts that must work for the design to succeed. If they both have a 75% chance of working, what is the overall chance of success?

6.58 A design has two major parts: component A has a risk of 1 and component B has a risk of 3. If component B fails there is an alternate component C with a better risk score of 2. (Note: Even though C is safer it may not be a better technical choice.)

(a) Construct a design decision tree.

(b) Find the overall design score using the risk numbers.

6.59 Over the life of a project the risk scores will decrease. Does this mean that the overall risk will also decrease?

6.60 Every project has tasks with higher risk. Two strategies include (a) work on the higher risks first or (b) do the small tasks first to leave time for the risky tasks. List advantages and disadvantages for each.

6.61 Assume you are part of a team developing antigravity boots. The team realizes that the project carries a level of risk, and decides to use two strategies to reduce overall risk. Which two strategies should be used?

6.18 Risk reduction with design alternatives

Large problems are the most difficult to solve when there are only a few poor alternatives. Eq. (6.1) in Fig. 6.25 shows how to calculate an average time for a design outcome. In other words, if, as a project manager, you made similar decisions, sometimes you would get lucky and projects would be done quickly, other times they would take longer. Similar equations can be written for cost and overall project success.

Consider the example project in Fig. 6.26, where there are three ways to make a new material based on temperature, affecting both cost and time. The result is a variation in time, cost, and probability of success. The question is: Which alternative should be tried first? In this case the decision is between cost and time. If the test sequence is ABC, then the average time would be 37.5 h and the average cost would be $23.15. If work goes well, the project would succeed on step A with a time of 10 h and a cost of $15. If A and B were tried and failed, the time would be 100 h and the cost would be $39. Statistically it is possible to complete the project in as little as 37.5 h, at a cost of $23.15. It is also possible to complete the project at a cost of $19.75, but requiring a time of 71.5 h. Strategically this allows the project time to be cut in half for a 15% increase in cost.

PROBLEM

6.62 After a design review the project team unanimously selects a hardware-based approach that will probably take 10 weeks and has an 80% chance of completion. The backup option is a software-based approach that will take 2 weeks and has a 60% chance of success. Calculate a mean time estimate for the task.

Where,

T_D = Average Time To Solution

T_i, P_i = Decision i takes T time and has a P chance of success

$$T_D = T_1 + (1 - P_1)T_2 + (1 - P_1)(1 - P_2)T_3 + \ldots \qquad \text{eqn. 5.1}$$

FIGURE 6.25

Assessing decisions.

Example,

Method	Time	Cost	Probability
A	10	15	50%
B	40	13	70%
C	50	11	75%

Decision A, B, then C

$$T_{ABC} = 10 + (1-0.5)40 + (1-0.5)(1-0.7)50 = 37.5$$

$$C_{ABC} = 15 + (1-0.5)13 + (1-0.5)(1-0.7)11 = 23.15$$

Decision A, C, then B

$$T_{ACB} = 10 + (1-0.5)50 + (1-0.5)(1-0.7)40 = 41$$

$$C_{ACB} = 15 + (1-0.5)11 + (1-0.5)(1-0.7)13 = 22.45$$

Decision B, A, then C

$$T_{BAC} = 40 + (1-0.5)10 + (1-0.5)(1-0.7)50 = 52.5$$

$$C_{BAC} = 13 + (1-0.5)15 + (1-0.5)(1-0.7)11 = 22.15$$

Decision B, C, then A

$$T_{BCA} = 40 + (1-0.5)50 + (1-0.5)(1-0.7)10 = 66.5$$

$$C_{BCA} = 13 + (1-0.5)11 + (1-0.5)(1-0.7)15 = 20.75$$

Decision C, A, then B

$$T_{CAB} = 50 + (1-0.5)10 + (1-0.5)(1-0.7)40 = 61$$

$$C_{CAB} = 11 + (1-0.5)15 + (1-0.5)(1-0.7)13 = 20.45$$

Decision C, B, then A

$$T_{CBA} = 50 + (1-0.5)40 + (1-0.5)(1-0.7)10 = 71.5$$

$$C_{CBA} = 11 + (1-0.5)13 + (1-0.5)(1-0.7)15 = 19.75$$

FIGURE 6.26

Example calculations for minimizing time and cost when options can fail.

6.19 Business strategy

When necessary try new things. Then learn from the mistakes.

Written vision and mission statements are used to define the purpose of companies and professionals. Normally these will describe the target customers and the types of products and services they provide. These are then used to guide decisions and set priorities, including design projects:

- Vision ("where we want to be")
  - to serve the customer
  - to have a major impact in the market
- Mission ("how we are getting there")
  - to provide coal-mining solutions

  - to find a stimulating job in aerospace control software design (personal)
- Goals ("the end of the path")
  - to have 50% share of subterranean mining equipment sales by 2020
  - to be a chief designer in 2 years (personal)
- Objectives ("parts of the path")
  - to lead the equipment market for mining equipment
  - to increase sales and production by 5% this month
  - to move into management positions (personal)
- Strategies ("how we will walk")
  - find three new customers
  - give demonstrations
  - build prototypes for test marketing

Larger organizations have a hierarchy that complicates strategic plans. The majority of the employees are near the broad base of the corporate pyramid (Fig. 6.27). Each management layer of the pyramid envelopes a larger number of employees. Business-wide missions and visions are developed by a CEO and approved by a board of directors. The business goals and objectives are developed by upper and middle management. The mission, vision, objectives, and goals are used to communicate the corporate priorities. As a communication tool, the strategy must be understandable, relevant, attractive, and applicable for all stakeholders. Executives will do this by incorporating input and assessment from employees, customers, communities, politicians, investors, and more.

Fig. 6.28 illustrates a more team- and customer-oriented view of a company and strategy. The purpose of management is to coordinate employees; the outer rings are the employees and customers who provide assessment feedback to management.

A strategic plan is developed to coordinate activities and written for communication. It is easy to forget the purpose and develop an elegant plan that nobody will take the time to understand. A good strategic plan will consider the following questions from employees:

- Is it easy to read? (Be clear and concise so that it can be read easily.)
- Is there a larger goal? (Describe what success means for everybody.)
- Where do I fit? (Show employees how they will be part of the success.)
- Why should I care? (Be rewarding for employees, to create motivation.)
- Can it be used? (Be relevant to daily activity so that it can be used.)

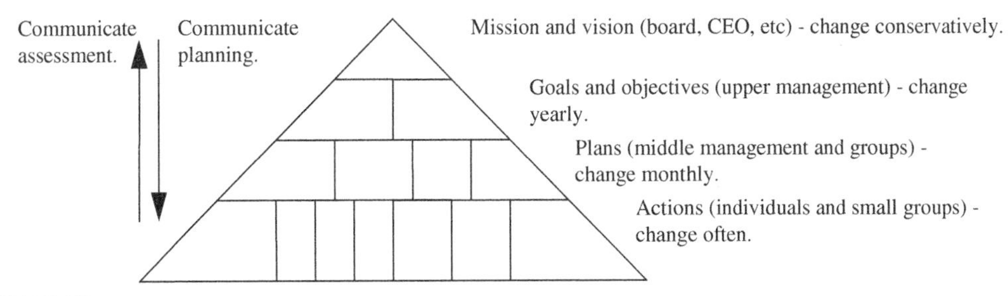

FIGURE 6.27

The planning pyramid (hierarchy).

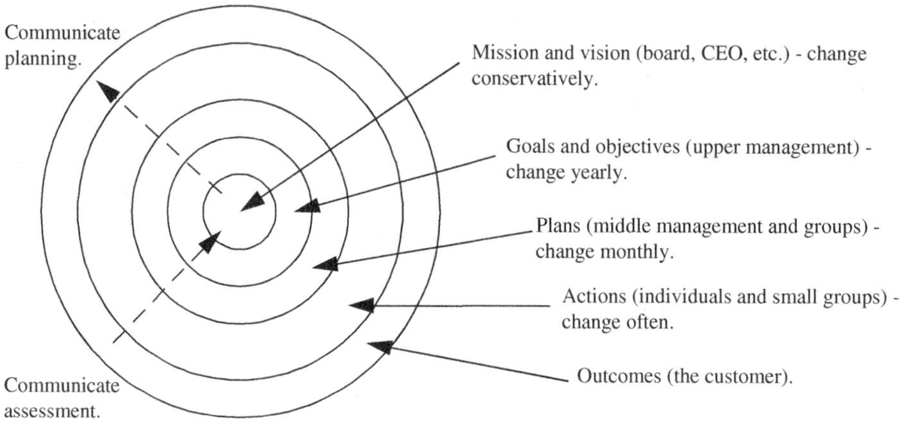

Communicate planning.

Mission and vision (board, CEO, etc.) - change conservatively.

Goals and objectives (upper management) - change yearly.

Plans (middle management and groups) - change monthly.

Actions (individuals and small groups) - change often.

Outcomes (the customer).

Communicate assessment.

FIGURE 6.28

The planning circle (team).

The SMART approach is used for goal development. Each goal is developed to satisfy the five criteria. A specific goal will clarify what is to be done and reduce confusion. Measurements are very important for evaluating progress. Unachievable goals will result in frustration. Irrelevant goals will be completed poorly, or not completed. If a goal is not trackable you have no idea if there is progress or if it will ever be reached.

S—Specific
M—Measurable
A—Achievable
R—Relevant
T—Trackable

Within a design project the plans should include milestones and objectives for tasks. At the end of the project there should be a deliverable that supports a corporate goal, an objective, or the mission. Fig. 6.29 shows how a plan is broken down into objectives.

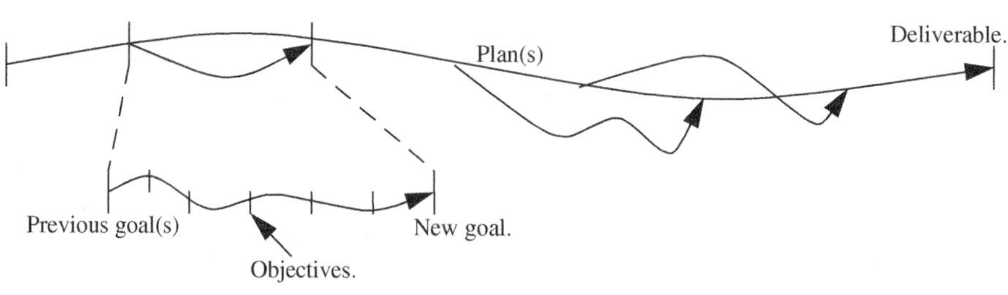

FIGURE 6.29

The planning hierarchy.

PROBLEMS

6.63 How do strategies guide decisions?

6.64 List three reasonable objectives and three reasonable strategies if you are running an ice cream stand.

6.65 Locate a statement of values, vision, and mission for Siemens.

6.66 Assume you are developing a plan for building a simple cheese sandwich. You have chosen the goals of (a) materials collected, (b) surface prepared, (c) sandwich assembled, and (d) workspace clean.

(a) Revise the goals using the SMART criteria.

(b) Add reasonable objectives to each of the goals.

6.20 Assessment and planning

Five-year-old risks reaped profits this year. What are the risks you are taking now?

The mission and vision for a company are updated every few years to address progress and changes in the business environment (Fig. 6.30). These are then used to set shorter-term objectives and goals on an annual and quarterly basis. Within these groups, more detailed plans are set to guide daily actions. The results of the actions are then assessed for effectiveness and consistency so that the processes in the company can continually improve.

Assessing the corporate goals should involve quantitative, qualitative, and intangible factors. Quantitative measures should be preferred, including projected and actual profits, market shares, employees, assets, and market surveys. Qualitative factors include new products, competition, new technologies, customers, and acquisitions. Intangible factors are important but unreliable. Examples include political

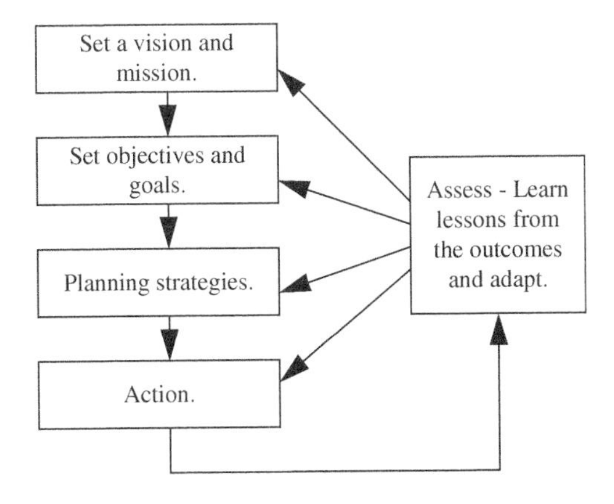

FIGURE 6.30

The large picture of strategy.

changes, research, predictions, customer loyalty, and customer preferences. At a minimum these issues need to be itemized and prioritized. Some helpful questions include:

- How have outcomes been measured in the past at the company and elsewhere?
- What are the major areas of growth and decline?
- What have been our strengths and weaknesses?
- What has changed recently?
- How have we changed?
- Are there any strengths or weaknesses tied to location, market, competitors, etc.?
- Are our expectations reasonable? What are their type and scale?
- How does our business operation differ from our competitors?
- Is the organization built around the mission?
- Does everything support adding value?
- What is the company approach: innovation, commodity, or other?
- What should the next product be?
- What can we do?
- What do we want to do?
- What is good for the company?
- Are bad decisions encouraged by misapplication or misunderstanding of metrics?

A direct method to capture a corporate position is strengths, weaknesses, opportunities, and threats (SWOT) analysis. This is normally done using groups of people from all parts and levels of the company. A set of key points is developed for the SWOT. After this is complete, the group can use the key points to develop new strategies for the company. These should support changes to the vision and mission statements. The following list is a very brief example of a SWOT analysis for a software company:

- Strengths
  - good at fabrication and testing
  - good software design
- Weaknesses
  - few solid customers
  - only one hardware designer
- Opportunities
  - underserved product markets
  - there is excess software design and production capacity
- Threats
  - there are alternative technologies we don't use
  - the cash reserves are very low

After the initial SWOT analysis the team works toward a new corporate strategy. Two possible strategies are listed as follows that show options that do, and do not, include electronic hardware design. A rash decision could be made but it is better to spend time and consider the implications of both. It is hoped that the outcome is something that receives enthusiastic support from all:

- Strategy A: Be the best designer of embedded system software
- Strategy B: Provide turnkey hardware and software-embedded solutions

The diagram in Fig. 6.31 shows some possible approaches to dealing with SWOT outcomes. Obviously the goal is to keep the company financially healthy and avoid risk. Although it is obvious, manage the threats and weaknesses, and build on the strengths and opportunities.

Creativity is an important part of the planning process that is difficult to organize. Some plan steps will be obvious from the assessments. Other plan steps will be found by combining ideas. Don't forget to "borrow" good ideas from other businesses. Occasionally the process will require some creativity. The following list offers some thought-provoking ideas to assist the process:

- Find low-hanging fruit.
- Don't take on too many jobs.
- Do what you do best.
- Focus on what you have, instead of new opportunities.
- The best way to predict the future is with the past.
- Try not to be everything to everybody.
- Focus is everything.
- Chicken or egg: Pick a point and start.
- Emotion will keep you on a path much longer than necessary.
- Using more steps and parts adds more chances of failure (i.e., complexity).

PROBLEMS

6.67 How is assessment related to the quote, "Those who cannot remember the past are condemned to repeat it" (George Santayana)?

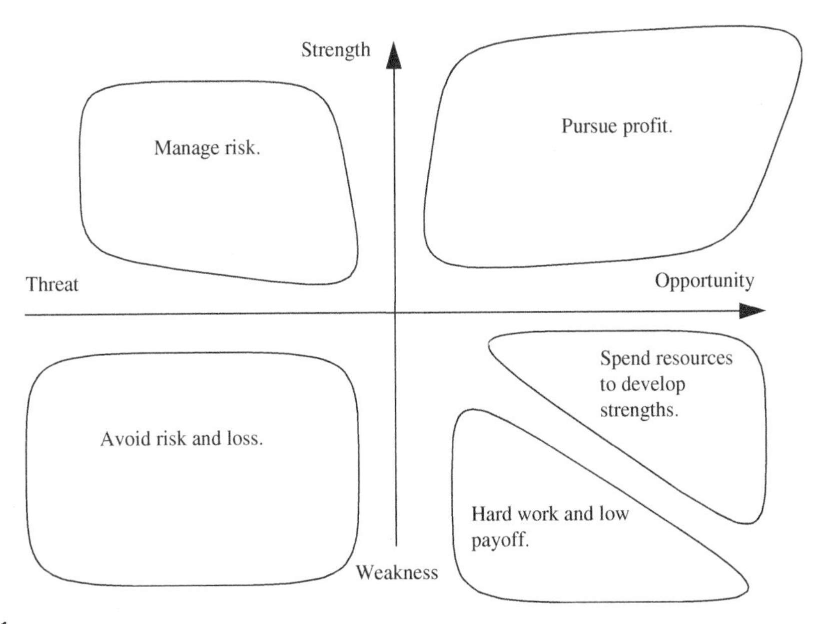

FIGURE 6.31

Dealing with the SWOT (strengths, weaknesses, opportunities, and threats) categories.

6.68 Consider the commercial space industry. In 2012, it included SpaceX and Virgin Galactic offering orbital payloads and high-altitude flights, respectively. Commercial interests are now considering asteroid mining and lunar trips. List 10 assessment points for the current commercial space industries business opportunities.

6.69 Develop a SWOT chart for a high school student deciding to apply for a job at a fast-food restaurant.

6.70 A company has been writing software to control clothes washing machines. There have been some delays and bugs that prompted a discussion about finding an outside software developer. Using reasonable assumptions, develop a SWOT analysis for the decision.

6.71 What are the major categories of project risks?

6.72 Develop a risk analysis for a technology product that has been announced, but does not have a scheduled release date.

Further reading

Baxter, M., 1995. Product Design; Practical Methods for the Systematic Development of New Products. Chapman and Hall.

Bennett, F.L., 2003. The Management of Construction: A Project Life Cycle Approach. Butterworth- Heinemann.

Bryan, W.J. (Ed.), 1996. Management Skills Handbook; A Practical, Comprehensive Guide to Management. American Society of Mechanical Engineers, Region V.

Cagle, R.B., 2003. Blueprint for Project Recovery—A Project Management Guide. AMCON Books.

Charvat, J., 2002. Project Management Nation: Tools, Techniques, and Goals for the New and Practicing IT Project Manager. Wiley.

Davidson, J., 2000. 10 Minute Guide to Project Management. Alpha.

Dhillon, B.S., 1996. Engineering Design; A Modern Approach. Irwin.

Fioravanti, F., 2006. Skills for Managing Rapidly Changing IT Projects. IRM Press.

Gido, J., Clements, J.P., 2003. Successful Project Management, second ed. Thompson South- Western.

Heldman, K., 2009. PMP: Project Management Professional Exam Study Guide, fifth ed. Sybex.

Heerkens, G.R., 2002. Project Management. McGraw-Hill.

Kerzner, H., 2000. Applied Project Management: Best Practices on Implementation. John Wiley and Sons.

Lewis, J.P., 1995. Fundamentals of Project Management. AMACON.

Marchewka, J.T., 2002. Information Technology Project Management: Providing Measurable Organizational Value. Wiley.

McCormack, M., 1984. What They Don't Teach You at Harvard Business School. Bantam.

Portney, S.E., 2007. Project Management for Dummies, second ed. Wiley.

Rothman, J., 2007. Manage It! Your Guide to Modern Pragmatic Project Management. The Pragmatic Bookshelf.

Salliers, R.D., Leidner, D.E., 2003. Strategic Information Management: Challenges and Strategies in Managing Information Systems. Butterworth-Heinemann.

Tinnirello, P.C., 2002. New Directions in Project Management. CRC Press.

Ullman, D.G., 1997. The Mechanical Design Process. McGraw-Hill.

Verzuh, E., 2003. The Portable MBA in Project Management. Wiley.

Verzuh, E., 2005. The Fast Forward MBA in Project Management, second ed. John Wiley and Sons.

Wysocki, R.K., 2004. Project Management Process Improvement. Artech House.

CHAPTER

Finance, budgets, purchasing, and bidding

7

7.1 Introduction

Professionals buy for needs, not wants.

Money, or cash, is a social contract for exchanging agreed values, but by itself the paper has no actual worth. Buying and selling involve trading money for a good or service. The monetary value of a product is subjective and includes factors such as material costs, labor expenses, sentiment, risk, taxes, functions, desperation, supply, reward, and so on. The strategic value of money is that it is "liquid" and can be conveniently traded for goods and services. To get cash one must have a thing that somebody else wants enough to give away their money.

Manufacturers and service companies use commercial expertise to combine other goods and services to create something with more value. After each sale the profit is the difference between the money received and the money spent. Profits can be used for things such as adding new commercial capabilities, taxes, personal rewards, philanthropy, and paying debts. A world-class company will maximize a sales price, minimize production costs, and use the profits wisely so that they will be able to continue producing.

7.2 Corporate finance

Given the importance of money in a company, it must be carefully managed. In companies there are many separate departments, including human resources, engineering, purchasing, accounting, and sales. The sales department maximizes the income, purchasing minimizes expenses, and accounting tracks the flow of money. As engineers our financial role is to specify purchases that will add value to the final products. Indirect costs include facilities, equipment, and product development. Direct costs include costs per product. Indirect costs are often divided into fixed and variable costs. The fixed costs will be spent, even if nothing is produced. The variable costs, such as machine maintenance costs, accumulate when more products are made, but cannot be directly related to each unit.

The direct costs are labor and materials, clearly tied to each product:

- Indirect, fixed: facilities, general production equipment (required even if there is no production)
- Indirect, variable: maintenance department (increases with overall production)
- Direct: components, product wages, single-job production equipment (costs associated with each unit)

Engineering Design, Planning, and Management. https://doi.org/10.1016/B978-0-12-821055-0.00007-4

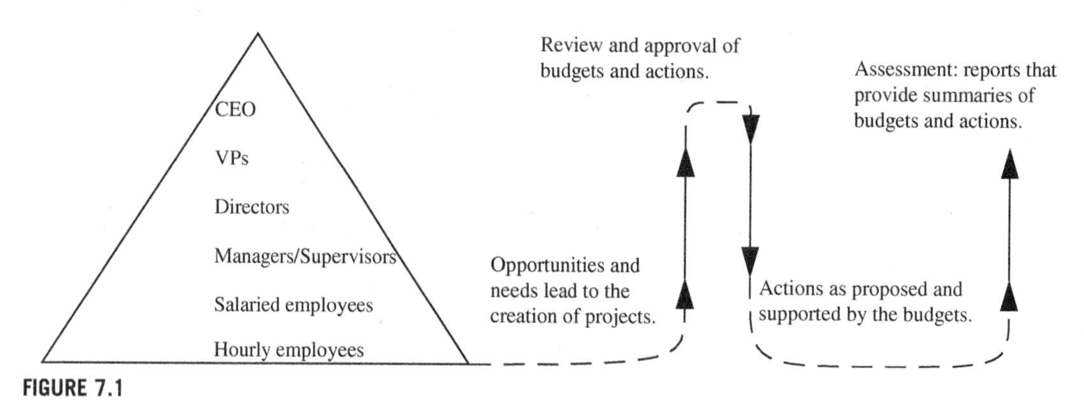

FIGURE 7.1

Financial reporting in a company.

Financial decisions generally follow a process of recommendation and approval. Projects begin with a recommendation of work and a budget (Fig. 7.1). If approved, the accounting department will track the work and give a regular report for the budget items and overall adherence. At an absolute minimum this is done yearly, but every quarter or month is more common. Computer-based accounting systems also allow more frequent but less accurate tracking. A budget is often subdivided into different expense categories, including materials, goods, services, travel, and labor for the project. As purchases are made, they are sorted into different budget categories. The difference between the planned and the actual expenses is called a variance. The regular financial reports generated by the accounting department include expenses, revenue, liabilities, and assets. Sometimes these reports will result in a change in project priorities and possible cancelation of projects.

7.2.1 Accounting

Fig. 7.2 shows a simplified financial model of a company. The bold arrows indicate money flow and the thinner arrows represent the flow of goods and services. Corporate assets are a combination of cash the company has, cash the company will collect, and the value of items it owns. Liabilities are what the company owes in terms of cash and items. The profit or net assets are the difference between the assets and the liabilities. When a company starts, and over time, it takes money from people, banks, stockholders, and others ("I" on the figure).

This increases the assets and liabilities but does not affect the profit. However, the company uses these assets so that it may buy components, combine them into more valuable goods, and then sell them. AR (accounts receivable) and AP (accounts payable) exchange goods for cash with suppliers and customers. Inventory is the collection of goods that are waiting to be sold.

From an engineering perspective our role is to maximize income while minimizing expenses. The main source of income is AR. There are many expenses, including payroll and AP, for the parts, materials, and equipment we use. With careful design and manufacturing we can reduce those costs, as well as inventory, waste, scrap, and operating expenses.

When dealing with senior management and other departments, keep in mind that this is how they see the company. Given the number of financial expenditures driven by engineering decisions, it is easy to understand how engineering can be easily associated with the liabilities. Many problems can be overcome by emphasizing that the projects add to the value of goods and services produced by the company and, as a result, increase the AR.

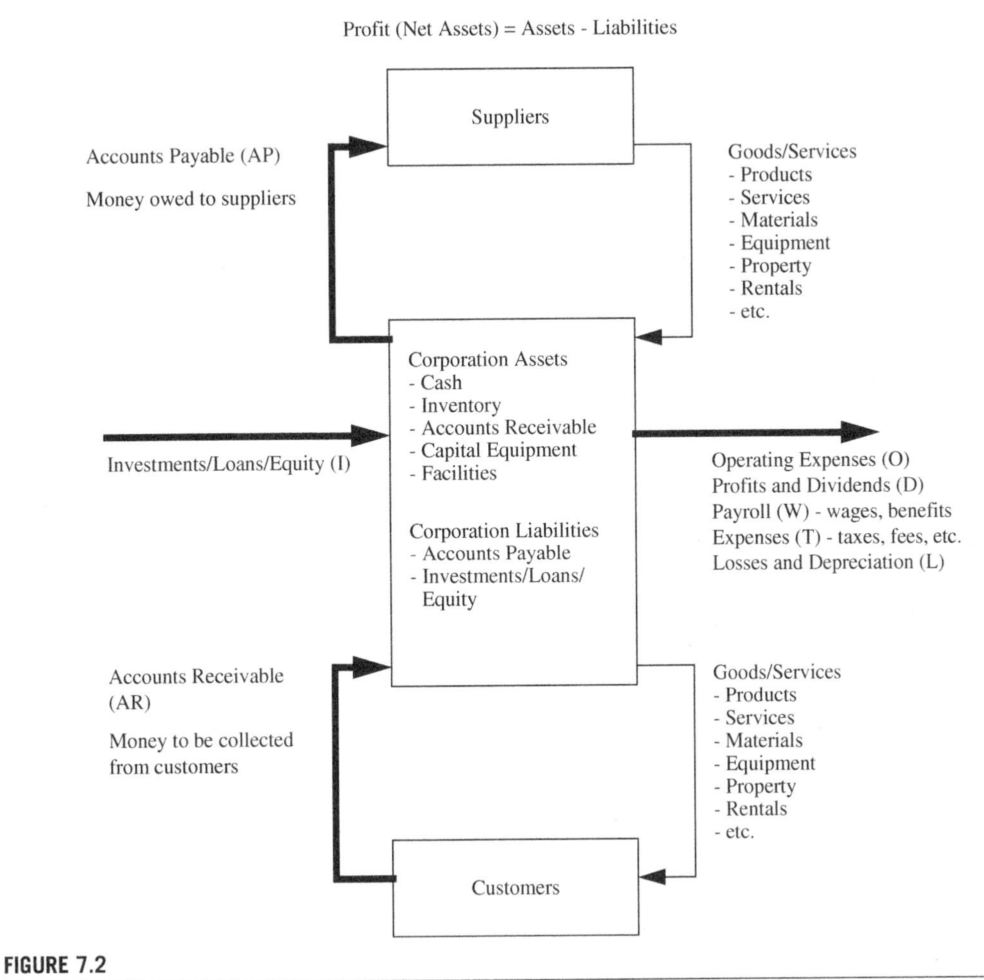

FIGURE 7.2

The flow of goods and money in a company.

PROBLEMS

7.1 Write an equation for cash flow in a business in which total cash equals zero.

7.2 Why are goods and services not assigned a cash value?

7.3 Value

Value is about perception, even with money. As designers and manufacturers, we are tasked with combining the value of many small things into something with higher value. As such, we need to be able to look at things for their value, not just the cost. Fig. 7.3 shows a simplistic scale for ranking value that is based on manufacturing perspectives. Determining the attributes of value for any

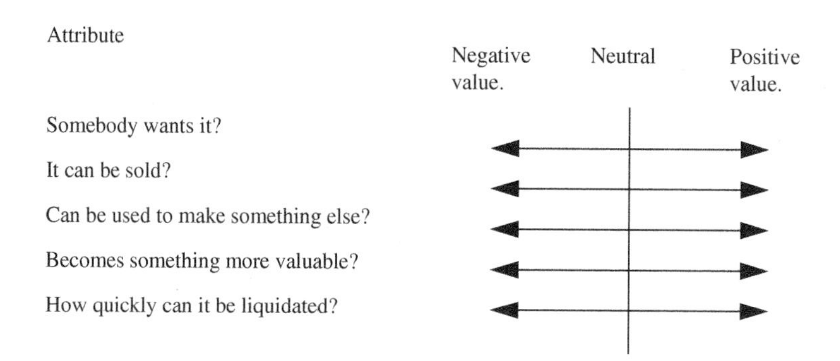

FIGURE 7.3

What is the value of a resource?

individual or business is a mix of understood and intangible factors. Each factor will have a positive or negative influence on the overall value. For example, a book with more topics might have more value, but a book with thousands of pages might be negative. Value is often easily related to finished goods, inventory, knowledge, existing capabilities, raw materials, equipment, customer lists, demand, and facilities. On the other hand, it can be difficult to assign value to intangible assets such as unproven concepts, goodwill, reputation, loyalty, new capabilities, and so on.

In general, money is the best resource in a stable, healthy economy. It is referred to as liquid capital because it can be easily converted into many resources or services from almost every sector of the economy. However, somebody with specific products must find specific buyers for their goods or services, thus constraining them to a very small part of the economy. The disadvantage of money, or liquid capital, is that it is constantly losing value because of inflation. So, the trick is to keep enough liquid capital to do what is needed at any point in time. The cash that is not needed is best placed in investments or activities that grow in value. Value-added activities in manufacturing focus on buying materials and combining them into something with a higher value. The purpose of budgets is to plan for the amount of cash needed, so that the rest can be invested when not needed.

Resources can be used as a generic term for value that is in a form other than money. These items can be directly converted to cash or can help increase the value of other items. A final product is a good resource because it can be sold and the value is translated into money. Something product specific, such as a solder mask or injection mold, is a much less liquid resource, but is still a resource, used to make circuit boards or shape plastic into usable parts. In designs and project proposals it is easier to justify resources that can be liquidated. This makes a strong case for simpler designs with fewer special parts and manufacturing steps. Addressing the issue of liquidity for larger projects will help managers assess the overall value of the project. Some examples of resources are as follows:

High liquidity

- Equipment
- Facilities
- Materials

Medium liquidity

- Business knowledge—finance, management, marketing, accounting
- Skilled labor
- Business systems, including information technology
- Intellectual property, including trade secrets and patents
- Technical expertise
- Design and testing equipment and facilities

Low liquidity (single-design projects)

- Production tooling
- Reputation
- Specialized parts
- Unfinished goods inventory

The value of a product is a combination of production costs and market factors determined by marketing and sales departments. For a successful product, these two elements are essential in forming the needs and customer specifications. If these groups do not work together, the result will often be a product that does not satisfy customer needs and is worthless. The production cost of the product represents all of the materials used in construction, production equipment, production labor, technical costs, shipping, management overhead, and much more:

Common customer values

- Understanding of use, key features, and clear functions
- Low cost overall relative to similar products
- A physical item, as opposed to software
- Brand recognition that provides trust, loyalty, status, etc.
- Emotional association with the product types, including personal history and social pressures
- Longevity, ruggedness, and the cost of replacement
- Urgency and desperation
- Early-to-market value for designs or features
- Novel or new designs

Development and production costs, from design and manufacturing

- Labor and related design expenses
- Acquiring new knowledge and expertise
- Short-term initial design decisions set overall design cost
- Long term, the cost of components and production is reduced
- Production, materials, parts, overhead, distribution, sales, profit
- Use of existing and underutilized resources

A few strategic items for resource planning are shown in the following list. Most of these are relevant when working with departments outside design and production groups. One particular concept that plagues many junior engineers and executives is sunk cost. Essentially, they consider money that has been spent as if it is still part of the decision-making process. Consider the design of a servo system. A hydraulic design has been underway for 15 weeks, and 4 more weeks are required to complete the design. A senior engineer found a new design using an electric motor that needs only 2 weeks for the total design. Which design should be used? Many would see the 15 weeks as an investment and choose to see it to completion. However, the correct course of action is to ignore the 15 weeks as sunk costs, and focus on the 2-week versus 4-week choice:

- Abundance versus scarcity: If a needed resource is difficult to replace, the cost will go up. In business terms, you want to use resources that are very plentiful. On the other hand, if your product is abundant in the marketplace, the cost will drop.
- Tangible versus intangible: The tangible value of a resource is preferred. Tangible resources include cash value or a material used to make a product. Intangible values are much harder to quantify and may never have a benefit. Intangible values include goodwill, image, or new opportunities.
- Goodwill: Goodwill is important in business, but make sure that everybody benefits, including you. Do not use goodwill to justify acceptance of a project at a loss.
- Free: If you give something for free, people will see it as worthless. Maintain the value of the product and brand. Many companies will hold the base price firm and negotiate using additional value goods and services at a discount.
- Prospect theory: The same value broken into smaller parts enhances consumer value. A single rebate of $120 per year will be valued less than $10 per month for a year.
- Liquidity: Apply a financial penalty to resources that are difficult to liquidate. If a project fails, you want to be able to sell the resources and move on to another project.
- Sunk costs: Money and resources that have been spent are gone. Do not make decisions based on money that has been spent, only the money and resources still needed.
- Depreciation: Over time all resources lose value because of tax laws, degradation, obsolescence, etc. If a resource is not needed, delay the purchase to save the money for other purposes. Don't buy a new desk if you can make the same amount of income with an old one.
- Maintenance: Some resources depreciate less if maintained. Base decisions for maintenance on a cost–benefit ratio.
- Opportunity cost: Once resources are allocated they are not available for new opportunities that arise. This should be considered a financial penalty.
- Brand value: A brand name takes time and money to develop and maintain. A brand loses value when it is used for products with different customer specifications or radically different design approaches.

PROBLEMS

7.3 A customer asks for a discount on a product in exchange for loyalty in future business. Discuss the implications in terms of real and tangible values.

7.4 Two options in sales comprise reducing the cost of a product or keeping the price fixed but adding other goods and services. What is the advantage of each?

7.5 What assets are liquid?

7.6 Depreciation encourages faster decisions to reduce cost, while opportunity cost encourages waiting for higher benefits. Put this in an equation form.

7.7 A cell phone service provider has decided to offer a rebate incentive of $100 to attract new customers. Is it better to offer the $100 rebate immediately or divide it into a rebate of $10 for 10 months.

7.8 Why do monopolies reduce the supply of product?

7.9 Give five examples of methods to add consumer value to a computer game without changing the software.

7.10 You open the company safe to find $100 worth of gold and $105 in cash. Which is worth more? Explain your answer.

7.4 Design and product costs

Figs. 7.4 and 7.5 show the typical cost-adding activities for products. For a consumer product there are many added steps for distribution and handling. Along the way, products are lost to breakage, samples, testing, warranty issues, and much more. It is very reasonable for a consumer product to have a materials/components-to-sales cost ratio of 5. A similar ratio is reasonable for specialty designs and equipment purchases. As a product moves from parts at a factory to a product in consumer hands it is accumulating costs and value. In a poorly designed system, there are many operations that add cost but do not add value. A value-added approach will carefully assess each operation for the product life cycle for value and cost.

Design activity adds value with some cost, as illustrated in Fig. 7.6. The curves shown in the graph are not universal, numerically accurate, or complete, but the concepts are useful for planning. Before a

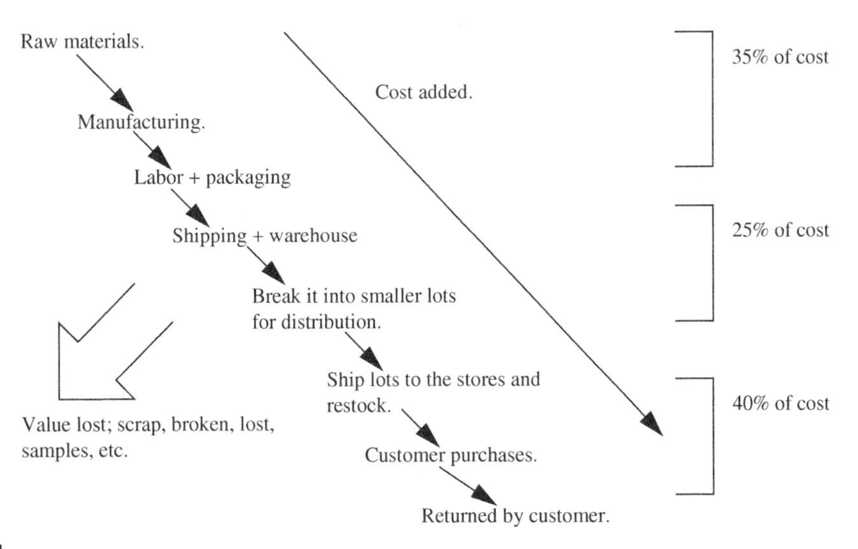

FIGURE 7.4

The cost of goods over the product life cycle.

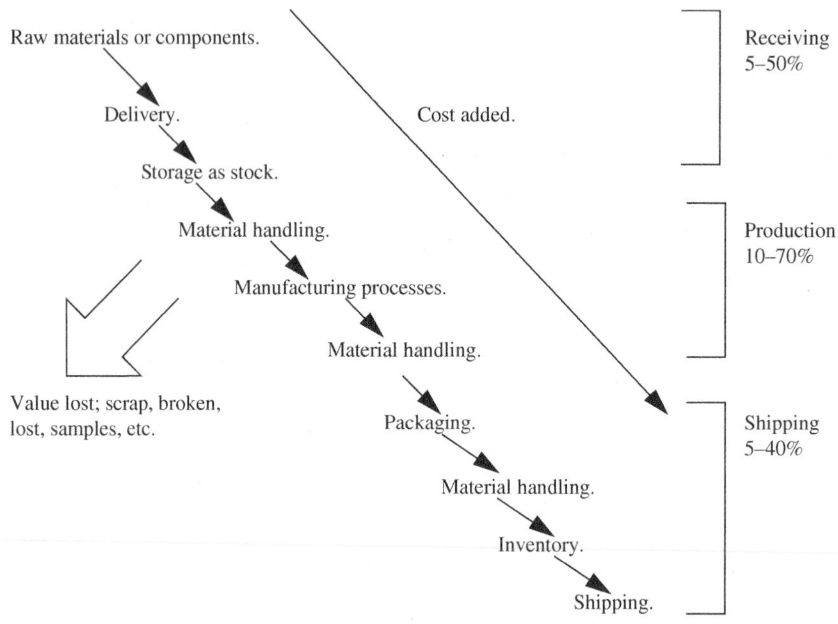

FIGURE 7.5

Adding cost in production.

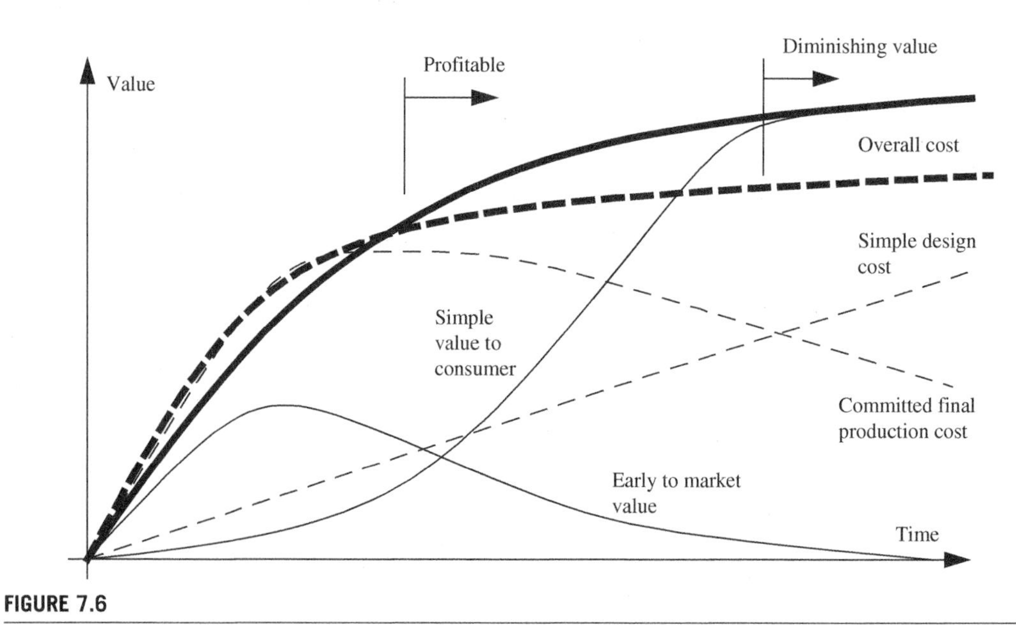

FIGURE 7.6

An example of incremental cost of design per part.

certain point, the design is simply not ready for market because (1) it will cost too much to produce, (2) many technical problems remain, and (3) it is lacking many features the customers want. However, a product that enters a market sooner can have a novelty advantage. Early in the design process, the technical decisions are very far reaching and valuable. With each generation of the product, the design matures, and the changes decrease and eventually become fine-tuning and optimizing.

Design activity may add value that the consumer is not willing to pay for. At the upper extreme of design work, there is so much design activity that there are diminishing returns because (1) the cost of design continues to increase, (2) the design work is focusing on features that many customers don't understand or want, (3) most of the production and material savings have been realized, and (4) similar products have already captured part of the market. A good manager will recognize when all of the "low-hanging fruit" in the design process is gone and resist unnecessary refinements. (Note: The design activities should continue until all specifications are met.) The key for a new design is to find the right balance to stop designing and start making. Sales-oriented businesses tend to rush the design process to get more new products on the market. Technically oriented companies tend to prolong the design process, to achieve precision.

Use quality, not precision, to decide when the design is complete.

The target market has an impact on the design approach, as seen in Fig. 7.7. A project to produce a single machine may use very expensive components to reduce development costs and risks. Projects for high-volume production should use more design time, to reduce component and manufacturing costs.

The activity of design creates a dilemma for project teams because it appears to add to the overall project time. However, a suitable amount of planning can reduce careless mistakes and surprises that could result in costly backtracking and waste. Fig. 7.8 illustrates two possible project approaches. In the "quality" approach, the team spends time planning the process before committing to any design decisions. Their efforts are effective, but superficially they seem to be falling behind. The other team quickly begins to order parts, makes design decisions, and shows apparent progress. As the design moves forward they discover that some components are not compatible, requiring additional design time. Then they discover that there was a misunderstanding about the customer needs. These and a few other setbacks slow the design process, and eventually debugging and building stretch out, as problems are identified and fixed.

If you kick a dead racehorse long enough it will cross the finish line.

Another argument in favor of extended planning is shown in Fig. 7.9. When money is actively being spent there is apparent progress, so there is pressure to move forward and begin spending money sooner. However, most of the eventual budget is set in the very early stages of design. Once the essential components and processes are set, the remainder of the design components will not vary much. Therefore, additional planning time at the beginning of the design can result in subsequent cost savings in the build and test phases of the project. This is more critical when designing for medium- to large-scale production.

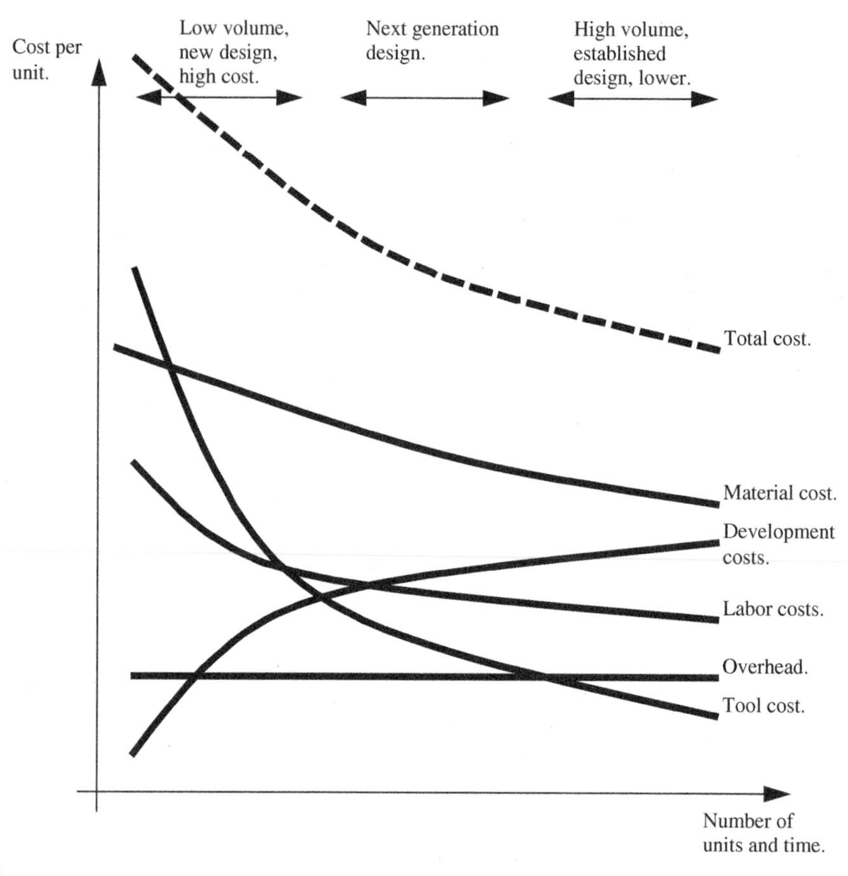

FIGURE 7.7

Production quantity cost reduction over time and quantity.

PROBLEMS

7.11 Suppliers and customers place different values on products. What value does shipping add for the customer and supplier?

7.12 Assume a company normally plans for a maximum variation of +5 each month. The finance office compiles the budgets to find the projected expenses. If the projected expenses are \$10,000 for November, then \$10,500 is kept in the bank account. Any remaining cash is invested to gain interest. Higher interest rates are available when the money is locked into bonds for a month or longer. If the expenses exceed the variance, then the business must borrow the extra money at a financial loss. If the budgets are below projections the extra money does not earn interest. Given this scenario, should a budget manager be rewarded for always being below budget?

7.13 Describe the trade-offs between longer and shorter design times. Draw a graph that relates manufacturing costs (y axis) to the design duration (x axis).

7.14 Is a low cash flow at the beginning of a project a sign of trouble?

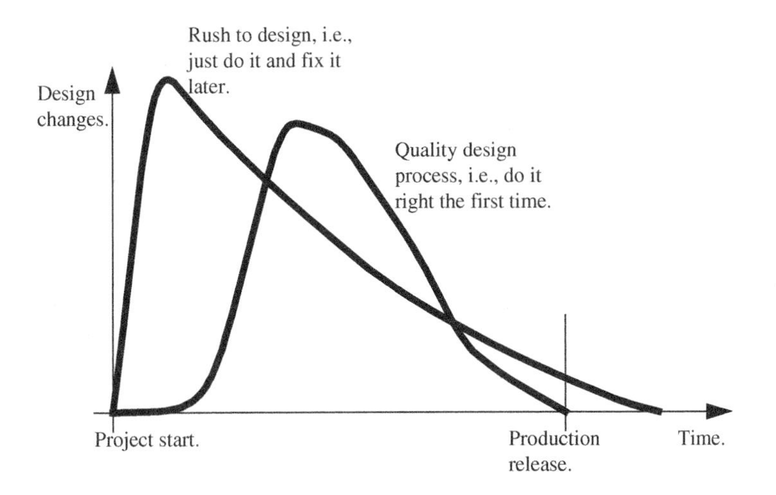

FIGURE 7.8

Shortcuts hurt.

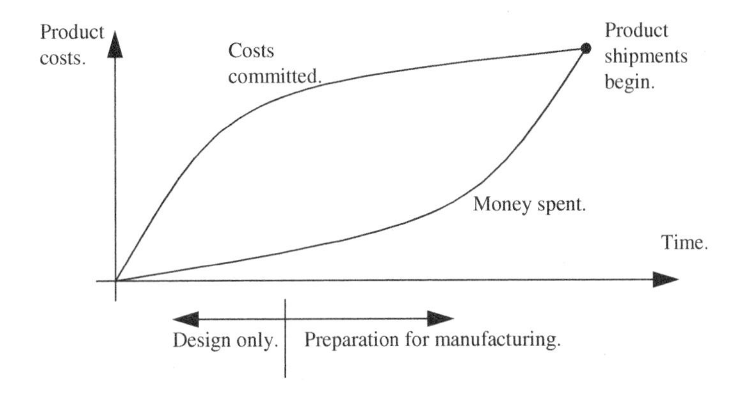

FIGURE 7.9

Actual project expenses and the estimated cost per production unit.

7.15 The materials for a new computer cost $200. What is a reasonable consumer price range? Consider all of the associated costs.

7.16 What is the cost of rushing through the design process?

7.17 Explain why a longer design stage is valuable when producing millions of products.

7.5 Project costs

During the design process, the teams will select components, equipment, and suppliers that will have an impact on the corporate cash flow. Given the importance of the projects to the future of any company, it is essential to consider and propose budgets for the finance department.

7.5.1 Budgets and bills of material

At the beginning of a project, the budget items and costs are estimated. As the project progresses, detailed budgets are developed. Each purchase and expense is tracked and added to the appropriate budget category. The following list shows a sample budget. This could accompany a project proposal that would justify each of the expenses. A budget in a proposal will often include a cash flow estimate, if the project duration is longer than 3 months:

- Electrical design department labor: $22,000
- Mechanical design department labor: $11,000
- Electronics, printed circuit boards, and components: $3000
- Rapid prototyping: $2500
- Injection molding tooling: $15,000
- Product artwork: $2000
- FCC radio emissions testing: $3500
- Patent attorney: $10,000

Labor is a component in all budgets, and it requires estimates before, and tracking during, the project. When the budget is prepared, an estimate sheet will be used (Fig. 7.10). Regular employee labor costs are normally divided between projects. Consider the example of a salaried engineer who earns $70,000 per year and receives another $30,000 in benefits. Last year, the engineer, worked 2000 h and recorded 1500 h of project work. The minimum labor rate for that engineer is $100,000/1500 = $67 per hour, but the company uses a figure of $80 per hour for general engineering labor. Hourly workers are paid specifically for the jobs they are working on. A machine operator who earns $20 per hour effectively earns $30 per hour with benefits. If he or she works on a job for 20 h, the labor is worth $600. Administrative labor, including a project manager, is often included in the overhead costs, or burden costs. Administrative costs can be added to hourly rates for employees or as a percentage of the total project budget.

Most projects will last months or years. Over that period the expenses and income will vary by time and project stage. Substantial income normally comes at the end of a project; expenses are greatest at the beginning and middle. For example, there may be a large expense for project equipment at the midpoint but no substantial income until the end of the project. Cash flow projections are used to set aside money for the purchases (Fig. 7.11). The company will keep some contingency funds available for the budget variances.

Task	Person	Hours	Rate	Period start.	Period end.	Monthly cost.	Actual
PCB design	Pere	150	$80	Jan	Jan	12,000	
Software design	Anya or Paul	200	$80	Feb	Mar	8000	
Housing design	Jacque	50	$80	Feb	Feb	4000	
Testing	Test employee	40	$30	Apr	Apr	1200	
Tooling	Machine shop	200	$70	May	July	7000	
Production	General labor	0.2/part		Sept			

FIGURE 7.10

An example of labor cost estimation.

A bill of materials (BOM) lists all of the parts required to produce or assemble some other device (Fig. 7.12) (see Resource 7.1). This has more parts than those listed in a budget because some of the parts will be work-in-process components. In other words, the original material has been worked on to produce new parts, but no purchase has been involved. A BOM is normally found on assembly drawings. Detailed budgets list all major components, minor components, prices, and quantities. A well-constructed budget should also list the catalog numbers, source/supplier, and status (e.g., not ordered, received, due 2 weeks, late 1 week). Consumables, such as bolts and wires, are normally listed under a "miscellaneous" heading. However, all other components should be listed and prices provided. If the components have been drawn from the engineering inventory, similar devices can be identified from catalogs, and those prices may be used.

Resource 7.1 A budget and BOM worksheet are included on this book's website: https://sites.google.com/site/engineeringdesignprojects/home/content/resources/budgetworkshee.

The true cost of any item is greater than the purchase price. This includes direct costs such as shipping, shipping insurance, customs duties, internal labor for handling, labor to install equipment, and so

Month/year	Expenses			Income			Notes
	Projected	Actual	Variance	Projected	Actual	Variance	
Jan 2012	14,500	13,923	577	10,000	10,000	0	First payment
Feb 2012	23,000	24,934	(1934)	0	0	0	
Mar 2012	46,000	44,482	1518	0	0	0	
Apr 2012	16,900			0			
May 2012	34,300			30,000			Second payment
June 2012	6100						
July 2012	1700			80,000			Delivery
Aug 2012	500						
Sept 2012	0			20,000			Withholding

FIGURE 7.11

Cash flow projection and report.

Item Description	Source	Quantity	Price	Date Needed	Date Ordered	Received	Paid
Plasma Cutter Xn8-C	Arcco plc.	1	12,899	May 5	Feb 15		
Robot Mini-9a 5 axis	Droidsnmore Intl	1	32,168	Mar 21	Jan 20	Feb 12	
Steel tube 30cm sq.	Metalemporium	200m	10/m	Feb 10	Jan 10	Jan 20	
PLC XJ-9, IO-combo	Sparkystuff Inc.	1	1290	Feb 10	Jan 10	Jan 13	Feb 11
Cabinet Nema 3, 1m	Sparkystuff Inc.	1	235	Feb 10	Jan 10	Jan 28	
Sensors - RF10-diff.	Spookyaction Ltd.	13	35/each	Mar			
Wiring Supplies	Internal store room		1000				
....							

FIGURE 7.12

A simple materials list.

on. Indirect costs include the space for storage, internal labor for storage, internal labor for restocking supplies, purchasing agents, accounting payments, initial time to order and track, and so on. When all of these are considered, a $10 part may need a $20 courier service, 1/2 h of accounting to pay the bill, and 1/2 h of receiving and delivery to an office. The final direct and indirect costs could easily become $60. A business-minded engineer considers these costs and tries to minimize them through careful planning and bulk ordering.

A project budget must be estimated before a project has begun. Three common approaches are outlined in the following list. An estimator with experience may be able to draw on knowledge of similar projects to identify required items and values; this is known as speculation. In top-down budget planning, the project is broken into stages and major components and then iteratively refined down to individual components. In bottom-up design, the key components are identified and then combined (rolled up) into larger budget items. Labor is a budget item for all stages of the project, and time estimates will be required. For product designs, the budget may also have separate prototyping and production estimates. An effective budget process will use a combination of all three of the approaches. The bottom-up approach will be most valuable when the material costs of a project are relatively high. When the component costs are low but the labor/development costs are high, the top-down approach will be more valuable:

- Top-down (parametric) budgeting: Get estimates from other divisions and managers and then augment with organization costs.
- Bottom-up (roll-up) budgeting: Get piece lists from designers and then augment with organization costs.
- Speculation: Use history and knowledge of other jobs to estimate costs based on the work breakdown structure.

Eventually the budget must be compiled into a single number and cash flow projection. The budget value will be used to prepare a customer quote or an internal project proposal. In both cases, a review is required to make sure there is a clear financial benefit. When a budget contains only direct parts costs and paid labor costs, it requires more assumption and estimates by managers. By adding more financial considerations, the budget is more likely to receive a positive review. The following list describes some of the issues that influence budget decisions. The best budgets and proposals will show a clear profit for the company at the end of the project:

- Some estimates of best-case and worst-case outcomes are very helpful, even if the worst case will result in a loss.
- Some managers will always provide less budget money than requested. This often happens when overestimates are common.
- Some managers will use budgets as maximum spending limits.
- In long-term budgets, money is allotted by time period. When a time period such as a month, quarter, or year is complete, the account may be closed and the money is returned to the corporate pool. This can be a problem if a large purchase is delayed to another budget period, when the money is not available.
- Contingency planning may allow special and justified requests that can exceed the budget. Contingency money is normally limited, so it is more difficult to convince a manager to use it.
- Some companies spend conservatively, leaving budget surpluses at the end of a month/quarter/year so that the money is not spent. Often this results in rushed purchases to avoid losing the money.

- A manager will look for a well-prepared budget or quote that includes project cost or quote = material costs + labor costs + overhead + profit (return on investment [ROI]) + risk costs + opportunity cost.

Once a budget is approved and the project begins, the expenses are tracked and compared with the budget. If the variance between projected budget and actual expenses is too large, positive or negative, the project may be reviewed. If the expenses are well below the budget, a manager should review the original project plans for other misestimates, missed steps, or changes in the work. When the expenses are far over budget, a project should be reviewed for new budget predictions and differences between the work plan and the actual progress. When a project is not following the original plan it may be revised and reviewed as if it is a new project proposal. The key concept is project control. If the budget and expenses match within a reasonable range, the future expenses can be predicted and the project is said to be in control. When the differences are unexpected or cannot be easily explained, the future budget cannot be predicted and the project is out of control. Some strategies that can be used for projects that are out of control include intervention, cancelation, or delay:

- Appoint a new manager and team.
- Lobby or negotiate for a higher budget.
- Abandon the project.
- Continue and accept the financial loss.
- Increase the frequency of reporting and review stages.
- Set aside contingency funds for the anticipated budget excesses.
- Break the project into stages, beginning with the most difficult tasks first, followed by others. An example would be a pilot study to develop a prototype, followed by another project to develop a production design.
- Break the project into core and optional parts. In this case the customer can select the components that will fit into the budget. This avoids the all-or-nothing scenario.

When preparing budgets, accuracy is important; a project that ends close to the projected budget is appreciated. A project that goes far over or under budget is poorly estimated. As project managers become more experienced at estimating, the variances should decrease. If there is a chance for budget variations, it is good practice to estimate these to provide a budget range. For example, a project that uses large quantities of electricity may be subject to varying utility costs. Previous variations in price or analyst predictions can be used to estimate the extremes and probable energy costs. The incentive to budget accurately is that a lower budget makes the project more likely to occur, and a higher budget makes profit more likely. A manager that overestimates budgets will be awarded fewer projects; a manager who underestimates may lose money.

7.5.2 Tracking budgets

For example, the cash-flow projections will be sent to the accounting department and relevant managers so that they can plan to have the needed money in the budgets (Table 7.1). The project manager can track the accounts and compare the expected with the actual expenses, to identify problems. It is normal practice to track the difference between actual and projected budgets, and when the difference is too great, an escalation procedure is initiated to examine the reasons for the variations in expenses.

Table 7.1 A sample cash flow table for a project.

Month	Monthly expenses	Total income	Financial position	Major event
Sep	10	0	(10)	Launch
Oct	20	0	(30)	PO received
Nov	55	0	(85)	Purchase of machine
Dec	20	0	(105)	
Jan	15	0	(120)	Design review
Feb	10	0	(130)	Production
Mar	100	280	50	Delivery
Apr	5	20	60	Final sign-off

PO, purchase order.

Table 7.2 A bill of materials worksheet.

Part number and name	Quantity	Source or supplier	Delivery date	Cost	Order status

Engineering groups and purchasing departments will use a BOM (Table 7.2). Early in the project this will have details for major purchases and estimates for various components. As the design progresses, more detail will be added and the purchasing department will use these details to purchase prototyping materials and negotiate for production volume purchases.

PROBLEMS

7.18 How can the salaries for project managers and CEOs be incorporated into project budgets?

7.19 Is the cost per hour for an employee the same as his or her salary?

7.20 Develop a draft BOM and monthly budget for the described design. The project is to develop a new cosmetic product within a 6-month period. The total budget is $26,000 for the design work and tooling. Bottle design prototypes will cost approximately $4000 and take a few weeks to produce and test. The final tooling for the bottle and cap will be $11,000 and take 6 weeks. An industrial designer will consult on the bottle shape and labels for $4000. The remaining funds will be allocated for engineering salaries.

7.21 Mini-case: The price isn't the cost

Tyrone required a sheet of acrylic plastic for a window on an electronics cabinet. A nearby plastic supplier had provided a quoted price of $55 for the sheet and $15 for delivery. When preparing to order the plastic, Tyrone checked an Internet supplier, PartsCo, and found that their cost was

only $40. Given the price that was $15 lower, Tyrone had the purchasing department order the material from PartsCo. The order was placed, and the sheet arrived a week later on a large shipping truck. The team thought nothing about the sheet, until they were entering purchase costs in the budget sheet.
In the past the team had purchased items from PartsCo and shipping costs were $5 to $30 depending on the time to delivery. They had not asked for a rush on the sheet so they expected a cost of no more than $15. The cost of delivery for the sheet from PartsCo was $50. After some investigation, they discovered that parts larger than a certain size or height-to-width ratio cannot be shipped using standard carriers. As a result, PartsCo used a special shipping service that is normally used for delivering much larger equipment. As is the normal practice, Tyrone gave the order to the purchasing department and did not order the part himself. The purchasing department verified his material choice and the purchase price. They simply assumed that Tyrone understood the shipping charges. What could Tyrone do differently next time to avoid the problem?
Does the saying "too good to be true" have any bearing in this situation?

7.6 Return on investment

If you are not using a resource to make money, you are losing money.

Inflation causes (liquid) money to constantly lose value. In addition, money that is not being used is missing an opportunity to grow in another investment (project). Both of these factors combined cause most companies to define an internal return rate in the range of 15% and up. For example, when justifying a project that costs $100, it must return $120 in profit in 1 year to have 20% ROI. If the company requires a 15% ROI the project would be a success. If the company requires a 30% ROI, it would be a failure. If a project lasts long enough, the time value of money must be considered; the future expenses and income are converted into present-day dollars. Given the nature of technology, and the changes in markets, most projects will have a maximum ROI period of 1 to 3 years.

A simplified relationship for calculating ROI is shown in Fig. 7.13. For each job i, there is a cost, C_i, and a bid, B_i. The ROI rate is set by management to cover additional costs of the project and other indirect company expenses and to generate profits. These other costs include everything from the sales department to the cleaning crew. For any job, the bid price should be the project cost, plus the ROI. At the beginning of a project, the values for C_i and B_i are estimates; by the year-end, they are calculated exactly using accounting and labor records. Factors that complicate setting the ROI are large changes in assets and liabilities.

If money is invested the value increases every year. If the money is spent it cannot be invested and value is lost. This creates an incentive to delay purchases and obtain payments sooner. For a project that stretches multiple years it can be difficult to compare present and future money values. The solution is to take all of the costs and compare them to a present, or future, value. For simple value-of money calculations, the rate of inflation can be used. For example, with a 2% inflation rate, what costs $1.00 in 2023 will cost $1.02 in 2024. Examples of present value, PV, and future value, FV, calculations are shown in Fig. 7.14.

Given that investments normally perform better than inflation, a company will use an internal rate of return (IRR) related to the normal returns on investments. For example, an investment in bonds may normally return 8% per year and the company chooses to use IRR = 8. Fig. 7.15 shows the equations for calculating FV and PV. In the example, a purchase of $1000 in 3 years requires only $751 in today's

Where,

ROI_i = Simple Return On Investment for job i.

B_i = Bid cost quoted for job i.

C_i = Estimated/Actual cost for job i.

n = Number of jobs in year.

R = Total income for the year.

E = Total expenses for the year.

$$ROI_i = \frac{B_i - C_i}{C_i} \qquad \text{eqn. 7.4}$$

$$B_i > C_i(1 + ROI_i) \qquad \text{eqn. 7.5}$$

$$ROI = \frac{R - E}{\left(\sum_{i=1}^{n} B_i\right)} + \text{inflation} + \text{profit} \qquad \text{eqn. 7.6}$$

FIGURE 7.13

Simplified bidding, costing, and return-on-investment relationships.

dollars, less than the $1000 for an immediate purchase. In contrast, spending the $1000 3 years earlier is the same as paying $1331 then. These calculations are used to take all of the project expenses, and income, and shift them to a common reference point in time.

PROBLEMS

7.22 If $100,000 were borrowed for 3 years at a 10% interest rate, how much would be due at the end of the loan?

7.23 Jenny borrows $200 with an annual interest rate of 9%. At the end of the first year, she repays $50. How much does she owe at the end of the second year?

7.24 Assume that land is appreciating at 4% per year (Note: This is equivalent to an interest rate). The company invests by buying $100,000 worth of land each year. After 5 years the company has spent $500,000. How much is the land worth after 5 years?

7.25 The IRR and ROI rates should be much higher than inflation. Why?

7.26 If $100,000 were borrowed for 3 years at a 10% interest rate, how much would be due at the end of the loan if $20,000 were repaid each year?

7.27 The sales engineering team estimates that a job will cost $15,000. The company uses an ROI rate of 20%. What is the minimum acceptable bid for the customer? What is the ROI if the bid is $25,000?

Where,

PV = present worth of the money (in today's dollars).

R_{A_j} = Annual revenues (income) for year j.

C_{A_j} = Annual costs (expenses) for year j.

j = j years in the future.

i = interest rate (fractional).

n = number of years for consideration.

FV = future worth of the money.

$$PV = C_0 + \sum\left[\left(R_{A_j} - C_{A_j}\right)(P/F, i, j)\right] \quad \text{eqn. 7.7}$$

$$(P/F, i, j) = \frac{1}{(1+i)^j} \quad \text{eqn. 7.8}$$

$$(P/A, i, n) = \sum(P/F, i, j) = \frac{(1+i)^n - 1}{i(1+i)^n} \quad \text{eqn. 7.9}$$

$$F = P(F/P, i, n) \quad \text{eqn. 7.10}$$

$$(F/P, i, n) = (1+i)^n \quad \text{eqn. 7.11}$$

A simple example of a $1,000 now, that will increase in value by 5% per year for three years. At the end of that time the equivalent sum is $1158.

$i = 5\%$ $\quad P = 1000 \quad$ $n = 3$

$$F = P(F/P, i, n) = 1000(F/P, 5\%, 3) = 1000\,(1 + 0.05)^3 = 1157.63$$

FIGURE 7.14

Present and future value calculations.

7.28 Assume the inflation rate is 3%. This year a chair is worth $500. How much should it cost in 1, and 2, years?

7.29 Assume the inflation rate is 3%. If the price for a chair is $500 now, how much should it have cost 2 years ago?

7.30 Prove the PV and FV calculations for a 3-year period.

7.7 Financial project justification

Break-even analysis looks for the match between project expenses and income (Fig. 7.16). The payback period is the time to the break-even point. For example, a project that costs $100,000, but generates an income of $40,000 per month, would have a break-even period of 2.5 months.

Internal Rate of Return (IRR) - The company value for cash. The value is always greater than inflation.

Discounted cash flow:

$$FV_i = PV_i(1.0 + \mathrm{IRR})^n$$

Assuming the IRR is valid the Future Value in "n" years.

$$PV_i = \frac{FV_i}{(1.0 + \mathrm{IRR})^n}$$

Note: These formulas are valuable when calculating values over multiple years. These formulas are not needed if using a spreadsheet with annual interest calculations.

For example

If a purchase of $1000 was made now, how much would it be worth in 3 years with an IRR = 10%?

$$FV_i = PV_i(1.0 + \mathrm{IRR})^n = 1000(1.0 + 0.10)^3 = 1331$$

If a purchase of $1000 was made in three years, how much would it be worth now with an IRR = 10%?

$$PV_i = \frac{FV_i}{(1.0 + \mathrm{IRR})^n} = \frac{1000}{(1.0 + 0.10)^3} = 751.31$$

FIGURE 7.15

The time value of money.

Where,

C_I = initial investment ($)

S_A = savings per year ($/yr)

N = payback period (years)

Break even analysis is the point where the costs and expenses are equal.

$$\sum \text{cash received} = \sum \text{cash spent} \qquad \text{eqn. 7.12}$$

Payback period is determined when there is a "break-even" point.

$$N = \frac{C_I}{S_A} \qquad \text{eqn. 7.13}$$

FIGURE 7.16

Break-even analysis.

When projects last years the time value of money becomes significant. To compensate, the future expenses and income are converted to present value, as seen in Fig. 7.17 In break-even analysis, the time period, n, is adjusted until the net present value (NPV) is equal to zero.

Multi-year projects are normally converted to present day dollars for even comparison. This is called the Net Present Value (NPV).

$$NPV = \sum_{i=0}^{n} PV_i = \sum_{i=0}^{n} \frac{FV_i}{(100\% + IRR)^i} = \sum_{i=0}^{n} \frac{(income_i - expenses_i)}{(100\% + IRR)^i} \quad \text{eqn. 7.14}$$

FIGURE 7.17

Net present value.

Where,

C_E = cost of new equipment.

I_S = revenue from sale of old equipment (salvage).

L_0, L_1 = labor rate before and after.

H_0, H_1 = labor hours before and after.

M_0, M_1 = maintenance costs before and after.

t = the break-even period.

$$CI = CE - IS \quad \text{eqn. 7.15}$$

$$S_A = (L_0H_0 - L_1H_1) + (M_0 - M_1) \quad \text{eqn. 7.16}$$

$$t = \frac{C_I}{S_A} \quad \text{eqn. 7.17}$$

FIGURE 7.18

Cost savings justification.

In many projects, the goal is improvement of a current process, design, and so on. For these projects there is often an initial cost for the new capital equipment minus the sales value from the old equipment, as shown in Fig. 7.18. The cost savings are calculated by comparing labor, time, and maintenance before and after the equipment change. The capital equipment purchased for improvements is usually a one-time cost at the beginning of the project, while the actual savings occur on a monthly basis. These projects usually use a break-even period for financial justification.

As equipment is used in production it slowly loses value. For example, a machine that is worth \$100,000, C_E, might lose \$25,000, D, in value over 3 years, n. After 3 years, the machine has a theoretical (salvage) value of \$25,000. For accounting purposes, the annual depreciation is claimed much like it was any material used in production. The \$25,000 depreciation from the machine is used to decrease the taxes paid by the corporation, as shown in Fig. 7.19. Corporate taxes are normally 50% of all profits, therefore the tax benefits of depreciation can be significant when justifying project budgets.

Given that assets wear at different rates, depreciation periods vary. For example, the depreciation period for a building could be decades, whereas the depreciation period for a laptop might be a few years. These time periods and rates are defined by government tax policies. One of the simplest calculation methods is the straight-line depreciation approach, where an equal value is subtracted each year.

Where,

A = after tax cash ($/yr)

B = before tax cash ($/yr)

D = depreciation of equipment ($/yr)

tax_{rate} = the corporate tax rate

$$A = B - T = B - (tax_{rate}C) = B(1 - tax_{rate}) + Dtax_{rate}$$ eqn. 7.18

$$D = \frac{C_E - I_S}{n}$$ eqn. 7.19

FIGURE 7.19

Basic depreciation and taxes.

Where,

D_j = The annual depreciation for year j

f_j = the depreciation factor for year j

$$D_j = C \times f_j$$ eqn. 7.20

Recovery % for a given depreciation period (US).

year	3 yrs	5 yrs	7 yrs	10 yrs
1	33.33	20.00	14.29	10.00
2	44.45	32.00	24.49	18.00
3	14.81	19.20	17.49	14.40
4	7.41	11.52	12.49	11.52
5		11.52	8.93	9.22
6		5.76	8.92	7.37
7			8.93	6.55
8			4.46	6.56
9				6.55
10				6.55
11				3.28

FIGURE 7.20

The US modified accelerated cost recovery system method.

Other methods include the US accelerated cost recovery system (ACRS) and the modified ACRS (MACRS), as shown in Fig. 7.20.

Although depreciation calculations are specific to tax codes and equipment types, the general principle will guide engineers doing project planning.

Smaller projects are often easy to justify without extensive calculations. However, projects that span years, or involve major capital purchases, should be discussed with the accounting and financing managers. Being aware of taxes, depreciation, salvage, and the time value of money will make you more effective as a project manager.

PROBLEMS

7.31 The NPV calculation converts all future expenses and income into current dollar values. Why is this a useful measure?

7.32 What is the difference between a payback period of 3 months and one of 1 year?

7.33 A machine was purchased for $100,000 and generates $20,000 per year as income. How many years would be required to break even, if the company charged a 10% internal interest rate?

7.34 A machine is purchased for $100,000 and the lender charges 10% for the use of money. What annual return is required for the machine to break even in 3 years?

7.35 A machine costs $100,000 and has a 5% yearly depreciation rate. Using the MACRS approach, indicate the deduction for each year, and the value if it is sold after 3 years.

7.36 A machine costs $100,000 and will be sold for salvage value in 3 years, for $30,000. The alternative is to lease a machine for $40,000 per year. If the company uses an interest rate of 10%, which option should be chosen?

7.37 An existing manual production line costs $100,000 to operate per year. A new piece of automated equipment is being considered to replace the manual production line. The new equipment costs $150,000 and requires $30,000 per year to operate. The decision to purchase the new machine will be based on a 3-year period with a 25% interest rate. Compare the present value of the two options.

7.38 A company purchases a solder reflow oven for $50,000 and uses a 3-year cost-recovery depreciation. How much will they deduct each year?

7.39 From a financial standpoint, purchasing equipment seems to be a zero net change in the company assets, where equipment replaces cash. In practical terms, equipment is something that is consumed, like ink in a printer. Depreciation is used to assign a cost to equipment based on standard aging. However, the NPV of the depreciation will be worth less than the purchase price. What is the NPV for the deductions of a company that purchases a solder reflow oven for $50,000 and uses a 3-year cost-recovery depreciation?

7.40 Engineers are not accountants, but they should understand basic accounting principles. Why?

7.8 Product life-cycle cost

The price of a product is only part of the cost of ownership and use. The following list provides a number of the expenses that go into the total cost of ownership and use. A designer should consider these elements. For example, an automobile customer who focuses only on the sale price will fail to recognize the expenses related to daily operation and repairs. Likewise, a consumer who purchases a large piece of equipment must be aware of the true costs. For example, a manufacturer that purchases a new computer-controlled milling machine for $100,000 should expect to spend another $50,000 over the life of the machine:

Infrequent

- Repairs
- Recalibration and recertification
- Retraining
- Upgrades and updates

Regular

- Consumables—ink, paper, disks, cleaning supplies
- Utilities—electricity, water, heating, fuels
- Labor—operators, maintenance
- Fees and licenses
- Insurance
- Annual tax depreciation
- Financing costs for larger purchases or leases

Fixed

- Purchase price
- End-of-life salvage price
- Delivery costs—truck, courier, personal automobile usage
- Installation costs—electrical, plumbing, construction, services, sewing
- Peripheral and support equipment—cases, covers, tests

PROBLEM

7.41 Estimate the lifetime cost of owning an automobile. Include price, insurance, fuel, licenses, maintenance, income from end-of-life sale, accessories, cleaning supplies, etc.

7.9 Business decisions

Change brings opportunity and risk. Without opportunity it is pointless to accept risk. It is also unwise to ignore risks while pursuing opportunities, a simple view of the risk–benefit spectrum is shown in Fig. 7.21. In an ideal world we would be able to be safe and have major benefits. If we are willing to take losses to be safe, then we are overly afraid of risk. To take risks with no benefits can bankrupt

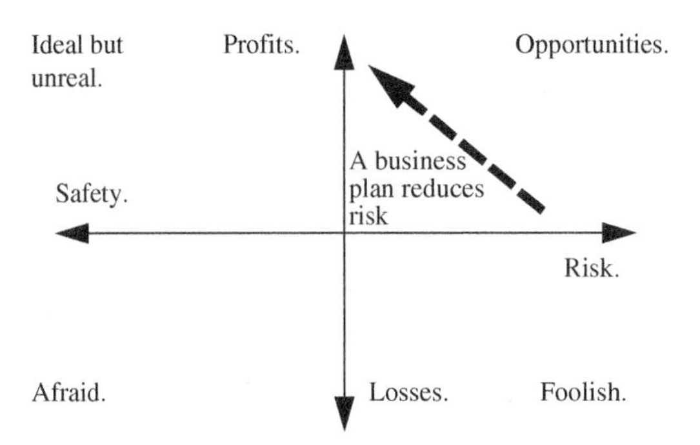

FIGURE 7.21

Risk–benefit spectrum.

your company. We normally expect to work in a situation in which there are benefits and risks. It is worth noting that everything contains a combination of benefits and losses that must compensate for risks. The trick to all of this is to recognize where you are on the spectrum. Benefits tend to be easy to recognize, while risk can be more elusive. Section 7.9 explores methods for identifying, quantifying, and managing risk. Only when the risks and benefits are understood are we able to make balanced decisions for project work.

PROBLEM

7.42 "You can't succeed without taking some risks" is a well-known saying. Which quadrant of the risk—benefit spectrum does it describe?

7.10 Purchasing

Don't inflate specifications in hope of getting better results.

Purchasing is more of a routine activity, done to acquire materials needed for daily engineering work. These can include components, evaluation kits, software, raw stock, fasteners, office supplies, and much more. Some of the terms related to purchasing are defined as follows:

- Supplies: These low-cost items are used for administration or production.
- Account: A supplier may have an account for routine purchases. This often has a total limit and the account balance is paid monthly.
- Reimbursed: When an employee submits a claim for money spent, he or she is reimbursed. Normally receipts are required.
- Preferred supplier: This is a supplier that does not require special clearance; it will often offer a discount for preferred customers.
- Original equipment manufacturer (OEM) or volume: Large purchases often require special arrangements, but will reduce costs. OEM parts are used in other products. For example, OEM software comes bundled with a computer; retail software is sold by itself.
- Prototype or small volume: These materials are available in smaller quantities at higher prices.
- Samples: Some materials are not available in small volumes but suppliers provide free samples for evaluation or design purposes. This is very common with special materials, electronic parts, and mechanical components. These often come with technical support from the suppliers.

For various reasons, the purchasing department is often separate from the design and manufacturing groups.

Some companies allow limited purchases directly by employees; however, larger items are directed through a purchasing agent. The agents are often very effective at ordering commodity and volume parts, but often have issues with specialty items. The typical purchasing process is outlined here with a few of the common pitfalls. Fig. 7.22 shows some of the procedures used by engineers when purchasing parts. Smaller purchases will often be made by an employee. The employee submits the receipts and is reimbursed for the amount spent:

(1) Smaller purchases can be made by engineers with company credit cards as personal purchases that are reimbursed. These often have limits in terms of cost and item types.

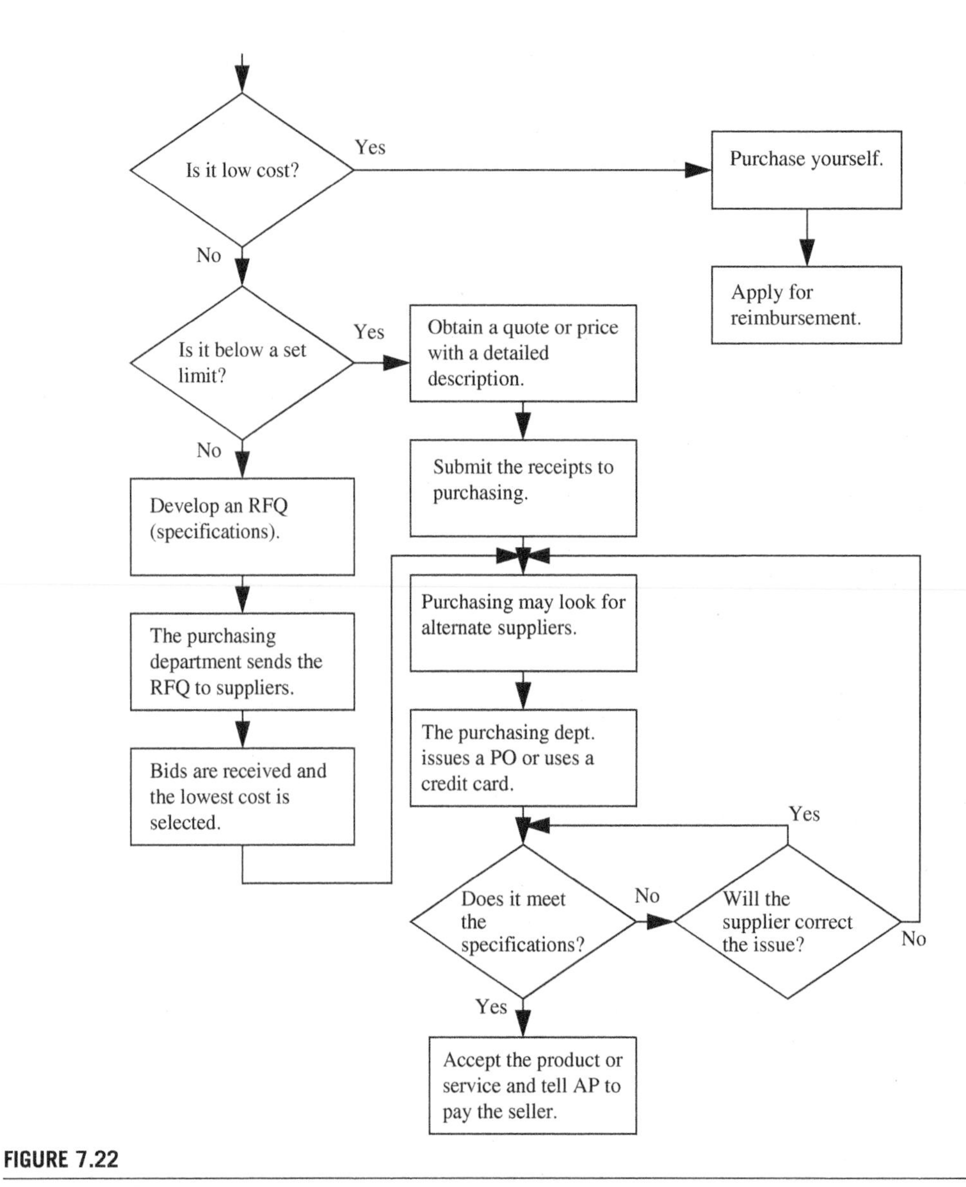

FIGURE 7.22

Purchasing an approved item. *AP*, accounts payable; *PO*, purchase order; *RFQ*, request for quotes.

(2) Requisition forms are normally used for larger orders. These forms typically ask for a preferred supplier, part numbers and descriptions, quantity, and other details. The ordering department indicates an account that is to be used to pay for the items. When these forms are incorrect, incomplete, or vague, the purchasing agent often makes assumptions. A typical problem is that a form says 10 bolts but the agent purchases 10 boxes of bolts.

(3) The purchasing agent sorts the forms by priority. Unless it is marked "urgent," purchasing agents may let orders wait for weeks. Even then, long delays are common. Assuming purchasing is a fast process is a costly mistake.
(4) The purchasing agent reviews the forms and looks for common items and suppliers. When an unusual supplier is specified, the agent may look for the same part at a known supplier. These changes may lead to nonequivalent substitutions. Using preferred company suppliers increases the chance of a successful order.
(5) The agent will then use the Internet, phone, or forms to place the order. Typographical errors are common in this process. Sometimes it is possible to enter orders on a supplier website (e.g., for electronic parts) and then have the purchasing department approve it.
(6) The supplier processes the order and prepares it for shipping. Naturally, mistakes will also occur in terms of items, quantity, and shipping. This includes shipping address. Many shipments arrive at the wrong location and sit for weeks unclaimed, sometimes a few feet from where they should have been delivered.
(7) The parts arrive and receiving logs, or shipping paperwork, are sent to purchasing/accounting. A wise engineer will check all parts as they arrive, comparing them to his or her own parts list. This is the last chance to easily catch purchasing mistakes. Logging these parts in the project BOM will reduce confusion and help track important components.

Every project has purchasing problems.

Many purchasing mistakes can be avoided by considering the following factors. Being careful and detailed will take more time before the order is placed, but will save substantial time and confusion after the parts arrive. After placing orders, or giving the request form to the purchasing department, you depend on others to satisfy your needs. If anything is not clear, there is a probability that it will be misinterpreted. Make it easy for others to do things correctly. Consider the example of a USB connector with a 10-digit manufacturer code that ends with an S (surface mount) or T (through hole) to indicate the mounting type. In haste, a production manager forgets to add the T to the end of the order, assuming that the supplier will send what they always do. A new employee at the supplier sees the code and assumes it is surface mount, the most popular component. To reduce inventory at the customer, the parts are scheduled to arrive the day before they are to be used. Obviously production is delayed because the wrong connectors need to be reordered because a T was not added to the end of a part number. The following list includes some hints that can be used to reduce purchasing problems:

- Use catalog part names and numbers exactly the way they are printed in the catalogs.
- Provide all supplier contact information, including the names of people.
- Highlight any special shipping instructions. When possible, include a name and phone number.
- Leave 2 extra weeks for purchasing to order the parts and an extra week for shipping.
- Develop a friendly relationship with the purchasing department so that they will call you if there are problems.
- Put your name and phone number on all paperwork.
- Double-check all of the details before sending the forms.
- Check all orders as they arrive. If they are late, look for them.
- Use the preferred suppliers and parts when possible.
- If possible, request quotes or create detailed orders yourself, to reduce purchasing errors.

PROBLEMS

7.43 An employee may be able to spend up to $400 and ask for reimbursement. Why do companies use these limits?

7.44 What is the advantage of a preferred supplier?

7.45 Engineers often send a purchase request to the purchasing department and receive something similar to what was on the request, but unsuitable. List three ways this can be avoided.

7.46 How does purchasing volume affect the negotiation between suppliers and customers?

7.47 A purchasing process that requires three bids has advantages and disadvantages. List three of each.

7.48 Mini-case: Creative purchasing
DefenseCorp does projects for a government that has limits for expenses. Restrictions on computer networking hardware limits purchases to a security-approved list. Permission to purchase equipment that is not on the list normally takes 6 weeks. One of the software engineers, Ned Flanders, decides that he needs a router for testing software. It will not pose any security threats or end up in the final design. He reviews the approved options and the least expensive is $1000 and has a shipping delay of 2 weeks. The type of router he needs is readily available at electronics stores for $40. A colleague tells him that in similar situations in the past, engineers have purchased similar items and submitted receipts for reimbursement. To avoid the restrictions, they list the expense as "office supplies." Should Ned use this approach to save time and money?

7.11 The supply chain for components and materials

This section assumes that the technical needs for a production component are already understood and specified; the remaining step is supplier identification. As discussed previously, the cost of a component or material is higher than the price for an individual part. Naturally, price is an excellent starting point for selection and comparison. Beyond the basic price are additional costs such as shipping, handling, insurance, and packaging. If more parts will be purchased in the future, the supplier must be able to provide similar components at a similar price. A component that is available from multiple suppliers and is always available is preferable to a component that is made by only one supplier in special batches by special order. Supply-chain management is the practice for formally planning and tracking the supply and cost for components. Managers and designers need to be aware of the supply chain when selecting components. Some elements of supply-chain management include:

- using fewer unique parts
- replacing custom parts with standard parts when possible
- fewer manufacturing steps
- using commodity parts when possible
- having alternate suppliers
- spreading the project risk between suppliers
- avoiding new designs when parts can be purchased
- Using suppliers with experience for risky designs

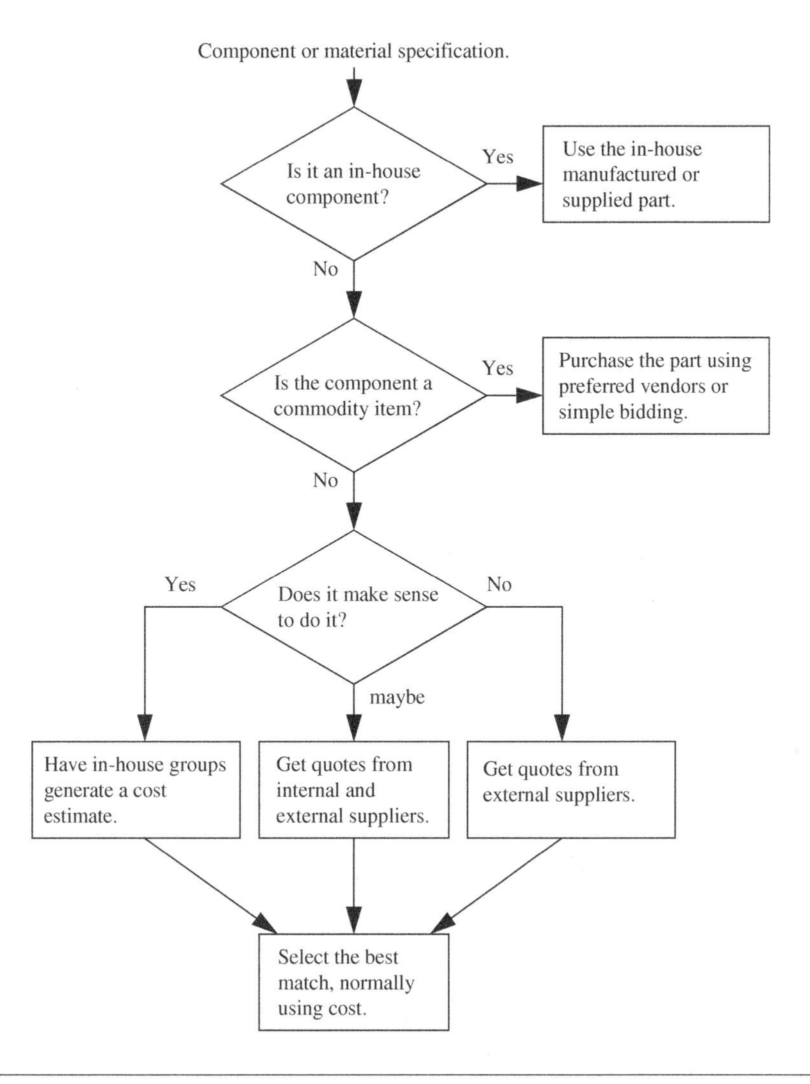
Component or material specification.
Is it an in-house component?
Yes
Use the in-house manufactured or supplied part.
No
Is the component a commodity item?
Yes
Purchase the part using preferred vendors or simple bidding.
No
Does it make sense to do it?
Yes
No
maybe
Have in-house groups generate a cost estimate.
Get quotes from internal and external suppliers.
Get quotes from external suppliers.
Select the best match, normally using cost.

FIGURE 7.23

Selecting project components.

- considering buying the tooling
- ensuring that the suppliers have reliable supply chains

A simple flowchart for considering purchases is shown in Fig. 7.23. Many companies have parts that they make, or stock, internally. If the company uses enough parts they can be bought in bulk to lower the cost. If the part is made within the company it should help decrease the overall cost and increase utilization of facilities.

Commodity items are excellent choices because they can be purchased from multiple competitors, to obtain lower prices. Most companies have arrangements with preferred suppliers for discounted purchases, but if problems arise they can easily change to another supplier. Specialty components require more consideration. If there is expertise, and free time, the parts could be made in-house. Otherwise external suppliers could be sought. In many cases a company will find that an outside supplier can provide components at a lower price than if they were made in-house. At the end of the process the cost should be low and the supply reliable.

"Outsourcing" is a term for using outside companies to do work that could be done in-house. This is not always the best solution but it can be valuable when (1) you do not have enough capacity to do the work, (2) the supplier costs are much lower, (3) the supplier has special expertise, and (4) there are strategic benefits. When outsourcing, you are delegating part of your work to another company. Working closely with the supplier will mean better results. The elements of a good supplier relationship include profits, technical support, clear specifications and tolerances, coordinated production, and communication. Some of the factors to consider when selecting suppliers are as follows:

- Intangibles
  - Reputation
  - First impressions
  - Eagerness and enthusiasm
  - Trustworthy independence in work
  - Personality
  - Self-awareness and vision
  - Dependence on limited resources or key people
- Business
  - Financial expectations
  - They provide something you need
  - They want some things you have
  - Single-source suppliers pose monopoly risks
  - Being a small customer to a large company cuts your purchasing influence
  - Corporate stability
  - Location and shipping issues
- Strategic
  - Previous work
  - Fits a clear need
  - A mutually beneficial relationship
  - Is innovation needed?
  - Is the work commodity in nature?
  - Corporate and individual philosophy

Fig. 7.24 shows an outsourcing decision path for a major task. It begins with a need, and eventually leads to a design project or decision to wait. If you cannot do the work well, a supplier may be able to do it at a reasonable cost. If you can do it, you may consider if there is enough benefit to doing it internally.

An extreme version of outsourcing is offshoring, which refers to work sent to other countries. In the most controversial cases this is done for cost reduction, but other reasons include unique skill sets and capabilities. An example of outsourcing would be a company that makes coffee makers. Design work is done at the corporate headquarters, located near Main Street, USA. The manufacturing, including packaging, is done by a supplier on another continent, to reduce labor costs. When a product shipment

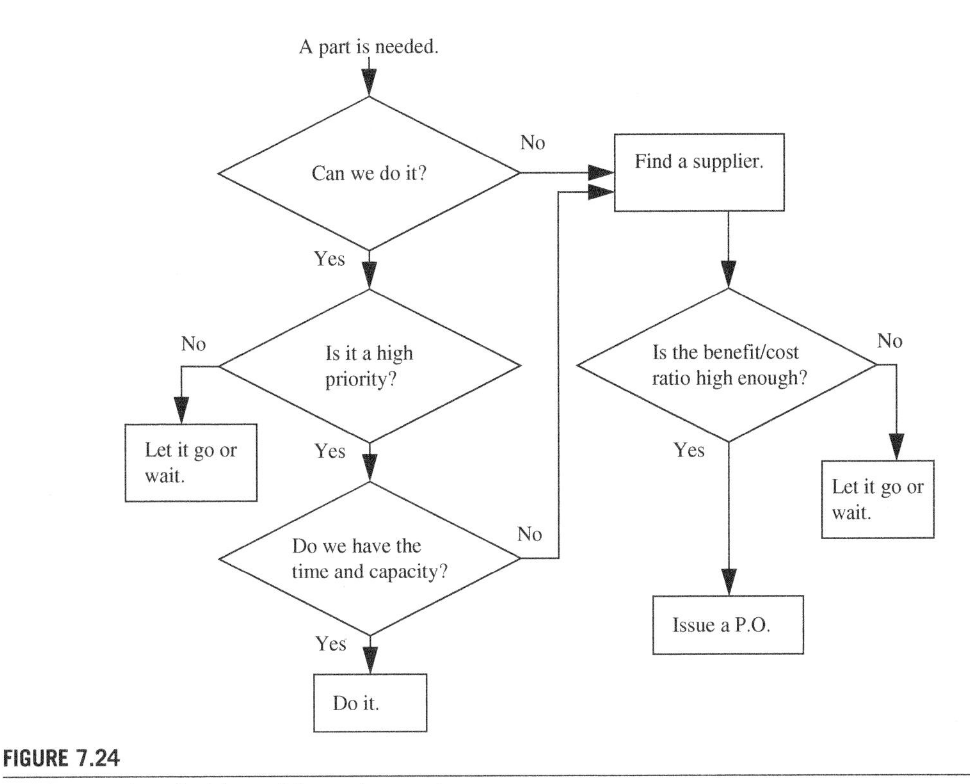

FIGURE 7.24

Build versus buy decisions. *PO*, purchase order.

is received, a testing lab in another country is used to verify the quality levels. When doing design projects, the team will often consider the components with higher costs and with special needs. Outsourcing may reduce or increase risks, delays, quality, control, image, and costs. It is strategic to consider that the supply chain is global and this is very unlikely to change. One business can benefit from a weak currency in another country, but also suffer because sales to that country decrease. Diversifying supply chains can be valuable in component sourcing. A case in point is the May 2011 earthquake and tsunami in Japan that disrupted computer memory production, threatening half of the world's supply. Customers that used the Japanese suppliers faced the prospect of higher-cost parts and halted production. When selecting suppliers, Japan would appear to be a very stable source, but in the global economy offshoring can also affect risk.

PROBLEMS

7.49 Assume a device uses 20 similar, but different, resistor values. One percent of all orders has a problem with delays, quality problems, or reduced quantities. What is the combined probability of problems? What would the probability of problems be if the 20 resistors were combined into two different types?

7.50 List three reasons a company might outsource a part it can build.

7.51 Mini-case: Four of eight
A machine was being designed to test vehicle braking systems. The design team at Specialmachines was well into the detailed design phase of the project. To increase accuracy they needed a set of low-friction bearings. There were a number of standard sizes available, but a supplier, Smoothco, was offering odd-sized bearings at a much lower price. Given that the machine was a unique design, Specialmachines decided to use the Smoothco bearings. After the detailed design was complete, a purchase order was sent to Smoothco for eight bearings. A couple of weeks later a crate arrived with six bearings, four were as ordered and two were only partially finished. A design engineer and the receiving manager spent time looking for a lost crate and checking the order paperwork. The engineer called Smoothco to ask about the odd order contents. The call was answered by a message saying that Smoothco was no longer in business, and gave another phone number to call. It was clear that the team would receive only four functional bearings.
Should the team use the four odd-sized functional bearings, and four normal-sized bearings? How could the team have minimized the risk of part purchasing? How would cost influence the decision?

7.12 Bidding

Bidding is a competitive process. Customers will state needs or specifications and ask for proposals and costs. A simple example could be a set of storage racks to hold a few tons of equipment. Multiple suppliers would review the customer needs. Each supplier would estimate the work required to make the shelves and develop a proposal and budget that satisfies the customer and ensures a profit. The key is to quote a price high enough to ensure a profit but low enough to beat the other bidders. To state the obvious, accepting jobs without profit will help put a company out of business. However, submitting higher bids will result in less work. Finding the right price point is a business issue that engineering must support.

Developing a bid, or quote, requires time and effort. Therefore the decision to submit a bid involves a few factors, including (1) the resources required to prepare the bid, (2) other opportunities, (3) likelihood of success, and (4) likelihood of profit. Fig. 7.25 illustrates a reasonable review process when preparing bids. The work should include a combination of customer specifications, concepts, and technical specifications. Most bids are submitted using budget estimates based on a technical specification.

It is unwise to submit a bid that does not provide a clear profit, or carries too much risk. However, somebody is bound to make a case for submitting a money-losing bid. Reasons for these include philanthropy, developing new business, using excess capacity, and obtaining expertise. In certain circumstances each of these can be valid reasons, but they require very careful justification. For example, some companies require that 5% of all profits are used for charitable activities. Sometimes projects are done to provide an example of capabilities, or increase awareness. A common example is a manufacturer that provides sample parts, design consulting, and production prototypes to a potential customer. Some nonfinancial factors to consider when submitting bids include the following:

- Intangibles
  - Customer reputation: Does the customer pay bills on time, argue during delivery, have a hands off attitude, micromanage?
  - Your reputation: Will it help or hurt your image?
  - What is the overall project risk?

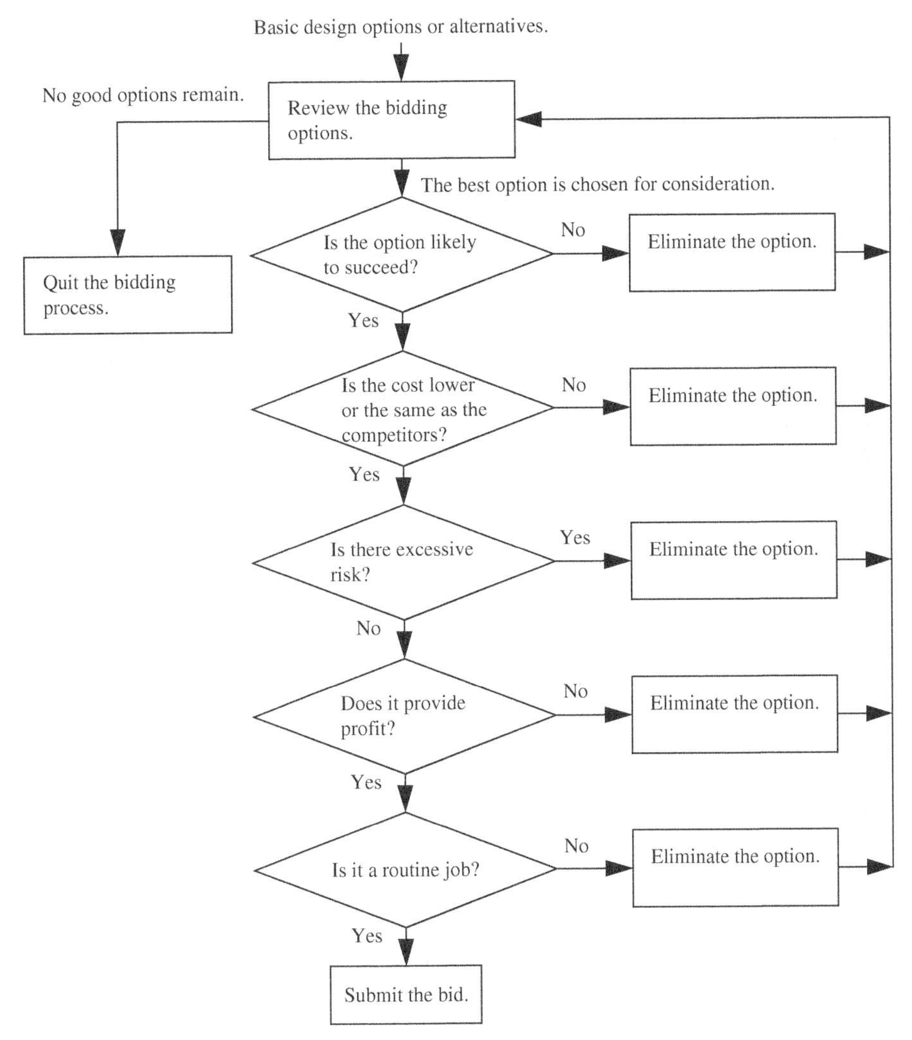

FIGURE 7.25

Bidding options.

- Are there specific technical risks?
- Is it consistent with company goals and objectives?
- Are there better opportunities?
- Does it create new opportunities?

- Killing points
  - Excessive risk (this may be handled with multiple project steps)
  - Unprofitable

- No free capacity
- Not our expertise
- Good points
  - Large profit and benefits
  - Easy with prior experience
  - Does not require any special effort

Branding is an excellent tool for considering discounted bids. A brand is a recognized value of a product or service. If the brand value is high, you can charge more for products and services. If it is poor, nonexistent, or negative, it can reduce the value of products and services. Consider a supplier that provides $10,000 worth of equipment for $5000, hoping that a customer will return for more. However, when the customer returns they want $100,000 worth of equipment for $50,000, based on the earlier deal. A better strategy would have been to charge the customer $10,000 for the first equipment, but offer them a $5000 discount on the larger order.

Increase brand value by:

- not offering discounts or sales
- providing other items free (devalue something else), instead of offering price discounts
- exchanging a thing of perceived value, such as membership status, etc.
- promotion
- association with higher-value products and brands
- being stable and trustworthy

Decrease brand value by:

- selling products at a lower cost
- giving products away
- creating the perception that the products are unpopular

The statistical variations in a project cost can be used to adjust budget estimates (Fig. 7.26). The process uses the probability of failures and the cost for each option to calculate a budget that will break even over a number of jobs. This approach is well suited to companies that do very few jobs and cannot afford the cost of a failure.

A company that does many jobs can handle budget variances by increasing the ROI. This is equivalent to self-funded insurance. Some useful factors to consider when preparing bids are as follows:

- The ratio of low-cost part price to the retail sales price 2–5, for volume, and 5–10 or more for custom.
- Research and development projects have high risk content and should not be contingent on success.
- High-cost part prices compared with retail sales prices are typically 2–10 times higher depending on the labor and design content.
- If retrofitting, inspect carefully for whether something is old, dirty, or poorly maintained. Working components can be worth more than new components that don't work.
- A manufacturer will typically make half to a third of the retail price before shipping and distribution.
- An ROI of 20–30% is reasonable. It may be higher for work with greater risk.

Aside: Consider a case where we have three methods to make a material.
Let's assume that we are quoting for the lowest cost. Then we would chose the sequence C, B, then A. In this case the low cost option was chosen and a 20% ROI was applied. This is a typical number applied by companies. The value varies by company.
In the example, the quote could be issued for $23.70, with the expected costs being around $19.75. If all goes well and the costs are $11.00 then the profit will be $12.70. If things go poorly the project could fail entirely (the customer does not pay) and the work costs $39.00. In this case a manager may decide that the risk of failure does not make the project worth taking. Of if it is taken then the Bid price would be increased to offset the increased risk. Other methods for assessing risk will be discussed later.

Method	Time	Cost	Probability
A	10	15	50%
B	40	13	70%
C	50	11	75%

Decision C, B, then A

$$T_{CBA} = 50 + (1 - 0.5)40 + (1 - 0.5)(1 - 0.7)10 = 71.5$$

$$C_{CBA} = 11 + (1 - 0.5)13 + (1 - 0.5)(1 - 0.7)15 = 19.75$$

$$C_{min} = 11 \qquad C_{estimated} = 19.75 \qquad C_{max} = 11 + 13 + 15 = 39$$

$$ROI_{given} = 20\%$$

$$B = \left(\frac{100 + ROI_{given}}{100}\right) C_{estimated} = \left(\frac{100 + 20}{100}\right) 19.75 = 23.7$$

FIGURE 7.26

Incorporating risk into bidding.

Some expenses for a project will be relatively obvious. Examples include key components such as a microprocessor, a programmable logic controller, a diesel engine, or a steel structure. Others will require some consideration of the project steps and variations. Some useful categories are shown in the following list.

Needless to say, talking to project engineers, production staff, and managers will reveal other costs. Typical budget items are also listed. The list can be reviewed when preparing a budget to ensure that the project needs have been considered. People who have prepared, managed, or quoted budget items will be a useful source of information. Casual conversations can provide a wealth of knowledge about problematic budget items:

- Components and materials
  - Look at all major items and break into physical pieces (parts, software).
  - Break major items into subparts, materials, etc.
  - Identify all of the off-the-shelf items with known costs.
  - Consider price fluctuations.

- Consider lead time and shipping costs.
- A contingency may be needed, such as an alternate supplier if the primary supplier fails.
- Consider rejected parts or products.
- Equipment required for project
  - Leased items
  - Available equipment
  - New purchases
  - Machine cost rate and utilization
  - Space on the production floor, in test labs, or in storage
- Shipping
  - Shipping and logistics
  - Customs and delays for international trade
  - Insurance for shipping and project failure
  - Taxes and import/export duties for international and global shipments
- Financial
  - Overhead
  - An allowance for fluctuations in currency values
  - Financing for large projects requiring bank loans or similar
  - Lost profits if other projects are delayed because of labor, equipment, or material shortages
  - Does a customer release part of the funds at any project milestones?
  - Payment details and delays
- Related expenses
  - Travel to the customer site
  - Customer entertainment
  - Fees for licenses, customs, testing, inspection
  - Training and consulting for technical issues
  - Printing documents, manuals, special labeling, signs, etc.
  - Prequote costs for prototyping, concept development, and proof-of-concept work
  - Testing and certification
  - Insurance for liability and recalls
  - Market studies and advertising
  - Samples
- Labor
  - Benefits for employees
  - Bonuses for employees
  - Look at all major work tasks and estimate labor time and cost
  - Estimate times for project stages
  - Training for staff and customers
  - Labor for non-project-related development
  - Paid sick days and vacations
  - Meetings related to the project
  - Inefficiencies such as task switching, problems with software, etc.
  - Training for new employees
  - Consultants or subcontractors
  - Other nontechnical staff

A final but very important note: The single most important task of a project is to estimate the budget. A simple oversight or miscalculation in a budget can be very expensive. Budget mistakes drive many companies out of business. Have somebody else verify the budget. Use common sense to review the budget. Take the time to do it right.

PROBLEMS

7.52 Why would a company refuse to discount a product?

7.53 When a company offers a less expensive version of its product, how is it changing the brand?

7.54 A wise manager will recognize when there are no good options and stop developing a bid. A poor manager will keep the process alive because of the resources "invested" already. What financial principle should be applied in this case?

7.55 There are two processes for making cheese. The low-cost process costs $10/kg to make cheese, but only 50% can be sold. The high-cost process costs $16/kg, but 80% can be sold. What should the production cost be for the cheese actually sold?

7.56 List 10 bidding factors a company could use if it delivered and installed trees for commercial businesses.

7.57 Write a general computer program to solve the following project costing problem. Test the program using the numbers provided. The program should accept the initial cost of equipment (C), an annual maintenance cost (M), an annual income (R), a salvage value (S), and an interest rate (I). The program should then calculate a present worth and the ROI.
Test values
C = 100,000
M = 20,000
R = 150,000
S = 10,000
I = 10
L = 3 years

7.58 Write a program that determines the ROI for a project given the project length, initial cost, salvage value, and estimated income. To test the program, assume that the project lasts for 36 months. The company standard interest rate is 18%. The equipment will cost $100,000 new and have a salvage value of $10,000. The annual income will be $50,000.

7.59 Consider the costs for the current and a proposed manufacturing work center. Use a break-even analysis to determine the payback period.
Current manual line

- Used 2000 h/year with 10 workers at $20/h each.
- Maintenance is $20,000/year.
- The current equipment is worth $20,000 used.

Proposed line

- The equipment will cost $100,000 and the expected salvage value at the end of the project is $10,000.
- Two workers are required for 1000 h year at $40/h each.
- Yearly maintenance will be $40,000.

Further reading

Baxter, M., 1995. Product Design; Practical Methods for the Systematic Development of New Products. Chapman and Hall.

Bennett, F.L., 2003. The Management of Construction: A Project Life Cycle Approach. Butterworth- Heinemann.

Cagle, R.B., 2003. Blueprint for Project Recovery—A Project Management Guide. AMACON Books.

Charvat, J., 2002. Project Management Nation: Tools, Techniques, and Goals for the New and Practicing IT Project Manager. Wiley.

Dhillon, B.S., 1996. Engineering Design; A Modern Approach. Irwin.

Fioravanti, F., 2006. Skills for Managing Rapidly Changing IT Projects. IRM Press.

Gido, J., Clements, J.P., 2003. Successful Project Management, second ed. Thompson South- Western.

Heldman, K., 2009. PMP: Project Management Professional Exam Study Guide, fifth ed. Sybex.

Heerkens, G.R., 2002. Project Management. McGraw-Hill.

Kerzner, H., 2000. Applied Project Management: Best Practices on Implementation. John Wiley and Sons.

Lewis, J.P., 1995. Fundamentals of Project Management. AMACON.

Marchewka, J.T., 2002. Information Technology Project Management: Providing Measurable Organizational Value. Wiley.

Portney, S.E., 2007. Project Management for Dummies, second ed. Wiley.

Rothman, J., 2007. Manage It! Your Guide to Modern Pragmatic Project Management. The Pragmatic Bookshelf.

Salliers, R.D., Leidner, D.E., 2003. Strategic Information Management: Challenges and Strategies in Managing Information Systems. Butterworth-Heinemann.

Siciliano, G., 2003. Finance for the Non-financial Manager. McGraw-Hill.

Tinnirello, P.C., 2002. New Directions in Project Management. CRC Press.

Ullman, D.G., 1997. The Mechanical Design Process. McGraw-Hill.

Verzuh, E., 2003. The Portable MBA in Project Management. Wiley.

Verzuh, E., 2005. The Fast Forward MBA in Project Management, second ed. John Wiley and Sons.

Webster, W.H., 2004. Accounting for Managers. McGraw-Hill.

Wysocki, R.K., 2004. Project Management Process Improvement. Artech House.

CHAPTER

8 Reliability and system design

8.1 Introduction

Failure can lead to damage. Minor damage is inconvenient and has minor costs. Major or catastrophic damage can include loss of life, loss of equipment, and much more. A normal design objective is to minimize the probability of major damage and find an acceptable likelihood and level of minor damage.

Reliability is a measure of availability. When a system fails, it is unavailable and no longer meets the specifications. A degraded system has partially failed but some of the specifications are still met. When the failure is catastrophic the design no longer meets any specifications.

Fig. 8.1 illustrates a simple trade-off between reliability and value. Any design will have some base cost level, even if it is entirely unreliable. Reliability can be increased easily with basic engineering or overdesigning critical parts. At some point the cost to raise reliability increases exponentially. For example, a car that is designed to last 1,000,000 km without repairs would cost much more than a standard automobile designed for 100,000 km. The result is the final designed cost that has a base level that rises rapidly as we strive for failure proof. On the other hand, a customer may expect a car to operate 50,000 km without maintenance, but if the value were 100,000 km, he or she would spend more. However, cars are normally discarded before 200,000 km so any higher reliability has no value.

PROBLEMS

8.1 What is the difference between failure and reliability?

8.2 Can a design be too reliable?

8.2 Human and equipment safety

Sure it works, but what happens when something goes wrong?

Safety is hard to define, but it can be defined in terms of danger, a function of likelihood, and damage. Fig. 8.2 is a visual model of the danger zone in safety space. A likely event can be expected to occur, while an unlikely event will occur rarely. Minor damage is inconvenient and will sometimes go unnoticed. Major damage can cause injury to humans and equipment. Catastrophic damage means loss of equipment or life.

Engineering Design, Planning, and Management. https://doi.org/10.1016/B978-0-12-821055-0.00008-6

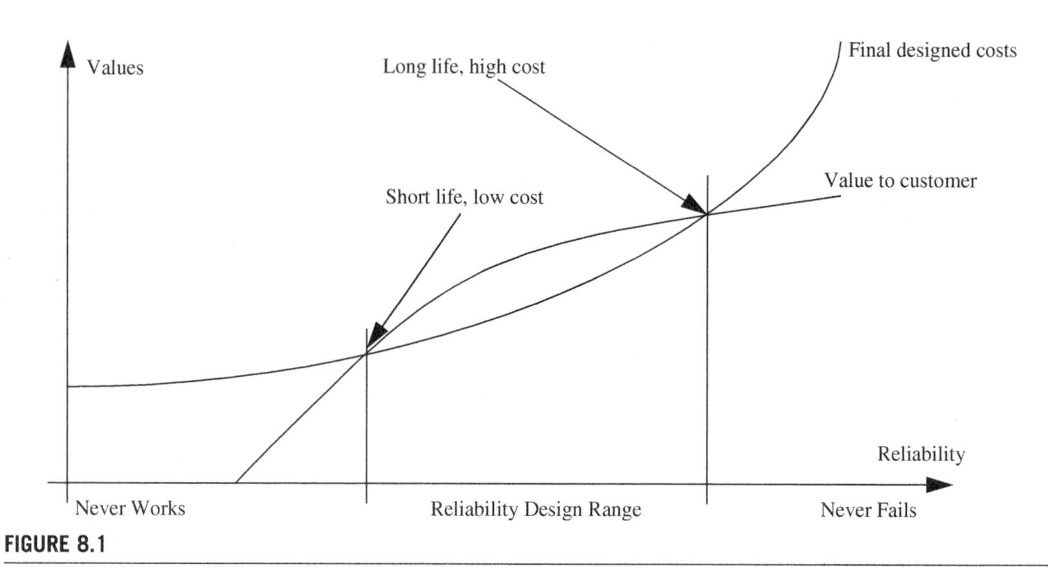

FIGURE 8.1

The design cost of reliability.

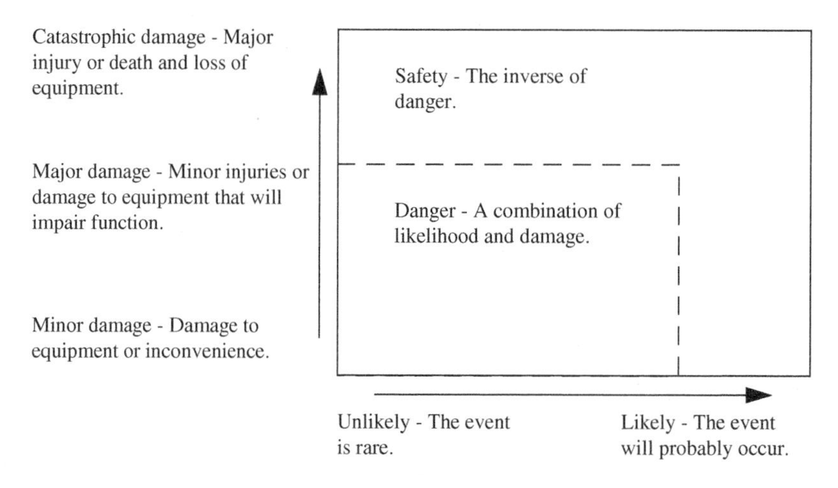

FIGURE 8.2

Danger zones.

For example, a car with minor damage may be scratched or require a new tire. Major damage may include body work, a new engine, a new transmission, or occupant injury. Catastrophic damage would result in scrapping the vehicle and maybe loss of life. On the other hand, damage to a computer might be a minor scratch, a major crack on a screen, or a catastrophic failure of the battery resulting in fire.

A risk assessment system was developed by the US military and is documented in MIL STD 882B System Safety Program Requirements (Fig. 8.3). The method uses a failure category, damage, and a failure frequency to assign a risk score. If a failure leads to minor damage it is ruled an acceptable risk. A catastrophic event that is very improbable is ruled negligible to avoid an unnecessary design focus.

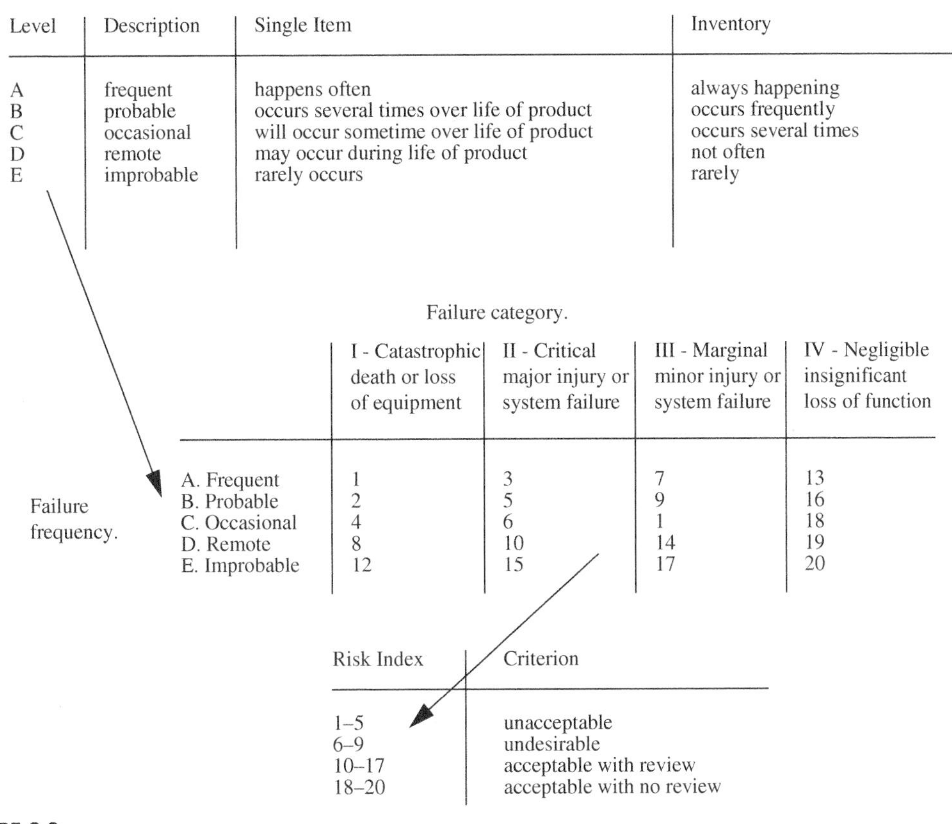

Level	Description	Single Item	Inventory
A	frequent	happens often	always happening
B	probable	occurs several times over life of product	occurs frequently
C	occasional	will occur sometime over life of product	occurs several times
D	remote	may occur during life of product	not often
E	improbable	rarely occurs	rarely

Failure category.

Failure frequency.	I - Catastrophic death or loss of equipment	II - Critical major injury or system failure	III - Marginal minor injury or system failure	IV - Negligible insignificant loss of function
A. Frequent	1	3	7	13
B. Probable	2	5	9	16
C. Occasional	4	6	1	18
D. Remote	8	10	14	19
E. Improbable	12	15	17	20

Risk Index	Criterion
1–5	unacceptable
6–9	undesirable
10–17	acceptable with review
18–20	acceptable with no review

FIGURE 8.3

Assessing design risks with MIL STD 882B.

On the other hand, events that are hazardous and likely should get more design effort, more manufacturing effort, more maintenance, and more care in use. If any design (or a component) is ruled unacceptable, it must be redesigned or modified to move it to the "acceptable with review" category (see Resource 8.1).

Human damage can be physical or mental (see following list). Some of the effects are immediate and obvious, including cut skin, broken bones, electric shock, burns, trauma, and crushed appendages. Other effects may occur slowly without an identifiable start; these include the effects of carcinogens, depression, repetitive stress injuries, and radiation exposure:

- Pinch points or crushing
- Collision with moving objects
- Falling from heights
- Physical strain
- Slippery surfaces
- Explosions or very loud sounds
- Electric shock
- Temperature/fire

- Toxicity: liquid, gas, solid, biological, radiological
- Trauma, depression, posttraumatic stress syndrome, phobias

Resource 8.1 See this book's website for safety standards: https://sites.google.com/site/engineeringdesignprojects/home/content/reliability.

Ideally, hazards are eliminated, but realistically, hazards should be minimized. Hazards can be reduced or eliminated using the techniques in the following list. When hazards cannot be eliminated, the likelihood, and thus danger, can be reduced with warnings. A static approach to warning is signs. An active approach could include flashing lights and sirens. In a practical sense, these should stop or redirect hazardous activities. Consider a room with a strong ultraviolet light used to kill bacteria. The hazard is temporary or permanent blindness caused by the invisible ultraviolet light. The likelihood of injury is reduced by a sign that reads, "Do not enter without UV eye protection." Another example would be a seat belt system in a car. A seat belt reduces the hazard of an accident and the warning sound makes it more likely the belt will be worn, thus reducing the danger.

Warnings

- Colors—red for warning, yellow for caution, and green for safe
- Horns, buzzers, sirens
- Lights
- Multiple senses to deal with deaf, color blind, blind, etc.
- Large easy-to-read signs
- Pictures on signs for people who cannot read English or are illiterate
- Markings on floors, equipment, controls, etc.
- Clear labeling when hazardous operation must occur
- A hazardous failure should be accompanied by warnings

Safeguards

- Barrier guards—physical blocks to separate operators and equipment; interlocks to disable the machine when the barrier is open
- Special protective equipment (e.g., earplugs)
- Hand pullbacks for pinch points
- Dead-man controls
- Presence-sensing devices
- Maintenance safety devices—additions (and procedures) that ensure that the machine is still safe, even though the normal safety equipment is disabled
- Passive safety devices (e.g., seat belts)
- Machine condition checks before steps move forward
- Vapor barriers and ventilation
- Extra protection for chemicals, including spills and splashes
- Fire protection for combustibles

Education

- Policies and procedures
- Worker training
- Supervision for safety practices
- Mandatory training for dangerous operations

Safety should be considered during the design stage. For established designs and methods, the safety issues are normally understood and regulated. For new technologies and methods, the safety issues must be audited.

Design tasks

- Do a closed-system analysis of the user and the design. List all of the energies, materials, and motions that cross the user–design interface boundaries.
- Identify dangers and work to reduce the hazard and likelihood to acceptable levels.
- Use failure modes and effects analysis (FMEA) for simple failures and fault trees for complex situations.
- Assess previous and competitors' designs and use customer feedback.

Professional practices

- Use workplace regulations from groups such as OSHA (US Occupational Safety and Health Administration) and the European Agency for Safety & Health at Work.
- Use product regulations from groups such as the NTSB (US National Transportation Safety Board) and the *European Civil Aviation Handbook*.
- Use industry standards such as UL, CSA, and CE.
- Use de facto standards.

Design parameters

- Test outside the design range.
- Use fail-safe operation.
- Make systems robust and error tolerant.
- The design should be easy to stop but require effort to start.
- Make it idiot and mistake proof—assume the user is untrained and unaware of the operation and dangers.
- Make points of failure tougher.
- Graceful degradation maintains operation at a lower performance level.
- Use fault detection and management.

Human contact

- Include guards for moving parts.
- Avoid sharp edges or points.
- Include guards for electrical hazards.
- Add grounding.
- Avoid health hazards (e.g., lead, choking, gases).
- Consider ergonomics.
- Make it easier for the user to be safe.

PROBLEMS

8.3 What is the difference between hazard and danger?

8.4 Explain why a warning should appeal to multiple senses.

8.5 For a disposable cup of hot coffee, list five hazards and the likelihood of occurrence. Use MIL STD 882B to rate the criticality of each of these events.

8.6 Based on your own experiences, list five ways disposable coffee cups tend to fail. How often does this happen? Put these in a table.

8.7 A kitchen knife is unsafe. The likelihood of damage is reduced by training that includes (1) do not cut toward your hands, (2) do not hold the knife by the blade, (3) cut on a stable surface, and (4) do not cut on regular table surfaces. The hazard of the knife is the sharp blade. Propose three ways to reduce the hazard (not the likelihood).

8.3 System reliability

Systems have many components that contribute to the overall functionality. When a component is used for only one function the failure affects only that function. For example, a car radio that fails affects only the audio functions. Failure of a component that works in parallel or sequence with other functions can have greater impacts. For example, a worn hydraulic pump reduces the pressure available for car brakes and increases the stopping distance. The interdependence of system components will determine how the reliability of each component influences overall reliability.

A design life cycle can be enhanced to improve safety (see Fig. 8.4). At the beginning the specifications are examined for reliability effects. In the detailed design stage the design uses good practices for increasing reliability. For higher-reliability systems the detailed design may include additional

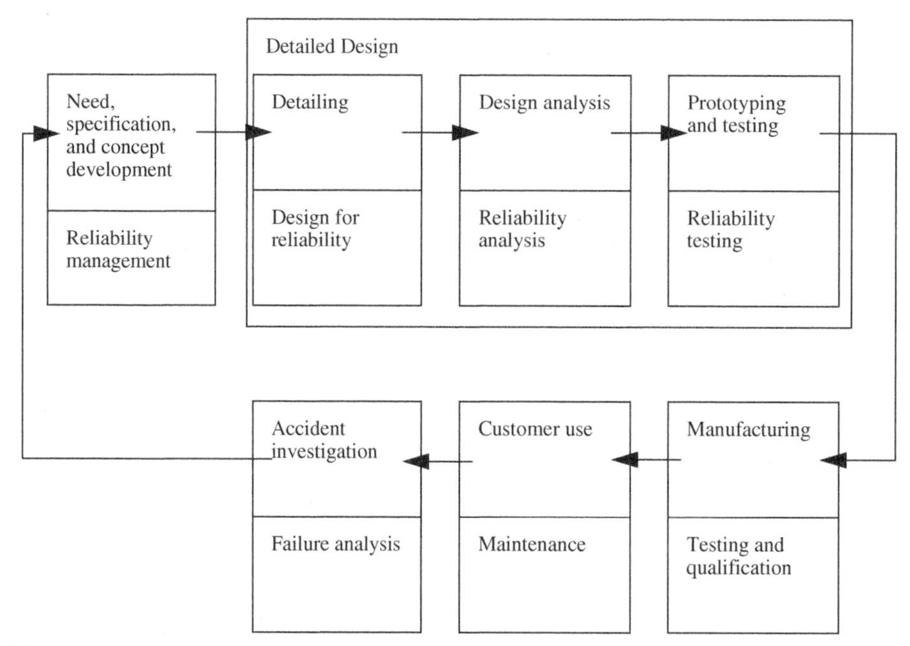

FIGURE 8.4

An example of a reliability design control procedure.

analysis and testing steps. During manufacturing the critical components must be monitored and tested for suitable performance. When used, the design should be monitored and maintained for reliability. Finally, when a device does fail it should be examined to determine the cause.

Reliability management looks at strategic business decisions as they relate to reliability. This can include decisions to accept, decline, or modify projects. Consider the example of a redesign of a truck. The previous truck model would be able to carry a load of 1000 kg for a life of 120,000 km. The customers want a new truck design that will carry 1200 kg for 150,000 km for the same price. Is it possible or are compromises necessary? Is the market still suitable? Can the lessons learned from the last truck model help increase the life of the new model with minimal cost? For another example, consider an office chair maker that makes a decision to produce a new seat that uses a mesh cover. Before making a commitment to the design there is substantial work to determine potential market size, technical risks, production costs, and more.

Design for reliability takes advantage of experience that avoids reliability problems. Consider the example of a truck maker that finds that 37 of all trucks carrying at least 900 kg had shock absorbers that failed at 60,000 km or less. Investigation shows that the failure occurred at a hydraulic seal because the rubber had aged. The designer then looks for a new material with a longer life.

Design analysis and prototype testing are very common. Consider an office chair with a new reclining mechanism. The new design is created on a computer-aided design system and then substantial time is spent using finite element analysis to predict failure. During this step the designer adjusts the geometry to reduce stress concentrations. Eventually a plastic mold is made and a few dozen parts are built for testing. The parts are put in machines that cycle the designs 100,000 times at the maximum load and then the parts are examined for failure. If the parts had worn or failed as predicted, the design moves to production, if not, design work continues.

Materials, processes, and methods vary once a product enters the manufacturing phase. Quality control methods are used to look for anticipated production problems. Unexpected production problems are identified with testing. A common practice is to take random samples of products from a manufacturing line and then test them until the point of failure: so-called destructive testing. For example, an electronics maker may take a computer power supply to a test lab for a "high potential" test in which it is exposed to a steady 600-V AC signal and a simulated lightning strike. Another common practice is to nondestructively certify or qualify each design by testing it after production. Consider an aircraft manufacturer that has a test pilot fly new aircraft and perform specific high-stress maneuvers before sending the airplane to a customer.

After delivery the customer chooses how the design is used. Maintenance may be part of the postproject activity. For example, an automobile manufacturer might recommend that a timing belt be changed every 100,000 km. Or, consider a company that sells radio transmission equipment; it may make a maintenance visit every year to adjust power and frequency settings. Reliability may also be maintained with practices such as fail-safe measures. For example, a coffee maker might have a fuse that will "blow" when the heating element draws too much current. This does lead to failure of the device, but it fails in a safe mode.

Products that fail in use are valuable sources of information. These can be inspected and evaluated from a technical standpoint to determine what caused the failure and how it failed. Consumer products are often returned for warranty repairs and industrial sales often have maintenance contracts. For example, an automobile maker may investigate a door handle design by buying a used car, looking at Internet discussions, reading reviews in magazines, doing customer surveys, and asking auto-part suppliers for the quantity of replacement part sales.

When any new technology reaches the design stage there are always unknown implications. As a result, any new design or technology carries risks with it. Designers will compensate by using higher factors of safety. Over time the design group will learn lessons and the design matures.

PROBLEM

8.8 When inserting batteries into portable electronics there are multiple points of failure. List the steps involved in inserting two AA batteries into a computer mouse. For each step list the problems that might occur.

8.4 Component failure

A component can be defined as a smaller self-contained piece of a bigger system, but there is some room for other interpretations. For example, in a radio the components include resistors, transistors, capacitors, inductors, integrated circuits, and more. A car radio is a component, as is the ignition control unit, speedometer, and headlights. When identifying components try to pick similar levels of functionality and group functions into subsystems to simplify analysis.

Component failure can mean many things such as the scenarios given in the following list. For our purposes we will assume that the component will not perform as needed. This could be either total failure or degraded performance. For example, a heater control system that will not keep an oven at 170°C will not kill all bacteria and has effectively failed:

- Temporary: predictable or unpredictable maintenance work; e.g., changing car tires
- Permanent: end of life; e.g., a tire fire
- Unpredictable: uncontrollable external influences; e.g., a flat tire
- Degradation: system wears or distorts over time; e.g., wear on a car tire or loss of pressure

The expected life of devices and components is normally available from the manufacturer in data sheets or from technical support groups. For products that follow legal requirements or elective standards these may refer to the minimum required life of the standard. When these numbers are not available they are found using life and stress testing. Some sample component lives are shown in the following list. For example, if a machine has a light bulb, it should fail at least 10 times before a million hours, or it should fail before 100,000 h of operation.

More than 10 failures per million hours

- Bearings; ball, roller
- Brakes
- Compressors and pumps
- DC generators
- Gears (complex); differential gears
- Light bulbs
- Solenoids and valves

More than one failure per million hours

- AC generators
- Heaters

- Motors
- Shock absorbers and springs
- Switches

Less than one failure per million hours

- Batteries
- Connectors
- Gears (simple)

PROBLEM

8.9 Find the expected life of a 100-W-equivalent light bulb using manufacturer specifications. This should include fluorescent, incandescent, and LED bulbs.

Reliability, R, is the fraction of components working at any point in time (Fig. 8.5). At the beginning of life all devices should be working, for a reliability of 1.0 or 100%. Over time more devices will fail and eventually no devices will work and the reliability will drop to 0.0 or 0%. Unreliability is the inverse of reliability. These numbers can be found by testing a number of components, N, simultaneously. When a component fails, the value of N drops by 1. This test continues until all of the units have failed. Test failure rates are often accelerated by intensifying conditions. For example, an electronic device can be tested by heating and cooling it rapidly, simulating years of daily temperature changes in a few hours.

A typical failure rate curve can be seen in Fig. 8.6. New products are generally more reliable, but that is also the point at which major problems arise. For example, a new laptop with a defect may appear to work well for a few days until overheating causes an early failure. Warranties are used to defend consumers again these types of defects. Engineering groups will look at warranty failures for manufacturing defects to determine fundamental design flaws. If the device makes it past this burn-in period it is likely to last for a long period before it starts to wear out. The expected life of a device should be the time at which wear-out failures are occurring. Devices that have been used until they wear out can be used to increase the life of a product.

Where,

$N(t)$ = the number operating at time t

$R(t)$ = the reliability, the portion surviving over the time $[t_o, t]$

$Q(t)$ = unreliability

$$Q(t) = 1 - R(t) \qquad \text{eqn. 10.1}$$

$$R(t) = \frac{N(t)}{N(t_o)} \qquad \text{eqn. 10.2}$$

FIGURE 8.5

Reliability rates for components.

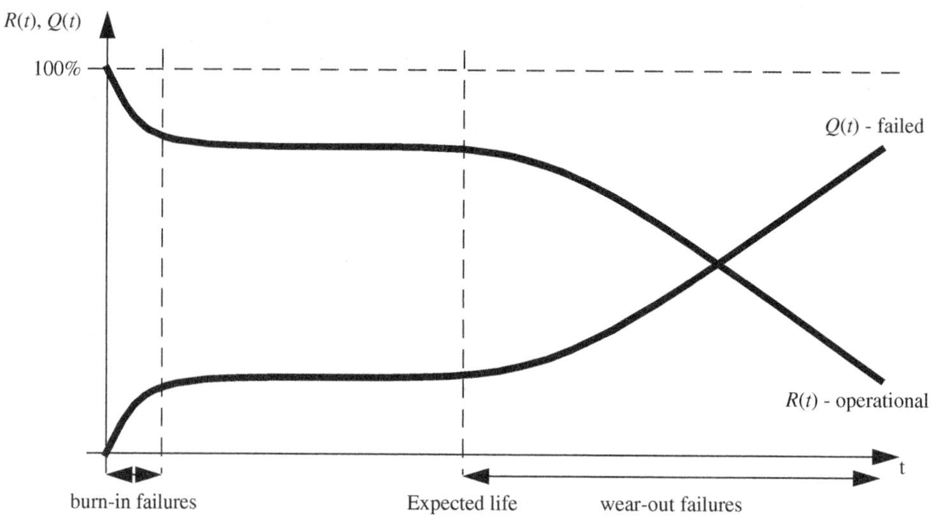

FIGURE 8.6

Sample curves for component reliability.

Where,

$\lambda(t)$ = the failure rate (units per period of time)

$Q(t)$ = unreliability= $1 - R(t)$

$$\lambda(t) = -\frac{\frac{d}{dt}R(t)}{R(t)} = \frac{\frac{d}{dt}Q(t)}{1 - Q(t)} \qquad \text{eqn. 10.3}$$

FIGURE 8.7

Failure rate.

Failure rate, λ, is the derivative of reliability (Fig. 8.7). In essence, it is the number of failures per unit of time. A positive failure rate means that devices are failing faster. A negative failure rate is unrealistic because it means that failed devices have begun to work again. When the failure rate is decreasing and approaching 0, the device is becoming more reliable. An increasing failure rate means that something is causing more failures. As in any engineering problem, it is important to know why the values are changing or constant.

The bathtub curve is named for its shape (Fig. 8.8). New designs fail at a high rate in their infancy. Once the burn-in is complete the failure rate is relatively constant. Failures during this period tend to be slow and steady. Eventually the device wears out and the failure rate increases. The useful service life is the bottom of the tub. An example of this curve would be a new cell phone. Sometimes these fail in the first day or two and need to be exchanged. Once the phone is in use it may fail as it is dropped and bumped. Eventually the battery does not charge and hinges break. Cell phone makers try to reduce the return rate by testing the phones for a few hours before packaging and shipping them to customers. This allows them to detect early failures. NASA uses a similar strategy when testing space components.

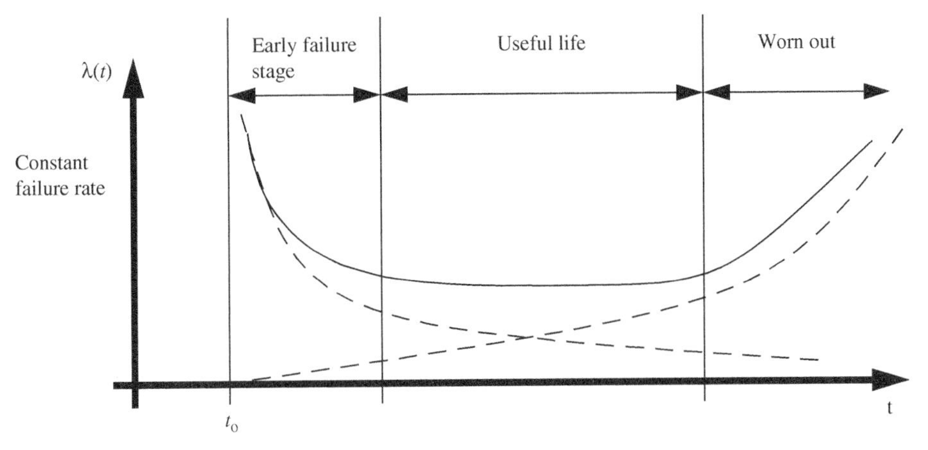

FIGURE 8.8

The bathtub curve.

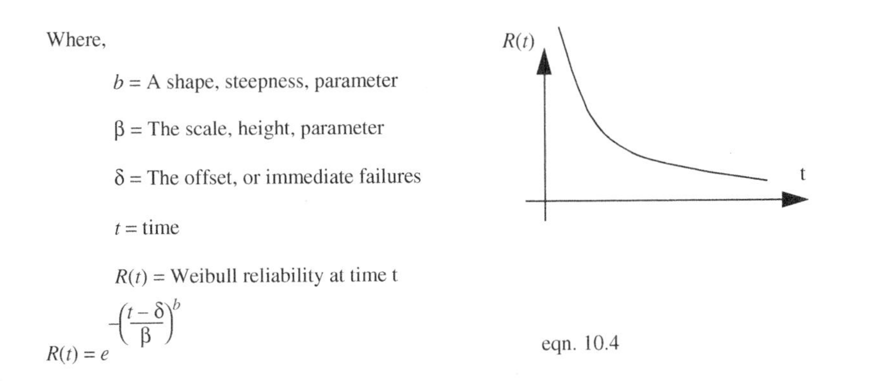

FIGURE 8.9

The Weibull distribution is used for infant component mortality.

The Weibull distribution can be used to model the infancy failures (Fig. 8.9). The natural number exponent is similar to the curve for the wear-out failures. The infancy failure rate will be very fast for most products and then the wear-out curve will be much more gradual. The parameters for the distribution can be found by fitting the function to initial test data. The function also includes an initial offset, δ, that compensates for devices that have failed before the first test. This function will be very useful for products that have inherent quality control problems. It is also possible to eliminate this function using burn-in testing.

Given the differing causes, the burn-in failures can be separated from the wear-out failures. When considering only the failure cause by wear-out the reliability curve can be approximated with a natural exponential distribution curve (Fig. 8.10). This derivation is based on the assumption that the failure rate is constant. This is reasonable for many products, but for high-precision estimates it should be verified for the design. In simple terms, a smaller failure rate, λ, means longer life and higher reliability.

Assume the failure rate is constant.

$$\lambda = -\frac{\frac{d}{dt}R(t)}{R(t)}$$

$$\therefore \frac{d}{dt}R(t) = -\lambda R(t)$$

$$\therefore \frac{d}{dt}R(t) + \lambda R(t) = 0$$

A standard first order homogeneous differential equation solution.

$$\therefore R(t) = e^{-\lambda t} + C$$

$$R(0.0) = 1.0$$

$$\therefore 1.0 = e^{-\lambda(0)} + C \qquad \therefore C = 0$$

$$\therefore R(t) = e^{-\lambda t} = \text{The Exponential Failure Law} \qquad \text{eqn. 10.5}$$

FIGURE 8.10

The exponential failure law.

PROBLEM

8.10 Explain the shape of the reliability graph.

The failure rate is a measure of reliability for a larger number of components. For an individual part, a designer needs to know the expected life of one component. The mean time to failure (MTTF) is a popular measurement. For example, if a designer wanted a design to last for 10,000 h he or she would select components with higher values such as 50,000 h. The MTTF derivation can be found in Fig. 8.11.

When a system fails it may be repaired and returned to operation. In those cases we can calculate a mean time to repair (MTTR). The repair rate, μ, is an estimate of repairs per unit of time. Like the failure rate and MTTF, the MTTR is the inverse of the repair rate (Fig. 8.12). When dealing with repairable systems the mean time between failures (MTBF) is a better choice than MTTF. Consider the example of a laptop that has an 11-month MTTF and 1-month MTTR, then the MTBF is 12 months.

Other interesting measures of reliability include:

- MCMT: mean corrective maintenance time
- MPMT: median preventive maintenance time
- MTTD: mean time to detect

$$\text{MTTF} = \int_0^{\infty} tf(t)dt$$

$$E[X] = \int_{-\infty}^{\infty} xf(x)dx$$

where,

X = a random variable

$E[X]$ = expected value

$f(x)$ = a probability density function

Given the probability density function and using integration by parts, we can find the relationship between the MTTF and reliability.

$$\text{MTTF} = \int_0^{\infty} t\frac{d}{dt}Q(t)dt = -\int_0^{\infty} t\frac{d}{dt}R(t)dt = \left[-tR(t) + \int R(t)dt\right]\Big|_0^{\infty} = \int_0^{\infty} R(t)dt$$

$$\text{MTTF} = \int_0^{\infty} R(t)dt = \int_0^{\infty} e^{-\lambda t}dt = -\frac{1}{\lambda}e^{-\lambda t}\Big|_0^{\infty} = -\frac{1}{\lambda}(0) - \left(-\frac{1}{\lambda}(1)\right) = \frac{1}{\lambda}$$

$$\text{MTTF} = \frac{1}{\lambda} \qquad \text{eqn. 10.6}$$

FIGURE 8.11

Mean time to failure derivation.

Where,

$$\mu = \text{the repair rate} = \frac{\text{number of repairs}}{\text{time period for all repairs}}$$

$$\text{MTTR} = \frac{1}{\mu} \qquad \text{eqn. 10.7}$$

$$\text{MTBF} = \text{MTTF} + \text{MTTR} \qquad \text{eqn. 10.8}$$

FIGURE 8.12

The mean time to failure and mean time between failures.

PROBLEMS

8.11 Why is reliability, R(t), a variable instead of a constant?

8.12 A printer ink cartridge is 40% reliable at 200 h. If 225,000 have been sold, how many will fail at 200 h?

8.13 Find the reliability and unreliability for the system approximated by the equation at time t = 100 h for:

(a) life = 1 h

(b) life = 100 h

(c) life = 1000 h

$$R(t) = e^{-\frac{1}{(life)}t}$$

8.14 An IT group at a large company monitors 10,000 computers using software. At 1000 h, a test of all of the machines found that 8950 computers were operating normally. Another test at 1010 h showed that 8975 were operating normally. What is the reliability and failure rate?

8.15 Angus works in the packing department, folding and then sealing boxes. When mistakes occur he must stop and correct the error before restarting. He agreed to participate in a time study that would help improve job cost estimates. A camera was installed that took video for a week. When the video was reviewed, a total of 27 failures were counted and they took a combined time of 6.23 h to correct. What are the repair rate and the MTTR?

8.5 System reliability

Systems are collections of components like those shown in Fig. 8.13. In this example there are multiple modules that contribute to function, in sequence. If any of the modules fails, the system fails. An example of this type of system would be an automobile with a fuel tank, fuel lines, ignition control, fuel injectors, spark plugs, valves, and so on. If any of these components fails the entire fuel system in the automobile fails. Each component in the system has an independent failure rate. The overall

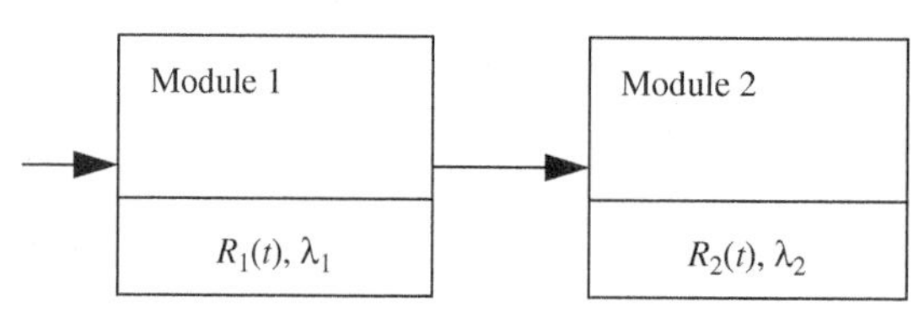

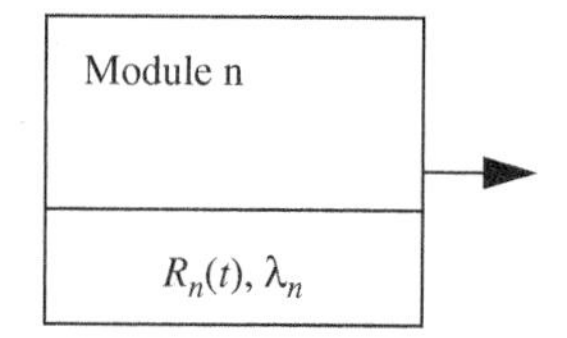

Where,

$R_s(t)$ = the reliability of a series system at time t

$R_i(t)$ = the reliability of a unit at time t

$$R_s(t) = (R_1(t))(R_2(t))\ldots(R_n(t)) = \prod_{i=1}^{n} R_i(t) \qquad \text{eqn. 10.9}$$

Now, consider the exponential failure law presented before. If each element in a system observes this law, then we can get an exact value of reliability.

$$R_s(t) = \left(e^{-\lambda_1 t}\right)\left(e^{-\lambda_2 t}\right)\ldots\left(e^{-\lambda_n t}\right) = \prod_{i=1}^{n} e^{-\lambda_i t} = e^{-\sum_{i=1}^{n} \lambda_i t} \qquad \text{eqn. 10.10}$$

FIGURE 8.13

Serial reliability configurations.

reliability of the system is found by multiplying the component reliabilities. For example, consider a flashlight that has been used for 10 h. The reliability of the bulb has dropped from 1.0 to 0.9. The reliability of the battery has dropped to 0.2. The total reliability of the flashlight is now $0.9 \times 0.2 = 0.18$. However, if the battery is changed the reliability becomes $0.9 \times 1.0 = 0.9$. This calculation illustrates the impact of the weakest part of a serial system. It also highlights the importance of preventative maintenance for unreliable components.

Systems can be made robust by adding components in parallel, as shown in Fig. 8.14. Consider the example of car headlights with a reliability of 0.8 at 10,000 h. If one headlight is lost, the other still functions, providing diminished but ongoing operation. At 10,000 h the chance that either bulb will fail is 20%. The chance that both will fail is $20\% \times 20\% = 4\%$, so the headlight system is 96% reliable at 10,000 h. A similar approach is used for aircraft with two engines. If one engine fails on takeoff, the other engine provides enough thrust for a safe landing, thus increasing the reliability substantially.

Some systems will use many components in parallel for reliability, business, or technology reasons. In these cases it may be necessary to have a portion of the system active for operation. The binomial operator can be used to determine the combined reliability (Fig. 8.15). Consider the example of a wind farm that has 12 turbines, but can run with as few as 8. At 20,000 h the reliability of each turbine is 0.8, therefore there is a 92% chance that at least 8 will be operating.

Up to this point the analysis has considered only parallel or series systems. To combine the elements, calculate parallel reliabilities first and then combine the results in the serial system.

During the design process an engineer will take care to differentiate critical and noncritical systems. With each of these systems the designer will consider the series and parallel effects on reliability. The designer also considers individual component failure rates. Adding higher-reliability components or using redundant (parallel) components will make the system more reliable. Some of the general design rules for reliability include the following:

- Shorten serial reliability chains.
- Use parallel, redundant, configurations for lower-reliability components.
- For unreliable components do a cost–benefit analysis for higher-cost and higher-reliability components.

As before, the MTTF, MTTR, and MTBF can be found using experimental or calculated values for the entire system (Fig. 8.16). When using reliability design methods, the MTTF for the individual system components should be combined mathematically and then compared experimentally to validate the models.

It is unusual to find a system with only one specification. As a result, it may be necessary to repeat these calculations for subsystems with different specifications.

PROBLEMS

8.16 Clyde bends boxes and then passes them to Maria, who closes and seals them. Clyde makes 2 mistakes per hour and Maria makes 1.5 mistakes per hour. How many times will they need to stop each hour to correct mistakes? What is the MTTF? If they require 0.01 h to correct the mistake, what are the MTBF and MTTR?

8.17 Maria and Clyde are now working in parallel, folding and sealing boxes. As before Clyde makes 2 mistakes per hour and Maria makes 1.5 mistakes per hour. Each mistake requires 0.01 h to correct. What are the MTTF, MTBF, and MTTR?

8.18 How are series and parallel reliability different?

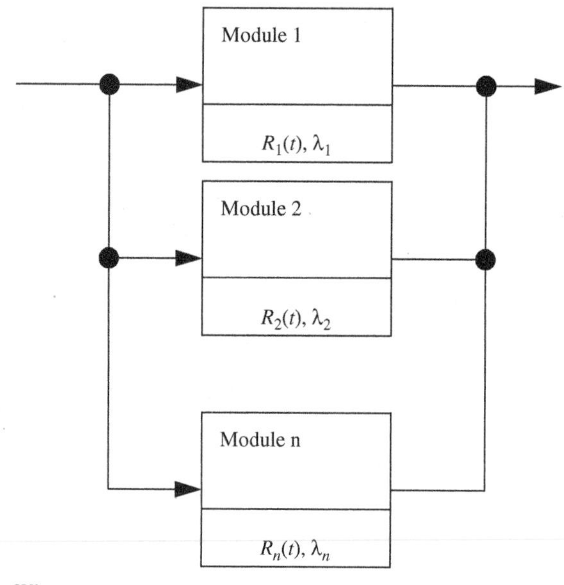

Where,

$Q_s(t)$ = the unreliability of a parallel system at time t

$Q_i(t)$ = the unreliability of a module at time t

$R_p(t)$ = the reliability of a parallel system at time t

$R_i(t)$ = the unreliability of a module at time t

$$Q_p(t) = (Q_1(t))(Q_2(t))\ldots(Q_n(t)) = \prod Q_i(t) \qquad \text{eqn. 10.11}$$

$$R_p(t) = 1 - Q_p(t) = 1 - \prod_{i=1}^{n} (1 - R_i(t)) \qquad \text{eqn. 10.12}$$

FIGURE 8.14

Parallel system components.

8.6 Passive and active redundancy

A triple-modular-redundancy (TMR) system is shown in Fig. 8.17. In this system there are three modules that do the same calculations in parallel. If everything is operating normally they should produce the same results. Another system compares the results from the three modules, looking for a mismatch. If the results from any module do not match the other two, a fault is detected and the majority of the two sets the control output. This system was used for the navigation computers for the space shuttles. The systems were designed and built by three independent contractors. Despite all of the testing, these

Where,

$R_{m;n}(t)$ = reliability of a system that contains m of n parallel modules

$R(t)$ = the reliability of the modules at time t

$\binom{n}{i} = \frac{n!}{(n-i)!i!}$ = the binomial operator (we can also use Pascal's triangle)

$$R_{m;n}(t) = \sum_{i=0}^{n-m} \binom{n}{i} (R(t))^{(n-i)} (1 - R(t))^{i} \qquad \text{eqn. 10.13}$$

Example: If at 20,000 hours the reliability of a turbine is 0.8 what is the chance that at least 8 will be running?

$n = 12 \qquad m = 8 \qquad R(20{,}000) = 0.8$

$$R_{m;n}(t) = \sum_{i=0}^{12-8} \binom{12}{i} (0.8)^{(12-i)} (1 - 0.8)^{i}$$

$$= \binom{12}{0}(0.8)^{12}0.2^{0} + \binom{12}{1}(0.8)^{11}0.2^{1} + \binom{12}{2}(0.8)^{10}0.2^{2} + \binom{12}{3}(0.8)^{9}0.2^{3} + \binom{12}{4}(0.8)^{8}0.2^{4}$$

$$= 0.927$$

FIGURE 8.15

Highly parallel system failure.

Where,

$A(t)$ = probability that a system will be available at any time

t_o = hours in operation over a time period

t_r = hours in repair over a time period

$$A(t) = \frac{t_o}{t_o + t_r} = \frac{\text{MTTF}}{\text{MTTF} + \text{MTTR}} = \frac{\text{MTTF}}{\text{MTBF}} = \frac{1}{1 + \frac{\lambda}{\mu}} \qquad \text{eqn. 10.14}$$

FIGURE 8.16

Mean times for complex systems.

systems were not ultimately proven until the first space shuttle mission. The approach helped to offset the critical risk associated with the guidance system.

If there is a random failure in any of the TMR modules, it will be outvoted and the system will continue to operate as normal. This type of module does not protect against design failures, where

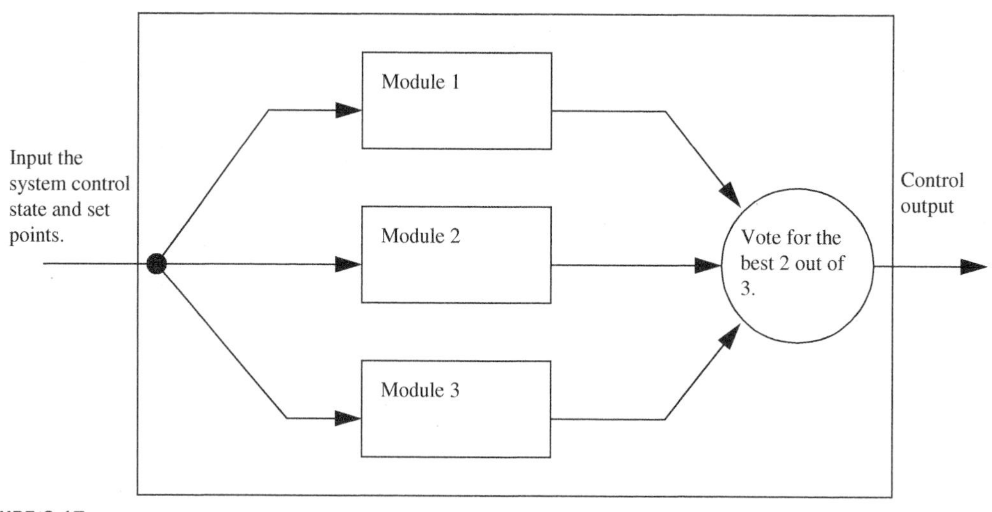

FIGURE 8.17

A triple-module passive-redundancy system.

all three modules are making the same error. For example, if all three had Intel Pentium chips with the same math mistake, they would all be in error and the wrong control output would be the result. (Note: This design problem occurred in one generation of Intel CPUs.) This module design is best used when it is expected that one of the modules will fail randomly with an unrecoverable state.

When hardware or operating system issues are expected, backup hardware can be prepared for "hot swap" (Fig. 8.18). In this approach, there is a monitor that looks for failure in the prime module.

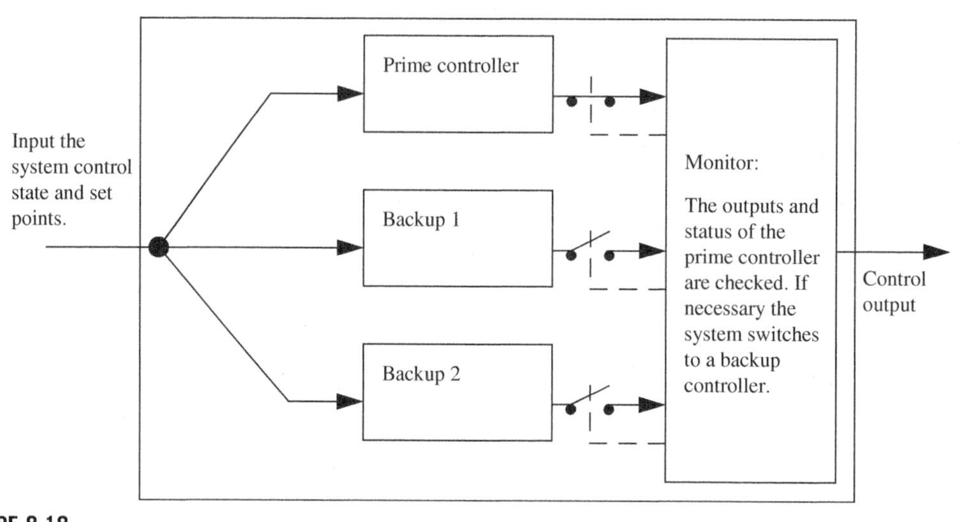

FIGURE 8.18

An example of an active redundant system.

In the event of failure the prime module is disconnected and the backup module is switched on to take its place. This approach is very common in mission-critical industrial control systems. The industrial controllers can self-detect software and hardware failures and can signal the next controller to take over. Another popular electronics approach is to use a watchdog timer. The software must reset the timer regularly, often every 1/10 s. If the timer is not updated, a fault condition is flagged. Sometimes all of the backup modules will run in parallel with the prime module, monitoring inputs but not able to change outputs.

This method depends on a careful design of the monitor module. As with the TMR modules, this system is also best used to compensate for complete module failure. If needed, this system can be used with analog electronics and mechanical components.

PROBLEMS

8.19 Four explorers are wandering through dense rainforest. Three explorers are assigned to navigating. The fourth explorer is leading the group and verifying the course. At the beginning of the trip all three are using maps and compasses separately to estimate the path. (a) The leader is listening to all three of the navigators and when two or three agree she takes that path. What type of redundancy is she using? (b) The next day the leader appoints one navigator to provide directions. When the navigator makes a mistake she cuts her off and moves to the next navigator. What type of redundancy is she using?

8.20 List three advantages and three disadvantages of redundant systems.

8.7 Modeling system failures

Estimating how a system will fail requires an identification of the failure modes. These are simple failures or more complex chains of events that may occur and lead to a total failure. For single-step failures, the numbers can be developed with FMEA. More complex analysis will consider chains of events that lead to failures.

Some components and systems have a history of failures that can be used as a starting point for identifying fault modes and causes. For example, the faults for a cell phone could be listed in a table, such as Table 8.1. Each of the specifications is listed with possible faults. The possible causes and effects are also captured. Beyond the stated specifications, there are others that are de facto or regulated. For example, there are a number of standards from groups that deal with the electrical, digital, mechanical, environmental, safety, and other requirements. This table is used to capture ideas, not to make judgments. Some of the faults, such as loss of signal, will occur, but being aware of when and how helps the design. Commonsense specifications should also be considered. For example, the phone should allow normal gripping; falling from a sweaty or greasy hand would be a bad fault.

Some faults are easy and obvious to identify, while others are more elusive. Group input will help to expand the list. Useful methods include brainstorming, Pareto charts, warranty returns, maintenance records, consumer discussion boards, and laboratory tests. Like brainstorming, it is important to identify all of the possible failures first. The negligible or unimportant cases can be left out of the following stages by decision, not negligence.

Table 8.1 Sample fault identification for a cell phone.

Specification	Fault mode	Possible causes	Possible effects
Range of 3 km	Cannot place call	Out of range Surrounded by metal Electromagnetic interference Electronics fault Software fault	Inconvenient Temporary
	Call is interrupted	Loss of signal Another incoming call Network error. Electronics fault Software fault.	Inconvenient Temporary
10 h talk time	Loss of power	Battery is out of power Battery is loose Electronics fault	Inconvenient Temporary
Temperature range −20°C to 60°C	Melting or fire	Cannot dissipate heat In insulating space No thermal shutoff	Loss of equipment Harm to user Permanent
	Stops working	Temperature above or below range Battery capacity lower at temperature	Inconvenient Permanent or temporary
IEEE 802.16m: 450–470, 698 −960, 1710 −2025, 2110 −2200, 2300 −2400, 2500 −2690, 3400 −3600 MHz	Outside range	Crystal PLL filter failure Narrowband filter	Legal violation Inconvenient Interferes with other devices
Common sense	Visible in normal light levels	Excess light Bad light sensor Back light fails	Inconvenient

8.7.1 Failure modes and effects analysis

The FMEA technique provides a rapid technique for analyzing and ranking failure (Table 8.2). Each major function of the device is considered for possible failure types, called modes. Technical knowledge of the device is needed to determine the modes of failure. The severity of a failure is described and assigned a score, S, where 1 represents trivial and 10 very damaging. The potential cause is identified and assigned a score related to how often it occurs, O, where 1 represents never and 10 always. The method of detecting and controlling the failure is described and assigned a score, D, where 1 represents always obvious and 10 impossible to detect. The risk priority number (RPN) is an overall danger score calculated by multiplying S, O, and D. Higher RPNs are design priorities. In the example shown in Table 8.2 the first case has a score of 200 and is the highest priority. A solution is proposed to make the failure less likely, reducing the overall risk score.

FMEA studies are relatively routine documents for engineers and can be constructed by groups in a relatively short time. However, these do tend to be bulky, and a product with a few dozen parts could result in 100 pages of documentation. The process for completing a detailed FMEA is outlined as follows:

Table 8.2 An example failure modes and effects analysis table for a consumer front door lock.

Item or function	Failure mode	Failure effect	Severity ratings (S)	Cause	Occurrence rating (O)	Control	Detection rating (D)	Risk priority number, RPN = S × O × D	Recommended action	Action to be taken	New S	New O	New D	New RPN
Latch	Cannot extend into strike plate on door frame	The latch does not engage and the door moves freely	8	Does not line up with strike plate and the latch cannot extend	5	Noise indicates when latch is extended	5	200	Provide additional adjustments	Yes	8	3	5	120
			8	Spring mechanism fails and the latch will not extend	3	Noise indicates when latch is extended	2	48	Add backup spring	No				
			8	Latch edge catches in mechanism and the latch cannot move	4	Noise indicates when latch is extended	3	96	Add rounds to prevent jamming	Yes	8	3	3	72
Handle	Does not retract latch	The door cannot be opened	7	Mechanism broken	4	The door cannot be opened even with handle motion	3	84	Improve slides	No				
	The latch retracts unless the user holds the handle		5	Spring broken	4	The handle hangs down and does not offer resistance	1	20	Increase spring strength					

- Item or function: Review the specifications, the concepts, and all system components to identify those that are critical. These must include safety, operation, enjoyment, and satisfaction considerations. Consumer product industries also require that these address aesthetic and finish details for visible parts.
- Failure mode: Each function and item should be reviewed for all of the ways it could fail. These range from common issues to the more remote. If an item is unheard of or causes no damage, it can be omitted from the list. When there are multiple failure modes, use multiple lines for the function.
- Failure effects: In the event of a failure, this is a reasonable estimate of the level of damage that will occur.
- Severity rating (S): This puts a numerical value to the severity of the failure effect. More severe effects are more important:
  1, no effect;
  2, very minor, the effect is rarely noticed;
  3, minor, most users notice the effect;
  4–6, moderate, the issue may cause irritation, annoyance, or impaired functions;
  7–8, high, there is loss of function and unsatisfied users;
  9–10, very high, the problem may result in loss of equipment and human harm.
- Causes: The cause gives an insight into the reason the failure has occurred. If there are multiple causes use multiple rows. These can be very specific and name components, materials, processes, and steps.
- Occurrence rating (O): A numerical score is used to estimate how often the failure may occur. Failures that happen often are more important:
  1, unknown, there are no known occurrences;
  2–3, low, there are very few failures;
  4–6, moderate, failures occur sometimes;
  7–8, high, failures occur often;
  9–10, very high, failures occur in almost all cases.
- Control: This is a description of how a failure can be detected and how it might be corrected or mitigated.
- Detection rating (D): This is a numerical score for how easily the failure will be detected so that it will reduce the effect:
  1, certain, the failure will always be noticed;
  2, almost certain, most users would notice the issue;
  3, high, users are likely to notice the problem;
  4–6, moderate, some effort is required to detect the failure;
  7–8, low, special effort is required to detect the failure;
  9–10, rarely, the fault may not be detected by normal, and special, procedures.
- Risk priority number (RPN): This is a product of severity, occurrence, and detection, where $RPN = S \times O \times D$. A higher score means that the failure is likely to go undetected, happen often, and have more negative effects. Higher scores are a higher priority for design.
- Recommended action: This is a technical description of actions that can be taken to reduce the RPN. These can focus on one or more of the severity, occurrence, and detection scores. These do not need to be done for every failure type but are essential for the higher RPN scores.

- Action to be taken: The recommended actions are considered for difficulty, effort, benefit, and RPN score impact. Those that provide the most improvement for the lowest costs are undertaken. Actions for unacceptably high RPN scores are also considered. Others are left for later.
- New S, O, D, and RPN: These are the updated values after the changes are used to verify the impact and identify remaining issues.

The 1−10 scoring scale can be replaced with firm numerical values if known to provide a better estimate of failure importance. This does require more accurate statistical data of occurrence, severity, and detection. For mature product lines and experienced designers, these numbers may be easy to estimate. For newer products, these may require prototyping, lab testing, and consumer testing. Sources of this information include:

- historical data for similar components in similar conditions
- product life-cycle values
- published values
- data from consultants
- experienced estimates
- testing
- warranty data
- market studies of customer opinions and complaints
- benchmarking competitor's products

The FMEA method looks at single-failure incidents, which is suitable for most designs. For example, an FMEA would consider the possibility of a car tire pressure failure. However, it would not consider that the flat tire causes a large force that breaks a steering component and the wheel is torn off. An FMEA analysis requires a systematic review of all of the system components and individual reliabilities (see Resource 8.2). The engineering value of the process is the identification of the components that are likely to fail and dangerous failure types, including the following:

- Analyze single units or failures to target reliability problems.
- Identify redundant and fail-safe design requirements.
- Identify single-item failure modes.
- Identify inspection and maintenance requirements.
- Identify components for redesign.

A few hours of time to create the study can identify problems that can lead to product failures that harm people and property. The business cost of failed products can be very high, including lost customers, warranty returns, fines, product recalls, and lawsuits.

Resource 8.2 An FMEA spreadsheet is included on this book's website: www.engineeringdesignprojects.com/home/content/reliability.

PROBLEM

8.21 Construct an FMEA for a disposable coffee cup.

8.8 Complex fault modeling and control

In a complex system unrelated components may interact in nonobvious ways. For example, consider a new laptop design that has an optional lighted keyboard. All of the system testing done with the base model showed that it became hot but heat dissipation was adequate. After reaching the market a number of the computers with backlit keyboards were found to burst into fire. After investigation it was found that at full processor load, and with the lighted keyboard on, the battery became hotter than measured in tests, the keyboard also generated more heat, and the thermal management system could not cool the system enough. The high temperature caused the battery to swell, burst, and catch fire. This type of failure would not be considered in a routine FMEA.

Fault trees can be used to define combined events in a system that lead to system failures. Fig. 8.19 shows examples of three possible causes and effects. The boxes with the causes include a brief description of the action that initiates the fault. The bottom half of the box indicates where the main source of the fault lies. In general, "design" means that the detailed design group has the most control over the issue.

Fault trees can be created for each specification or failure type. An example of a fault tree is shown in Fig. 8.20. In this example the fault occurs when the phone will not turn on. Each of the arrows indicates a different cause that could create the fault. Some of the conditions are ANDed because they must all occur for that fault to happen. The fault tree in the example shows a couple of ways that dropping the phone could lead to the fault. This tree can be used to estimate probability of failure, identify candidates for testing, and set design priorities. The fault tree also indicates the likelihood for each event. This is equivalent to the unreliability for components, Q. Although this example uses constants, these values could be defined as values measured at the end of a year in use.

The likelihood of failure for this tree can be calculated using the statistical relationships in Fig. 8.21. The ANDed probabilities are simply multiplied. The ORed probabilities are assumed to be independent, so they may both occur at the same time. If the probabilities were mutually exclusive

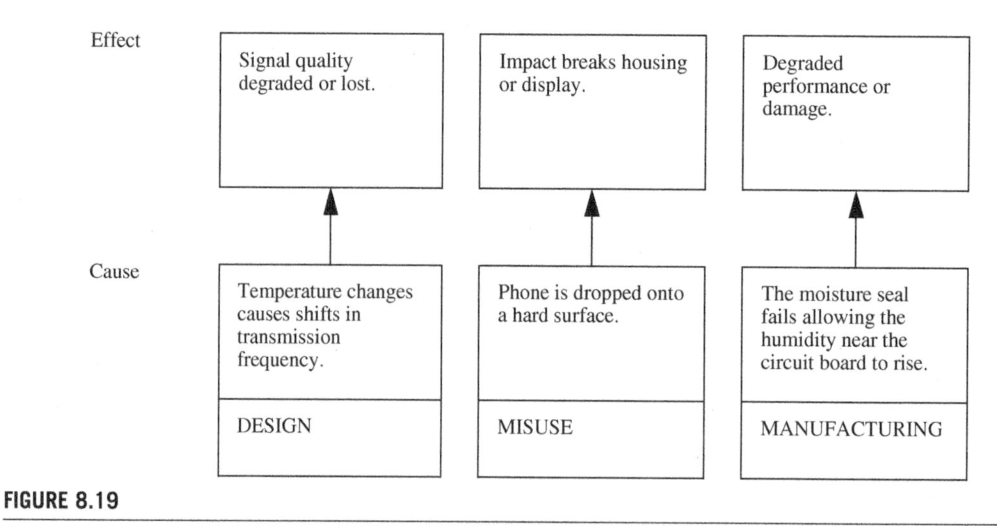

FIGURE 8.19

Single-step failure cause and effects.

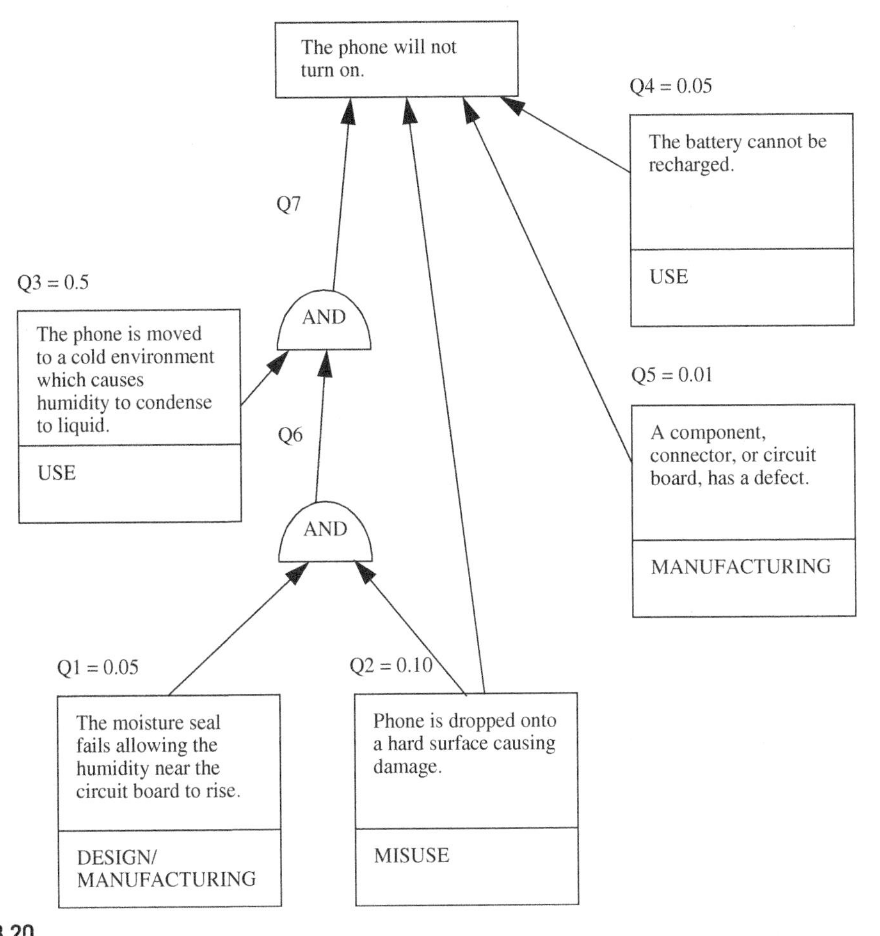

FIGURE 8.20

A sample fault tree.

they would not be able to occur at the same time. The calculations for the ORed faults are done two at a time to simplify the calculations. The likelihood of failure in order is $Q_2 = 0.1$, $Q_4 = 0.05$, $Q_5 = 0.01$, and $Q_7 = 0.0025$. This suggests that the design priority should be (1) case dropping, (2) battery life, (3) manufacturing defects, and (4) moisture seals. In fact, dealing with case breakage alone would resolve two-thirds of the failures.

An event tree looks at the sequence of steps leading to a failure (Fig. 8.22). This approach is more visual than a fault tree, and provides an alternative method for validating a fault tree. In this tree all of the initiating events are placed in double boxes. The intermediate boxes show factors that contributed to the overall failure of the device. As mentioned before, the trees expose sources of failure but it is up to the designers to select the problems to be solved. In this case, better connectors are easily identified, while battery usage may be resolved with a replaceable battery.

For AND conditions:

$$Q_{combined} = Q_1 Q_2$$

For mutually exclusive OR conditions:

$$Q_{combined} = Q_1 + Q_2$$

For independent OR conditions:

$$Q_{combined} = Q_1 + Q_2 - Q_1 Q_2$$

$$Q_3 = Q_1 Q_2 = 0.05(0.1) = 0.005$$

$$Q_7 = Q_3 Q_6 = 0.5(0.005) = 0.0025$$

$$Q_{27} = Q_2 + Q_7 - Q_2 Q_7 = 0.10 + 0.0025 - 0.10(0.0025) = 0.10225$$

$$Q_{45} = Q_4 + Q_5 - Q_4 Q_5 = 0.05 + 0.01 - 0.05(0.01) = 0.0595$$

$$Q_{2457} = Q_{27} + Q_{45} - Q_{27} Q_{45} = 0.10225 + 0.0595 - 0.10225(0.0595) = 0.1557$$

FIGURE 8.21

Fault probability calculation rules and example continued.

PROBLEM

8.22 What is the probability that this system will fail?

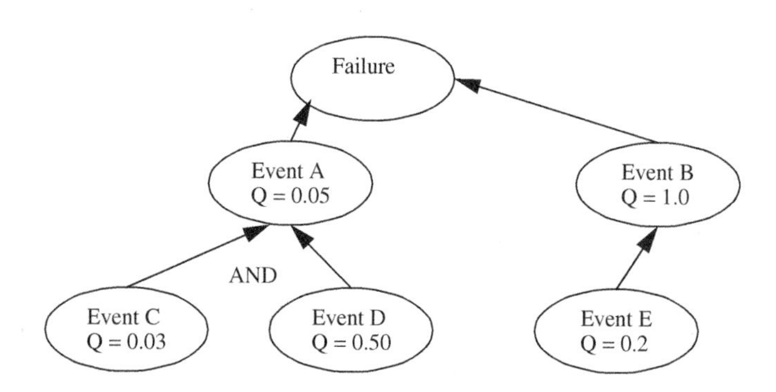

8.9 Designing reliable systems

System reliability is highly subjective. For example, a set of brakes on a roller coaster should fail in an engaged state, to safely stop the trains. Even if they don't slow the train, the riders will generally be safe.

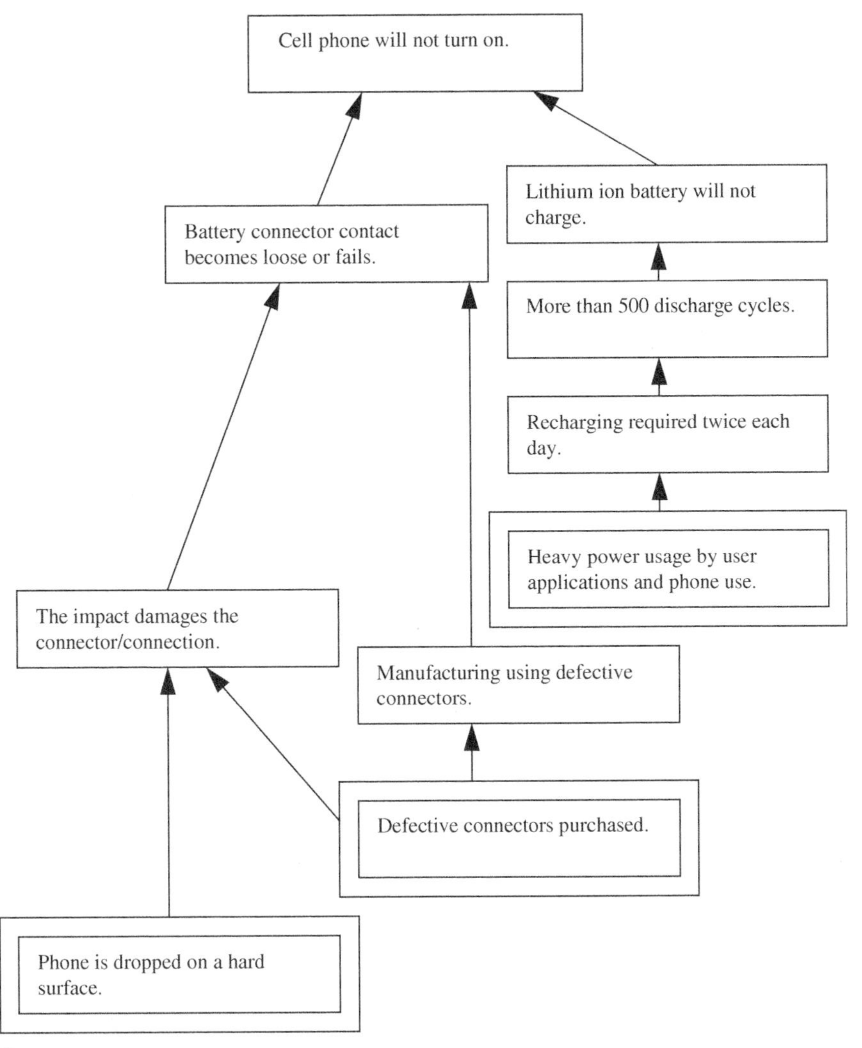

FIGURE 8.22

A sample event tree.

However, if a set of brakes on a car fails, then it may not stop, or it might come to a sudden halt and slide off the road. In both cases, the brakes perform a similar slowing function, but the car application is unforgiving. Some of the key terms for describing reliable systems are as follows:

- Rugged: a system that will resist faults;
- Robust: a system that will recover when faults occur;
- Failure tolerant: in the event that one system component fails, the entire system does not fail;
- Reliable: the probability that a system operates through a given specification;
- Available: the probability that the system will be available at any instant required;

- Dependable: a system that has a high reliability for important functions;
- Mission critical: a system that must operate even in the event of failures;
- Catastrophic: failures could lead to major damage and injury;
- Maintainable: a system that can be returned to operation with maintenance;
- Unstable or sensitive: a system that is easily faulted;
- Fault coverage: the probability that a system will recover from a failure.

Reliability objectives, and constraints, should be selected for each design and application. As with any expectation, excessive reliability adds cost and complexity to the design process. In addition, reliability does not always mean longer life and operation. For example, a rearview mirror in a car should break when excessive force is applied. This is done for passenger safety during collisions. The design criteria may call for a target breakaway force of 50 N. By contrast, a part that breaks simply will have a lower cost, but more expensive repairs may be required. Another magnet-based system may be reattached with no special effort, but at a higher cost.

If a mission-critical system is designed for maximum availability, the design objectives include increasing reliability and recovering from faults when they occur. Some of the strategies employed to increase the availability are given in the following list. Consider an electronics and software system for braking freight trains. If the system fails the results can be catastrophic. Each of the critical braking components in the system should have backup components. Lower-reliability components should also have backup components. If the system does fail, the electronics should disengage so that a manual braking system can be used. The computer in charge of braking should have ways to reset itself or switch to another control computer if problems are detected:

- Backup: a secondary system that can be used to replace the primary system if it fails;
- Fail operational: even when components fail, the system continues to operate within specifications;
- Prime: a main system that is responsible for a task;
- Redundant: secondary systems that run in parallel with the prime and will be able to hot swap if the prime fails;
- Time critical: the system has a certain response time before a failure will occur.

Fail-safe systems are preferred when possible. When a problem occurs the system should move to a low activity state. If power or control is removed from a system it should fail safely. For example, an elevator system should not drop or rise when power is lost. If the elevator door is open the elevator should not move. If the elevator is not at a floor the doors should not open. These types of designs require that the designer select components and structures that will be safe if they do not have any power or if they are cut from the system during operation. Safing is a process whereby a system that has failed is shut down appropriately (i.e., actuators halted, brakes applied, or whatever is appropriate to the situation). Safing paths often include the following:

- Equipment
  - Braking equipment
  - Removal of power to actuators
  - Consideration of complete power failure
  - Operator control should be available, even when automated systems are in place
  - Multiple safing paths should be available
- Operator training and decision-making
  - Safing procedures

  - Attempt to manually repair
  - Ignore
- Software and electronics
  - Checksums
  - Parity bits
  - Software interlocks
  - Watchdog timers
  - Sample calculations

Software and electronics warrant special discussion. There are formal methods for software design, often required for the military, transportation, aviation, and medical industries. In general, the software must be deterministic, meaning that all possible scenarios are considered, including errors and error recovery. For example, consider a military system in which there is a good chance part of the system will be forced to fail by impact or explosion. In these cases the software must compensate and maintain maximum functionality.

Electronics are also prone to variations in voltage levels and electromagnetic noise. To compensate for data transmission errors, parity and check bits are used to detect errors in communication. Checksums can be used for blocks of data, and gray code can be used for detecting errors in sequential numbers. The quantity of redundant hardware can be halved by doing the same calculation twice at different points in time on the same processor. If the calculation results are compared and found to be different, a transient fault would be detected. This can be important in irradiated environments where data bits can be flipped randomly. Software redundancy involves writing multiple versions of the same algorithm or program. All of the algorithm versions are executed simultaneously. If a separate acceptance algorithm estimates that the primary version is in error, it is disabled and the secondary version is enabled. This continues as long as remaining modules are left. Fault-tolerant systems are designed so that there is a mechanism for detecting failure and enabling alternate system components. Common components in these systems include the following:

- Monitoring systems: Check for system sanity or failure of systems to report failures.
- Emergency control functions: These are functions that switch control when faults are detected. In some cases this might include human intervention and be triggered automatically. These systems are intended to eliminate or reduce the effects of a failure.
- Time-outs: When an event or process does not respond in a reasonable time, a fault condition is flagged.
- Tethers: Systems are tied together physically or electronically to ensure presence or location.
- Operator inspection: Tools are provided for operator-detected faults in the event of system failure.
- Remote monitoring: Operators in a remote location can monitor and correct problems using vision systems, networking, warning lights, sirens, etc.
- Redundant connections: Multiple wires and paths, network paths, and multiple radio frequencies will ensure that one damaged connection will not fault the system.
- Connection failure detection: If a connection is disrupted the condition can be detected. A system that uses 4 and 20 mA for true and false will be broken if the current drops to 0 mA. In mechanical systems this could be a loss of hydraulic pressure.
- Echoes and acknowledgments: If the original message or command is sent back it can be verified or the receiver can send back a message that the command was received and processed correctly.
- Approved operation: Systems may require a review before operation to detect potential faults.

- Critical sensors: Additional sensors are added and monitored for dangerous components and systems.

There are thousands of regulations, standards, and professional practices for designing safe systems. These normally have similar steps, including (1) identify the hazards, (2) determine severity, (3) determine likelihood, (4) identify unacceptable dangers, and (5) modify the designs as needed. For example, a fire safety standard would identify various fuels, flammability, probability of ignition, reasonable estimates of injuries, and methods for removing or modifying the materials. MIL STD 882B was presented earlier as another method for risk assessment. A similar methodology is used in the electronics and software safety standard IEC 61508/61511 (see Resource 8.3). The standard covers electrical, electronic, and programmable electronic systems. Three categories of software languages are covered by the standard:

- FPL (fixed programming language): a very limited approach to programming. For example, the system is programmed by setting parameters.
- LVL (limited variability language): a language with a strict programming model, such as ladder logic in programmable logic controllers (PLCs).
- FVL (full variability language): a language that gives full access to a system, such as C.

Resource 8.3 IEC safety website: www.iec.ch/functionalsafety or www.engineeringdesignprojects.com/home/content/reliability.

PROBLEMS

8.23 What is the difference between sensitive systems and tolerant systems?

8.24 Aircraft landing gear is a critical system; if it does not work the result is catastrophic. Assume the system is controlled by a computer in the cockpit that drives an electric motor in the wing wheel well. List 10 features and steps that would increase the safety of the system.

8.10 Verification and simulation

After a program has been written it is important to verify that it works as intended, before it is used in production. In a simple application this might involve running the program on the machine and looking for improper operation. In a complex application this approach is not suitable. A good approach to software development involves the following steps in approximate order:

(1) Structured design: Design and write the software to meet a clear set of objectives.
(2) Modular testing: Small segments of the program can be written and then tested individually. It is much easier to debug and verify the operation of a small program.
(3) Code review: Source code is reviewed by the programmer and the design group.
(4) Modular building: The software modules can then be added one at a time and the system tested again. Any problems that arise can then be attributed to interactions with the new module.
(5) Design confirmation: Verify that the system works as the design requires.

(6) Error proofing: The system can be tested by introducing expected and unexpected failures. When doing this testing, irrational things should also be considered. This might include unplugging sensors, jamming actuators, operator errors, etc.
(7) Burn-in: This is a test that lasts a long period of time. Some errors won't appear until a machine has run for a few thousand cycles or over a period of days.

Program testing can be done on machines, but this is not always possible or desirable. In these cases, simulators allow the programs to be tested without the actual machine. The use of a simulator typically follows the basic steps listed below:

(1) The machine inputs and outputs are identified.
(2) A basic model of the system is developed in terms of the inputs and outputs. This might include items such as when sensor changes are expected, what effects actuators should have, and expected operator inputs.
(3) A system simulator is constructed with some combination of specialized software and hardware.
(4) The system is verified for the expected operation.
(5) The system is then used for testing software and verifying the operation.

PROBLEMS

8.25 Why is simulation useful for system reliability?

8.26 Reducing a larger system to modules simplifies design and testing. Why?

8.27 Fans can be used to control electronics, including CPUs. If the fan fails the result could be catastrophic. Develop options to make the fan safer by (a) reducing the chance of failure, (b) reducing the impact of failure, and (c) detecting the onset of failure.

8.28 Sometimes parts will make noise, change color, become loose, or bend before failure. Explain how this reduces the danger.

8.29 List common sources of failure in the design process.

8.30 How does the bathtub curve relate early- and late-life failures?

8.31 What is the relationship between the bathtub curve and the Weibull distribution?

8.32 What is the purpose of device "burn-in"?

8.33 What is the failure rate if 15% of all devices are working at 4000 h.

8.34 If a unit cannot be repaired, would you use MTBF, MTTR, and/or MTTF?

8.35 A design must have an MTTF of 20,000 h.

(a) The device is made with two identical units connected in series. What is the required MTTF for each unit?

(b) The device is made with two identical units connected in parallel. What is the required MTTF for each unit?

8.36 A system component has an MTTF of 10,000 h, but the customer requires 30,000. There are 2 design options to be analyzed.

(a) If one component can be added in parallel, what is its required MTTF?

(b) How many of the 10,000 MTTF components would be needed in parallel to meet the 30,000-h MTTF?

8.37 Two components are used in series. One component has been in use for 5000 h and has a reliability of 0.60. The second component was replaced 1000 h ago and has a reliability of 0.85.

(a) What is the combined reliability?

(b) What is the reliability if both components are 0.85 reliable?

(c) What is the reliability if both components are 0.60 reliable?

8.38 A ship will still operate if 6 or more cylinders in a 10-cylinder engine are working. Each cylinder has an MTTF of 8000 h. What is the reliability of the engine at 10,000 h?

8.39 A traffic light has three $30 traffic light modules. Each module has an MTTF of 30,000 h. The labor to replace the three modules is $50 during routine maintenance. However, if any one of the light modules fails, an emergency crew must replace the three modules at a labor cost of $600. A standard policy is to replace the modules at regular intervals before they have failed. What should the replacement interval be to minimize the costs?

8.40 A PLC-based control system has three parallel control modules. Each module has an MTTF of 20,000 h. The monitor has an MTTF of 50,000 h. What is the combined reliability of the system?

8.41 What is the probability that the system shown in the diagram will fail?

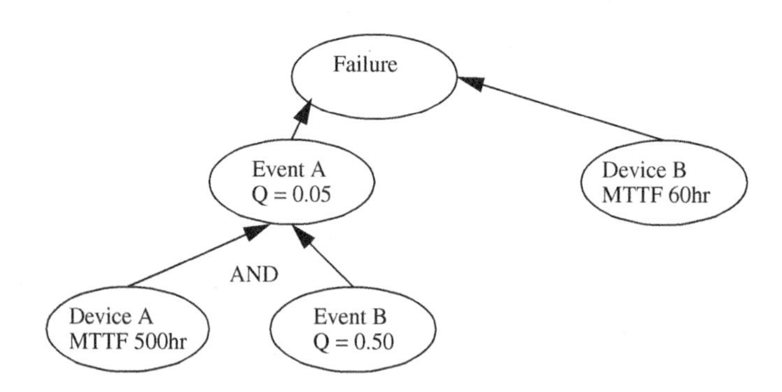

8.42 Develop a list of five catastrophic, five major, and five minor faults for a car tire.

8.43 Standards define the reliability of devices. Find a published standard that lists minimum requirements for a consumer product.

8.44 Case: The Challenger

The explosion of the Challenger space shuttle was the result of management failures at NASA and Morton Thiokol and technical weaknesses. (See Challenger accident report PDF file on course website: www.engineeringdesignprojects.com/home/content/accident-reports.) List five events in which the chain of events could have been easily changed to prevent failure. The essential timeline is as follows:

1972: Morton Thiokol was chosen to design the solid rocket boosters (SRBs) based on a modified Titan III rocket. One major change was an O-ring seal along the rocket body that was made longer and a second O-ring that was added to provide a redundant seal.

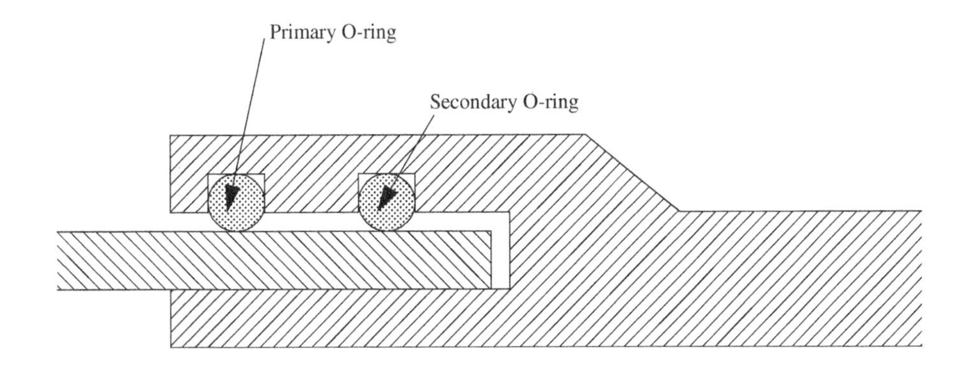

1977–78: An engineering test showed that under pressure the joints rotated significantly, causing the secondary O-ring to become ineffective. Morton Thiokol management chose to accept the risk.

1980: The O-ring joint was listed as 1R on the critical item list (CIL), which indicates possible catastrophic failure. The "R" indicates redundancy because of the second O-ring. There were 700 other items on the CIL.

1982: The space shuttle was declared operational. During the flights problems were identified and assigned a tracking number to start problem solving. The O-ring problem was noticed but not assigned a number. Eventually the problem was noticed and the CIL rating was changed from 1R to 1 to indicate that there was no backup. Morton Thiokol paperwork was not updated and it still listed the seals as 1R. When pushed to recognize the change, Morton Thiokol disagreed with the criticality change and went to a referee procedure.

1984: The O-ring erosion during launches became a significant issue. NASA asked for a review of the asbestos putty used to reduce the heat effects on the O-rings. Morton Thiokol responded that the putty and O-rings were failing sooner because of the higher-than-needed testing pressures, and this was confirmed with tests. It said it would investigate the effects of the tests.

January 1985: The space shuttle was launched at the coldest temperature in the history of the program. The cold stiffened the O-rings and prevented them from deforming and sealing. After booster rocket recovery, the O-rings were examined and showed the greatest degradation of all flights.

January to April 1985: The flights continued and the issues with the O-rings persisted. The launch temperature and the O-ring condition were positively correlated. Morton Thiokol acknowledged the problem but stated that the second O-ring would ensure safety.
April 1985: During a flight the primary O-ring did not seal and the secondary ring had to carry the pressure. The secondary O-ring was showing degradation and would have eventually failed. The near failure of the backup resulted in a committee decision to set a minimum temperature for launches. The report was distributed within NASA and Morton Thiokol, but there were questions about who received a copy and if they read it.
July 1985: To prevent a disaster, a Morton Thiokol engineer recommended that a team be set up to study the O-ring seal problem.
August 1985: Morton Thiokol and NASA managers briefed NASA headquarters on the O-ring problems, with a recommendation to continue flights but step up investigations. A Morton Thiokol task force was set up.
October 1985: The head of the Morton Thiokol task force complained to management about lack of cooperation and support.
December 1985: One Morton Thiokol engineer suggested stopping shipments of SRBs until the problem was fixed. Morton Thiokol management wrote a memo to NASA suggesting that the problem tracking of the O-rings be discontinued. This led to an erroneous listing of the problem as closed, meaning that it would not be considered as critical during launch.
January 1986: The space shuttle Challenger was prepared to launch on January 22; originally it had been scheduled for July 1985 and was postponed three times and scrubbed once. It was rescheduled again to the 23rd, 25th, 27th, and then 28th. This was a result of weather, equipment, scheduling, and other problems.
January 27, 1986: The shuttle began preparation for launch the next day, despite predicted temperatures below freezing (26°F or −3°C) at launch time. Thiokol engineers expressed concerns over low temperatures and suggested NASA managers be notified (this was not done). A minimum launch temperature of 53°F had been suggested to NASA. There was no technical opinion supporting the launch at this point. The NASA representative discussing the launch objected to Thiokol's engineer's opinions and accused them of changing their opinions. Upper management became involved with the process and "convinced" the technical staff to withdraw objections to the launch. Management at Thiokol gave the go-ahead to launch, under pressure from NASA officials. The shuttle was wheeled out to the launch pad. Rain had frozen on the launch pad and may have gotten into the SRB joints and frozen there also.
January 28, 1986: The shuttle director gave the OK to launch, without having been informed of the Thiokol concerns. The temperature was 36°F or 2°C.
11:39 a.m.: The engines were ignited and a puff of black smoke could be seen blowing from the right SRB. As the shuttle rose the gas could be seen blowing past the O-rings. The vibrations experienced in the first 30 s of flight were the worst encountered to date.
11:40 a.m.: A flame jet from the SRB started to cut into the liquid fuel engine tank and a support strut.
11:40:15 a.m.: The strut gave way and the SRB pointed nose cone pierced the liquid fuel tank. The resulting explosion totally destroyed the shuttle and crew.
11:40:50 a.m.: The SRBs were destroyed by the range safety officer.

8.45 Mini-case: Space units

In December 1998, the US Mars Climate Orbiter was launched and began the trip to Mars. In September 1999, the satellite reached Mars and was preparing to enter a permanent orbit. This process involved adjusting the path to be tangential to the desired orbit circle. The rocket was slowed so that when it was near the tangential point, the satellite velocity matched the orbital velocity. If the satellite was too slow, or too close to the planet, it would crash. If the satellite was too fast, or too far from the planet, it would miss. As the Orbiter approached the planet, four 22-N side thrusters were available to adjust the approach angle and a 640-N braking engine was available to slow the Orbiter from interplanetary speeds to orbital speeds. The ground crew, on earth, monitored the trajectory of the satellite and used it to calculate the burn times for each of the engines to achieve orbit. There was a radio delay of several minutes, so the team uploaded the calculated values and then waited. Normally the satellite would have done the calculated burns, achieved orbit, and then sent a status update. For this satellite the status update never came. Needless to say, there was some concern over the $327 million failure. The executive summary from a report on the failure follows (NASA, 1999). (See MCO accident report PDF file on the course website: www.engineeringdesignprojects.com/home/content/accident-reports.)

This Phase I report addresses paragraph 4.A. of the letter establishing the Mars Climate Orbiter (MCO) Mishap Investigation Board (MIB) (Appendix). Specifically, paragraph 4.A of the letter requests that the MIB focus on any aspects of the MCO mishap which must be addressed in order to contribute to the Mars Polar Lander's safe landing on Mars. The Mars Polar Lander (MPL) entry-descent-landing sequence is scheduled for December 3, 1999.

This report provides a top-level description of the MCO and MPL projects (section 1), it defines the MCO mishap (section 2) and the method of investigation (section 3) and then provides the Board's determination of the MCO mishap root cause (section 4), the MCO contributing causes (section 5) and MCO observations (section 6). Based on the MCO root cause, contributing causes and observations, the Board has formulated a series of recommendations to improve the MPL operations. These are included in the respective sections. Also, as a result of the Board's review of the MPL, specific observations and associated recommendations pertaining to MPL are described in section 7. The plan for the Phase II report is described in section 8. The Phase II report will focus on the processes used by the MCO mission, develop lessons learned, and make recommendations for future missions.

The MCO Mission objective was to orbit Mars as the first interplanetary weather satellite and provide a communications relay for the MPL which is due to reach Mars in December 1999. The MCO was launched on December 11, 1998, and was lost sometime following the spacecraft's entry into Mars occultation during the Mars Orbit Insertion (MOI) maneuver. The spacecraft's carrier signal was last seen at approximately 09:04:52 UTC on Thursday, September 23, 1999.

The MCO MIB has determined that the root cause for the loss of the MCO spacecraft was the failure to use metric units in the coding of a ground software file, "Small Forces," used in trajectory models. Specifically, thruster performance data in English units instead of metric units was used in the software application code titled SM_FORCES (small forces). A file called Angular Momentum Desaturation (AMD) contained the output data from the SM_FORCES software. The data in the AMD file was required to be in metric units per existing software interface documentation, and the trajectory modelers assumed the data was provided in metric units per the requirements.

During the 9-month journey from Earth to Mars, propulsion maneuvers were periodically performed to remove angular momentum buildup in the on-board reaction wheels (flywheels). These Angular Momentum Desaturation (AMD) events occurred 10–14 times more often than was expected by the operations navigation team. This was because the MCO solar array was asymmetrical relative to the spacecraft body as compared to Mars Global Surveyor (MGS) which had symmetrical solar arrays. This asymmetric effect significantly increased the Sun-induced (solar pressure-induced) momentum buildup on the spacecraft. The increased AMD events coupled with the fact that the angular momentum (impulse) data was in English, rather than metric, units, resulted in small errors being introduced in the trajectory estimate over the course of the 9-month journey. At the time of Mars insertion, the spacecraft trajectory was approximately 170 km lower than planned. As a result, MCO either was destroyed in the atmosphere or reentered heliocentric space after leaving Mars' atmosphere.

The Board recognizes that mistakes occur on spacecraft projects. However, sufficient processes are usually in place on projects to catch these mistakes before they become critical to mission success. Unfortunately for MCO, the root cause was not caught by the processes in place in the MCO project.

A summary of the findings, contributing causes and MPL recommendations are listed below. These are described in more detail in the body of this report along with the MCO and MPL observations and recommendations.

Root Cause: Failure to use metric units in the coding of a ground software file, "Small Forces," used in trajectory models

Contributing Causes:

(1) Undetected mismodeling of spacecraft velocity changes

(2) Navigation Team unfamiliar with spacecraft

(3) Trajectory correction maneuver number 5 not performed

(4) System engineering process did not adequately address transition from development to operations

(5) Inadequate communications between project elements

(6) Inadequate operations

- Navigation
- Team staffing

(7) Inadequate training

(8) Verification and validation process did not adequately address ground software

MPL Recommendations:

- Verify the consistent use of units throughout the MPL spacecraft design and operations
- Conduct software audit for specification compliance on all data transferred between JPL and Lockheed Martin Astronautics
- Verify Small Forces models used for MPL

- Compare prime MPL navigation projections with projections by alternate navigation methods
- Train Navigation Team in spacecraft design and operations
- Prepare for possibility of executing trajectory correction maneuver number 5
- Establish MPL systems organization to concentrate on trajectory correction maneuver number 5 and entry, descent and landing operations
- Take steps to improve communications
- Augment Operations Team staff with experienced people to support entry, descent and landing
- Train entire MPL Team and encourage use of Incident, Surprise, Anomaly process
- Develop and execute systems verification matrix for all requirements
- Conduct independent reviews on all mission critical events
- Construct a fault tree analysis for remainder of MPL mission
- Assign overall Mission Manager
- Perform thermal analysis of thrusters feedline heaters and consider use of preconditioning pulses
- Reexamine propulsion subsystem operations during entry, descent, and landing

Given the summary, select two MPL recommendations that should be a high priority. Explain your choice.

Reference

NASA, November 10, 1999. Mars Climate Orbiter Mishap Investigation Board Phase I Report.

Further reading

American Institute of Chemical Engineers, 1992. Guidelines for Hazard Evaluation Procedures: With Worked Examples, second ed.

Cooper, D., Grey, S., Raymond, G., Walker, P., 2005. Project Risk Management Guidelines: Managing Risk in Large Projects and Complex Procurements. Wiley.

Dhillon, B.S., 1996. Engineering Design; A Modern Approach. Irwin.

Dorf, R.C. (Ed.), 1993. The Electrical Engineering Handbook. IEEE Press/CRC Press, USA, pp. 2020–2031.

Leveson, N., 1995. Safeware: System Safety and Computers. Addison-Wesley Publishing Company Inc.

Rasmussen, J., Duncan, K., Leplat, J., 1987. New Technology and Human Error. John Wiley & Sons Ltd.

Ullman, D.G., 1997. The Mechanical Design Process. McGraw-Hill.

CHAPTER

Communication, meetings, and presentations

9

9.1 Introduction

Communication is a tool.

Engineers need to communicate effectively. Formal and informal methods of workplace communication include writing, oral presentations, and meetings. These are directed from a speaker or a writer to another. The receiver of the communication is often called the audience, but we will use listener. Fig. 9.1 shows a simple model of knowledge transfer that can be used when analyzing communication. Motivation is the reason for communicating, such as to send an invoice to a customer, announce the birth of a child, or a provide a letter of reference for employment. When the motivations do not match expectations, the reader may misunderstand, ignore, or overreact. The content of communication should include a suitable amount of detail to satisfy the motivation.

Common problems include too much or too little detail, irrelevant information, and information that is too technical or too simple. Formats range from formal reports to simple comments in a hallway. If the format does not match the listeners' expectations they will have to make special efforts to understand, leading to frustration and misunderstanding. Transmission can be critical to ensuring the message arrives and is received. For verbal conversations this may be a simple nod of the head. For formal items, such as contracts, this might require a courier service that tracks delivery times, a receiver signature, or a written response from the receiver. For legal matters, transmission requirements might be very specific. In simpler cases, this model is excessive. For new communication, and problem solving, this model can be used to identify and solve problems. For example, when negotiating project milestones with a customer, apply the model for approval at each of the stages.

PROBLEMS

9.1 Use the speaker/listener model to describe the following communication types. Specifically state motivation, content, format, and transmission:

(a) A phone call to a salesperson

(b) A presentation of a new design proposal to a customer

(c) A funny email to a colleague

Engineering Design, Planning, and Management. https://doi.org/10.1016/B978-0-12-821055-0.00009-8

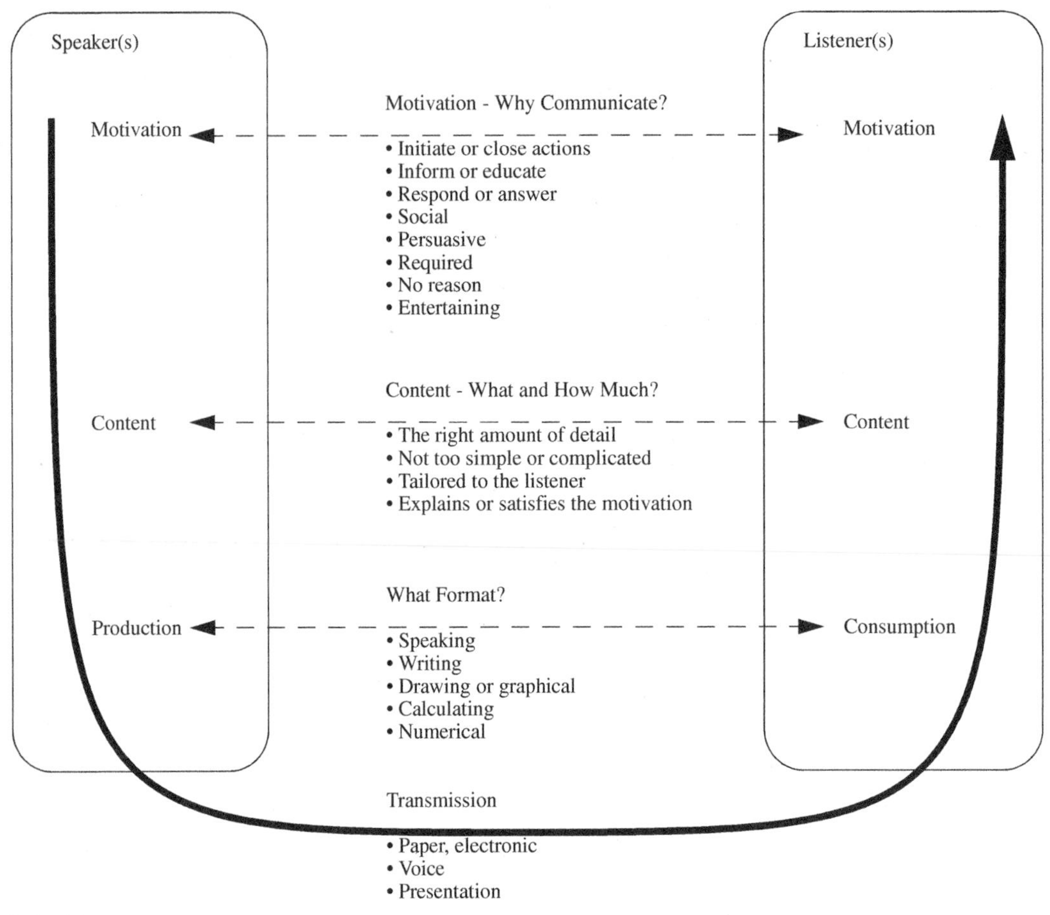

FIGURE 9.1

A simple model of communication.

9.2 Mini-case: A memo from the accounting department

A very brief memo was received from the MassiveCorp accounting department. It was so succinct it could win an award for brevity. It read, "The purchasing policy now requires that all food purchases be approved by supervisors. Food purchases above $20 per person must be approved by the VP of finance." The motivation for this memo was to inform employees about a new company policy. The content was the new policy with details important for anybody purchasing food. The memo format was quite suitable for this type of message and the requirement could also be added to the expense reimbursement forms. The memo could be sent on paper or by email, based on normal practices at MassiveCorp. If the reimbursement forms were changed, then the updated forms could be sent by email to all managers or employees.

Vlad, a project engineer, received the memo and remembered the details for his visit to customer sites. Sonja, a supervisor in shipping, received and read the memo. At the next department meeting, she used her new purchasing authority to buy lunch for each of the employees in the department. When she submitted the expense claim form it was denied. After some emotional questions, she discovered that the memo applied only to supervisors that already have food budgets for entertaining customers.

Address the communication mistakes with reference to motivation, content, format, and transmission that would have avoided this problem.

9.2 Speakers/writers and listeners/readers

When communicating you are the performer and the audience. The audience is watching, listening, reading, and more to see what messages you provide and what questions you ask. At some point you give the stage to somebody else. If you present to the audience they will be engaged and interested. This analogy is very effective as we take roles as speaker/listeners, reader/writers, presenter/audience, lecturer/class, and so on.

Audience is a broad term used to describe the person(s) you are communicating with. The other person also needs to know what type of audience you represent. As an example, a phone call between two engineers will contain more technical content than a call to the sales department. An engineer may want to focus on technical content, whereas an accountant may want to focus on tracking expenses. Everybody has preferences for communication. Examples include email versus phone messages versus text messages versus meetings versus memos. Some audiences want more background information; others want the communication to be brief. Select a target audience for your message and frame it for effective communication. To understand the audience, consider what content they expect and how they prefer to get it. If you are the audience, help the speaker understand what content you need and how you prefer to get it. Some questions to prompt this process include:

- What do the listeners say they want?
- What do the listeners really want?
- Why can't the listener do the task?
- Why is the listener interested in you?
- How do the listeners expect to use your work?
- Which of the following methods of operation are preferred?
    - Trust (verbal) versus formal (written)
    - Detailed versus strategic
    - Firefighting versus working ahead
- How will your work fit into the listener's systems?
- What is the budget?
- What is the schedule?
- What has happened before in this, or a similar, project?
- What existing resources or previous work can be used?
- What is the schedule?

Crude audience categories for business people are given in Fig. 9.2. Larger companies normally have more people involved in decision-making, more structured and defined business processes, and a need to understand risks and benefits. On the other hand, an individual inventor may accept risks easily, make decisions on the spot, and have fewer business expectations. Some general (but not binding) definitions of various groups are described in the following list:

- Inventors: Inventors typically have very little business knowledge, have very few resources, are in a vulnerable position, and need help. Inventors sometimes have idealistic approaches that exceed

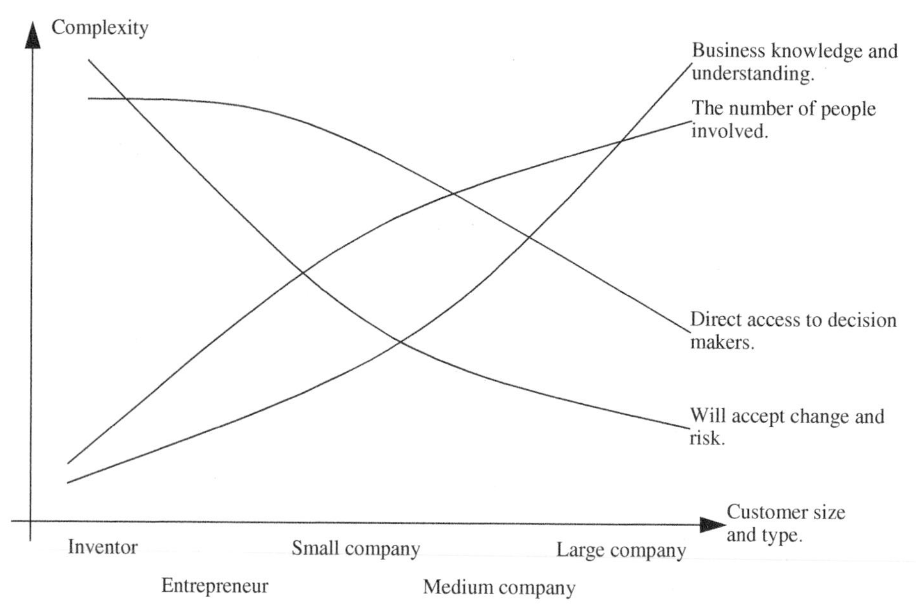

FIGURE 9.2

Generalized communication factors as a function of business maturity.

reasonable practice. What you say and what they hear will often mismatch. Always repeat details and decisions to ensure they agree.

- Entrepreneurs: Entrepreneurs have variable levels of business knowledge, but limited resources. By their nature entrepreneurs are adaptive and willing to change. The free-flowing environment can make it hard for them to settle on a fixed design.
- Small companies: These people have done some good things and are probably growing. In a small company people trade roles freely and procedures tend to be informal. They should be sure they are saying the same thing the listener is hearing.
- Medium/large companies: These are distinguished by established markets, structures, and business knowledge. People in medium and large companies may need to adapt to company-wide systems to be effective. Procedures and roles can be very narrowly defined.

The key to understanding the audience is to find out what they know and want. Communicate to those perspectives, but adapt as you get to know them more.

PROBLEMS

9.3 If you watch Internet videos you are the audience. How can you become the performer?

9.4 Assume you are writing an email to ask for technical help with software.

(a) Write five questions for the email.

(b) Indicate how the reader would interpret each of the questions.

9.5 Consider a computer saleswoman with corporate accounts and retail sales, both having different needs and wants. List five objectives each for corporate and consumer customers.

9.6 Is it always true that individuals from larger companies have more business knowledge?

9.7 List three reasons an inventor would approach you.

9.3 What are you saying?

As a presenter you are responsible for making yourself understood. It is not enough to recite information and expect understanding. A simple fact stated from your perspective requires that the listener think more, thereby leaving opportunity for misunderstandings. If you stop to think and then relate the presentation to the audience, they are more likely to understand correctly. The key concept for all communication is to be clear and concise. Say the important things and then reinforce them. Avoid the vague. Omit less important information. State your reasons for communicating.

In technical terms our listeners and readers come with some anticipation of structure. For example, a problem-solving process is shown in Fig. 9.3. Written problem solutions will normally follow these

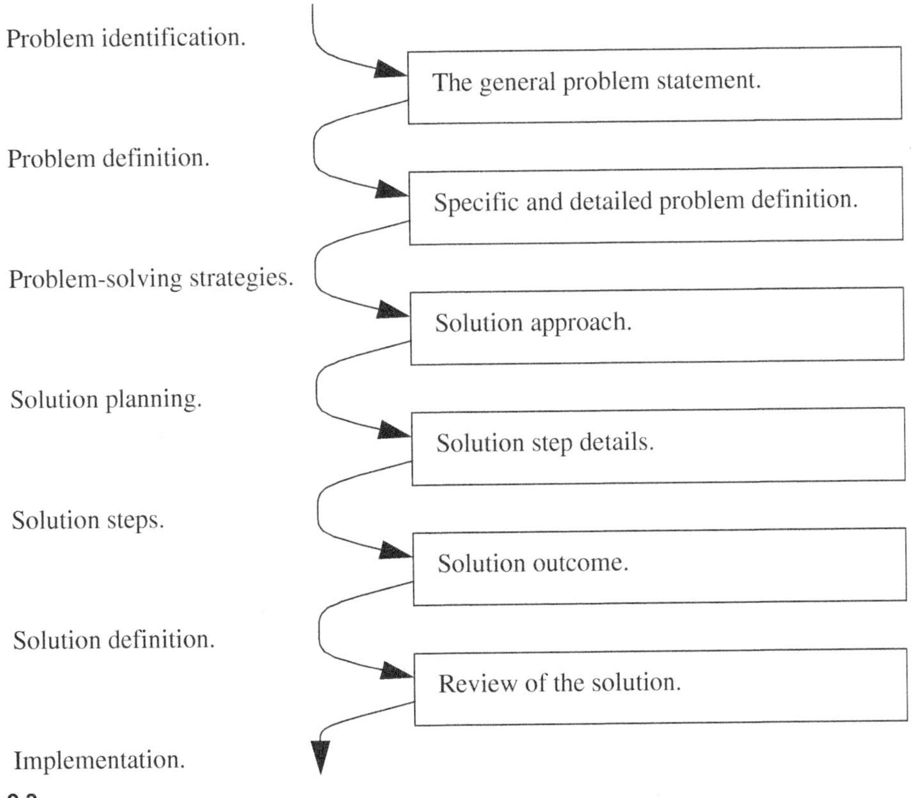

FIGURE 9.3

A procedural model of problem solving.

steps to ensure a logical and justifiable result. If you follow these steps you will appeal to a process that other technical people will understand, thus reducing the amount of work required for comprehension.

In technical work, information tends to be highly structured, as illustrated in Fig. 9.4. For example, a design has a motivation, or subject, such as the design of a new coffee cup. The implementation uses new approaches for insulation and sealing, called topics. These require new materials and shapes, or concepts. The final design contains many details. One way to relate these intellectually is with a

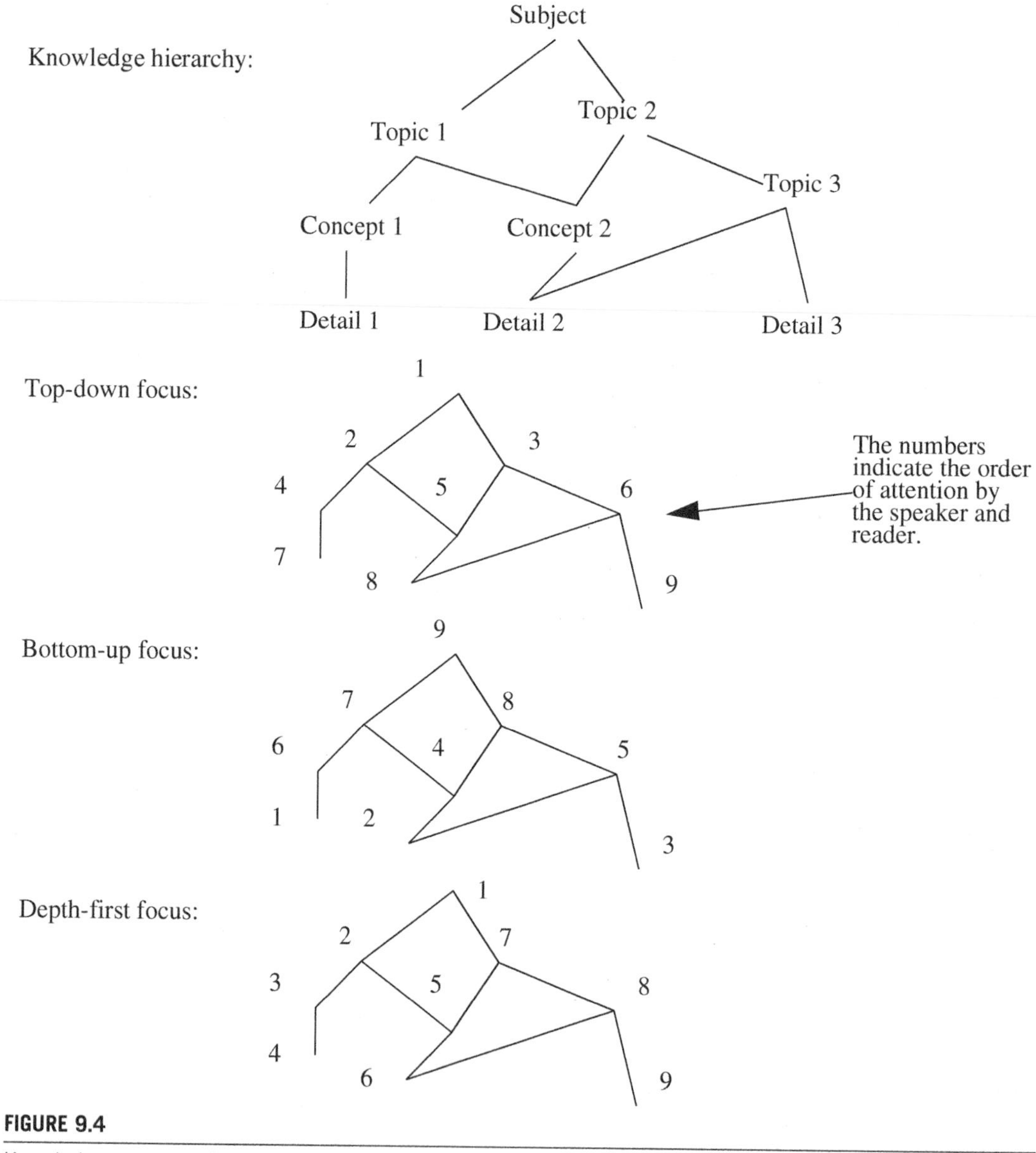

FIGURE 9.4

Knowledge sequence for communication.

knowledge hierarchy tree. In a design report, these may be communicated top to bottom with increasing level of detail. A listener must absorb everything in sequence to understand the details. In a bottom-up approach the listener is presented with a large number of details that must be remembered before they are eventually related. In a depth-first approach, one subject—topic—concept—detail is presented and explored in depth before the next topic. This creates more variety, but entire sections of the details are withheld until the end. All three of these presentation methods have advantages and disadvantages. A good practice is to concisely summarize the work with the top-down approach so that people understand what is expected. After this, use the depth-first approach for the presentation and then bottom-up for the summary, or vice versa.

When you are the speaker or writer, a good approach is to present in stages, possibly starting with a top-down summary followed by a listener response. This can then be followed with bottom-up or depth-first details. A critical mistake is presenting too many details at once and overwhelming the listener. A speaker who never gets to the point can be equally frustrating.

Other effective tools for presenting knowledge are outlined as follows:

- Analogies: Taking a difficult concept and describing it with an analogy can help understanding and recall.
- Examples: An example of techniques and applications can be useful for understanding.
- Concrete: Using exact descriptions will be more effective. For example, you could say a "portable computer" or "laptop."
- Active: Active voice is better for listeners and readers. However, it can be a negative when trying to convey impartiality or detachment.
- Action: When possible, describe what actions are required to lead to a result. For example, instead of "write a report," say "write a report containing …."
- Entertain: Limited diversions, such as jokes, can increase audience interest and empathy.
- Visualize: Pictures, hand gestures, casual sketches, etc., can help communicate with a concrete and visual audience.
- Abstract: Some listeners prefer abstract concepts over details. Providing these as a parallel to details can reach more of the audience.
- Engage: Develop a way for the audience to actively apply the knowledge you have presented.
- Echo: As a listener it is a good idea to repeat what the presenter has said. Likewise, you should expect your audience to echo your words.

PROBLEMS

9.8 Is encyclopedia information presented in a top-down or bottom-up style?

9.9 What is the difference between subjects, topics, concepts, and details?

9.10 What communication tools can be used to convey the impression of impartiality?

9.11 How can you verify that what was said and what was heard are the same?

9.12 What must occur before solving a problem?

9.4 Critical listening and reading as the audience

The ability to acquire, retain, and use information is often cited as one of the most valuable professional skills. Some physicians receive criticism for asking a couple of questions, making a diagnosis, and then moving to the next patient. Other physicians receive praise because they allow the patients to describe their concerns before making a diagnosis. The result is often a diagnosis that is more correct and suited to the patient. The outcome is a happier patient who is more likely to follow the prescribed treatment and return to the same doctor. When working with others, we will have casual interactions and conversations but at some point we will need to have detailed discussions. Let each speaker talk before trying to provide answers. While someone is talking you need to build a mental picture of what he or she is saying and what he or she wants. If you do not understand, ask questions. When done listening, restate what he or she has said to be sure that you understand what was said.

Then act.

Effective listening and reading are learned skills. You must be patient and try to absorb all details, delay making decisions or judgments, critically analyze the information, and fill in the gaps. This process is complicated because of the differences in communication styles and effectiveness. Some speakers/authors will provide too much information; others will say very little. A speaker who gets off topic can be redirected with a question. Encourage quiet speakers by showing interest and asking questions.

An abstract view of the listening process is shown in Fig. 9.5. When somebody begins to give you information, begin by collecting details and trying to fit them to what you know. If there is anything you didn't know or disagree with, make a mental note (or a note on paper). Continue to listen and think

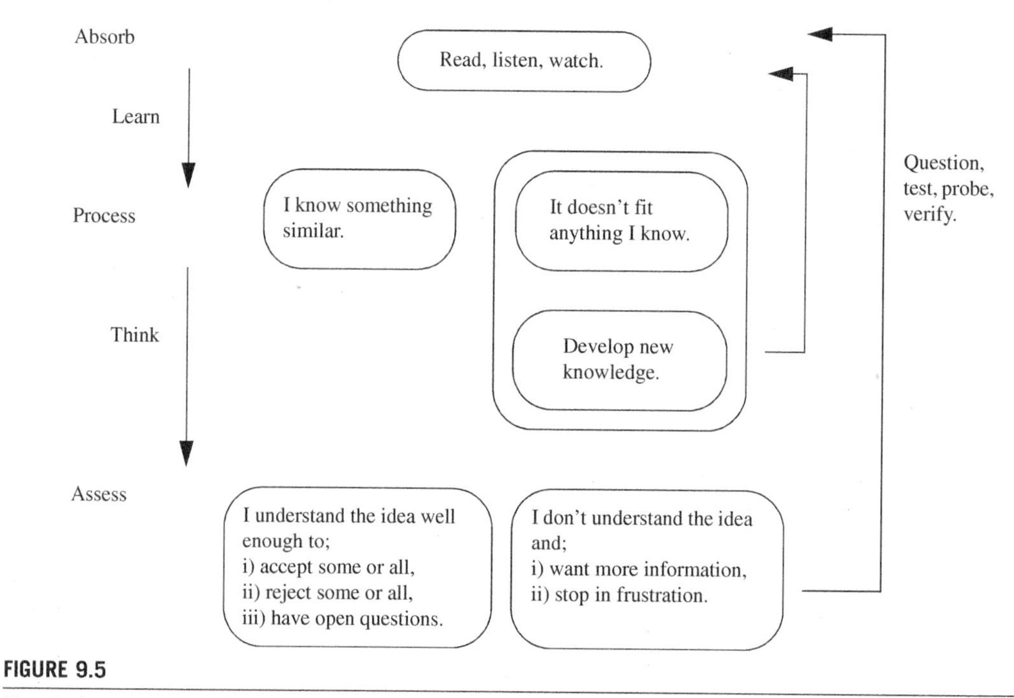

FIGURE 9.5

Knowledge assessment in critical thinking.

until the person has finished. Review your mental notes to look for inconsistencies. If things seem to be incomplete, or don't make sense, then ask questions. It is important to be quiet and listen while you are processing and assessing.

Don't act until you understand the other view.

Fig. 9.6 shows a more specific critical thinking process that can be used for problem solutions. If the writer or speaker has used a problem-solving approach like that shown in Fig. 9.6, the presentation will be easily analyzed. If the information is presented in a different order, the listener has to work harder. Regardless, it is the listener/reader's obligation to answer each of these questions in turn.

Secondhand knowledge can be unreliable. This is because each speaker says something one way, but a listener interprets it another. Verification of knowledge will reduce the number of costly mistakes created by misunderstandings. A very simple and effective form of verification is paraphrasing or echoing. Essentially, when the other speaker is done you summarize what you understand using your own words. An example of this process is given in the following list. Consider the example of a quick conversation in the hallway in which you are told "The parts for the machine are coming tomorrow." The next day you visit the loading dock to get the parts, but they are not found. After clarification, you learn that the speaker meant the supplier was shipping the parts tomorrow. A simple verification would have been "I will get the parts tomorrow."

(1) Absorb, process, assess, verify.
(2) Repeat the critical information using terms such as "I believe you said ...," "You think ...," "You feel ...," etc.
(3) Restate what was said but in words and terms that you commonly use. Be careful not to be confrontational, judgmental, or defensive.
(4) When done, ask something like "Have I heard that correctly?," "What am I missing?," etc.
(5) Wait for a response. If it is defensive, or harsh, find out why.

Some poor practices are described in the following list. It is very common for the speaker to be unaware that he or she is doing these things, so, as always, prudence is the best approach:

- Quiet: A lack of communication can be an indicator of problems. Warning signs include being too busy, avoiding, or working on another task.
- Half truths: Enough is said to imply an incorrect conclusion. For example, "I did not ask for a bribe" does not mean "I did not take a bribe." These statements are often overly specific. In these cases repeat back the meaning you took and ask for confirmation.
- Signal-to-noise ratio: The speaker talks for a while and it may be complete gibberish. The main point is then slipped in quickly or skipped altogether. If at any time you feel confused, this is probably the case. Just say "I think I might have missed it, but can you tell me ...?"
- Changing the subject: This occurs when you feel like you still have something to say but it is no longer relevant. It can be resolved with a statement such as "To stay on topic ...," or "To finish the last topic ...," or "Before we move on ...."
- Glossing over details: If the level of depth or style somebody uses changes, you should ask why. This often indicates a different level of understanding or willingness to communicate knowledge.
- Agreement: One or more times somebody will ask for agreement on something obvious. Once you are agreeing, he or she will follow with a controversial statement and expect you to agree reflexively. Watch for somebody making statements and expecting only a "yes." You are in control after he or she finishes and before you say yes, so pause.
- Babbling: There is no clear point to make.

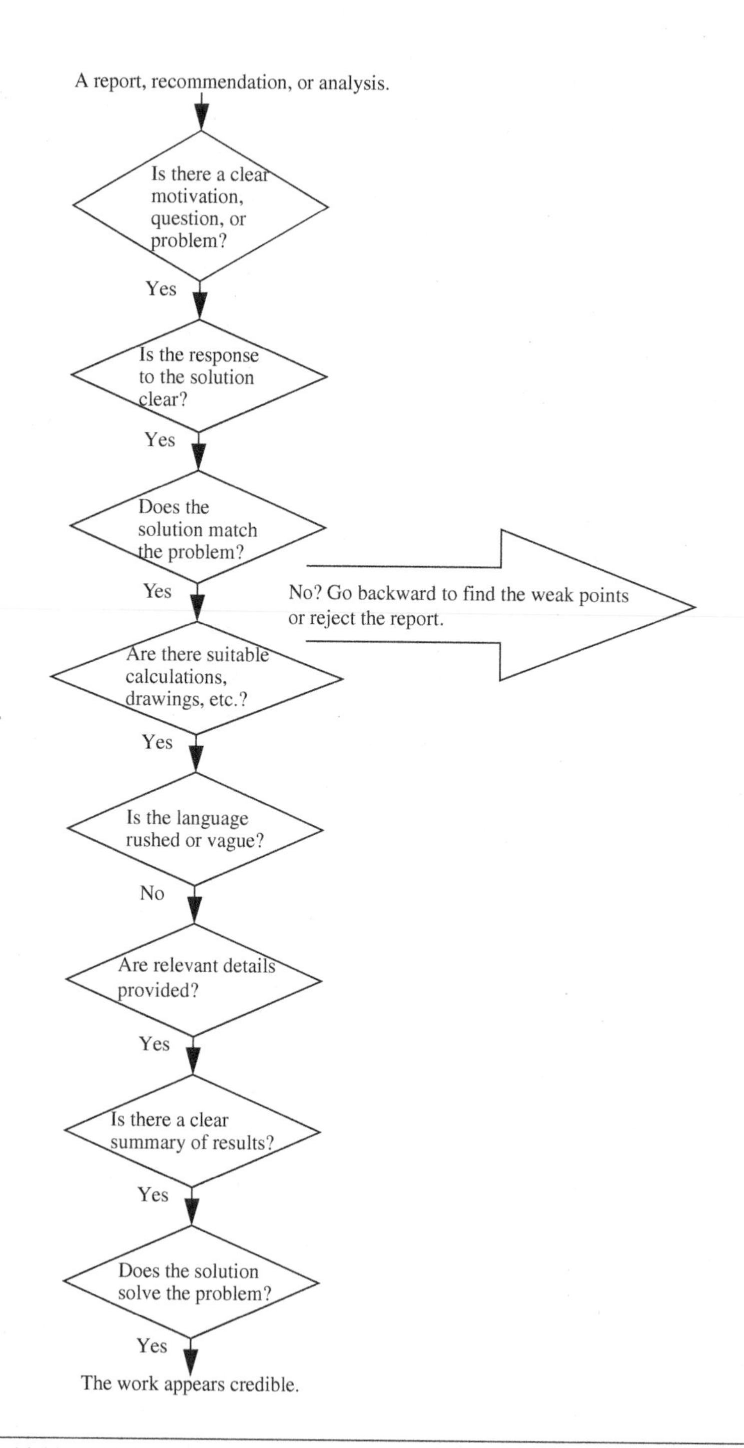

FIGURE 9.6

A general critical-thinking process for assessing technical reports and proposals.

- Overloaded: Trying to say too much at once.
- Not reinforced: Emphasize a point by restating it or getting a suitable response.
- Premature decisions: Making a decision before somebody has finished xtalking.
- Not related: An inability to relate the point to the listeners' needs or perspectives.
- Emotion: Strong reactions eclipse the message.
- Judging: Making a decision/judgment/evaluation about a statement or person based on a single statement.
- Power: The power, or rules, for communications are not understood. This is a major problem when punishment and reward are involved.

PROBLEMS

9.13 Effective communication: List five reasons people may not communicate effectively. Develop strategies to overcome each of these.

9.14 Why is thinking an important step in listening?

9.15 What can you do if somebody wants to ask a question but makes a statement instead?

9.16 Critical listening: In a group of four to six, have a discussion on the relationship of the environment and the development of new energy sources (or another complex topic). Discuss these issues for approximately 5 min. Once the discussion is done, have everyone answer the following questions individually.

(a) Create a list of all of the points you heard during the discussion.

(b) Who was ignored? Which opinions?

(c) Were you ignored? Which opinions?

(d) Compile a complete log of the discussion using all the lists developed in (a).

(e) What opinions did you miss?

9.17 To explore listening and interpretation:

(a) Find one or more partners.

(b) Take turns and make one verbal statement about something new or controversial.

(c) Have the others in the group listen and then write down their interpretation of each statement. (Note: Do not merely copy what was said.)

(d) Collect the written interpretations for your statement.

(e) Review each to indicate where they match and don't match what you said or meant.

9.18 Find another engineer, and one of you assume the role of presenter, the other that of the audience. The topic is the kinetic energy of a moving particle ($E = 1/2\ mv^2$). The listener should take notes while the presenter talks. At the end of the presentation, both of you review the notes for completeness and agreement. How would this be different if done with a salesperson?

9.5 Interpersonal communication skills

9.5.1 Verbal communication

The majority of working communication is verbal. In-person or phone communications are fast ways to transfer information and make decisions. The issues with verbal communication include (1) frequent misinterpretation by the listener, (2) written communication takes precedence, and (3) you do not have time to consider your answers. Some common methods used for verbal discussions are listed as follows. Many of these will not be appreciated until a problem occurs:

- Follow-up: If the content of a discussion is important, capture the content in an email or memo and send it back to the other person "for the record." This also gives them a chance to respond if there is disagreement.
- Cover your assets: Keeping a record of important discussions can be valuable if any problems arise later. Very detailed note-takers will keep track of the date/time/place/people as well as the topic of discussion and outcomes.
- Notes: Taking notes while talking can be the best way to make sure that nothing is forgotten.
- Echo: Repeat what you have heard so that the other person can verify.
- Ambush: Do not raise unexpected questions and expect a good response. Also, if you feel pressured to respond, stop until you can respond carefully.
- Recording: Using recording devices can capture discussions for a more formal record. Be aware that secret recording is illegal in many places.
- Quality: When somebody is rushed or distracted he or she will sometimes give incorrect answers or forget what he or she has said. Don't count on these answers.

In informal discussion you will face many unexpected events. The ability to deal with these is commonly called "thinking on your feet." This ability is good because you are able to make high-quality decisions quickly; however, when you have to think on your feet you are being forced to respond. Being good at thinking on your feet is like being good at dodging bullets. It is a good skill to have but you don't want to use it every day. Some strategies for dealing with these situations are:

- Screen each decision with maturity.
- Can you do it? How well? Boring can be good.
- What is in it for you and the listener (win-win)?
- What does the other person want? Ask questions, then listen.
- Don't commit to anything until you have had time to consider.
- Don't let somebody rush you into bad decisions; slow things down.
- Write everything down.
- Try to get the listener to give you a solution.

Telephone calls are a fact of business. When you call somebody, say who you are and ask if they have some time to talk. If they do, then respect their time, tell them why you called, and get to the discussion quickly. At the end of the call summarize the outcome. If you do not reach them, leave a message using the sequence of steps in the following list. Although these seem simple, it is very frustrating to receive a rambling message, from somebody you don't know, who quickly mumbles their phone number 3 min into a message. Make it easy for somebody to call you back with the answer you need:

(1) First, identify yourself and your company/group, etc.
(2) Give your phone number clearly.
(3) Concisely say why you are calling and if you want something.
(4) Restate your name and phone number.

PROBLEMS

9.19 Why is it important to follow a meeting with a written summary of the discussion?

9.20 List three probable outcomes if you ask somebody to make a major decision without preparation.

9.21 Give an example of thinking on your feet.

9.22 If you feel pressured to make a decision, what should you do?

9.23 Why is it important to repeat information and decisions?

9.6 Casual written communication

Email, text messages, and notes are mainstays of modern business. However, being written, they are on the border between verbal conversations and formal commitments. Email is particularly troublesome for communication because something that was written in haste cannot be "unsent." Mistakes and unintended meanings can be difficult to correct. The most memorable anecdotes involve angry emails sent quickly and regretted seconds later.

Email comes in many forms, including information, approvals, requests, jokes, personal, spam, bills, and reminders. Many professionals will receive hundreds of emails per day, so it is important to make yours easy to read and respond. Some basic tips for composition are given in the following list. The worst cases are often emails from somebody you recognize, without a descriptive title, where the email never gets to the point. If an email takes more than a few seconds to categorize it will often end up in a list to read later. When you receive a few of those from the same person they tend to be ignored immediately:

- Consider why you are writing the email, e.g., question, information, request, etc.
- If possible, put your critical ideas in point form.
- Say what you need/want at the top, not the bottom.
- Get to the point quickly.
- Proofread at least twice.
- Use a spell-checker.
- Avoid incomplete sentences or vague wording.
- Attachments can be hard to read on portable devices such as phones. If possible, copy the important text to the email body.
- In any form of communication ask questions as separate points or paragraphs. When they are bunched together or tightly packed they are easily missed or ignored.
- Do not mix simple and complex issues in the same email. Send one for a fast response and another for a later list.
- You are more likely to get reader attention if you provide something interesting.

Personal email gets more attention. When anything is sent to a large group it is much easier to ignore. An email is much more likely to be read and get a response if it is sent to one person and his or her name is in the body of the email. If the reason for the email is obvious, in the subject line, it will often get appropriate attention. Reasonable conventions for email distribution include:

- Use cc if the message needs awareness, closure, or record keeping.
- Use bcc for large lists to avoid "reply to all" spam.
- If using lists, review the recipients, especially for sensitive information.

Even though email is "semiformal," you should save everything business related. Put these items in folders or use labels and store them in an archive. Being able to search old emails can be a great source of information and a record of decisions. Some mail programs will permit automatic email sorting and prioritizing. You can make this process easier for everybody if the subject for an email includes the subject or project name at the beginning, such as "Project Mouse: optical sensor selection." Items that are urgent can begin with "[Urgent] PR required." Items that do not need a response could be labeled "[FYI] Shell RP is complete."

PROBLEMS

9.24 What are three benefits of having an action-oriented title for an email?

9.25 If an email has multiple points what order should be used?

9.26 When is it better to break one email into separate emails?

9.27 When should you use cc and bcc on email?

9.28 List 10 ways you can modify an email to reduce the effectiveness.

9.7 Selling

It is hoped that you enjoy time spent with the businesspeople that you meet, but at some point you will need to do some formal business. In a very abstract sense, business communication is driven by the need to create and add value. Any two professionals will be looking for a beneficial outcome. This comes in many forms, but the most applied form is sales. In an engineering context engineers will be dealing with, or providing technical support for, salespeople. The general procedure for sales is outlined in the following list. After an initial assessment, there is an offer of some good or service, followed by determining a value, and agreeing to an exchange. The best results in sales negotiations will happen when there is mutual benefit:

(1) Ask about problems and needs.
(2) Listen and learn about the customer needs.
(3) Consider what you can offer and how they will understand it.
(4) Describe the options you can offer, outline the best versus the acceptable solutions.
(5) Get the response and consider.
(6) If there is no acceptance, look for a chance to negotiate.
(7) Look for a chance for agreement or follow-up activities.

PROBLEMS

9.29 What will probably happen if you present a solution before you have listened to the customer?

9.30 Putting your solution in terms the customer can understand will help him or her visualize the purchase. How can you learn what is important to the customer?

9.31 Assume you are the customer talking to a supplier representative and he or she is presenting a solution that does not solve your problem. List three strategies you can apply to lead the representative to a solution you will use.

9.8 Praise and criticism

Critics often give the best advice.

Praise and criticism are natural parts of everyday communications. They are how we share our likes and dislikes. Receiving praise can be very important and can reinforce good efforts. Receiving criticism can be very difficult when it feels personal. Of course, it is much easier to give criticism than to receive it. In the best work environments praise and criticism are shared for mutual personal growth. Dysfunctional environments will have only praise, or criticism, or neither. Other signs of dysfunction include communication in only one direction or connecting criticism with threats. Some of the good and bad signs are described as follows:

Bad
- An optimist refuses to give negative feedback.
- A pessimist gives only negative feedback.
- Somebody will give criticism freely but object to any personal criticism.
- Somebody with more power ties criticism to threat.
- Somebody with less power is threatened by criticism.
- Criticism is used to force social conformity or to create a monoculture.

Good
- There is a balance of praise and constructive criticism.
- The praise is heartfelt.
- The criticism is impersonal and constructive.
- There is freedom to criticize and praise.
- Criticism is seen as an opportunity, not a history.

  Criticism can be very difficult to accept, especially if it seems personal. Needless to say, the worst reaction is to be angry and defensive. Attacking the critic will only prevent him or her from being honest in the future and possibly ruin professional relationships. In truth, it is very hard to find people who will give you honest feedback. Critics give us a more impartial source of self-assessment. When criticism is good it can identify opportunities for self-improvement. Misguided criticism often has altruistic intentions that offer insight. Using criticism constructively is the mark of a mature professional. Some strategies available when receiving criticism include:

- Being defensive
  - Remember that you will tend to hear it as more negative than it is.

- Always assume it is offered constructively.
- If somebody offers strong opinions it is because he or she cares. Sometimes these come with personal attacks.
- Take notes if it is complex. This also gives you some time to think about it.
- Stay calm, breathe deeply, and take your time.
- Thank him or her for the honesty.
- Remember that nobody starts a day by saying "I want to be evil today." The other person truly believes he or she is right from his or her own perspective.

- Responding
    - Don't argue with criticism; accept it as an opinion that is owned by somebody else.
    - Ask for clarification and what he or she expects from you.
    - If you can't resolve it, then "let it go" and "walk away."
    - Do not make excuses or use scapegoats for protection.
    - If you made a mistake, admit it. Apologize if it has caused harm or inconvenience.
- Using it
    - Use criticism as a source of unbiased, outside perspective of yourself.
    - Consider the accuracy of the criticism.
    - Look for something valuable in the criticism.
    - If you do not agree with the criticism consider what the other person thought. Is there anything you can learn from his or her comments?
    - Seek out and listen to people who will give genuine criticism. It is much easier for people to smile and tell you everything is fine.
    - Encourage environments where criticism can be given freely. You can do this by encouraging criticism of yourself, leading by example.

To paraphrase Robert Burns, "What a gift it would be, to see ourselves as others do." Believing we are doing our best is natural, but we all have much to learn. When we share our insights with others it is a way to give them that other perspective. It is an act of kindness. As such, it is important to offer it in the best way possible:

- Keep it soft.
    - Avoid being personal (use passive voice). For example, replace "you made many mistakes" with "many calculations were incorrect."
    - Humor is highly subjective and should be avoided.
    - Do not mix criticism with personal insults or attacks.
    - Allow the other person to respond and then acknowledge his or her views. He or she needs to be heard.
    - Do not ask for excuses or demand acceptance.
    - Consider how you would feel if you were criticizing yourself.
    - Use the Socratic method and ask questions that lead the listener to your conclusion.
    - Watch the body language for signs of stress or anger.
- Make it useful.
    - Be constructive by offering possible solutions.
    - Offer an alternate course of action that would resolve your concern.
    - Be concise and clear.

Sincere praise and appreciation reaffirm our contributions. When you provide sincere praise to others you will reinforce their good works, and it is more likely they will do the same. When you receive praise you should appreciate the honesty. At the simplest level, praise can be a brief compliment. More elaborate forms include written letters, meetings, public announcements, and formal awards:

Giving praise

- Be detailed.
- Focus on the positive outcomes.
- When you feel good about someone's efforts let him or her know.
- Being personal (active voice) is fine and will increase the benefit. Emphasize the use of "you" and similar words.
- Less personal praise would be "the drawings looked good," whereas encouraging praise might be "your work on the drawings was very detailed."

Getting praise

- Thank the other person.
- Consider what he or she liked.
- Give credit to others as appropriate and do not take credit for work done by others.
- When you get constructive criticism, express your appreciation.

PROBLEMS

9.32 Relative to giving criticism, discuss the phrase "You will catch more flies with honey than with vinegar."

9.33 What is a reasonable ratio between praise and criticism?

9.34 Why is third-party, passive voice better when offering criticism?

9.35 A good practice is to respond quickly to good work with genuine, detailed, praise. What can happen if praise is overused?

9.36 List five advantages and disadvantages of pessimism and optimism.

9.37 Write three good responses to the criticism "I cannot believe how many mistakes you made."

9.38 Rewrite the following criticisms to be more constructive.

- **(a)** I hate that color.
- **(b)** Your company has a terrible reputation.
- **(c)** Why are you always late?
- **(d)** Tell me why I shouldn't fire you.
- **(e)** That was a total waste of time.

9.39 Should you use "I" and "you" when praising somebody?

9.40 Is criticism bad? Explain.

9.9 Saying yes, maybe, or no

Whether it is by agreeing, offering, or volunteering, young professionals easily make the mistake of assuming too much responsibility. Saying yes to a simple request seems to offer career opportunities while also pleasing others. Saying no definitely results in immediate disappointment. Agreeing to every request can result in an overwhelming list of tasks. Some tasks will be done to satisfaction; other tasks will fail to be completed as wanted or on time. When you fail, it is more disappointing than saying no. Say yes only to tasks that you know you can do well with your given time and resources. If you are unsure say "I will let you know later." Some other strategies for saying no or "maybe" are:

- Delay the response by deferring to somebody else: "I don't have the authority to make that decision, I will need to check."
- Ask for time: "I will need some time to consider the options." This requires a follow-up with a conscientious "no."
- Be direct: "That would not provide the benefits I need." "At this point I can't justify the time/cost, etc."
- Be passive: "There are not enough resources to make that possible."

The default answer to any request should be "no" unless you are certain you can provide satisfactory results. Saying you will do something that you will not be able to do well is a waste of effort for everybody. If you think there is a chance, say "maybe" instead. As a warning, many business people will test you by asking you to do something a little frivolous or ridiculous. A rushed yes or no response will undermine their trust.

PROBLEMS

9.41 What are three strategies to use if you feel pressured to make a decision?

9.42 What are the disadvantages of pressuring somebody into a decision?

9.43 Mini-case: Yes-man

Larry meets Lefty after work at a casual restaurant. Lefty says he has been having trouble bonding carbon nanotubes to an aluminum substrate for their new Leisure Suit project. Larry says he has looked for synthetic rubbers before and might be able to help. Larry is pretty sure he could get through the whole specification problem in 2 h. However, the next day Larry looks at the catalogs and data sheets and finds only 14 adhesives recommended for nanotubes and none of those is rated for aluminum. In frustration he calls Al at LizardChem to ask about alternatives. Al tells Larry that there are two adhesives he thinks may work but they have not been used with nanotubes and aluminum bonding before. LizardChem sends samples of the adhesives, and they are tested. One of the adhesives works very well but is very expensive. Larry sends his results to Lefty, who is disappointed that the only option is so expensive. Instead of the 2 h he had expected, Larry has spent 23 h, and his gold–silicon bonding project is now almost a week behind. Lefty says Larry will have to catch up on his own time because he had not authorized that much time.

Did Larry produce successful results? State reasons for, and against, the success of the adhesive specification work. At what points in the process could Larry have made other choices?

9.10 Answering questions

People dislike the question phases of presentations for the wrong reasons. Some things to remember about questions are shown in the following list. In general, it doesn't really matter how a question is asked. The objective is the same: to get information. Your task is to provide relevant answers. A simple flowchart for replying is given in Fig. 9.7:

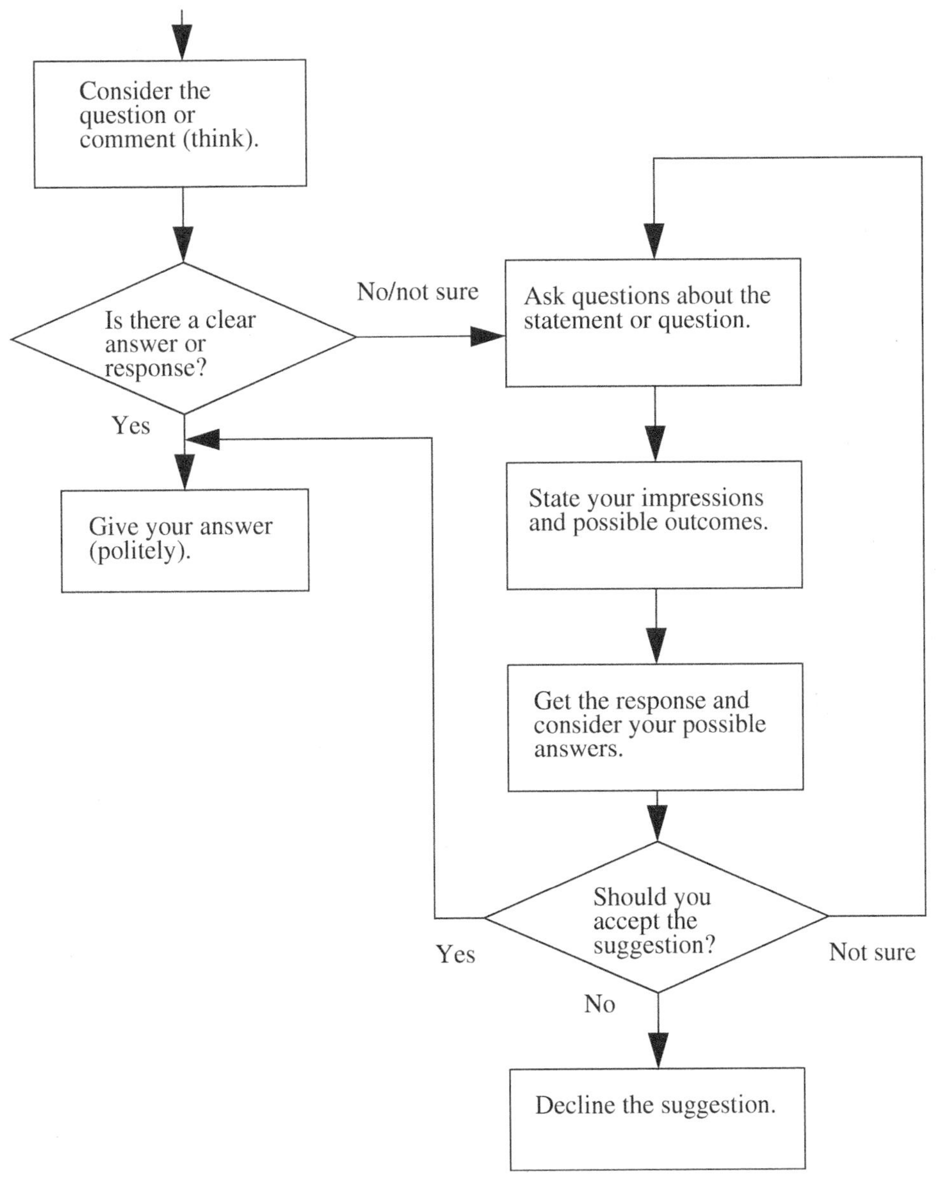

FIGURE 9.7

Responding to suggestions.

Quantity of knowledge

- More answers are better, but you cannot know everything.
- If you don't know something, respond "I am not sure," or better, "Let me see if I can find the answer."

Confusing questions

- People asking questions will often provide clarifications, if you ask for them.
- People making statements sometimes don't have questions. When they are done you can say, "Thank you," or, "Can you suggest how I could use the information?"
- If it doesn't make sense, ask them to repeat their question.
- Do not assume that everybody heard and/or understood everything you said.

Difficult questions

- You get difficult questions only because people care about the results of your work.
- Too much detail in the presentation is overwhelming. Questions mean that you have started the right discussions.
- People asking questions can have as much anxiety as when they are the presenter.
- A presenter should never become defensive. Deal with harsh questions politely.

PROBLEMS

9.44 Does a presenter need to know everything?

9.45 How could a speaker respond in the following circumstances:

(a) He or she does not know an answer.

(b) He or she is very confused about the question.

9.46 Assume you are presenting and experiencing some speaker anxiety. Is it possible that somebody asking a question might feel the same anxiety? Explain.

9.47 What could you do if you ask for questions and somebody makes a statement?

9.48 Pretend you were giving a presentation about the basic requirements for a bicycle. How would you respond to the following questions?

(a) "Why would I ride a bike?"

(b) "How high is the tallest bicycle?"

(c) "I have ridden bicycles for decades and don't know why you would want brakes on the back wheel."

(d) "You have said we need 10 gear speeds; what about a variable speed transmission?"

(e) "Why not store the braking energy to help propel the bike?"

9.11 Meetings

If you want something, start by giving.

Meetings are a time for shared communications and decisions in a group. They must have a clearly stated purpose that is the basis for an agenda. All participants should understand the expected outcome and run the meeting toward those goals. All discussions in a meeting should lead to some action; items that are simply for information can be distributed other ways. Meeting actions need to be recorded with (1) what led to the decision, (2) what action will be taken (or if no action is required), (3) who will do it (if more than one person, assign a leader who is responsible for it), (4) when it must be complete, and (5) what "done" means. Some pointers to help with meetings are the following:

- It is a good practice to send reminders before a meeting and send follow-up summaries after.
- Start and end on time. If people know you will start late they will arrive late next time.
- Do not let complaining take over the meeting. If it is a regular occurrence, set aside 10 min at the beginning for complaints and then change to business.
- Cancel meetings if they are not productive. This sends a message that meetings are to add value.
- If a discussion item involves only part of the group, have them discuss it outside the meeting and report back.
- Stick to the agenda and push new topics to the end if time permits.
- Do not discipline people in public meetings.
- For each item on the agenda, have an outcome. Note that "being informed" is not an outcome.
- The meeting leader should initiate, and finalize, a discussion, but other people should talk in between.
- Make sure that if somebody is at the meeting he or she will contribute beyond simple approval.

The cost of meetings tends to be hidden because it is not tied to a single project and often does not appear on budgets. But if in each 8-h day a team spends 2 h in meetings, they have spent up to 25% of their available work time. Therefore it is essential for meetings to be well run and as brief as possible. Meetings are often inefficient because they do not have a clear focus or execution involving the group. Typical problems are as follows:

- Late: not starting on time
- Overtime: going past the appointed time
- Unplanned: no clear set of business
- Action: not action oriented
- Social: the meetings are considered a time for social interaction
- Regular: meetings that are held because of the time instead of need
- Monopoly: a few people dominate the meeting
- Information: the meeting is used to transmit routine information
- Missing stakeholders: some critical people are missing for agenda items

Resource 9.1 See Appendix A, Section A.4, for a meeting-planning checklist.

PROBLEMS

9.49 Do casual conversations have a part in meetings? Explain.

9.50 Give five examples of action items.

9.51 A meeting includes six employees with hourly billing rates of $90. Five minutes are needed before and after the meeting to collect items and move between offices and the meeting room. How much does a 15-min meeting cost? How much does a 60-min meeting cost?

9.12 Purpose and procedures

It costs money and time for employees to sit in a meeting. Don't spend money to have them listen. Pay them to contribute.

The basic process of planning and holding meetings shown in Fig. 9.8 is to plan the meeting and use an agenda to verify that the purpose is well understood. Keep the meeting focused on actions and ensure that the meeting outcomes are well understood. A meeting is normally run by a single leader whose duty is to make sure that the meeting follows an agenda and communicates the results with written minutes. A common practice is to have another person take minutes in the meeting so that the leader is not distracted.

Formal meetings are normally held for decision-making, progress review, project reviews, and approval. These meetings are driven by agenda items and the outcomes are decisions and action items. Some standing items will appear on every agenda, for example, a weekly review of the budget and customer orders. New and old agenda items should follow a process to avoid ill-formed or frivolous agenda items. A reasonable method for managing an agenda is shown in Fig. 9.9.

Normally, meetings occur face to face, but other forms of meetings are becoming more common. These follow the same basic rules but typically require a bit more management to be successful. The main challenge of web- or phone-based meetings is that normal facial and body language cues that accompany in-person meetings will be missing. The following list offers a set of rules for phone- and web-based conference calls. It's important to set up and test the technology ahead of time:

- Use web clients if also sharing documents.
- The host should arrive early and welcome guests.
- When talking, pause often and wait for responses.
- Ask for comments often.
- Look or listen for people who aren't talking and ask them for responses, too.
- Noise is always a problem and "mute" is a cure.
- Provide a summary and ask for questions before ending any agenda item.
- If anybody is dominating the call you can often repeat his or her point to recognize his or her input and then ask for the next person to speak.

A typical meeting is documented with minutes. A sample is shown in Fig. 9.10. Tracking who is, and is not, present makes it clear who needs to be informed. Approval of previous minutes ensures that everybody agrees to what is written. Noting the date, start time, and end time helps track when information was exchanged and decisions were made. Identifying action items makes it clear who is to do

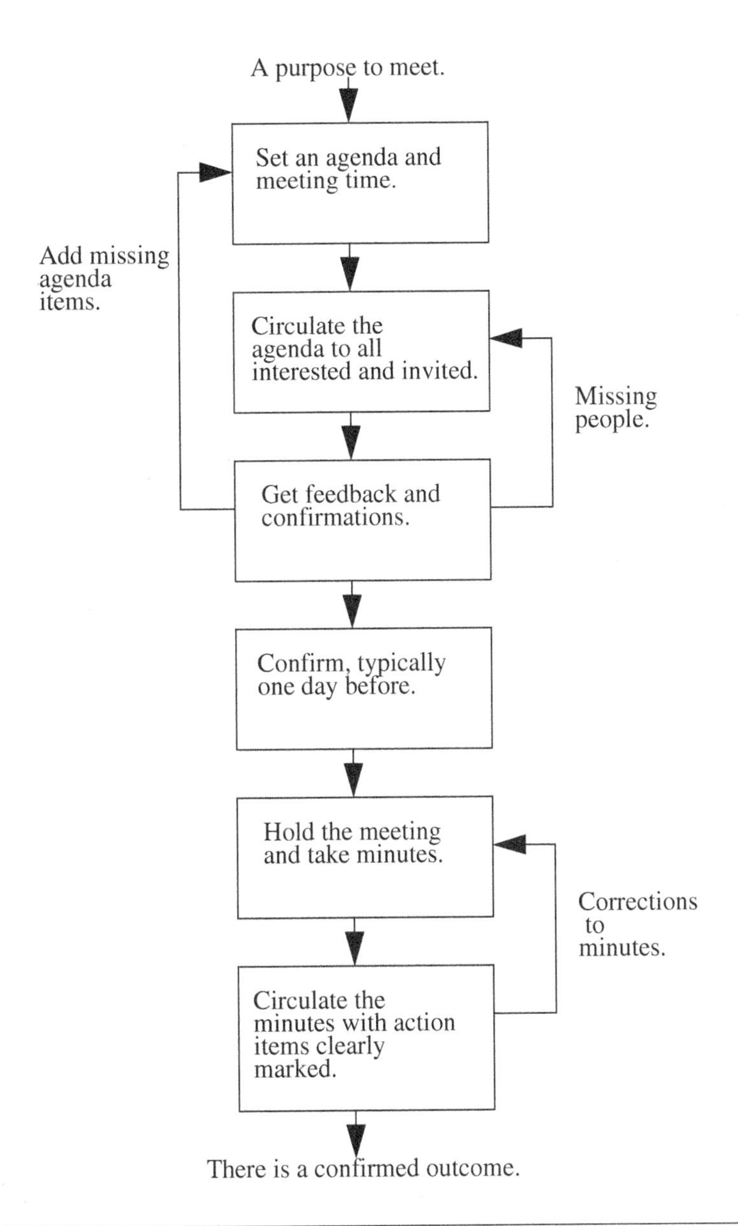

FIGURE 9.8

Arranging meetings.

what. A common mistake is to bypass action items and assume that people understand what they should do after the meeting. A newer approach to keeping action items is to use a spreadsheet with listed items, decisions, action items, people, and times (see Resource 9.2).

Resource 9.2 A spreadsheet for minutes is included on this book's website: www.engineeringdesignprojects.com/home/content/resources.

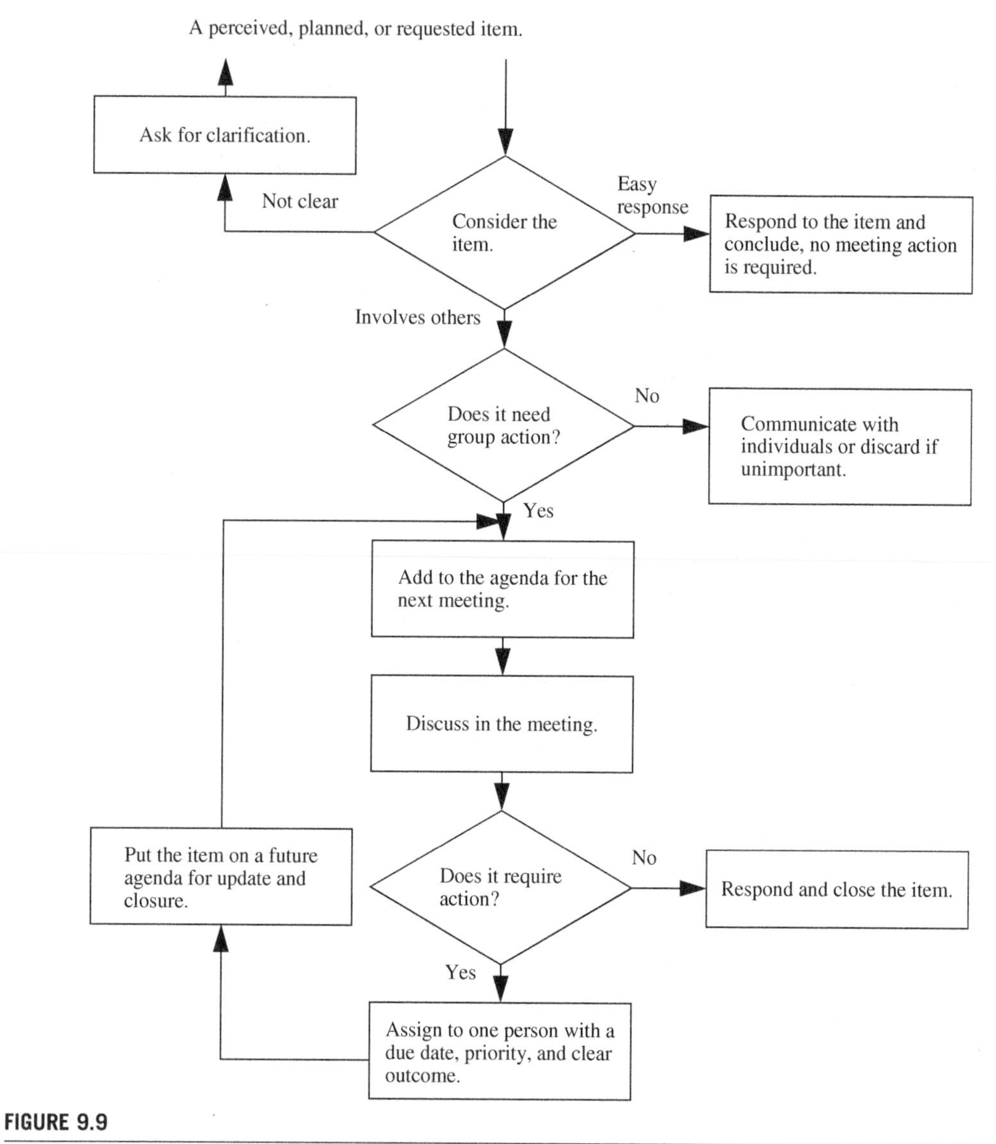

FIGURE 9.9

The life of an agenda item.

PROBLEMS

9.52 List five benefits of using an agenda for a meeting.

9.53 Why would an agenda refer to old action items?

9.54 List five benefits of meeting minutes.

Project Review Meeting - June 12, 2016
attendees: Mary Contreery, Jack Sprate, Peter Rabert
regrets: Bernie Toste

1. Began at 1:30pm
2. Approval of minutes from May 21.
3. [Mary] review of project progress.
   - purchase and delivery of spaghetti bender
   - packaging station designed and in construction
   - project is ahead of schedule except for drying station.
4. [Peter] Site preparation issues.
   - concrete work is done.
   - electrical and water services to be done in one week.
5. [All] Engineering changes
   - modify the batch size to 100 kg and cycle time to 20 minutes
   - change the packaging station house to stainless steel instead of polypropylene.

Action: [Jack] ensure packaging station material change.

6. Decision to meet again July 15.
7. Adjourned 2:20 pm.

FIGURE 9.10

Sample meeting minutes.

9.13 Customer and supplier meetings

There are a variety of reasons for meeting with customers, suppliers, and others outside your project group.

These may include:

- review of preparation work
- approach to project timeline
- scheduling
- what comes next
- research to learn what you don't know
- develop a preliminary system
- develop a list of questions
- write a draft of a functional specification
- refine the functional specification
- assign agenda items to one person
- ask sponsors about prototype supplies or money; indicate a proposal will be provided
- start considering tests and prototypes that are needed

In meetings where decisions are made, there should be some agreement to exchange resources, services, or money. If you do not prepare for a meeting, you could make some costly mistakes. There is nothing inherently negative about this process; everyone wants to get the most value for money spent. Fig. 9.11 shows the general sequence of steps for preparing and running a meeting. In simple terms, you should pick a range of acceptable outcomes before the meeting. During the meeting you should work toward your best outcome, but compromise toward your poorest acceptable outcome. If none of the outcomes are acceptable, get an idea of what they would like and end the meeting. Do not accept

Prepare for a meeting.
Prepare an ideal outcome and minimum fall-backs. List the unknowns.
A clear agenda.
Could Include;
• Written documents
• Presentations
• Verbal Reports
• Demonstrations
• Summary of past meetings
Meet with others and start with a discussion of unknown items.
Discuss ideal outcomes for both parties.
Unsure.
Discuss differences.
Too many issues.
Both agree.
Common ground exists.
Accept positions and adjourn meeting.
Concede some goals on both sides to find a compromise.
Discuss possible outcomes including another meeting.
Both agree and get some benefits.
The discussion is exhausted.
Accept positions if better than minimum fall backs and adjourn meeting.
Do not commit to anything and end the meeting.
Follow up in writing.

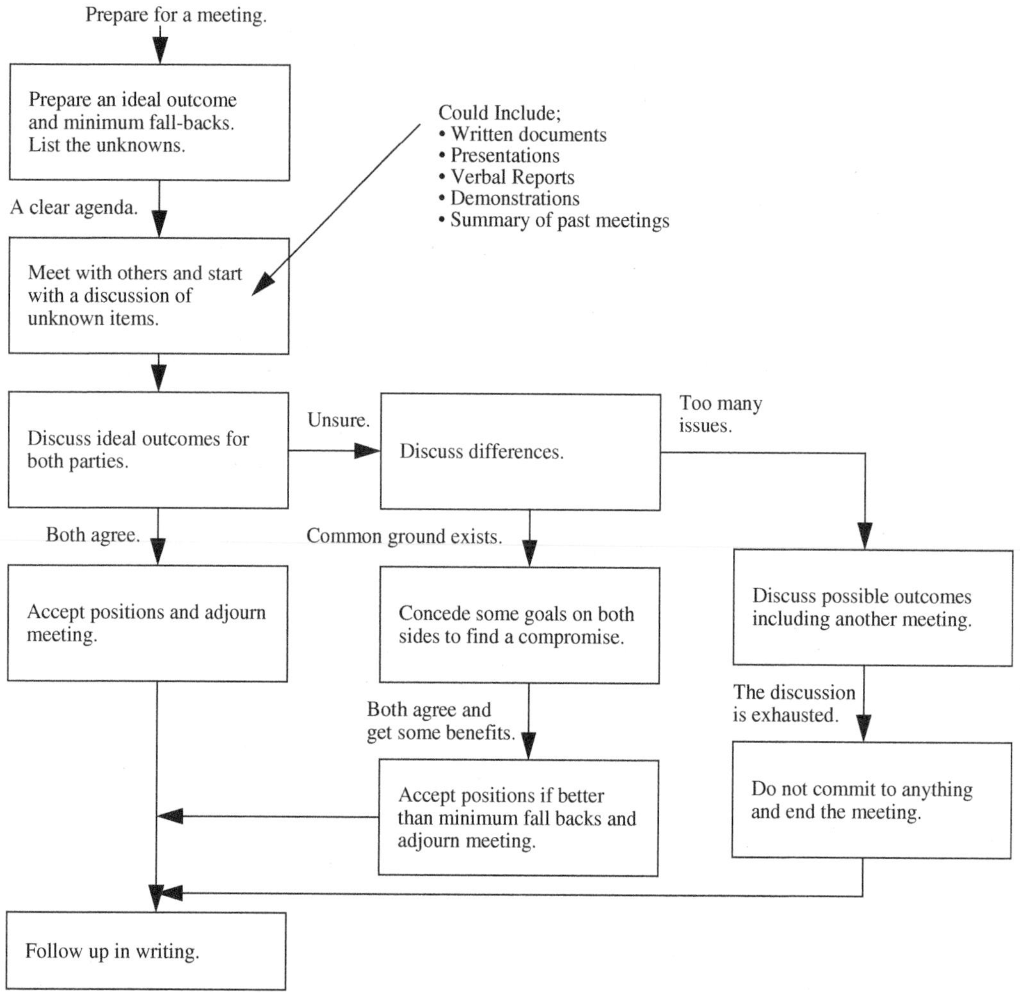

FIGURE 9.11

A reasonable approach to negotiating meetings.

new outcomes that you have not considered. When you come to an acceptable agreement, stop. You may do other things in these meetings, but don't confuse the cost-related outcomes with the other items discussed.

If you are determined to make poor decisions, you can use the process in Fig. 9.12. To do a truly terrible job, you should show up to a meeting to see what the others propose. If their proposal sounds good enough, you accept it on the spot. If they do not know what they want, you should propose a reasonable offer. The offer must be good enough for you but be so attractive that they will accept it. After you have reached an agreement with the customer, there is a temptation to "sweeten the pot" by throwing in a few extras. However, it is pointless to offer more cost and effort after the agreement has been made. Professionals using these approaches are dangerous as suppliers and customers.

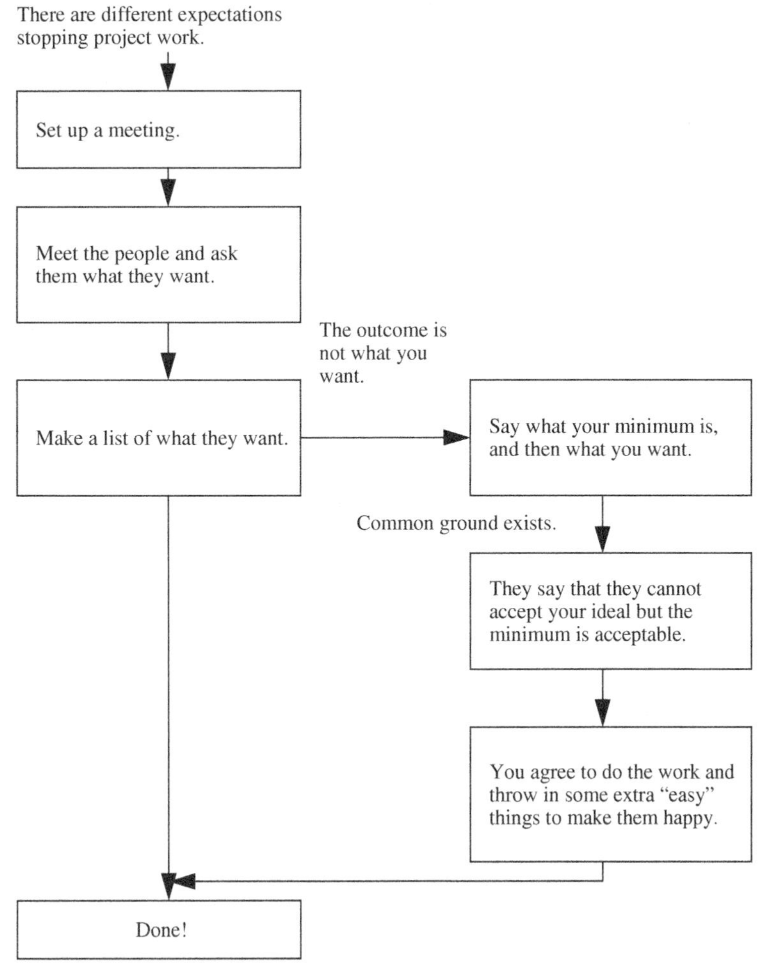
There are different expectations stopping project work.
Set up a meeting.
Meet the people and ask them what they want.
Make a list of what they want.
The outcome is not what you want.
Say what your minimum is, and then what you want.
Common ground exists.
They say that they cannot accept your ideal but the minimum is acceptable.
You agree to do the work and throw in some extra "easy" things to make them happy.
Done!

FIGURE 9.12

A poor approach to meeting, requesting, and bargaining.

Sales work is an early form of meeting where someone offers services, hoping to start some business relationship. A wise sales team will know what their standard product line and costs allow. If things move swiftly they may be able to make deals in the first meeting. If not, it may take additional meetings. In the early stages these meetings will have more of a free-flowing form. If any decisions are to be made, make sure they meet previously agreed sales guidelines:

- Screen each decision with maturity.
- Can you do it? How well? High pay for simple work is great, depending on your perspective.
- What is in it for you and them (win-win)?
- What do they want? Ask questions, then listen.
- Don't commit to anything until you have had time to consider.

- Write everything down.
- Before the meeting:
  - List who is involved (customer, management, technical staff, clerical, others).
  - What are their motivations and yours, both positive and negative?
  - Develop a strategy that gives everybody what they want (win-win).
- Avoid "strong arm" techniques when bargaining.
- Poor bargaining examples:
  - "I will pay you so you will do what I want."
  - "If you don't, I will sue you."
- Good bargaining examples:
  - "I want this and it will benefit you, too."
  - "Tell me what you need to help you say yes."
- Don't be afraid to walk away from negotiations if there is not enough immediate benefit.
- You are better off turning down something than doing a poor job.
- Money already spent does not count.
- Try for a win-win situation. For the best long-term outcome everybody should get something out of the process.
- Know what you want to achieve and what is ideal, acceptable, and unacceptable.
- If something new comes up, ask for time to consider; making decisions on the spot is very risky.

PROBLEMS

9.55 List three benefits of not preparing for a meeting.

9.56 List three advantages and disadvantages of making major decisions in meetings.

9.57 Discuss the statement "Discussions are not successful unless you are a clear winner."

9.58 What should you do if a customer proposes a solution you have not considered before?

9.59 When are multiple negotiation meetings required?

9.60 Develop an agenda for a project review meeting with the customer during the conceptual design phase.

9.14 Presentations

Would you listen to yourself talk for 30 min uninterrupted?

There are many presentation types, but the three we will focus on are (1) projected slides, (2) whiteboards, and/or (3) verbal. Consider a design review meeting in which the presentation begins with prepared slides followed by questions. The slides are well prepared but the answers for questions are given on the whiteboard or verbally. For any of these methods to be effective, preparation time is required. Each approach requires different skills.

Common presentation types include:

- Formal presentation: well-prepared slides followed by questions
- Webinar: like a formal presentation, but delivered over the Internet
- Lecture: well-prepared slides and notes presented using a projector and a whiteboard
- Information meeting: notes are prepared and presented verbally to a group, often including handouts

Fig. 9.13 shows a useful method to guide the process of presentation preparation. The key steps in this process are to understand what you need to present and the outcome you want. You also want to choose a format for the presentation so that you can edit the content later. The content can then be copied from other sources. It is hoped that most of the work has been done ahead of time and you need only to add some pictures and text for clarity. The trick at this point is to carefully choose the essential content, but not everything. Critical review is required to correct errors and ensure that there is sufficient detail, but not too much.

Resource 9.3 Sample presentation https://engineeringdesignprojects/home/content/presentations.

PROBLEMS

9.61 Why is it important to begin a presentation with a discussion of motivation?

9.62 Can a meeting be part of a presentation? Explain.

9.15 Presentation motivation

There are many purposes for giving presentations, including lectures, seminars, workshops, technical updates, design reviews, project launches, and so on. Regardless of the reason it is critical that the purpose for the presentation be clear. The following list presents some questions that can be asked to give focus to the process. It is best to keep each of the answers to a single sentence to force clarity in the process:

- Who is the audience for the presentation: decision makers, participants, and other stakeholders/audience?
- What do you want to get out of the meeting: decisions, approvals, agreement, or something else?
- What does your audience want to get out of the meeting?
- What do you have to offer them?
- What do they already know?
- What are their values?

Purposes for engineering presentations might include:

- Project review: to complete a phase of project work
  - Presenter purpose: to show the work done and be allowed to continue
  - Audience purpose: to ensure that resources are being used well
- Progress: a regular meeting to monitor ongoing activities
  - Presenter purpose: to seek feedback on project progress, good and bad

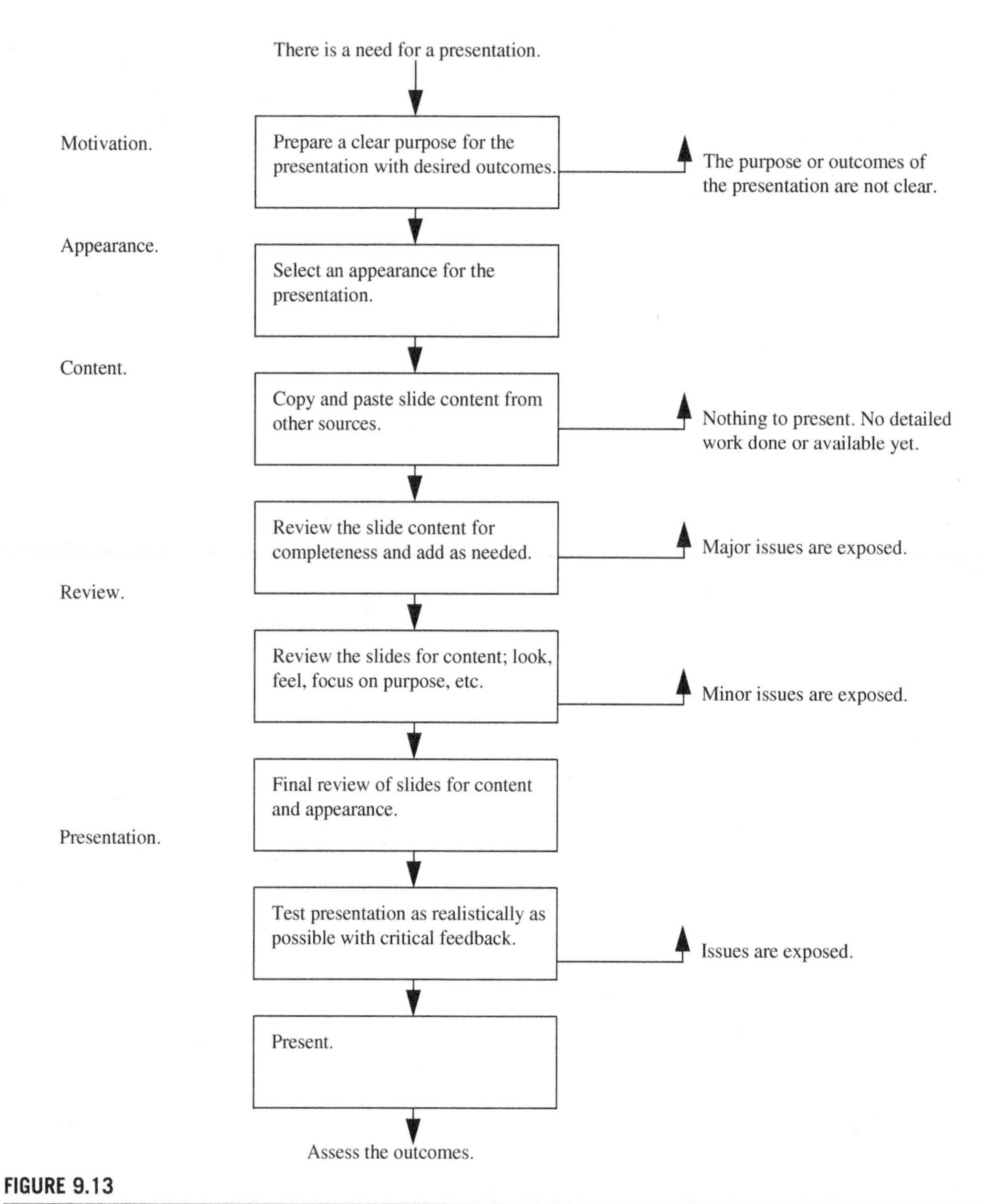
There is a need for a presentation.
Motivation.
Prepare a clear purpose for the presentation with desired outcomes.
The purpose or outcomes of the presentation are not clear.
Appearance.
Select an appearance for the presentation.
Content.
Copy and paste slide content from other sources.
Nothing to present. No detailed work done or available yet.
Review the slide content for completeness and add as needed.
Major issues are exposed.
Review.
Review the slides for content; look, feel, focus on purpose, etc.
Minor issues are exposed.
Final review of slides for content and appearance.
Presentation.
Test presentation as realistically as possible with critical feedback.
Issues are exposed.
Present.
Assess the outcomes.

FIGURE 9.13

Preparing a presentation.

- Audience purpose: to be able to influence the project work
- Project launch: a meeting held at the beginning of a project or project phase
  - Presenter purpose: to build enthusiasm and engagement by stakeholders/audience
  - Audience purpose: to get information about the project and learn how they will support it

PROBLEMS

9.63 What might happen if you do not know the audience for a presentation?

9.64 Create five one-line motivation statements for meetings.

9.16 Content

If you don't understand it, it probably doesn't make sense to anyone else.

The content to be presented should normally be completed while doing the work; the presentation is assembled using these resources. If content is being created for the presentation only, there is a question about adequate preparation. If you already have technical documents, presentations can be easy to assemble by cutting and pasting. The content of a presentation can be organized in a number of equally good ways:

- Purpose: The presentation should start with a purpose and everything should support that.
- Copy: If you have written documents, begin by cutting and pasting the major presentation details.
- Organize: Structure the content to create and then answer questions from the audience.
- Summaries: Stop often to review material and put it in context.
- Overviews: Keep the audience aware of what you will be talking about next.
- Pragmatic: Stay focused and realistic. Avoid overstatements and wild claims.
- Brief: Each major idea should begin with a single-sentence summary, followed by detail.
- Edit: Do not try to present all of the details. Focus on the bigger picture first.
- Condense: Gather ideas together into more concise pieces.
- Surprises: Don't keep information back to be revealed later. Start with the outcome and then show how you got there.
- Review: The content should be reviewed for logical flow.

When laying out a presentation, use the critical-thinking process and remember this is how people will probably be considering your work. An example of a strategic presentation layout is given in Fig. 9.14. The sequence of the slides is designed to help the listener ask questions on a higher level and then the slides will address the questions. Naturally a presentation cannot contain every detail, and people will disagree or be confused; this can be addressed in the question session.

Some good practices to follow as you assemble the materials are the following:

- Purpose: Consider each slide and how it supports the purpose of the presentation; it should be obvious and clear.
- Lines: Line drawings are always clearer than pictures.
- Formats: File formats are important. For crisp line drawings use PNG or GIF. For photographs use JPG.
- Text: Use sound bites that are less than one line.
- Figures: Use figures that complement the words when possible to reach more of the audience.
- Graphs: Replace tables with graphs or charts.
- Software: Hyperlinks can simplify and speed up external links to software and to websites.
- Relevant: Related visuals are good (i.e., pictures). Unrelated visuals are confusing.
- Relate: Emphasize what the audience wants to know.
- Questions: Anticipate questions and prepare responses.

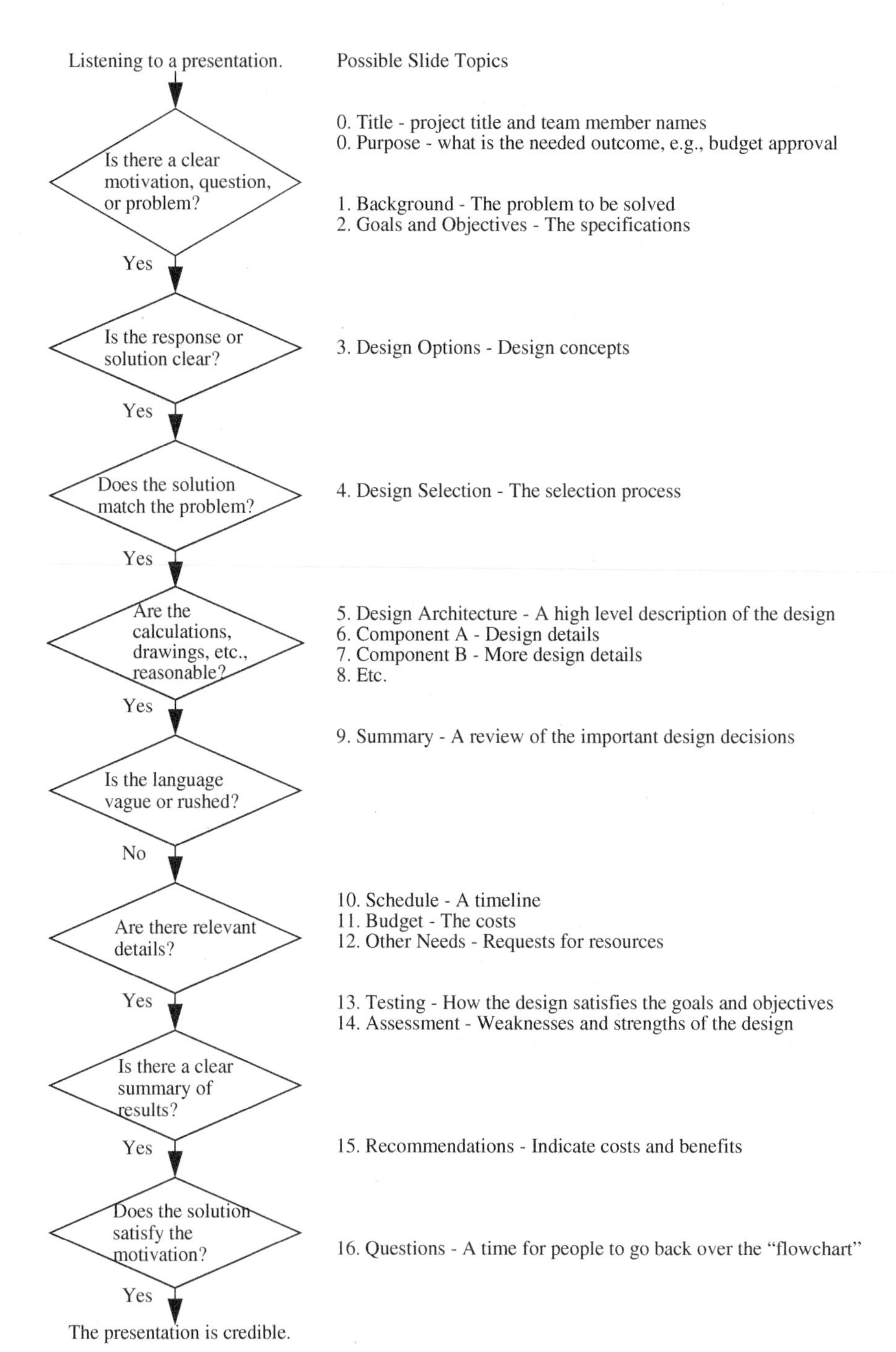

FIGURE 9.14

Keeping your audience on track.

Having an overall model for the presentation can help organize the material. If possible, use a common standard approach that is understood by the group. If none is available, consider organizing the topics logically using one of the following categories:

- Chronological: when things occur
- Hierarchical: in a structure such as main projects and subprojects
- Classification: category
- Spatial: the physical arrangement
- Sequential: in steps

PROBLEMS

9.65 Create five examples of graphical slides for a new car body design.

9.66 Should mechanical or printed circuit board (PCB) drawings be used at the beginning of a presentation? Explain.

9.67 Provide an example of a presentation topic well suited to a bottom-up information structure.

9.17 Presentation appearance and effectiveness

The presentation appearance should be chosen before content is added. Most presentation software has a variety of tested themes, but you could always create your own. In simple terms, the appearance of the slides will have an impact on how well the audience perceives your work and how well they understand the presentation. The following points can be used when preparing the slides as a whole. These visual details are best appreciated from a distance. Stand back and flip quickly through the slides:

- Theme: Use a consistent theme, including colors, backgrounds, fonts, and layout.
- Highlight: Selectively highlight a few important details using underlines, colors, bold, italic, large fonts, boxes, etc.
- Pages: Do not number the slides (e.g., 26 of 216).
- Layout: Use a consistent position for the title and text on the slides.
- Colors: Use a few easy-to-read colors that do not clash or fade into the background.
- White space: Cluttered slides are hard to read and can be overwhelming.
- Fonts: Use consistent font types. Avoid mixing fonts.
- Sounds: Use sounds very little, and only when you need to grab attention.
- Size: All fonts should be 16 point or bigger.
- Short: Use single-line descriptions. Anything longer than one line will not be read. Aim for 30 words per slide.
- Enlarge: Smaller details in figures, tables, equations, graphs, and numbers will not be visible.
- Pace: A pace of 1 min per slide is reasonable. Have a very good reason for each slide that deviates.

A trial run of a presentation can help to expose problems and opportunities for improvement. There are a number of common presentation issues, such as those shown in the following list. When reviewing a presentation, listen and take notes. Review your notes during and after the trial presentation. You should also ask the audience to repeat what they heard and what seemed fuzzy. Practically, there will be minor issues with a number of the following topics, but selecting two to four major issues to correct will provide the best results:

- Purpose: The purpose is evident through the whole presentation.
- Titles: The title for each slide should be relevant and brief.
- Spelling: Use spelling and grammar checkers.
- Small: Look at equations, graphs, tables, and figures for details that are too small. Printed handouts are a good option.
- Animations: These often fail in presentations. If it is essential, use an external video viewer. Otherwise use screen captures.
- Software: External software does not always respond appropriately. Screen shots are better.
- Data: Too much data will overwhelm the audience. Highlight the key values. Provide handouts if essential.
- Equations: Keep these to an absolute minimum—never more than one per slide.
- Visibility: Good photographs can be hard to see in presentations. Use software to brighten the picture.
- Network: Internet connections are notoriously unreliable. It is better to use screen captures when possible.
- Time: Shorter presentations have more impact. Longer presentations lose the audience.
- Pace: The time required for each slide should be relatively equal.
- Flipping: If you plan to refer to an earlier slide, copy it, or the content, to a newer slide.
- Confusion: Overly complex information should be simplified.
- Jargon: Uncommon technical terms and acronyms should be minimized and defined when used.
- Consistency: Watch punctuation, capitalization, tense, and structures for consistency.

Tools for modern presentations can improve the overall experience for the presenter and audience, but at the cost of increased complexity and points of failure. A few of the common technologies and some tips are shown in the following list. The key is to be familiar with your equipment (e.g., laptop or tablet) and the equipment in the presentation room (e.g., projector and lighting):

- Familiarity: Presenters unfamiliar with new equipment, software, or location will fumble while trying to make things work. Usually, the troubled presenter receives help from somebody in the audience. The worst case is the presentation fails or is plagued by technical problems. This can be avoided by preparation.
- Video connections: For video projects to work properly the following things must all occur. If any are not correct it will not work:
  - The computer video output is active—this often requires keystrokes on laptops, sometimes much more.
  - The computer is plugged into the correct video projector port and you have the correct cables going to the right places.
  - The video projector is configured to use that video input; most projectors have multiple video inputs.
  - The projector is capable of displaying the video output from the laptop; a high-resolution screen is often a problem.
  - The entire desktop screen may not be visible on the video projector. With some screen resolutions, the sides can be cut off.

- Laptops and other computers:
  - Backup—Make a backup of the presentation on a USB stick in case your laptop runs out of power, gets lost, won't connect, etc.
  - Video—Most projectors have VGA connectors; if you need something else get an adapter.
  - Resolution—Do not count on a screen resolution higher than 1024×768.
  - External—Learn how to turn on your laptop video output (e.g., Fn–F8 on a PC).
  - Mirrored—Laptops can mirror the screen or add a second screen. Both work, but you need to know your laptop.
  - Connections—Check for network connections, if needed.
  - Power—Find an outlet or keep your battery charged.
  - Eye candy—Turn off screen savers and pop-up notifications (such as email or chat clients).
  - Sound—There are frequent problems with sound. Connections are risky and may not be available. Consider taking your own speakers if sound is essential.
  - Remote—A wireless presenter or mouse can reduce the walking to and from the computer.
- Fun: A laser pointer can be helpful.
- Light: Keep the lights bright enough so that the screen is visible. Watch for glare.
- Loud: Use a microphone for a noisy or large room.
- Assistant: If doing anything special, have somebody else run the computer so that the presenter is not trying to run software.
- Offsite: Video conferencing is possible, but there will be problems. Have support.

PROBLEMS

9.68 How many slides should be presented in a 30-min period?

9.69 List five problems that could occur with computer-based presentations.

9.70 How many seven-character words could fit on a standard PowerPoint, or equivalent, slide?

9.71 Occasionally, slides must have fine detail that is hard to read. What strategy should be used to solve the visibility problems?

9.18 Presentation style

Being nervous is natural but should not overwhelm you. It is critical to remember that people sitting in the audience have been where you are. When you get nervous the audience does not get angry or critical. They don't want to make you nervous. Speaking to an audience is not the same as talking to a person. When talking to an individual you will get a few cues such as nods, smiles, and verbal responses. However, when people sit in an audience these cues often go away. When talking to a group, blank faces do not mean they are not listening or caring. You cannot rely on facial expressions.

Excellent presenters have spent many years working on their personal style. Naturally this is a process that never ends. There are a number of elements to consider when developing a personal presentation style:

Essential
- Clear—Speak clearly and loudly so that you can be heard at the back of the room. Consider microphones in large rooms.
- Rehearse—An end-to-end dress rehearsal in the presentation room with a critic will expose many problems early.

Relaxing
- Natural—Your natural personal style will work best. Do not assume personas.
- Breathe—If you are feeling rushed or anxious, pause, breathe, pause, breathe again, pause once more, then start again.
- Space—Organize the space for your supplies and movement so that you don't trip, stumble, or drop anything.
- Slow—Take your time. Fewer, but carefully chosen, words can be more effective.
- Posture—Don't lean on things; stand straight and keep your hands out of your pockets.

Relating to the audience
- Reaction—Sometimes you will get clues from the audience to speed up and slow down.
- Focus—Avoid visual distractions that draw eyes from you. This includes screen savers, other people, open blinds, etc.
- Flow—Speak and then pause, so that the audience can absorb details.
- Look—Keep your focus on the audience. Scan the room, look at people from the front of the room to the back corners.
- Interactive—If the audience is passive, their attention will drift away from you.

Good practices
- Flexible—Things will go wrong. Adapt and move on.
- Reading—It is fine to refer to the slides, but do not read from them.
- Repeat—Questions and comments from the audience should be repeated for clarity.

An entertaining presentation can help people focus for longer periods and retain more knowledge. To do this you need to transform the audience from passive listeners to active participants. One way is to simply pose a question for them to consider. Another is to make the audience part of the presentation. Professors often do this in lectures by asking students to answer questions or come to the board. A judicious mixture of the following methods can enhance a presentation. However, if overused the message can be lost or the presentation becomes annoying.

Tactile

- Borrow—Use something from an audience member.
- Things—Have a few things to pass around the audience, but not too many.

Intellectual

- Jokes—Good, but pick with care.
- Comics—Artwork that is related to the topic and/or is humorous.
- Pictures—Photographs are engaging, and more so when they contain people.
- Task—Give the audience something to contribute based on the talk. For example, they can write notes on a card.
- Attention—Consider the attention span to be a couple of minutes for each topic (much like the length of an Internet video such as on YouTube).

Interactive

- Questions—Ask questions of the audience as a whole and individually.
- Pause—Use short breaks during which you look around, ask for questions, or say something in a more informal way.
- Closeness—Walk toward the audience and change the physical distance.
- Eyes—Make eye contact for a few seconds or more. Talk to that person.
- Personal—Notice and compliment something somebody has with him or her.
- Games—Have the audience stand and stretch.
- Move—Walk, motion, gesture, but don't pace or use overly repetitive movements.

PROBLEMS

9.72 Everybody has had speaker anxiety at least once. A majority of speakers are anxious every time. The audience is normally very sympathetic and supportive when a speaker is anxious and nervous. Describe a time you were listening to an anxious speaker and how you felt.

9.73 List five methods to interact with the audience and keep their attention.

9.74 Why does breathing relax a nervous speaker?

9.75 List 10 presentation or speaker elements that you have seen in the past and would like to use.

9.19 Harmful and deadly presentations

The following lists some minor issues that presenters will work to overcome. Experience often resolves these:

- No presentation objectives: If you don't know what your audience should do at the end of your presentation, there is no need for you to present. Knowing your objectives is the key to developing an effective presentation.
- Poor visual aids: Visual aids are designed to reinforce to your audience the main points of your presentation. Without effective visuals, you are missing a key opportunity to communicate with your audience.
- Ineffective close: Closing your presentation is extremely important. It is when you tie up your presentation and spell out what you want your audience to do. A weak close can kill a presentation.
- Mediocre first impression: Audiences evaluate a presenter within the first 120 s of the beginning. Presenters who make a bad first impression can lose credibility with their audience and, as a result, diminish their ability to effectively communicate the information in the presentation.
- No preparation: The best presenters prepare for every presentation. Those who prepare and practice are more successful in presenting their information and anticipating audience reaction. Practice does make perfect!
- Lack of enthusiasm: If you are not excited about the presentation, why should your audience be? Enthusiastic presenters are the most effective ones around.

- Weak eye contact: As a presenter, you are trying to effectively communicate with your audience to get your message across. If you don't make eye contact with the members in your audience, they may not take you or your message seriously.
- No audience involvement: The easiest way to turn off your audience is by not getting them involved in your presentation. Use audience involvement to gain their "buy-in."
- Lack of facial expressions: Don't be a zombie. Effective speakers use facial expressions to help reinforce their messages.
- Sticky floor syndrome: There is nothing worse than a speaker who is glued to the floor. Be natural and don't stay in one place.

Some presenters do some terrible things to the audience without realizing what they have done. This normally happens when they break a few of the basic rules and get caught up in what they are doing. The good news is that most people are not aware they are doing it, and will change once they notice. Some of the classics are:

- Demonstrations are fraught with delays, mistakes, and failures.
- The presentation uses every available feature, including sound, lights, and animations.
- The screen saver is hypnotizing.
- The presenter can't find the on button. Know your equipment!
- The presenter uses microtext, which is unreadable when the font is too small.
- The presenter does not make eye contact, mumbles, is too quiet, etc.
- Regular paragraphs are cut and pasted as if they are on paper.
- The presenter uses mathematical derivations or large equations, which are almost impossible to present on slides.
- A reader will turn to the screen and read the text verbatim.
- A constant droner will fill every gap in the presentation with "ah," "um," "OK," etc.
- A fiddler plays with objects and travels the room.
- Jedi Knights use laser pointers on the audience.
- Caffeine addicts use laser pointers to exaggerate small jitters.
- Slide flippers jump forward and backward to find slides.
- File hunters go looking for lost files on a hard drive while the audience watches.
- Zombies stay up all night to prepare.

PROBLEMS

9.76 Create a list of five presentation skills you need to strengthen.

9.77 Give three advantages and three disadvantages of having physical demonstration units.

9.78 List five presentation or speaker elements you have seen that you found distracting or confusing.

9.79 Prepare a 10-slide presentation on how to run effective meetings.

9.80 What are the three C's of communication?

9.81 Describe reasonable audience expectations for:

(a) a presentation on a new technology

(b) a testing laboratory presenting a quality-control failure

(c) a presentation to a customer for a design change

Further reading

Bacal, R., 2004. The Manager's Guide to Performance Reviews. McGraw-Hill.
Bryan, W.J. (Ed.), 1996. Management Skills Handbook; A Practical, Comprehensive Guide to Management. American Society of Mechanical Engineers, Region V.
Carnegie, D., 1981. How to Win Friends and Influence People. Simon-Schuster.
Dhillon, B.S., 1996. Engineering Design; A Modern Approach. Irwin.
Heldman, K., 2009. PMP: Project Management Professional Exam Study Guide, fifth ed. Sybex.
Heerkens, G.R., 2002. Project Management. McGraw-Hill.
Malandro, L., 2003. Say It Right the First Time. McGraw-Hill.
McCormack, M., 1984. What They Don't Teach You at Harvard Business School. Bantam.
Nelson, B., Economy, P., 2005. The Management Bible. Wiley.
Pritchard, C., 2004. The Project Management Communications Toolkit. Artech House.
Ullman, D.G., 1997. The Mechanical Design Process. McGraw-Hill.
Wysocki, R.K., 2004. Project Management Process Improvement. Artech House.
Zambruski, M.S., 2009. A Standard for Enterprise Project Management. CRC Press.

CHAPTER

General design topics

10

10.1 Introduction

Know your tools.

Each engineering discipline has well-developed tools. Across all of these there are many general topics applied to all design types.

10.2 Human factors

Engineers will often focus on the technically difficult design factors and overlook the small but important details. The result is often an unhappy customer saying, "They should have thought about that." There are a number of approaches to ensure the design satisfies the users. The simplest is to imagine how the customer would use the device. It is critical to think about the minute details such as where the user will look, bend down, push buttons, move, and so on. Consider developing physical or simulated models of the design and then physically moving through the use processes, much like a performance by a mime. Look for steps that are complex, tiresome, or awkward.

10.2.1 User interaction

Technical design is function, artistic design is intent.

A user script, such as the one given in the following list for an oven, can be quite long; however, it allows the designer to think through the steps. The example below is at a relatively high level. Each of these steps could be subdivided further. For example, the "open door" steps should include how the user grabs the handle on the door, the pulling or pushing forces, where the user releases the handle, and proximity to the heat sources. This analysis would raise questions about ergonomics, user experience, safety, mechanical design, and more:

Installation

(1) Install 220-V AC single-phase 40 A with ground.
(2) Ensure at least 80-cm opening.

Engineering Design, Planning, and Management. https://doi.org/10.1016/B978-0-12-821055-0.00010-4

(3) Floor must be level within 4 cm.
(4) Remove oven from shipping box.
(5) Position in front of space.
(6) Plug in and look for lights.
(7) Move into place.
(8) If problems occur, use the troubleshooting guide.

Baking

(1) Open the door and check for contents.
(2) Use the start button to turn the oven on and the up/down buttons to change the temperature to 220°C.
(3) Wait until the temperature is close to 220°C.
(4) Open the door far enough that it latches open.
(5) Place the food inside while avoiding hot surfaces.

Cleaning

(1) Notice excess dirt or smells when in use.
(2) Open the door to ensure the oven is empty.
(3) Close the door and turn on the oven clean function.
(4) Wait until the clean cycle turns off.
(5) Inspect the oven.

Storyboards provide a visual approach to analysis (Fig. 10.1). In this example some of the major actions are identified. The design team could ask how each setting could be better for the user. For example, is the computer too hot when it is on the legs? The transition between states is also critical. For example, how is the computer removed or returned to the case? A detailed approach storyboard might focus on a computer program that uses the keyboard and mouse.

Flowcharts are useful when describing sequential processes (Fig. 10.2). In this example the steps for oven cleaning are described. Each of the steps represents a design interaction with a user.

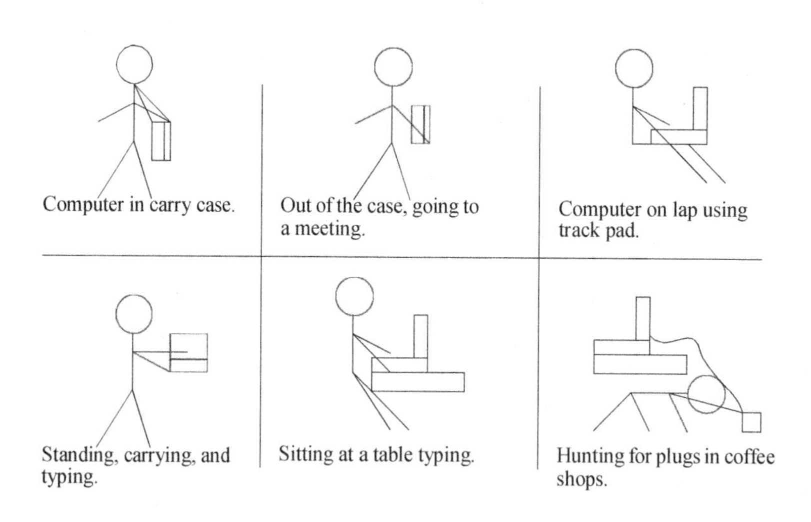

Computer in carry case.
Out of the case, going to a meeting.
Computer on lap using track pad.
Standing, carrying, and typing.
Sitting at a table typing.
Hunting for plugs in coffee shops.

FIGURE 10.1

A storyboard for laptop user interaction modeling.

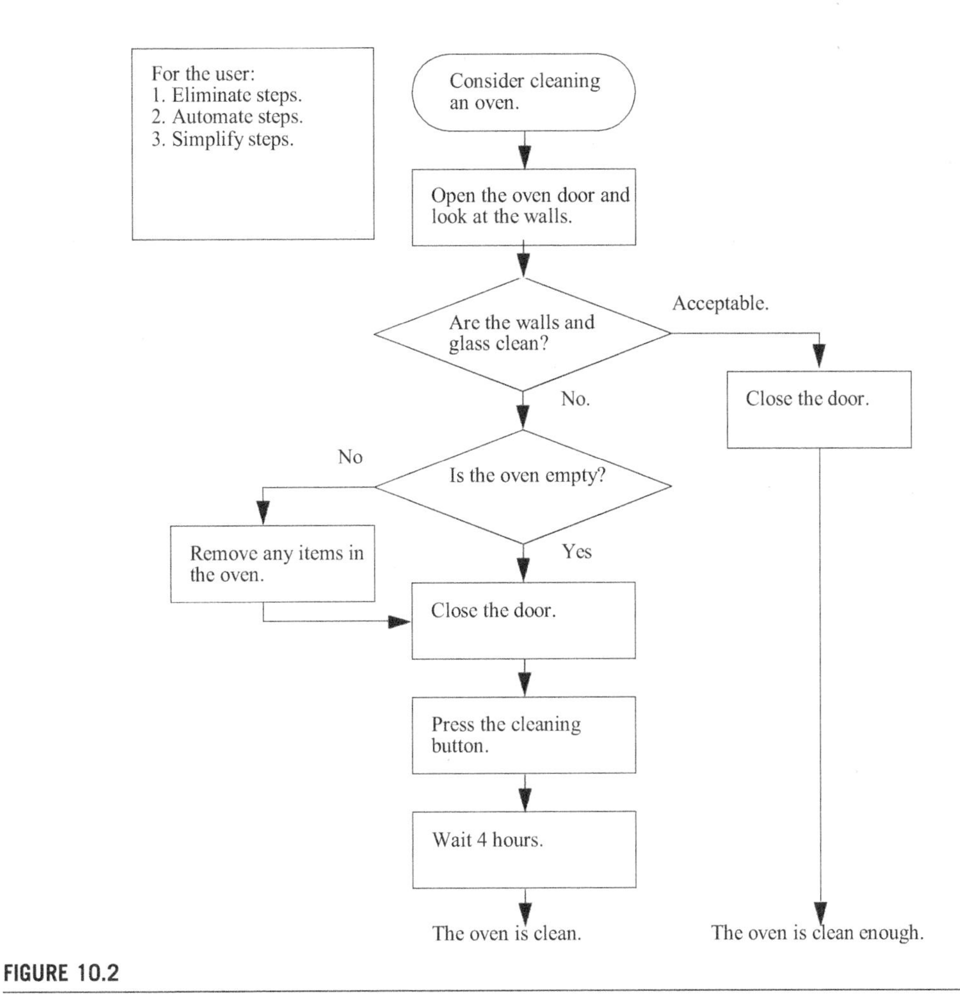

FIGURE 10.2

Flowchart for modeling user operation sequences.

Eliminating steps makes it more convenient and enhances the user experience. Simplifying or automating steps are also possible options. For example, the user process would be much simpler if a sensor was used to detect excess dirt and a second sensor detected items in the oven. If the oven is empty and dirty it could turn on automatically.

Modes of operation, or states, are convenient for describing system behavior (Fig. 10.3). The state diagram in the example is for a standard laptop power control. It makes the user options and design options very clear. It also allows standardization, such as holding the power button for 15 s turns off the computer. The state diagram also exposes problems; in this example, if the "boot is complete" case does not occur, the system will be locked in booting. The assumption for a state diagram is that only one state, a bubble, is on at a time, and true transitions, the arrows, will move to the next state.

Industrial designers, by degree, are educated to consider the user interaction and aesthetics in product design. They can serve as excellent members of the design team or as consultants. Industries that use industrial designers have much happier customers.

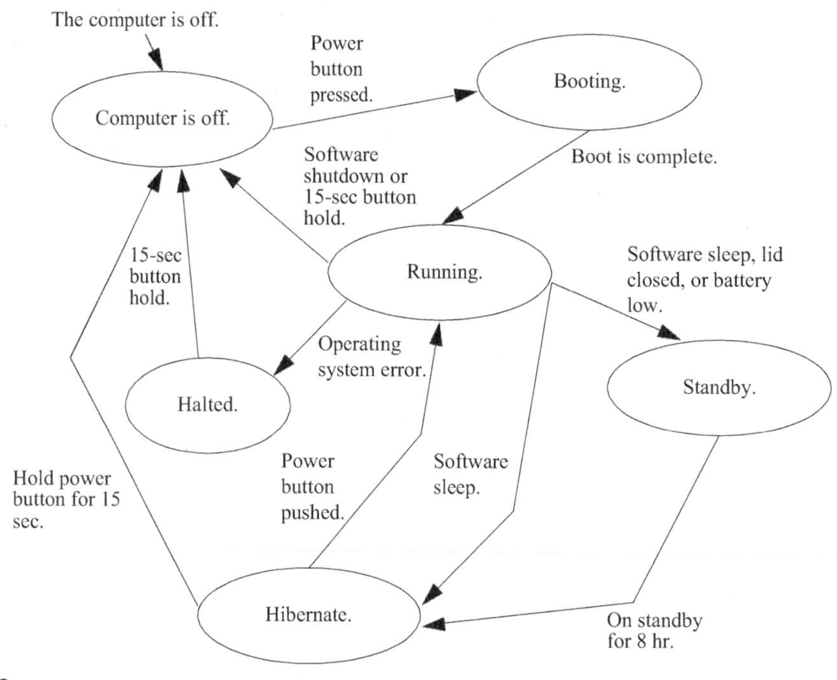

FIGURE 10.3

A state diagram for user option modeling.

PROBLEMS

10.1 Write a simple script for a user pouring milk into a cup.

10.2 List three advantages of a script, a storyboard, a flowchart, and a state diagram.

10.3 Consider a microwave oven that does not have a clock.

- **(a)** Draw a state diagram for a microwave oven.
- **(b)** Draw a flowchart for a user boiling a cup of water.
- **(c)** Draw a storyboard for boiling the water.

10.2.2 Ergonomics

Intellectual and physical requirements define user effects and affects. Sight, smell, taste, sound, and touch define human input. These are converted quickly to impulsive reactions and slowly to cognitive processing. If necessary, the body uses intent and motion to cause an action. In common use, ergonomics refers to allowing people to work effectively with no effort or harm. A higher-level goal for ergonomics is to have a system that does not require any effort and works intuitively to achieve a goal. The basic tenets of ergonomics include the following:

- Do not exceed physical limits and cause damage.
- Relax: Don't overwhelm senses and cause an impulsive reaction.
- Make it easy for the brain to understand the message.
- Do not require prolonged or exaggerated exertion.
- Work should require minimal control and effort.

In the workplace, ergonomics has become synonymous with worker safety. Notable examples include repetitive stress injuries, hearing loss, asbestos, and carcinogens. As a result, workplace ergonomic issues are regulated by the government and the courts. As a professional you are obligated to ensure that your designs can be manufactured without ergonomic harm. This can be done during design by identifying ergonomic problems and planning solutions. From a business perspective, ergonomics means lower costs and higher morale, as it is easier and safer to work. Nonergonomic designs result in longer cycle times, more quality problems, and minor and major injuries. The major sources of industrial ergonomic issues are rushed designs, short-term cost savings, and inexperience. Design for assembly (DFA) is one of many processes that are used in design to lower assembly cost, time, and effort.

Consumers seek out products that are more ergonomic. Nonergonomic products are normally uncomfortable or dangerous. If the product is less comfortable than a competitor's the customer may be lost. Ergonomic issues that can lead to minor or major injuries often lead to recalls and lawsuits. Consider a computer mouse chosen by a consumer considering the comfort. The button locations are too far right and the index finger is strained slightly when it clicks. Approaching 1,000,000 clicks, the wrist muscles are strained and carpal tunnel syndrome develops. (Note: One click every 5 s, 6 h per day, 5 days per week, 52 weeks per year is 1,123,200.) Testing and consumer feedback are essential when developing consumer products.

Manufacturing ergonomics is focused on preventing injuries first and increasing productivity second. The limits are often defined in terms of maximum forces, positions, sound levels, light levels, and support equipment. Typical ergonomic problems in manufacturing are listed in the following, along with possible solutions:

- Stress
  - Issues
    - Strain: unneeded strain on worker (e.g., hunching over)
    - Efficiency: unnatural motions will slow production
    - Cumulative trauma disorders: muscle strain injuries (lifting 30-lb packages all day)
    - Repetitive stress injuries: repeated motions (e.g., carpal tunnel syndrome in the wrists)
  - Solutions
    - Training for proper lifting methods
    - Rearrange operation locations and sequence to reduce unnatural motions
    - Use special lifting equipment
    - Use ergonomically redesigned equipment (e.g., computer keyboards)
    - A work area 30–50 cm in front of the torso will work well
    - Provide support for extended reaches, carrying heavier loads, and twisting motions
    - Provide rests and spring returns for the users
    - Soft contact surfaces reduce fatigue
    - Keep wrists straight and elbows resting or hanging
    - Allow adjustments for comfort

- Information overload/confusion: excessive, inappropriate, or a lack of detail (e.g., fighter pilots, air traffic controllers)
  - Redesign displays to be clear with a minimum amount of good information
  - Use of color to enhance pictures and text
  - Simplify controls to the minimum needed
  - Provide multiple-sense cues such as textured knobs to reduce operator looking
- Eye strain: fine focus or bad lighting
  - Adjust the lighting
  - Use magnifying lenses or cameras for small details
  - The work is straight ahead, to reduce eye strain
- Noise: direct hearing or annoyance (e.g., piercing tones, just too noisy)
  - Special hearing protection equipment
  - Redesign work spaces to reduce noise reverberation
  - Redesign equipment to reduce sound emissions
  - Sounds below 80 dB for constant noise and below 100 dB for short durations
- General
  - Comfortable temperature and humidity
  - Isolate lower-frequency vibrations to prevent motion sickness
  - Isolate high-frequency vibrations to prevent loss of sensation and nerve damage

A user interface is the boundary between equipment and machine or process. These range from complex user interfaces for software to a simple shape on a pair of scissors. The following list focuses on user experience and complements the safety ergonomics list. The listed items include (1) having the design provide information the way the user expects and can use easily and (2) having the design directly accept user direction in a natural form. In abstract terms, a good design will simply feel like an extension of the user:

Measures
- Learning and training time before use
- Retention of operation knowledge
- Error rates
- Error severity
- Operation speed
- Perceived satisfaction and comfort

Accommodation
- Works for everybody regardless of:
  - Missing sense—touch, sight, smell, color, taste, sound, etc.
  - Physical limitation—fine motor skills, force, reach, height, dexterity
  - Limited cognitive abilities
  - User mistakes
- Adaptable
- Tolerant

Expectations
- The interface matches both the task and the user
- Builds on prior knowledge of other products
- Requires no learning time

- Provides rapid feedback to confirm actions and changes
- The user is always aware of what is happening in the system through the interface
- Does not need the user to adapt to the system
- Gives the user "joy"
- Easy to stop and start at all times
- Comfortable to use
- Easy to reach (e.g., stop switches)

Techniques

- Poka-yoke, idiot proof, error proof
- Script user expectations
- Provide error detection
- Standardization
- The interface should internally align user expectations to machine requirements
- Visually obvious placement, appearance, and labeling
- Consistent appearance and theme
- Simplify
- Lighting

Methods for implementing interfaces

- Use of senses such as touch, smell, sight, sound, and taste
- Controls include switches (push, toggle, touch, etc.)
- Sliders, knobs
- Typical displays include indicator lights, touch screens
- Audible tones (buzzers, beeps, bells, etc.)

Note: The topics in this safety section are posed from a human perspective, but they also apply to equipment in general, as addressed in Chapter 8.

Resource 10.1 MIL STD 1742F (1989) on this book's website: www.engineeringdesignprojects.com/home/content/human-factors.

PROBLEMS

10.4 List 10 ways an office desk could be made nonergonomic.

10.5 List 10 methods for warning users.

10.6 Assume a poor workspace design has resulted in a number of work injuries. An engineer developed a proposal with two options: (a) redesign the work cell for $8000 or (b) train workers for $1000 and budget for additional sick days. Provide three benefits of each option.

10.7 What are five features of a great ergonomic design?

10.2.3 Law

Criminal law and civil law differ in terms of intent, application, and standards. Criminal courts focus on punishment, and the standard is "beyond a reasonable doubt." In civil cases, the court decides with a "balance of truth" and provides compensation for the wounded. Civil courts may find both parties at fault and provide partial compensation for damages. Problems resulting from ethical engineering projects rarely have any criminal implications but sometimes result in civil lawsuits. Three areas of civil law most important to engineers include contract law, torts, and intellectual property.

Liability is a measure of responsibility for damage by the defendant to the plaintiff. Damages can include personal injury, financial loss, disclosure of secrets, damage to property, failure to satisfy a contract, and so on. As engineers and companies we are required to provide "reasonable care" to those who will come in contact with our products or work. Reasonable care is somewhat arbitrary but is normally set by the current standards of practice. For example, if playground equipment normally has 30-cm rails for safety, designing with 26-cm rails could be considered unreasonable. Luckily, there are many government, industry, and de facto standards that can be used as guidelines for reasonable care. One example would be the flammability of cloth in car seats, as defined by government regulations. Industry standards include electrical insulation for equipment to prevent accidental electrocution. De facto or ad hoc standards include computer button placement to avoid repetitive stress injuries. In some cases we will do work that requires "great care" in excess of "reasonable care." In these cases, failure to meet these standards will also make us liable.

It is a simple fact that the products of our work will age and eventually fail. A warranty is a guarantee that the work will not fail within a reasonable amount of time. A warranty also defines the actions needed in the event of failure. For consumer products these warranties are typically 30 days to 1 year when voluntarily offered by companies. Regional and national governments often have laws that set minimum warranty periods to protect consumers when buying automobiles, houses, electronics, and much more. In specialty projects the warranties and liability can be defined differently in contracts as long as they do not violate local laws. Sometimes there is no formal warranty definition, but there is an implied warranty. Simple examples include food and clothing purchases that do not show obvious defects until use.

When the product of our work does not meet the standard of reasonable care or results in damage, we are said to be negligent. If the issue is the result of a mistake or oversight, the negligence is minor and the compensation is often the same as the damages. If the mistake or oversight is intentional or clearly violates professional standards, then it is gross negligence. In the case of gross negligence the court may also award compensation plus additional punitive compensation. For example, if an engineer makes a calculation mistake that leads to cracking in office chair legs, the customers cannot expect more than the cost of replacing or repairing the chair. If the engineer had reduced the chair legs' strength below industry standards to reduce costs, he or she may have to pay the cost of the chair plus additional compensation to the plaintiff. In clear cases, the plaintiff or defendant may receive money to cover reasonable legal costs.

Compensation for damages is awarded based on responsibility and the ability to pay. As an example, a consumer hurt by a playground slide may sue the slide maker, slide installers, and playground owner. The court might assign 35% negligence to the adult who fell off, 30% to the slide manufacturer who made 26-cm slides, 15% to the installer who purchased a substandard slide, and 20% to the playground owner who allowed adults to play on the child-sized equipment. The plaintiff's injuries resulted in medical bills and loss of employment totaling $500,000. The plaintiff, installer, and owner combined only have $100,000 in assets. The remaining $400,000 would have to be paid by the manufacturer if they could afford it. The result will be that a plaintiff will sue many defendants to increase the chance that all damages will be paid. In cases in which a defendant is not found responsible he or she may be able to recover legal costs from the plaintiff.

When damages have occurred, the soon-to-be plaintiff is required to mitigate the damages. Mitigating damages means taking reasonable actions to stop the damages from worsening. For example, if a car tire becomes flat the driver should stop driving to mitigate the damage. Driving for another hour might damage other car parts. The court would say that the damages are the tire, but the other car parts were damaged because the driver did not mitigate the damages.

Contracts exist when there is an agreement to exchange value. Written contracts are best, but implied and oral contracts can also be enforceable, though not as easily. Contracts are written as a combination of terms to define a variety of critical details, including those in the following list. Although the legal language can seem odd, it is very specific. When a lawyer writes or reads a contract he or she is careful to look for specific clauses (numbered paragraphs). Both the presence and the lack of specific terms can be critical. Good contracts will outline details of the final deliverables, methods for resolving disputes, dates and procedures for delivery, and warranties, as follows:

- Involved parties
- Intellectual property ownership and transfer of project materials
- Confidentiality and disclosure procedures
- Design requirements
- Delivery dates
- Conflict resolution requirements
- Exchange of value: money for design work
- Required schedule
- The process for ending the contract work
- A process for altering or voiding the contract
- "Acts of God" clause
- Breach of contract and possible results
- Assignment or transfer of liability during and after the project
- Arbitration

In larger projects, a contract may include subcontractors who are doing part of the work for the main contract holder. In these cases the main contract may outline the roles and relationship between the contractor and subcontractors. This arrangement is very common in the construction industries.

When one or more of the parties to a contract violate one of the terms it is called a breach of contract. Many contracts include an arbitration process that defines how suspected breaches should be resolved. Arbitration is a process that involves an impartial third party such as a lawyer or engineer to hear the case and decide an outcome without taking the issue to court. This provides a faster and less expensive resolution to many problems. If these clauses exist in contracts they may prevent going to court, unless they are against local laws. When a contract is breached the plaintiff will sue the defendant. The court will hear the case and assign blame and damages accordingly. These outcomes often include financial compensation, orders to stop work, and termination of the contract.

When the plaintiff is a member of the general public with no direct relationship to the defendant, the case is called a "tort." These cases often claim that the negligence of the plaintiff led to some sort of personal damage to the defendant. There are often multiple defendants for the payment issue mentioned before. In class action cases, there are a large number plaintiffs joined as a group. A simple example of a tort is a person sitting on a chair that broke and caused an injury. An example of a class action suit is a defect in automobile tires that leads to 2000 automobile accidents.

Designers will sign multiple nondisclosure agreements (NDAs) as they meet with many customers. These agreements have terms that restrict what information can be shared for some period of time. Some of the standard clauses are given in the following list. A variation on the NDA is the noncompete agreement. When hired, most technical staff will be asked to sign a document that does not permit them to leave the company and continue the same work. If you have secret information, such as a new invention or product, you should have suppliers sign an NDA before providing details. If asked to sign NDAs on behalf of your company you need to observe company policy:

- Introduction: Agreement to agree
- Section 1: A definition of what is considered secret
- Section 2: You or your people cannot share the secrets
- Section 3: Preagreement knowledge is not secret
- Section 4: The secret is owned, as in property
- Section 5: The agreement does not give property rights
- Section 6: The reason for the agreement is secret
- Section 7: You get access to new secrets for 1 year, but you must keep the secrets for 5 years
- Section 8: New agreements cannot replace this one
- Section 9: The contractual clauses may be enforced only if they are legal
- Section 10: An invalid section does not invalidate the agreement
- Section 11: This agreement replaces all previous agreements
- End: Signature to agree to terms

Given that engineers often generate and use strategic knowledge, these agreements provide protection for employers. A typical set of terms would prevent a former employee from working for a competitor for 1 to 5 years. Lawsuits do occur between former employers, former employees, or new employers. When presented with a noncomplete agreement, consider having it reviewed by a lawyer.

Intellectual property is becoming increasingly important. Patents, copyrights, and trademarks provide legal recognition of rights. They do not provide instant legal relief. If a patent is violated within a period of about 20 years, the holder can sue the violator to stop use and pay damages. A patent does have great value in determining the balance of responsibility, but it is not guaranteed. Practices to prevent and keep patent rights include using NDAs before applying for patents, notifying suspected violators, keeping design and meeting logs, and so on.

Copyright mainly protects artistic works such as images, videos, and text for a period of many decades. Trademarks are unique logos, names, and decorations. As an example, a design project for an alarm clock might violate a patent for a color display, the alarm sound might violate a music copyright, and the outer case may be too similar to a trademark design.

There are a number of laws and regulations that are specifically directed to engineering design. Violating these often leads to major fines and legal sanctions. One example is the regulation of radio-wave frequencies, in the United States done by the Federal Communications Commission. Another example is the regulation of automobile designs to ensure passenger and environmental safety. The European Commission has formally developed a set of regulations to cover the numerous members of the European Union. In other regions such as North America, countries tend to collectively develop or adopt existing standards. A very small number of the regulation areas can be seen in Table 10.1. Needless to say, any designer must be aware of these regulations, design to meet them, and test to ensure compliance.

Table 10.1 Regulation types and local requirements.

Regulation of certification type	United States	European Commission	Canada
Workplace health and safety	OSHA, NIOSH	EU-OSHA	CIRB, HMIRC
Engineering school accreditation	ABET		CEAB
Public safety	UL	PPE, REACH, CPD/CPR	CSA, CCPSA
Food and drugs	FDA	EFSA, EMA	CFIA
Transportation	NHTSA, FAA	EMSA, EASA, ERA	TSB
Communications	FCC	BEREC	CRTC
Environment	EPA, CAA, CWA, HSWA, PPA, TSCA	EEA, ECHA	CEAA

There are many horror stories about frivolous lawsuits and excessive awards. In practice these will happen at times for less sensational reasons. Many lawsuits can be settled out of court to save legal fees, delays, and negative publicity. Doing all work with best efforts, good intentions, and due diligence is the best protection against legal issues. After all, nobody starts a workday by saying, "Let's do a mediocre job today, even if it kills somebody."

PROBLEMS

10.8 What is a standard of reasonable care?

10.9 What may happen if an engineer does not follow a standard in design?

10.10 Is a warranty always 1 year? Explain.

10.11 Why do lawyers sue everybody connected to a defective product?

10.12 List five ways to mitigate damages if a fuel tank begins to leak.

10.13 How is a tort related to a contract?

10.14 What are five typical clauses found in a contract?

10.15 When would you ask somebody to sign an NDA?

10.16 What is the difference between a regulation and a de facto standard?

10.2.4 Sustainability and environmental factors

Leave things better than you found them.

Sustainability is often defined as an overlap of three factors (Fig. 10.4). Engineering designs both solve and create problems in all three areas. A design is economically sustainable if it is profitable, ensuring that somebody can earn a living doing the work. Environmentally sustainable designs do not degrade

FIGURE 10.4

Elements of sustainability.

the environment; they leave it the same or better than it was found. Socially sustainable designs build and strengthen communities. By contrast, designs that are not economic will result in losses and failed products. Designs that are not environmentally sound will result in loss, or damage, to natural resources (e.g., toxic waste). Unsocial designs degrade the quality of life using practices such as "sweatshop labor." A truly sustainable design will make positive economic, environmental, and social contributions. A design that has only one or two of the factors will eventually fail or result in harm. Economic factors are a natural part of engineering work, but we must not forget the environmental and social factors.

Designs can support social needs in a number of ways. Devices that allow handicap usage include more human factors in the designs. Manufacturing techniques that allow products to be produced locally allow communities to profit from their own work. Designs that provide better housing, sanitation, clean water, abundant food, and education help lift the standards of living. Reducing the chance of injury improves enjoyment of life. Some key questions to ask are:

- Who will benefit from the design?
- Will, or could, the design have a negative impact on anybody?
- Does the design offer convenience?
- Does the design allow new opportunities?
- How will somebody use the design?
- What are the social trade-offs for the design?
- Does the design improve quality of life?

Engineering designs can reduce the cost of purchase and ownership for a product. When the cost is low enough it can be purchased easily, with a profit for the designer and manufacturer. If the design is not profitable it will be viable only as a hobby, social cause, or government program. Engineers can develop new technologies and methods to reduce design costs and make products more viable.

ISO 26000 is available for companies looking for guidance in sustainability. (Note: It is not used for certification, like ISO 9000.) The standard provides seven core areas with related issues. These issues become very important when doing international and global trade. It is simply unethical to move business operations to another country for the sole purpose of avoiding legal, social, and environmental rights. However, business is often drawn to other countries for lower labor costs, labor supply, and natural resources. Ethically we are obliged to leave the environment and people unharmed. Morally we are driven to improve the quality of life of the people we touch. The ISO 26000 core subjects and issues provide guidance for reviewing daily and strategic business practices:

Core subject: Organizational governance
Core subject: Human rights
- Due diligence
- Human rights risk situations
- Avoidance of complicity
- Resolving grievances
- Discrimination and vulnerable groups
- Civil and political rights
- Economic, social, and cultural rights
- Fundamental principles and rights at work

Core subject: Labor practices
- Employment and employment relationships
- Conditions of work and social protection
- Social dialogue
- Health and safety at work
- Human development and training in the workplace

Core subject: The environment
- Prevention of pollution
- Sustainable resource use
- Climate change mitigation and adaptation
- Protection of the environment, biodiversity, and restoration of natural habitats

Core subject: Fair operating practices
- Anticorruption
- Responsible political involvement
- Fair competition
- Promoting social responsibility in the value chain
- Respect for property rights

Core subject: Consumer issues
- Fair marketing, factual and unbiased information, and fair contractual practices
- Protecting consumers' health and safety
- Sustainable consumption
- Consumer service, support, and complaint and dispute resolution
- Consumer data protection and privacy
- Access to essential services
- Education and awareness

Core subject: Community involvement and development
- Community involvement
- Education and culture
- Employment creation and skills development
- Technology development and access
- Wealth and income creation
- Health
- Social investment

PROBLEMS

10.17 List the ISO 26000 issues that would be important if workers were being chained inside a building during working hours to prevent theft.

10.18 What makes a design sustainable?

10.19 Assume you are trying to implement sustainability practices in a new labor market. You have allocated money for social development, environmental cleanup, and workforce development. Your employer tells you to put all of the money into worker development because their wages will pay for the other things. List five compelling arguments to keep the original budget.

10.2.5 Engineering for our environment

In order: eliminate, reduce, reuse, recycle, new.

Of all disciplines, engineering has the greatest impact on the environment. This includes the materials and energy used in production, the efficiency of products, the waste during use, and the end-of-life disposal. For example, we may decide to use a cell phone circuit board component that costs $0.01 less, but generates an extra 10 g of carbon dioxide and requires an additional 100 g of mined ore. If there are 10,000,000 devices made, the cost savings is $100,000, but the resulting waste is 100,000 kg of carbon dioxide and 1,000,000 kg of tailings. In this example it is very easy to focus on the cost savings of $100,000 but a responsible designer will also consider the environmental impacts.

Environmentalism is not new, but as our understanding of the ecosystems expands, so do our efforts to protect the planet. Pollution such as human waste, soot, sulfur dioxide, chemicals, refrigerants (e.g., Freon), smog, and asbestos have been identified and regulated or banned over centuries. Some governments have been very proactive in environmental reforms, including the European Union, Canada, Japan, and the US state of California. Regardless, numerous government agencies are involved with regulating materials such as those in the following list. The regulations are enforced using economic incentives, warnings, restrictions, fines, and prison sentences. These agencies often include resources and initiatives to provide information and help industry solve environmental issues:

Chemicals

- Benzene
- Carbon tetrachloride
- Chloroform
- Cyanides
- Dichloromethane (methylene chloride)
- Inorganic arsenic
- Methyl ethyl ketone

- Methyl isobutyl ketone
- Phenol
- Tetrachloroethylene
- Toluene
- Trichloroethylene
- Trichloroethane
- Vinyl chloride
- Xylene(s)

Heavy metals
- Beryllium
- Cadmium and compounds
- Chromium and compounds
- Copper
- Lead and compounds
- Mercury and compounds
- Nickel and compounds
- Zinc and zinc oxides

Product families
- Fuels and lubricants
- Paints and coatings
- Batteries
- Raw materials processing (e.g., ores, natural resources, agricultural)

Ecology
- Endangered species and biodiversity
- Global warming and ozone layer depletion
- Groundwater
- Migration paths and breeding grounds/waters
- Waterways and food chains
- Biologically active chemicals and life forms

Climate and air
- Fine suspended particulates (e.g., asbestos, ash, dust)
- Combustion (e.g., coke oven emissions, coal/oil/gas boilers)
- Greenhouse gases (e.g., carbon dioxide, Freon)
- Hydrocarbons
- Nitrogen oxides
- Photochemical oxidants
- Sulfur oxides

Others
- Noise and vibration
- Radioactive solids and liquids
- Waste from mining or similar activities
- Waste heat and light
- Solid waste (e.g., ash, packaging, consumables, scrap metals, scrap plastics)

A challenge for environmentalism is that it can add cost to designs. As a result, a business that is more environmentally friendly can have a cost disadvantage. One response to this issue is consumer education that promotes environmentally responsible products. The ISO 14000 and 19011 certification standards have been developed to provide some level of environmental responsibility. A more effective approach is to legislate minimum standards so that all manufacturers have the same environmental cost penalty. An excellent example was the EU directive to eliminate electronic solder that uses lead; this is denoted by the RoHS marking on circuit boards. Solder that uses lead is easier to use, less expensive, and less prone to grow "tin whiskers." Since the standard has been set, lead-free solder is used in most products. Without the legal requirements, manufacturers would be at an economic disadvantage to use RoHS standards, but a universal ban eliminates the competitive advantage.

As governments have chosen to intervene in commercial affairs for environmental benefits, new laws and regulations have been developed. These include a variety of criminal and civil penalties dictated by laws and regulations. For example, in an extreme case a company could be tried in criminal court for knowingly dumping toxic waste. Although it is difficult to send a company to jail, large fines and court orders are reasonable.

Normally employees inside a company are immune from legal actions against the company, but in the case of design engineers there are many environmental laws that apply directly to them. There are a wide variety of laws, agencies, and organizations that influence manufacturing and consumer products:

Government
- EEA—European Environmental Agency
- EPA—Environment Protection Agency (USA)
- OSHA—Occupational Safety and Health Administration (USA)
- NIOSH—National Institute of Occupational Safety and Health (USA)

Voluntary
- UL—Underwriter Laboratories (USA)
- CSA—Canadian Safety Association

Every design has an environmental impact. As designers we need to (1) identify the problems, (2) minimize and eliminate problems, and (3) look for ways to improve the environment. Identifying the problems begins with an audit of environmental factors. A sample of a simple audit form for a paper cup is shown in Table 10.2. As with any audit, it is difficult to decide where it starts and ends. In this case the analysis starts at the paper mill. Each component and process has an impact on the environment. In general, it is very difficult to quantify environmental impacts.

A graphical approach, product life-cycle analysis, focuses on a product or design as it moves from concept and raw materials through to eventual retirement, as shown in Fig. 10.5. This diagram focuses on the different major inputs to the process from the perspective of the manufacturer. The steel for the knives, forks, and spoons can come from scrap metal, and at the end of the product life it will be recycled. Some components, such as the manufacturing equipment, can also be recycled. Probably the most environmentally unfriendly process is the coating. In this process, electricity is used with environmentally hazardous chemicals often containing heavy metals. This process is essential for preventing corrosion and rust, and many controls are in place to minimize environmental damage. Electricity is required to make and run the manufacturing equipment. Although easy to ignore, each kilowatt-hour will require some sort of fuel that will become a pollutant. For an environmental audit to be more accurate it should include the environmental impacts contributed by outside sources. A manufacturing example includes the following:

Table 10.2 A simple environmental audit spreadsheet for a paper coffee cup.

Factor	Element or process	Consumes	Produces	Environmental impact	Reducing impact?
Raw paper	Paper pulp	Trees Handling labor Equipment grinders Fresh water	Waste bark, wood, brackish water		Add recycled paper
	Bleach				
	Presses	Electricity, equipment, labor	Wastewater, waste		
	Heaters	Fuel oil	Heat humidity		
Adhesives					
Forming process					

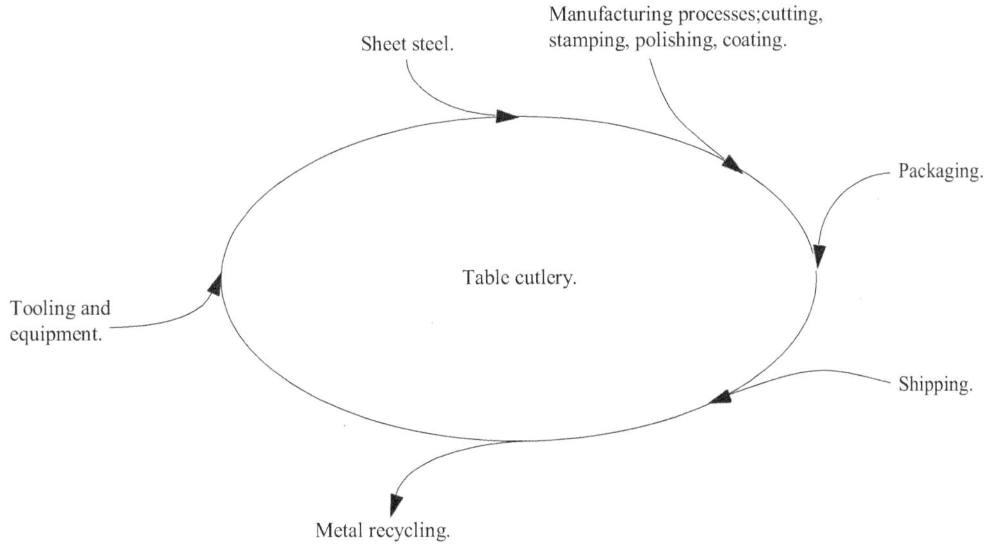

FIGURE 10.5

A sample life-cycle analysis diagram.

(1) Life-cycle problems
  (a) Extracting raw materials often results in damage to the environment.
  (b) Purifying raw materials produces by-products and requires energy and other materials.
  (c) Shaping materials into useful forms also produces by-products and requires energy and other materials.
  (d) During the life of a product there is upkeep, maintenance, and consumption.
  (e) A product must be discarded at the end of its life.

(2) Life-cycle solutions
  (a) Use less (eliminates 1a, 1b, 1c, 1d, and 1e).
  (b) Reuse when possible (eliminates 1a, 1b, and 1e).
  (c) Recycle (eliminates 1a and 1e).

(3) Pollution
  (a) Air-based exhaust
  (b) Runoff to waterways
  (c) Stored toxic dump
  (d) Stored solids
  (e) Discharges/waste (gas, liquids, solids) from production processes
  (f) Energy/fuel utilization in production
  (g) Aging of the product (decay, inert, toxic, etc.)
  (h) Energy/fuel efficiency in use

It is difficult to objectively assess environmental impact. A solid engineering approach to assessing impact is to use quantifiable metrics wherever possible. Some of the current metrics include:

- carbon dioxide, and similar pollutants, emissions by mass (kg, lb)
- solid wastes by volume or mass (m^3, kg, lb)
- energy usage by volume, mass, or energy
- liquid wastes, toxic and benign, by volume
- toxic wastes by volume, mass, and type
- airborne wastes by volume, mass, and toxicity

Many companies have embraced environmental missions. The ISO 14000 certification standard was developed to formalize the process. A variant of the ISO 14000 standard is the ISO 19011, which adds the ISO 9000 quality control standards. The standard requires processes and documentation that include audits of products to track wastes, recycled/reused materials, energy consumption, and similar elements. The standard encourages the inclusion of environmental impact in the conceptual and detailed design phases, in particular, adding environmental impacts to the budget and project control processes.

The standard priorities for design are (1) eliminate, (2) reduce, (3) reuse, (4) recycle, and then (5) create new materials and parts for designs. For example, a design with minimal packaging is more environmentally friendly than a similar design with larger packaging. The tooling, materials, energy, and handling efforts are reduced. Design factors that will benefit the environment include the following:

- Materials
  - Eliminate hazardous materials.
  - Reduce or eliminate pollutants and waste, including gases, liquids, and solids.
  - Reduce the use of materials overall.
  - Prefer commonly recycled metals such as steel, iron, aluminum, and copper.
  - Ensure plastic parts are clearly marked with the standard recycling symbols.
  - Use recycled materials, including glass and paper.
- Reusable components
  - Use parts from older versions.
  - Parts can be used elsewhere.
  - The product can be renewed or upgraded.
  - Separate durable and short-life components.
  - Make the design easy to disassemble.
  - Make the design easily repairable.

- Upgrade, downgrade, or discard
  - Best: Prefer materials/parts that can be used in the same form with no processing.
  - Good: Materials/parts that can be put in the same or better form with processing.
  - Poor: Materials/parts that must be downgraded in reuse and recycling applications.
  - Avoid: Materials/parts/assemblies that can only be discarded as scrap, such as:
    - mixed metal parts
    - coated parts
    - hard to separate parts because of connections that are glued, riveted, etc.
    - ceramics
    - thermoset plastics
- Customer specifications, concepts, and technical specifications
  - Fewer parts and materials
  - Long life
  - Reusable or recyclable
  - Easy to separate and sort materials
  - Uses fewer resources and less energy
- Resource consumption
  - Minimum energy and utility usage
  - Minimal consumables (e.g., refillable ink cartridges instead of disposable)

An important part of the reuse and recycling steps is the ability to separate and sort materials. It can be very difficult to separate materials that are screwed, welded, and glued together. It is easy to recycle a product that will divide into separate materials by removing or breaking fasteners. For example, a good design would require the removal of four screws to separate a television. Coincidentally this also tends to reduce assembly costs. In practice, separating materials can be very difficult. In some cases it requires intensive manual labor, such as removing copper from electric motor windings. The other extreme is devices that shred the materials and then separate the metals and other materials. To recycle automobiles, the toxic materials (e.g., oil, batteries, and fuel tanks) are removed first. The engine block may also be removed for separate processing. The car is then passed through a shredder that reduces it to pieces a few centimeters or less in size. These small pieces are then sorted into steel, "fuzz," and heavier scrap pieces.

The highest-quality materials for recycling are presorted by material type. When the materials are mixed they must be separated using automated and manual processes. Luckily, steel is easily sorted using a magnet. Certain metals such as aluminum are easily identified by color and mass. Paper is easily identified and removed, and it can be sorted into different grades. Glasses are also easily identified and removed. Plastics are more difficult to sort because the appearance and mass are similar. To help with this process, plastic pieces should be marked with the symbols in Fig. 10.6.

PROBLEMS

10.20 Why is reuse better than recycling?

10.21 Create an environmental audit spreadsheet for an orange.

10.22 Why is noise considered pollution?

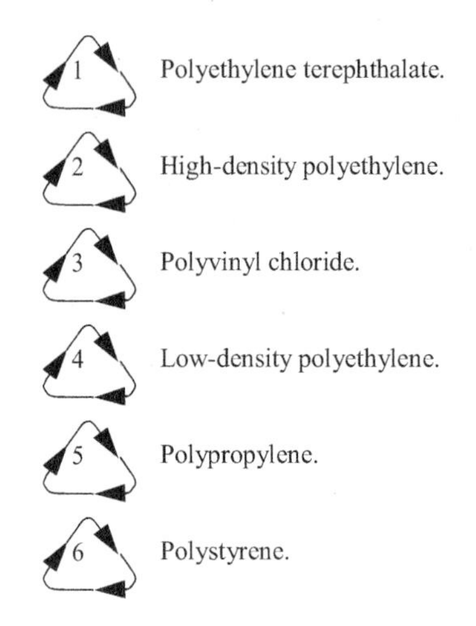

FIGURE 10.6

Plastic recycling symbols.

10.23 What are five rules for making assemblies easier to recycle?

10.24 Suggest 10 ways to redesign a car to lessen the life-cycle impact it has on the environment.

10.25 Which environmental factors will affect workers?

10.26 What factors would need to be considered when measuring noise in a manufacturing plant?

10.27 Discuss the two main organizations that deal with environmental issues. Who do they serve? What are their main environmental concerns?

10.2.6 Design for X

When a design is easy to manufacture, the result is a better product that costs less. Experienced designers understand and use knowledge of processes to improve the design of parts, assemblies, and entire products. The family of design for X (DFX) techniques has been developed to capture the knowledge of experts, giving designers guidelines for analysis and redesign. There is no single body of "design-for" techniques, but a few are listed in Table 10.3. DFA and design for manufacturing (DFM; sometimes called DFMA) are the two most popular.

In a traditional corporate structure, the Design Engineer (a corporate title) completes a design and then sends it to the Manufacturing Engineer (another corporate title). The design engineer sets the product details and the manufacturing engineer designs the processes and tools to make it. Once the design reaches the manufacturing engineer, many of the manufacturing decisions are predetermined. In other words, the design of the product is finished before the plans to make it are created. The possible outcomes include (1) the design works well, (2) the design features are difficult and costly to produce, (3) the manufacturing designer changes features to make it work, or (4) the manufacturing and design teams resolve the issues. Another option is concurrent engineering, in which the design and manufacturing engineers work together through the design process, as shown in Fig. 10.7. Concurrent engineering doubles the number of problems being solved at the same time, increasing the complexity

Table 10.3 Design for X types.

Design for X acronym	Meaning
DFA	Design for assembly
DFD	Design for disassembly
DFEMC	Design for electromagnetic compatibility
DFESD	Design for electrostatic discharge
DFI	Design for installability
DFM	Design for maintainability
DFM(A)	Design for manufacturing
DFML	Design for material logistics
DFP	Design for portability (software)
DFQ	Design for quality
DFR	Design for redesign
DFR	Design for reliability
DFR	Design for reuse
DFS	Design for safety
DFS	Design for Simplicity
DFS	Design for sustainability
DFT	Design for test

of the process and the labor required. However, the outcome of concurrent engineering is a product that is much easier to manufacture well, with less labor in tooling design and production planning. Sequential engineering reduces the complexity of each stage to just functional or manufacturing design. Therefore, sequential engineering requires less design labor but additional manufacturing design time. In practice, most engineering departments will use a mixture of both techniques, and as designs and designers become more mature the need for concurrent approaches will be reduced. Concurrent design approaches are evident when there are frequent design meetings with people from all stages of the design process, ranging from marketing to shipping.

Product designers who use DFX methods will add time to the design process but will reduce the time for manufacturing design and production problems. Designs that have been through a DFMA process are notable because they have:

- shorter production times
- fewer production steps
- smaller parts inventory
- more standardized parts
- simpler designs that are more likely to be robust
- lower cost and maintenance tooling

PROBLEMS

10.28 What is DFMA?

10.29 How does DFX add extra steps to design?

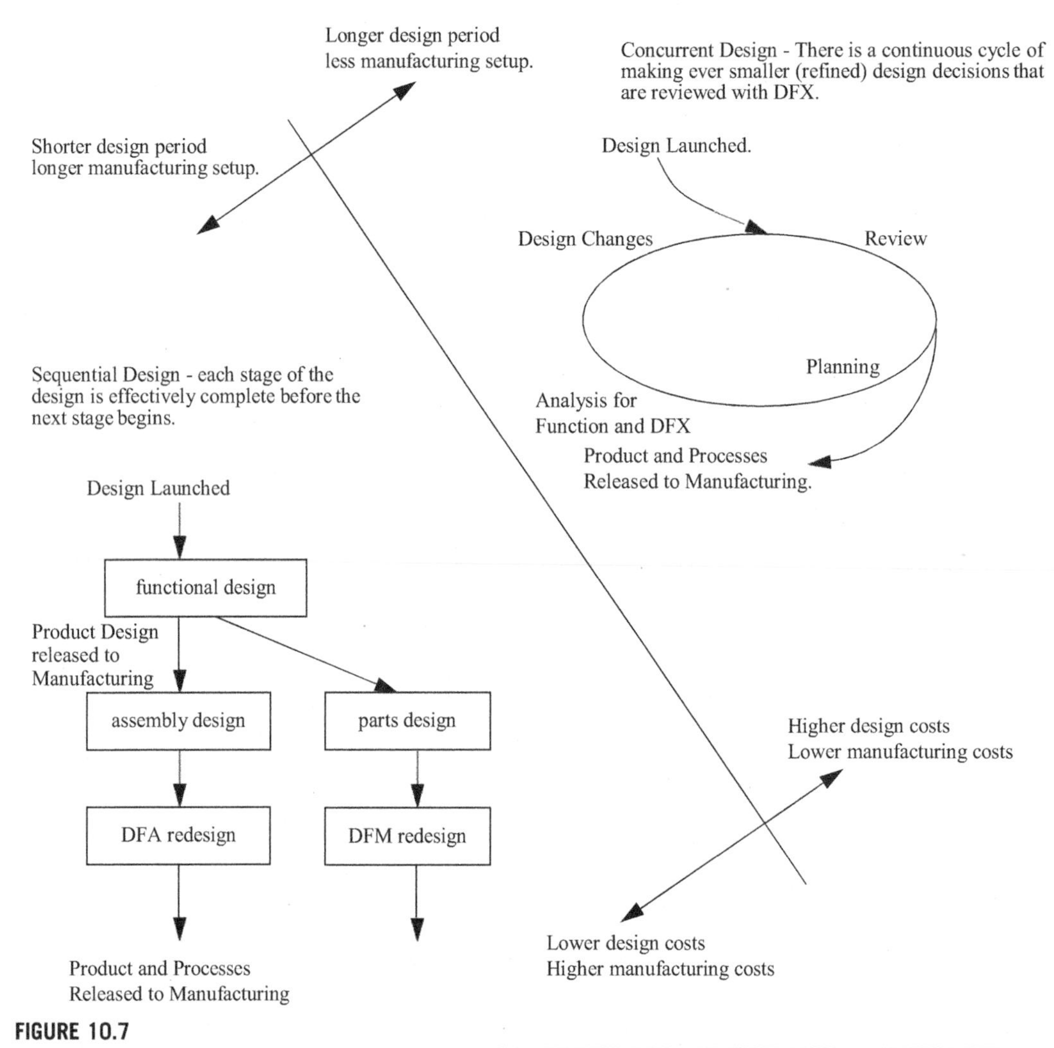

FIGURE 10.7

Application of design for X (*DFX*). *DFA*, design for assembly; *DFM*, design for manufacturing.

10.30 What problems do DFX methods solve?

10.31 What is the difference between sequential and concurrent design?

10.3 Quality

Precision is a lofty goal. Quality is a realistic objective.

Tighter tolerances require more precision, but past a certain point the consumer will not notice or value the added effort. Fig. 10.8 shows how an increase in precision has value for the consumer, but past a

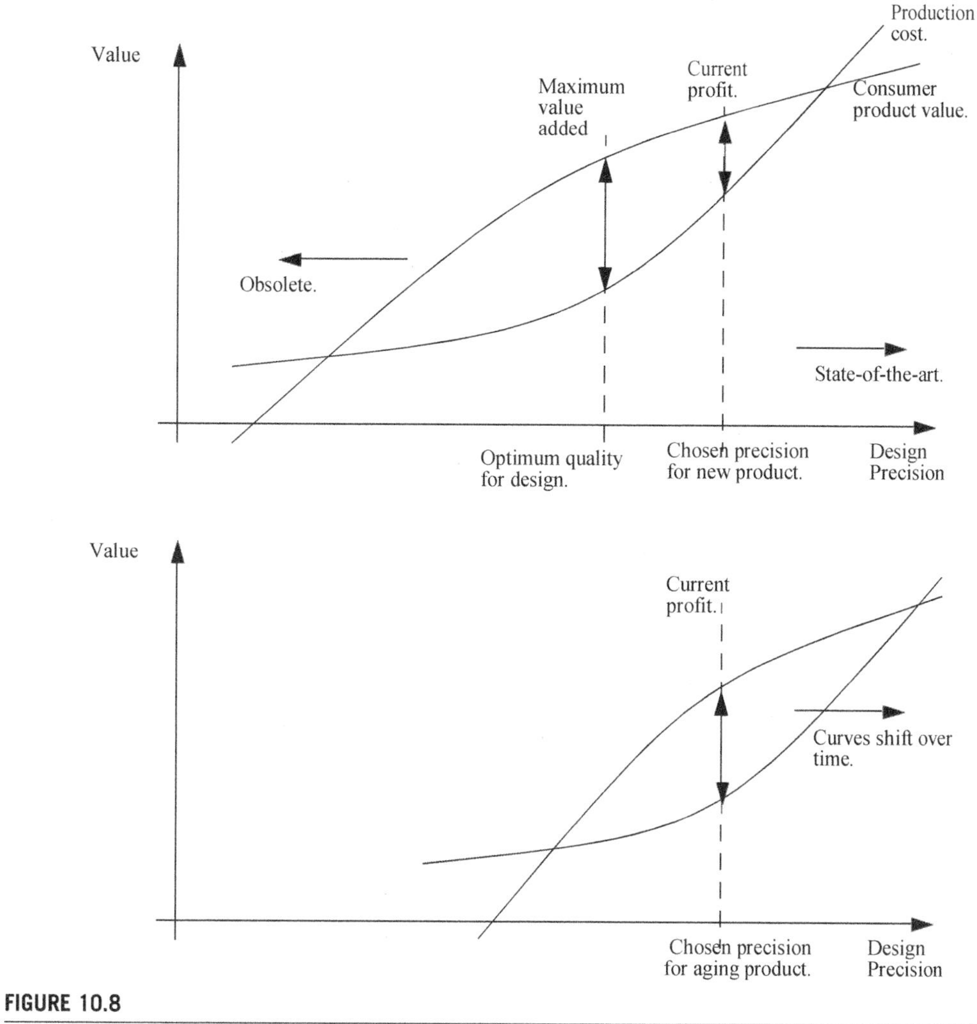

FIGURE 10.8

Selecting design precision.

certain point the increased precision does not result in an increased customer value. Consider the space between pixels on a laptop screen. If the pixels vary by 0.1 mm, the consumer will notice the variations. If the pixels vary by less than 0.01 mm, the user may never notice. An accuracy of 0.001 mm would offer no tangible benefit to the consumer. As the precision of a feature increases, the manufacturing costs will rise quickly. The state-of-the-art defines where the cost curve begins to accelerate exponentially. The widest space between the curves indicates maximum profitability. As technology develops, these curves shift to the right, while a specific product remains fixed at one precision value. Over time the profitability changes until consumer value is so low the product becomes unprofitable.

The key lesson of Fig. 10.8 is to select enough precision to maximize the profit area as the curves shift over time. The quality-control issues determine how close the actual precision matches the specified precision.

$$\text{Quality} = \frac{\text{Specified}}{|\text{Specified} - \text{Actual}| + \text{Specified}} 100\% \quad (10.1)$$

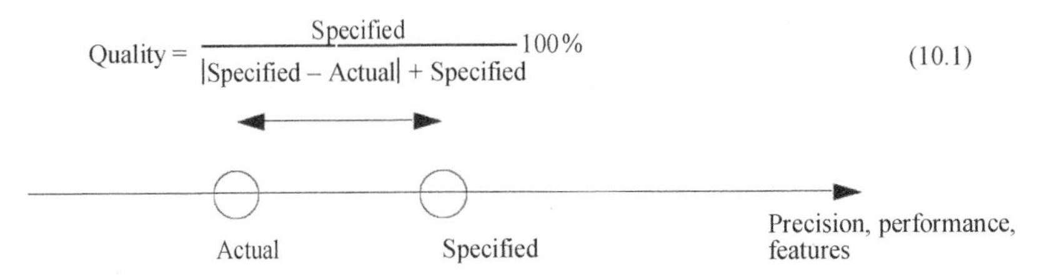

Say the specification called for a count of 200. The actual count was 180. Therefore the quality is:

$$\text{Quality} = \frac{\text{Specified}}{|\text{Specified} - \text{Actual}| + \text{Specified}} 100\% = \frac{200}{|200-180| + 200} 100\% = 90.9\%$$

Say the specification called for a count of 200. The actual count was 220. Therefore the quality is:

$$\text{Quality} = \frac{\text{Specified}}{|\text{Specified} - \text{Actual}| + \text{Specified}} 100\% = \frac{200}{|200-220| + 200} 100\% = 90.9\%$$

FIGURE 10.9

The relationship between quality and specifications.

Quality is a measure of how well a design meets but not necessarily exceeds the customer specifications.

Something that exceeds the specifications can be as much of a failure as a design that does not meet the specifications. A simple metric of quality is an inverse of the distance between the specified and the actual measures, as shown in Fig. 10.9. In the figure there is a specification of 200 candies in a bag. The machine fills the bags by opening a chute for 0.6 s. If the candies flow freely, the bag fills faster and has more than 200 candies. If the candies are slightly jammed, the final bag content is less than 200. The average candy count is used for the actual. The quality calculation (Eq. 10.1 in Fig. 10.9) puts a quantifiable metric to the deviation of the actual from the specification. In both cases the actual is 20 counts away from the target, so the quality is 91%.

The specifications define a performance target for the design and manufacturing of the feature. There are a number of factors that have an impact on the actual performance. Some of these are designed differences between the actual and the specified performance. For example, the designers know that the target is 200 but anything less is unacceptable, and therefore choose a design value of 202 (Fig. 10.10). Random variations expand the value of 202 to a range from 200 to 204. These variations arise from differences in candies, manufacturing processes, bags, and other difficult-to-predict factors. In this example the segment of the normal curve that is below 200 is filled in black. If the standard deviation is 0.7, the number of bags below 200 should be less than 3 in every 1000. The location and spread of random distributions are chosen by the engineers.

In the candy example, a value below 200 constitutes a failure. If parts are checked for quality control there are a number of common alternatives for the rejected parts. One extreme is when there is a single rejected part the entire batch is discarded. This option is very costly. The loosest approach is to recognize the issues but ship the product anyway. In this example of candy bags, underfilled bags could lead to fines or lawsuits, but do not cause any serious risk. Automobile companies use repair shops for

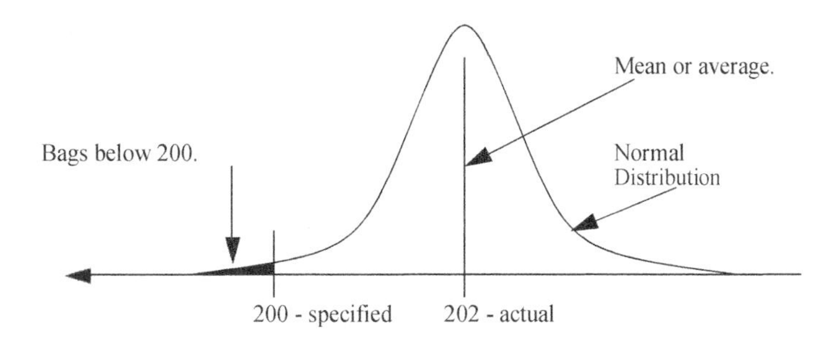

FIGURE 10.10

Statistical variation of actual feature values compared with specifications.

cars that do not pass inspections. Chicken-egg producers have little control over the chickens laying the eggs, but they sort the raw products to meet specifications. Eggs are sorted by sizes that command different prices. Some options for quality failures are as follows:

- Rework or repair the device to meet specifications.
- Discard the device or devices.
- Accept the parts and fix the problems later.
- Downgrade (lower returns); e.g., try to get the largest chicken eggs, but when smaller eggs are produced, sell for less.

Quality-control systems focus on achieving the precision of the specifications. Success in achieving or failing quality goals comes at every stage of the engineering process and beyond. A few of the common foci are given in the following list. Designers will set precision and quality goals that the rest of the company must be able to meet. A design that includes purchased components that are state of the art will make purchasing and management more difficult. On the other hand, a design that is approved by people across the company is much more likely to reach quality goals:

Design

- Specifications—Clarify what is important in a product; the rest is second priority.
- Design—Consider specifications, standards, and tolerances; keep it simple; evaluate production capabilities, safety, models, life testing, and engineering changes.
- Design—Products must match specifications; any more, or less, is a waste of time and resources. For example, if a chemical company "tunes" its production for a certain impurity in a raw material, a sudden improvement might hurt its product quality.
- Review of specifications—Make sure the specifications describe the product well and the specs are useful to customers.

Manufacturing

- Process inspection—Ensure conformance to the specifications and problem correction.
- Manufacturing engineering—Consider processes, equipment, standards, and layout.
- Manufacturing supervision—Ensure good employee attitude and training.
- Suppliers—Inspect incoming materials, parts, supplies, and equipment.
- Packaging and shipping—Consider packing materials, documents, delivery, and the environment.

Other business functions

- Marketing—Be aware of customer standards, current market, competition, liability, government standards, independent lab standards, customer surveys, and dealer and store surveys.
- Purchasing—Select materials and components, evaluate suppliers (rating, distances, etc.) and single/multiple suppliers, and follow up on rejected goods and schedules.
- Product service—Ensure adequate install, repair, and part supply.
- Support—Ensure support from the top (CEO and board) and adequate funding, staffing, training, and evaluation.

Quality control programs normally focus on manufacturing. At every substep of manufacturing, value is added by labor, parts, equipment, and the cost of business. When a design does not meet a specification, it may be scrapped or corrected. Either way, the design has accumulated value. The worst case is to add a critical defect to a design early in manufacturing and continue to add costs to the part, only to have it fail at the end of production. In manufacturing, you want to find and correct mistakes when they occur. This principle is universal, including rejected semiconductors, warped beams, and allowable contaminants in a biomass. The general manufacturing approaches to quality control include:

- Gating: Examine final product only for pass/fail. It is not very effective for correcting problems, and rejected parts cost more.
- Design of experiments (DOE): Various process parameters can be varied (e.g., speeds, feeds) and the effects can be examined to determine the best settings for a process. This process is more proactive.
- Statistical process control (SPC): During production, parts are measured and variations are monitored. Machines are adjusted to prevent poor part production. This process is more reactive.
- Part acceptance: These processes accept or reject parts after production. Typical techniques are go/no-go gauges, visual inspection, and so on. Rejected parts can be used to correct processes, much like SPC methods.

As previously mentioned, quality control is a company-wide concern; Fig. 10.11 shows a few tools used by engineers. Early in the design process design for quality is used to select design attributes that are easier to inspect and control. DOE identifies better production parameters. During production, SPC

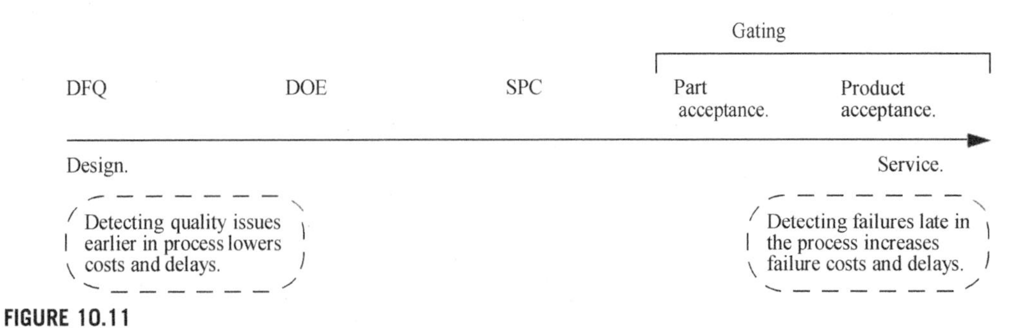

FIGURE 10.11

Engineering-focused quality control. *DFQ*, design for quality; *DOE*, design of experiments; *SPC*, statistical process control.

examines processes for consistency. Part and product acceptance is used to ensure that quality goals have been met, as opposed to SPC, which looks for unexpected process variation.

PROBLEMS

10.32 Is higher precision better? Explain.

10.33 What is the difference between part acceptance and quality control?

10.34 How would a gating process be used to sort eggs by size?

10.35 Generally, why should production try to meet specifications and not exceed them?

10.36 The graph shows two curves that relate the cost of a product to the expected value.

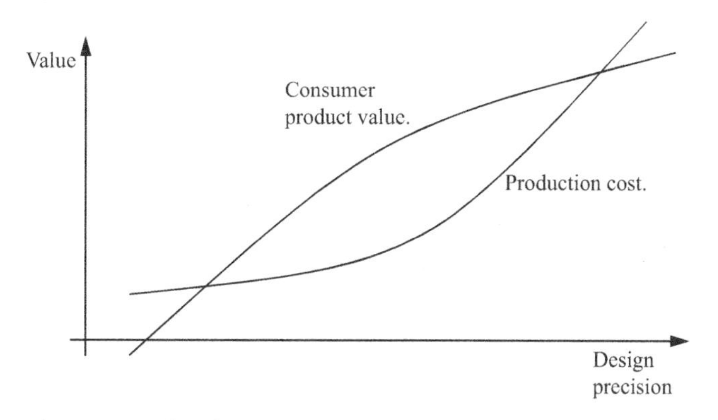

(a) What are the sources for the two curves on this graph?

(b) How can this graph be used when setting engineering specifications?

10.37 List some material and process variables that can affect quality.

10.4 Identification of problem causes and control variables

Outcomes and behaviors are the result of complex interactions. Sometimes a single cause dominates an outcome and is easy to identify and adjust. For example, the mass of a bucket is governed by the mass of the contents. However, there are always other factors that influence the outcome. For example, the bucket may have small holes that retain solid materials but allow liquids to drip out, so liquid mass drops over time. The holes may be halfway up so the bucket loses only the top half of the liquid. Or, the bucket has a removable handle that will change the mass. When considering causes there are two fundamental types:

- Chance: normal, or natural, variations that occur in a system (e.g., dice)
- Assignable: controllable parameters such as material, process parameters, operator skill, etc.

Knowing the causes of problems also reveals ways to control the system. The control variables are those items you can change to change the outcome. Once these are known, other techniques can be used to quantify the effects, including theoretical modeling, DOE, trial and error, Pareto diagrams, statistical analysis, and experience.

A basic approach to identifying causes is to simply list everything that comes to mind. A group can take this process farther. This approach will identify a majority of the issues. There are other techniques that can be used to guide the process and organize the results, including cause-and-effect (CE) diagrams.

PROBLEMS

10.38 List five examples of random processes.

10.39 Is it possible to explain truly random variations?

10.4.1 Cause and effect diagrams

In 1943, Kaoru Ishikawa developed the CE diagram, an example of which is shown in Fig. 10.12. In the figure, we are looking for the factors that affect driving time. The major causes selected include traffic, vehicle, distractions, and speed limit. The major causes are not always clear and it is up to the individual or group creators to select these. In addition, equally valid CE diagrams may look entirely different, but the important factor is to identify causes. For each of the major factors the contributing factors are added with side arrows, and side arrows can be added to these. For example, under the main heading of "vehicle" the top speed is indicated as a factor. The top speed is influenced by motor power and vehicle aerodynamics. In this example, three items are identified as significant

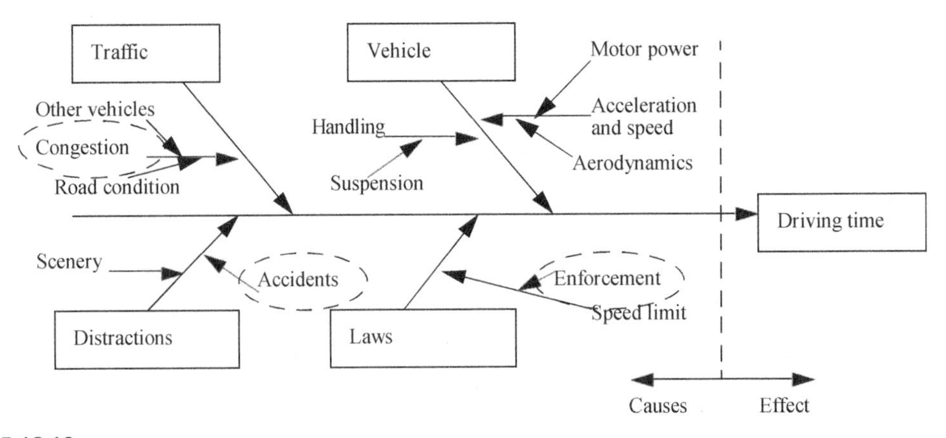

FIGURE 10.12

An example of a cause-and-effect, Ishikawa, or fishbone diagram. Note: When constructing the diagram, consider factors that both reduce and increase the effect. Any cause can be subdivided into finer factors. Not all of these will be significant but they should be noted.

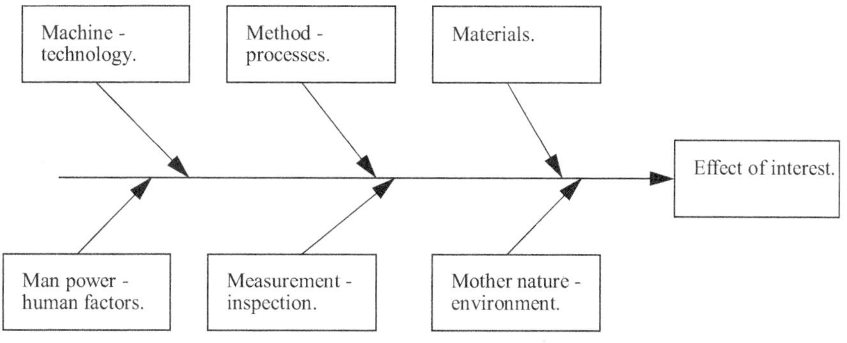

FIGURE 10.13

Cause and effect diagram example for manufacturing.

factors and a design team would focus on these first. Looking ahead, these could be three factors for a DOE test.

The factors identified in the CE diagram may not be significant, but the diagram allows them to be recognized and then ruled out systematically. In other words, it is better to identify all possible sources and then rule some to be negligible. A simple mistake would be to leave them off because they are probably not important. Once the diagram is complete, the branches can be traced back to identify significant causes. The significant causes are used to find the control variables.

A CE diagram for manufacturing is shown in Fig. 10.13 with six factors commonly used for manufacturing analysis. Like the example of travel time, alternative major causes can be developed as needed.

A CE diagram should be generated for the situations listed below. If it is created and shared with the technical team it will provide a common frame of reference for identifying and solving problems. Ideally it will be posted in a public location and updated as needed:

- Reaction to a problem has occurred, and the source(s) must be identified.
- During design, a CE diagram is developed to relate causal factors to specification performance.
- A CE diagram is developed during planning for a robust design by identifying usage factors.
- A CE diagram is constructed during planning for manufacturability.

PROBLEMS

10.40 Develop a CE diagram for the process of making cookies. Use the manufacturing CE example.

10.41 How many of the CE factors will be important?

10.4.2 Pareto analysis

The Pareto principle states that 80% of the problems are the result of 20% of the causes. To this end, a relatively simple chart is used to highlight problems. Fig. 10.14 is an example of an application. The

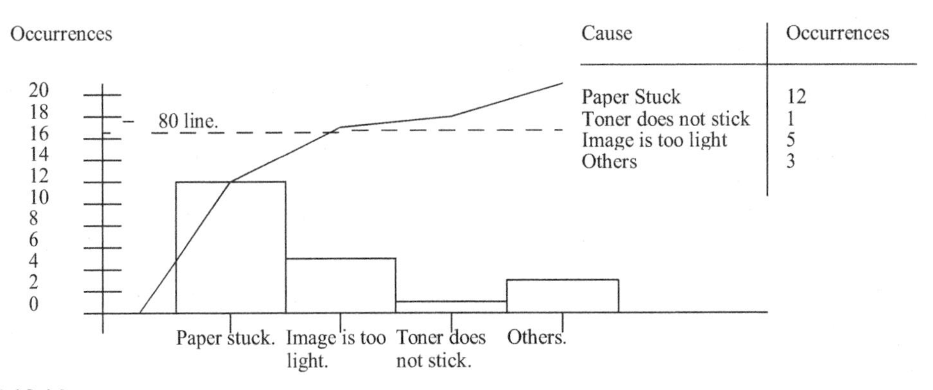

FIGURE 10.14

A Pareto diagram for poor photocopies.

Pareto chart is normally preceded by a CE diagram. In this example, the engineers identified the three issues that will have the greatest benefit. The process begins by watching for unwanted effects and then tracing the problem back to a cause. These values are listed in a table and then plotted in a bar chart in order of size, with the largest on the left. A cumulative occurrence count is also added to illustrate the priorities. In this case, "paper stuck" and "image too light" are the two problems within the 80% line and deserve the engineering attention. Although simple, this technique can quickly focus attention. The problem it solves is when a group gets too focused on less important problems.

PROBLEMS

10.42 (a) Draw a fishbone diagram for the production of cookie dough. The quality to be measured is the ratio of chocolate chips to dough per cubic meter. Note: The components are weighed separately and then mixed together in a large tub. (b) Select the most reasonable causes from (a), make up a tally sheet, fill it with some data, and draw a Pareto chart. You must consider that there are three different operators that may do the weighing and measuring.

10.43 Consider the CE diagram for painting a house in (a). Data were collected and used to construct the Pareto chart in (b). The diagram reveals that "missed spot" is the most common problem. What factors in the CE diagram could result in the "missed spot" problem?

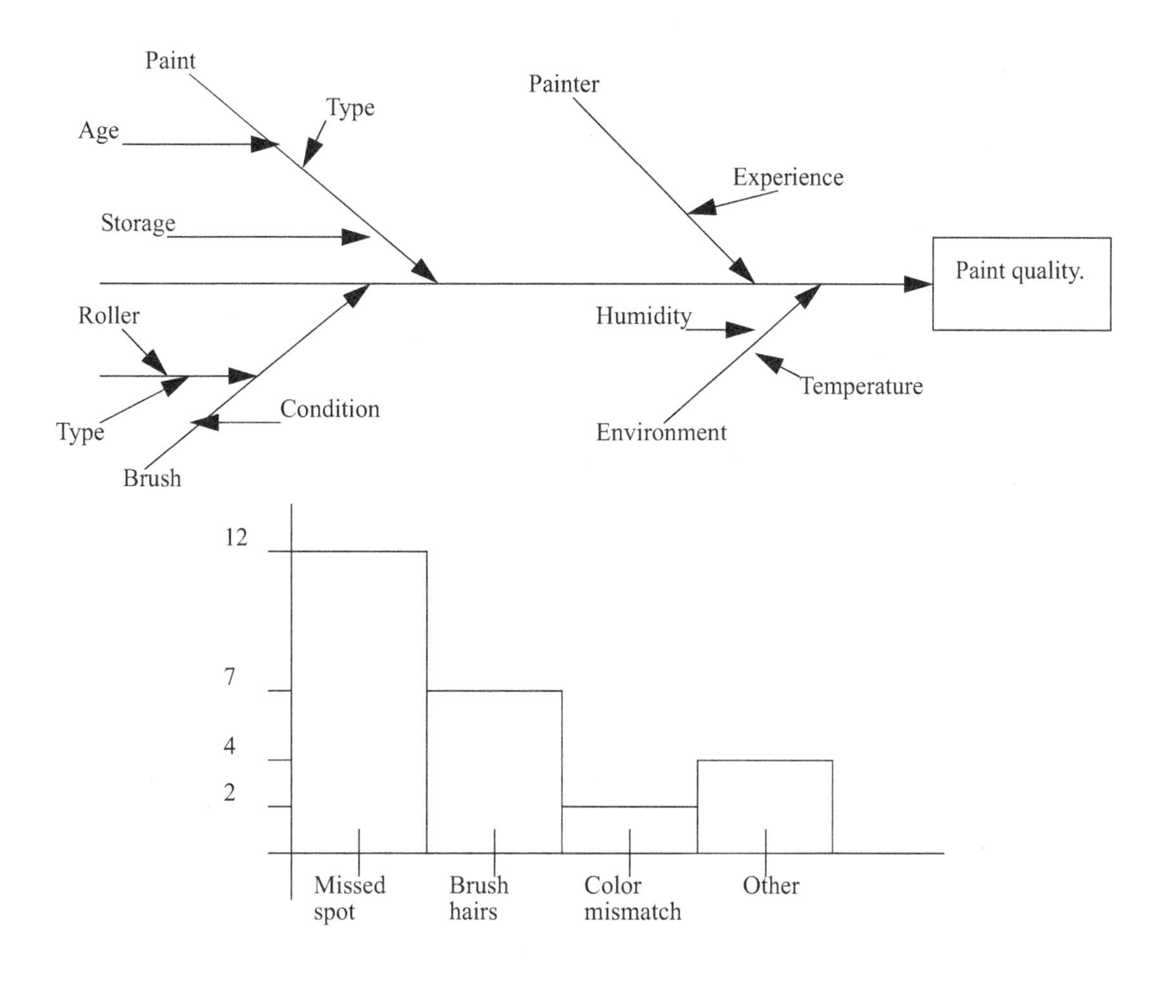

10.44 Draw the Pareto diagram for the data in the table. The data indicate the number of reported errors made when taking fast-food orders by telephone.

Day	Operator		
	Tom	Dick	Harry
Monday	12	8	3
Tuesday	9	7	7
Wednesday	7	9	9
Thursday	8	4	2
Friday	21	9	24
Saturday	28	12	9

10.4.3 Experimentation

Experimental methods are used for expected cause and effect relationships. At a minimum these may prove that causal factors do, or do not, have an effect. Engineers typically also look to quantify relationships using graphs, tables, and equations. Ideally we use the scientific method: (1) develop a hypothesis, (2) develop and conduct an experiment, and (3) analyze the results for proof of the hypothesis. Some examples of hypotheses are shown in the following list, with some of the enhancements typical of engineering work. The hypothesis is critical because it drives the process of impartially designing the experiment, conducting the tests, and analyzing the results:

- The device will operate for 1000 h at 120% of the rated load.
- Increasing the pressure will increase the flow rate.
- A thickness of 2 mm will provide a heat dissipation of 10 W.
- Regenerative braking is possible using rare-earth magnets.
- The temperature will increase with calcium content.
- Cracking in the springs is a factor of carbon content, quenching temperature, and heating temperature.

It is natural to have a bias in engineering work. The bias normally comes from personal ownership of ideas, a desire to reduce complexity, an attempt to reduce work, and personal preference. To overcome this we have been educated to (1) develop a realistic test for the hypothesis, as opposed to the preferred solution, (2) ensure consistent conditions for each test sample, (3) use controlled samples or tests, (4) isolate experimental variation, and (5) collect data or results that can be measured impartially.

The last important experimental step is to analyze the data relative to the hypothesis, as shown in the following list. Simpler approaches will look for some difference calculated with percentages. More mature analysis will consider statistical confidence and use methods such as Student's t test. When the results of an experiment are clear, there will be a clear positive or negative relationship between the causal input variables and the output effect variable. When there is no major change, the results may be ruled inconclusive. Normally, inconclusive results are not used to eliminate a hypothesis. For example, a hypothesis that adding eggs would change the bread texture is actually true, but if the experiment called for only 5 mL of egg in a 2-kg mass the results would be negligible. When the results are inconclusive you must decide to stop or redesign the experiment:

- Positive or negative relationship
- Quantitative or qualitative relationship (possibly graphs, equations, tables, observations)
- Interaction between inputs
- No significant relationship

Simple experiments have a single cause and effect. When there is only one input and one output variable the experiment is simpler to develop. As more inputs are added, the process requires more planning. For example, a set of two values that each range from 0 to 100, in increments of 1, would require 10,201 tests to evaluate all of the possible combinations. A wise experimenter would use larger intervals to reduce the test time but still obtain useful results. For example, using values of 0, 50, and 100 would reduce the number of trials to 9. If an experiment has multiple effects or output variables the analysis is repeated for the effects of the inputs on each single output.

A sample engineering experiment is shown in Fig. 10.15. The purpose of the experiment is to increase the yield, or decrease the percentage of defective parts. The experiment starts with an

Effect: The quenching process is resulting in cracking in steel springs. Sometimes the yield is as low as 60 (i.e. 40 failures out of 100 parts.)

Causes: A CE diagram identified carbon and quenching temperatures as important.

Hypothesis: The quantity of cracked springs is a function of the carbon content of the steel and the pre-quenching temperatures of the steel and oil.

Procedure overview: Run multiple trials and vary the i) steel carbon content (C) from 0.5-0.7, ii) pre-quench steel temperature (S) from 1450–1600°F, and iii) oil temperature (Q) from 70-50°F. Each condition will tested with four batches of 100 springs. The uncracked springs will be counted.

Trial 1a: C = 0.5, S = 1450°F, Q = 70°F (The base-line for comparison.)
Acceptable Parts: i) 72/100, ii) 70/100, iii) 75/100, iv) 77/100
Average yield: 73.5

Trial 1b: C = 0.5, S = 1600°F, Q = 70°F (Elevated steel temperature.)
Acceptable Parts: i) 78/100, ii) 77/100, iii) 78/100, iv) 81/100
Average yield: 78.5

Analysis 1: A higher value of pre-quench steel S = 1600F produces a higher yield rate so it will be kept higher for the following trials. Therefore the following trials will use S = 1600F.

Trial 2a: C = 0.7, S = 1600°F, Q = 70°F (Higher steel carbon content.)
Acceptable Parts: i) 77/100, ii) 78/100, iii) 75/100, iv) 80/100
Average yield: 77.5

Analysis 2: A higher carbon content in the steel C = 0.7 was a small but negative change. Therefore following trials will use C = 0.5

Trial 3: C = 0.5, S = 1600F, Q = 50°F (Cooler quench bath temperature.)
Acceptable Parts: i) 79/100, ii) 78/100, iii) 78/100, iv) 83/100
Average yield: 79.5

Analysis 3: The was a small positive increase in yield with a cooler quench bath temperature.

Conclusion: Increasing the pre-quench steel temperature increases the yield about 5. A colder quench bath increases the yield by 1. A lower carbon content increases the yield by 1. A yield of 79.5 was obtained with C = 0.5, S = 1600°F, Q = 50°F.

FIGURE 10.15

A one-factor-at-a-time experiment example. *CE diagram,* cause and effect diagram.

identification of the variables and a clear idea of what needs to be tested: the hypothesis. Although brief, the procedure lays out the critical variables. Reducing the number of experiments is important because this type of experiment probably takes hours per batch and thousands of dollars in materials and labor. To do this the experiment starts with three values for parameters C, Q, and S. Four separate batches are run and the yield values are averaged. Given the variation between batches, the choice to run four batches is reasonable. Trial 1a serves as a basis for comparison for the next trials. In trial 1b the temperature, S, is changed and the positive effect noted. At this point the experimenters decides to hold S at the new higher values for the experiments. The following two trials change one of the variables and note the effect. The last two trials were less significant and it would be possible to argue they are negligible. The conclusion refers back to the hypothesis. At this point the engineers would need to consider the following questions:

- Is the increase in yield enough to stop experimenting?
- Could other values produce higher yields? Possibly S = 1550°F–1650°F?
- Would C and Q have more impact if S stayed at 1450°F?
- Are there other variables that could be adjusted?
- Is an 80% yield a reasonable maximum expected value, or is more possible?
- Were any batch yield values artificially higher or lower because of errors or other factors?

PROBLEMS

10.45 Why would a design engineer conduct an experiment?

10.46 What is the scientific method?

10.4.4 Design of experiments

DOE is a technique for reducing the number of tests required to determine the effects and interactions of variables. In the experiments, there is some sort of measurable outcome such as the power of an amplifier, chemical purity, or surface finish. The variables cause some effect on the outcome variable. When there is no clear relationship between the input variable and the output variable there is a need for testing to find an optimal solution. Consider a recipe for cookies to be developed by food and manufacturing engineers. They begin with the recipe and a CE diagram for taste. Some early testing and Pareto analysis suggest that the major factors are moisture content, baking time, and sugar content. To determine the better production parameters they vary the three factors and then use taste testers to produce numerical scores. They decide to use three values for each variable, and determine that they would need to do 27 separate tests to find the best combination of all three. However, using DOE would allow similar results with fewer tests.

The procedure for applying DOE to quantify cause and effect problems is shown in Fig. 10.16. Naturally, the process begins with some recognition of causes and effects that can be quantified or measured. Expected values are used to design the experiment and collect data. The results will indicate which variables have a positive or negative impact on the output. The results will also show which variables have some form of interaction. If the results verify or provide useful input values, the experiment can end or be refined. It is also possible that the results do not show the desired values. Sometimes this occurs because there are causes that have not been considered. Other times factors are assumed to have an influence that is actually negligible.

The DOE technique begins with a user-specified set of factors, limits, and responses. The factors are the input variables that will be changed. The levels are the input variable values to be used for testing, including the maximum, minimum, and optional intermediate points. The response must be the effect that is to be measured. Without the DOE method it would be necessary to try all of the value combinations or arbitrarily select some tests to skip. The DOE method uses a carefully selected set of

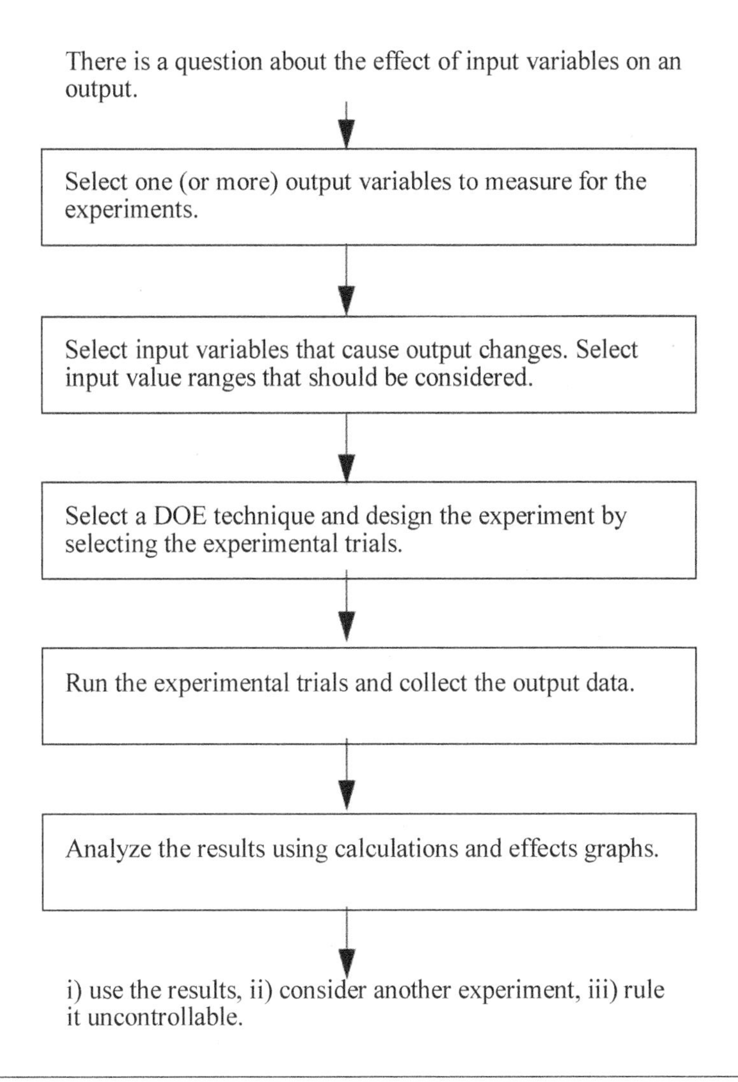
There is a question about the effect of input variables on an output.
Select one (or more) output variables to measure for the experiments.
Select input variables that cause output changes. Select input value ranges that should be considered.
Select a DOE technique and design the experiment by selecting the experimental trials.
Run the experimental trials and collect the output data.
Analyze the results using calculations and effects graphs.
i) use the results, ii) consider another experiment, iii) rule it uncontrollable.

FIGURE 10.16

The design of experiments (*DOE*) process.

factor changes to reduce the number of tests, and a set of calculations isolates the effect of single variable changes. In effect, the method cuts down the number of tests while providing reasonable single-variable and interaction results. The DOE-specific steps are as follows:

(1) Identify process variables (inputs) and dependent variables (outputs). Outputs should be continuous values.

(2) Select discrete values for the inputs. The most basic approach is to pick a high and low value for each.

(3) Create a data collection table that has parameters listed (high/low) in a binary sequence. Some of these tests can be left off (fractional factorial experiment) if some relationships are known to be insignificant or irrelevant.

Trial Run	a. (S)	b. (C)	c. (Q)	Yields from four batches	R = Yield (avg)
1 (___)	1450	0.5	50	73, 76, 75, 76	75.0
2 (a)	1600	0.5	50	79, 78, 78, 83	79.5
3 (b)	1450	0.7	50	72, 70, 71, 69	70.5
4 (ab)	1600	0.7	50	81, 80, 79, 80	80.0
5 (c)	1450	0.5	70	72, 70, 75, 77	73.5
6 (a c)	1600	0.5	70	78, 77, 78, 81	78.5
7 (bc)	1450	0.7	70	75, 73, 75, 73	74.0
8 (abc)	1600	0.7	70	77, 78, 75, 80	77.5

Note the binary sequence with values at the max (high) and min (low)

FIGURE 10.17

Design of experiments example for spring yield.

(4) Run the process using the inputs in the tables. Take one or more readings of the output variable(s). If necessary, average the output values for each of the experiments.
(5) Graph the responses, varying only one of the process parameters. This will result in curves that agree or disagree. If the curves agree, the conclusion can be made that process variables are dependent. In this case, the relationship between these variables requires further study.
(6) Calculate the effects of the process variable change.
(7) Use the results of the experiment to set process parameters, redesign the process, or design further experiments.

The example in Fig. 10.15 is repeated in Fig. 10.17. The terminology for DOE experiments is *n*-factorial, and in this case there are three factors, making this a 3-factorial experiment. Please recall that the example was to increase the yield rate for carbon steel springs that have been stamped. The process has three variables that are varied between high and low, thus giving eight possible combinations. Values have been added for the four additional trials. As before, the yield for each trial run is measured using four batches to calculate an average yield.

After the test data have been collected, the effects graphs can be drawn (Fig. 10.18). These graphs provide a visual presentation of effects. Each line on the graph represents two trials, with and without a factor. In this example, the four lines on the left are for tests with the change in factor a, or S, the temperature of the steel before quenching. The middle lines are for the change in b, or C, the carbon content. The right side is for the change in factor c, or Q, the quench temperature of the oil. In this example it is clear that factor a has a consistently strong effect on the result, or yield rate. Factors b and c are not as clear and some of the trends are opposite. In general terms, the greater the slope, the greater the effect. Lines straight across mean no effect. Lines that cross indicate that the relationship is not as simple and clear. Again, the consistently positive slope for factor a indicates that it is the dominant controlling factor.

The effects can also be expressed numerically, as shown in Fig. 10.19. These equations are specific to the 3-factorial experiment, but other experiments have a similar form. Essentially, these equations are the averages of effects with and without each factor. As seen in the graphs, the effect of factor a is very large. The effects of factors b and c are much smaller. Therefore the conclusion for the experiment would be that the temperature of the steel before quenching is a major factor and the results for carbon content and quenching oil temperature are less conclusive.

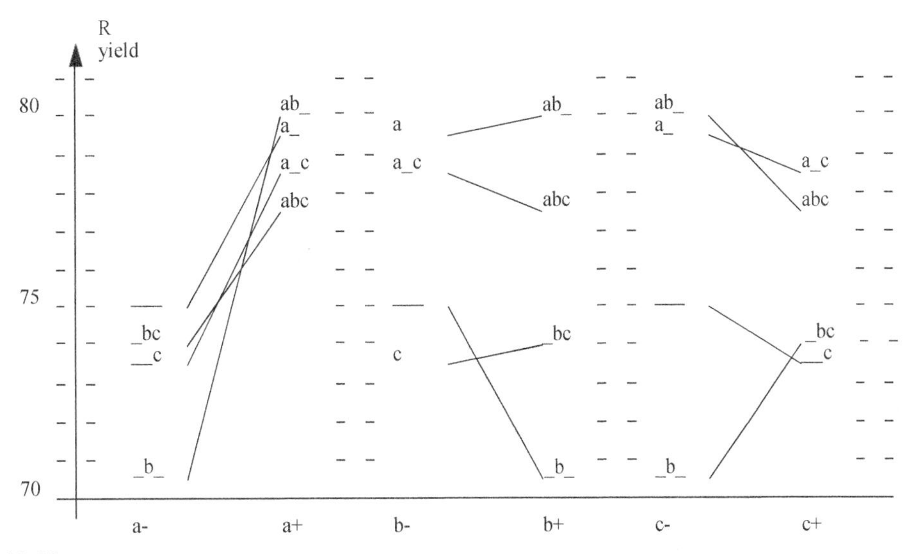

FIGURE 10.18

Design of experiments effects graphs.

Main Effect = Average at High − Average at Low

R = Yield (avg)

R1 = 75.0
R2 = 79.5
R3 = 70.5
R4 = 80.0
R5 = 73.5
R6 = 78.5
R7 = 74.0
R8 = 77.5

$$\text{Main Effect of a} = \frac{R_2 + R_4 + R_6 + R_8}{4} - \frac{R_1 + R_3 + R_5 + R_7}{4}$$

$$= \frac{79.5 + 80.0 + 78.5 + 77.5}{4} - \frac{75.0 + 70.5 + 73.5 + 74.0}{4} = 5.625$$

$$\text{Main Effect of b} = \frac{R_1 + R_2 + R_5 + R_6}{4} - \frac{R_3 + R_4 + R_7 + R_8}{4}$$

$$= \frac{75.0 + 79.5 + 73.5 + 78.5}{4} - \frac{70.5 + 80.0 + 74.0 + 77.5}{4} = 1.125$$

$$\text{Main Effect of c} = \frac{R_1 + R_2 + R_3 + R_4}{4} - \frac{R_5 + R_6 + R_7 + R_8}{4}$$

$$= \frac{75.0 + 79.5 + 70.5 + 80.0}{4} - \frac{73.5 + 78.5 + 74.0 + 77.5}{4} = 0.375$$

FIGURE 10.19

Design of experiments effect calculations.

PROBLEMS

10.47 How can DOE help an engineer improve a process?

10.48 You have collected the data shown in the table as part of a 2-factorial experiment for making slushies. There are two process variables you control, a quantity of sugar and a quantity of salt that are added to the water. These modify the freezing temperature of the slush. Draw effects graphs and calculate the effects of changing the parameters. State whether they are dependent or independent.

Sugar (g)	Salt (g)	Freezing temperature (°C)
40	3	2
40	5	−4
60	3	6
60	5	7

10.5 Statistical process control

A good manufacturing process is predictable.

Consistency is the key to quality in manufacturing. Each process or machine will have a target setting and some amount of random variation. In addition, the machine settings will drift. For example, an oven is a standard piece of industrial equipment for applications including heat-treating steels, soldering circuit boards, baking cookies, and separating oils. In these applications, there are target temperatures for the process, but also a tolerance band for operation. The oven temperature control can be quantified statistically using an average and standard deviation. In an ideal environment the oven temperature can be kept within a few degrees. The standard deviation describes the random variations caused by gusts of cold wind, temperature of incoming parts, variations in gas flow, and so on. The average, or mean, is the target temperature, but over time this will drift if the burners become dirty, the temperature sensor degrades, or the oven develops air gaps. In manufacturing, we want to track these changes so that we can control the processes and get predictable and consistent production outputs.

SPC is a method for monitoring the statistical variations in a process. The word "control" suggests that this method makes adjustments to a process, but this is not the case. SPC only indicates if the process is statistically predictable and therefore "in control." The normal distribution is at the heart of the SPC method. As parts are manufactured, samples are taken and measured. These measurements are then added and compared with historical values. If the statistical processes don't match, the process is called "out of control" and then stopped.

Fig. 10.20 shows a set of measurements from a production process. Four samples were measured at five different times. When the first samples were taken at 12:15, the average was 11.5. The sampling process is a combination of when samples are taken and how many measurements are taken for each sample. In this case, four measurements are taken for each sample. The time between samples may be longer for slower or more stable processes. For fast-moving or fast-changing processes the samples may be more frequent; 100% inspection systems are becoming more common in automated quality-control systems where every part is measured. When deciding the time and the parts to sample it is critical to be irregular. For example, always selecting the parts from the top of a pile may mean that you miss defects in parts that are put into the pile first. Another example is always sampling parts at 2:00 p.m., so the operators put aside some good parts for you to measure.

To consider these samples from an engineering perspective, the candy packages varied in count from 9 to 15. The average varied from 11.5 to 13.5. The numbers do seem to change randomly so

Machine: Candy Packaging Machine
Date: June 19, 2019
Operator: I. B. Fule

Time	Samples (Xi)	avg (X)
12:15	12, 15, 10, 9	11.5
2:45	13, 14, 13, 14	13.5
3:15	14, 13, 10, 11	12
5:30	11, 12, 9, 10	10.5
6:00	12, 15, 13, 12	13

FIGURE 10.20

Sample process data.

it does not seem to have a pattern. Is this good or bad? Without knowing the specification or history of the process, we cannot say. For example, the bags may contain candy with alphabet letters. The count is for the number of candies with the letter shape "S." If the specification says that each bag should have 5 to 15, then we are in compliance. However, two bags were measured to have 15, meaning it is probable that we will exceed or have exceeded the upper limit.

SPC charts are used as a visual interpretation of the statistics that is easy to update and interpret. The data from the previous sample are plotted in Fig. 10.21. The Xbar, upper control limit (UCL), and lower control limit (LCL) are based on the historic data set. These are updated slowly over time, but the current data points are plotted on the graph and compared with these immediately. The Xbar value is the average of all samples ever taken on the machine. The UCL and LCL represent ± 3 standard deviations for all readings ever taken. The simplest rule is that if any of the sample points fall outside the control limits, the process is out of control and it is stopped. The random probability of this occurring is 3 in 1000, making it very unlikely that the process is behaving randomly. In the example chart, the reading of 13.5 is at the limit, and the reading of 10.5 is outside the limits. The process is not

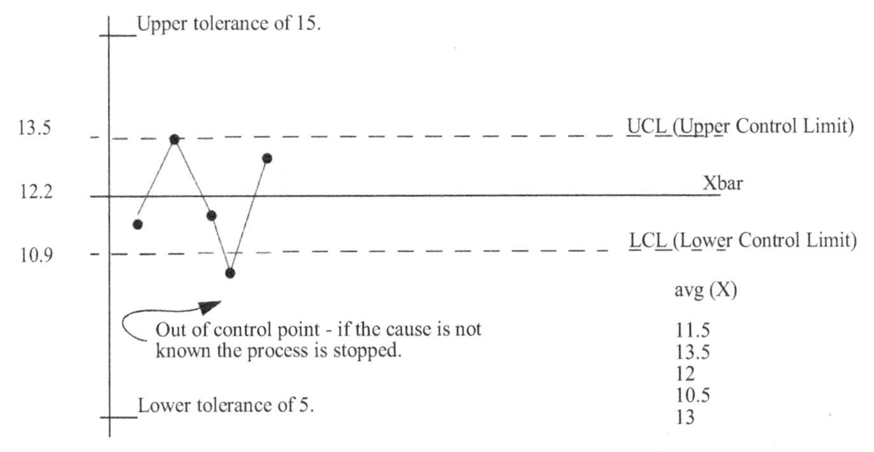

FIGURE 10.21

A sample control chart.

behaving normally and must be stopped until the problem is identified and corrected if necessary. It is worth noting that the UCL and LCL should be inside the tolerances to ensure that specifications are met, resulting in a higher quality. In this example, the value of 10.5 is within the tolerance range, but outside the normal operation for the machine. However, an inconsistent machine can produce parts outside of the specifications.

There will be a separate SPC chart for every job on every machine. At the beginning of every job there is a setup process in which parts are made for testing, not for use. As the process begins to produce consistent parts, sample measurements are taken and used to set the Xbar and UCL/LCL for the SPC chart. As the job continues, more samples are taken and the chart is adjusted. Large events may require that the old SPC data be put aside and a new chart started. Some of these events include new tooling, design changes, material changes, equipment changes, recalibration, and so on.

Sampling should be done to maximize randomness and reduce patterns of time, operators, shifts, breaks, vacations, and so on. Selecting groups of parts for samples is commonly done in one of the following two ways:

- Instant-time method: At predictable times pick consecutive samples from a machine. This tends to reduce sample variance and is best used when looking for process-setting problems.
- Period-of-time method: Samples are selected from parts so that they have not been presented consecutively. This is best used when looking at overall quality when the process has a great deal of variability.

Samples should be homogeneous, from the same machine, operator, and so on, to avoid multimodal distributions. Table 10.4, taken from US MIL STD 414, suggests sample sizes based on the total lot size. For example, if we have a batch of 200 coffee cups we should use a sample size around 25 in total. If we choose to take five samples every time, we would need to return five times to collect all of the samples. The time and selection of the cups should be random. Sometimes these sampling processes are destructive and the sampled parts must be discarded. Other times the tests are nondestructive and the parts are returned to the lot. The cost of destructive testing can sometimes influence the testing methods and process design.

Keep in mind that the mean and control limits are how the machine behaves, not what you specify. You should specify tolerances that are larger than the UCL and LCL of the SPC process. If one

Table 10.4 MIL STD 414 lot and sample sizes.

Lot size	Sample size
66–110	10
111–180	15
181–300	25
301–500	30
501–800	35
801–1300	40
1301–3200	50
3201–8000	60
8001–22,000	85

tolerance is only 3.0 standard deviations from the mean, then you can expect to have 3 failed parts per 1000. If your tolerance is 6.0 standard deviations, the failure rate will drop to fewer than 4 parts per 1,000,000. For well-known processes and machines, the accuracy and tolerances will be obvious. Otherwise, test runs or DOE methods can be used to find the capabilities of the machine.

PROBLEMS

10.49 Are the UCL and LCL the same as the tolerances?

10.50 Does SPC monitor the part quality or the process control?

10.51 Is it important to use a consistent measurement time when sampling parts?

10.52 Given the data set 4, 4, 7, 9, 10, 6, 8, calculate the:

(a) mean

(b) mode

(c) median

(d) standard deviation

(e) variance

(f) range

10.5.1 Control chart calculations

Calculations of the mean/average and the standard deviation, or range, are at the heart of the SPC method. There are two primary charts: the Xbar chart tracks changes in the mean, and the other tracks changes in the standard deviation or the range. In particular, the chart for standard deviation is called the s chart, while the R chart is for range. Practically, the R chart is usually chosen over the s chart because it requires fewer and simpler calculations with similar results. The center lines and control limits for the charts are found using the sampled data; an example is provided in Table 10.5. In this example there are five measurements taken at a time. The average, range, and standard deviation are calculated for these five readings only. So the fifth data sample has a mean value of 1.52, a range of 0.17, and a standard deviation of 0.067. Each sample would be (1) added to the control chart immediately and (2) eventually used to update the mean and control limits for the chart.

The mean and control limits for the Xbar chart can be calculated using the equations in Fig. 10.22. For this chart we are using the seven sample values in the example. The mean and standard deviations are calculated and then used to find the limits by adding and subtracting three standard deviations. The data points have been added to this chart. All lie close to the mean, with none outside the control limits, so the process appears to be in control. For convenience, a few more sample points may be added to

Table 10.5 Statistical process control data samples.

n	Date and time	X0	X1	X2	X3	X4	Xbar	R	s
1	Oct 12, 10:02	1.50	1.62	1.72	1.46	1.63	1.59	0.26	0.105
2	Oct 12, 14:56	1.77	1.48	1.52	1.54	1.52	1.57	0.29	0.116
3	Oct 12, 17:21	1.63	1.59	1.53	1.64	1.60	1.60	0.11	0.041
4	Oct 13, 9:43	1.57	1.51	1.65	1.56	1.77	1.61	0.26	0.101
5	Oct 13, 15:27	1.42	1.59	1.51	1.57	1.51	1.52	0.17	0.067
6	Oct 14, 12:46	1.45	1.71	1.48	1.44	1.67	1.55	0.27	0.129
7	Oct 17, 10:18	1.46	1.68	1.74	1.56	1.56	1.60	0.28	0.108
...									
n−1									
n		a	b	c	d	e	(a + b + c + d + e)/5	Max(a,b,c,d,e) − min(a,b,c,d,e)	SD(a,b, c,d,e)

this chart before updating the UCL and LCL values. Although we will not discuss it here, there are some methods to approximate the standard deviation using the range.

An R chart can be used to estimate changes in the standard deviation, or randomness, as shown in Fig. 10.23. As before, the range values are averaged for the center line and then the control limits are three standard deviations above and below. In this graph the points are within the control limits and the amount of randomness remains the same. The alternative to the R chart is the s chart. The process to develop the s chart is identical to that for the Xbar and R charts. Note that the R and s sample values are not identical, but they are proportionally similar. So choosing an R chart will make the process simpler for shop-floor calculations at a cost of some accuracy.

If any sample point falls outside the control limits, the process is out of control and the process should be stopped until the issue is resolved. Sometimes a point lies outside the range for a known reason that has been corrected, such as a new operator in training. In the cases in which the sample is known to be a one-time problem, the point can be discarded and production can continue (Fig. 10.24). If the cause is less clear and is likely to occur again in the future, the process should be stopped and the problem resolved.

The possibility of having a point outside the control limits is low: 3 in 1000. There are other cases that are unlikely if the variation is random. Fig. 10.25 shows a case in which the patterns are not varying as much as they should. This visual form of analysis makes use of zones between the first, second, and third standard deviation. It is unlikely that seven samples in a row would be in zone C. An example of causal issues might include operator mismeasurement, tool changes, changes in process parameters, and so on. It is also rare to find a series of six samples that increase in one direction. This suggests that the process is drifting in one direction, possibly with a loose setting, temperature rise, or tool wear.

$$\bar{X} = \frac{\sum_{i=1}^{n} \bar{X}_i}{n} = \frac{(1.59 + 1.57 + 1.60 + 1.61 + 1.52 + 1.55 + 1.60)}{7} = 1.578$$

$$\sigma = \sqrt{\frac{\sum_{i=1}^{n} (\bar{X}_i - \bar{X})^2}{n-1}} = 0.0325$$

$$UCL = \bar{X} + 3\sigma = 1.578 + 3(0.0325) = 1.676$$

$$LCL = \bar{X} - 3\sigma = 1.578 - 3(0.0325) = 1.481$$

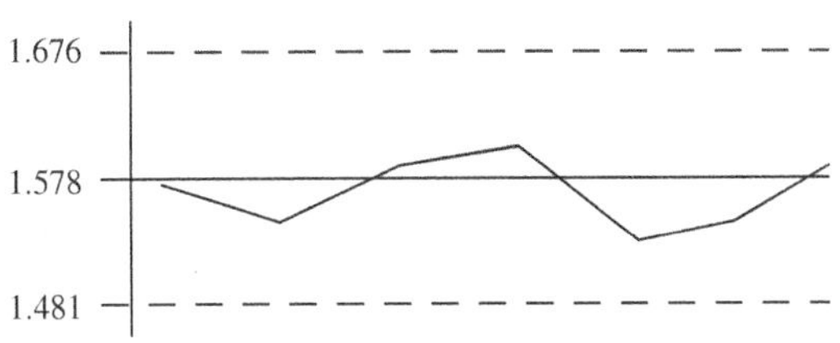

1.676

1.578

1.481

FIGURE 10.22

Sample Xbar-chart calculations. *LCL*, lower control limit; *UCL*, upper control limit.

Other trends that are a little more difficult to identify are shown in Fig. 10.26. It is less likely for a sample to fall in zone A or zone B. Multiple sequential points in these zones are very unlikely and indicate a shift in the sample mean. For example, the probability of one point in zone A is under 3. The probability that two samples in a row are in zone C is under 0.06. These trends suggest that something has occurred to shift the process in one direction. For example, a fixture might have been hit, shifting it a short distance in one direction, or an operator might have adjusted a process speed or cycle time.

Designers must be aware of SPC methods because they are ubiquitous in manufacturing. When there is an issue with a design there is a probability that it has occurred in manufacturing. Being able to interpret the SPC charts allows a designer to quantify variations within the specifications. For example, the values from SPC charts can be used for tolerance stack analysis. In addition, a designer and manufacturing engineer must be able to quantify what accuracy and variation can be expected from a manufacturing process. This can be done by analyzing SPC data from similar jobs run on the machine before.

$$\bar{R} = \frac{\sum_{i=1}^{n} R_i}{n} = \frac{(0.26 + 0.29 + 0.11 + 0.26 + 0.17 + 0.27 + 0.28)}{7} = 0.2343$$

$$\sigma = \sqrt{\frac{\sum_{i=1}^{n} (R_i - \bar{R})^2}{n-1}} = 0.0675$$

$$UCL = \bar{X} + 3\sigma = 0.2343 + 3(0.0675) = 0.4368$$

$$LCL = \bar{X} - 3\sigma = 0.2343 - 3(0.0675) = 0.0318$$

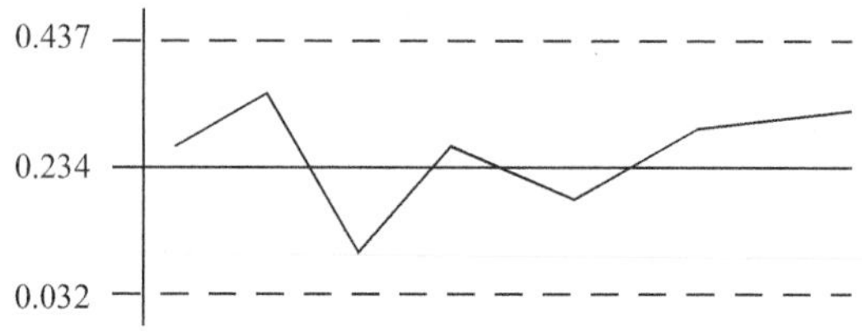

0.437

0.234

0.032

FIGURE 10.23

Sample R-chart calculations. *LCL*, lower control limit; *UCL*, upper control limit.

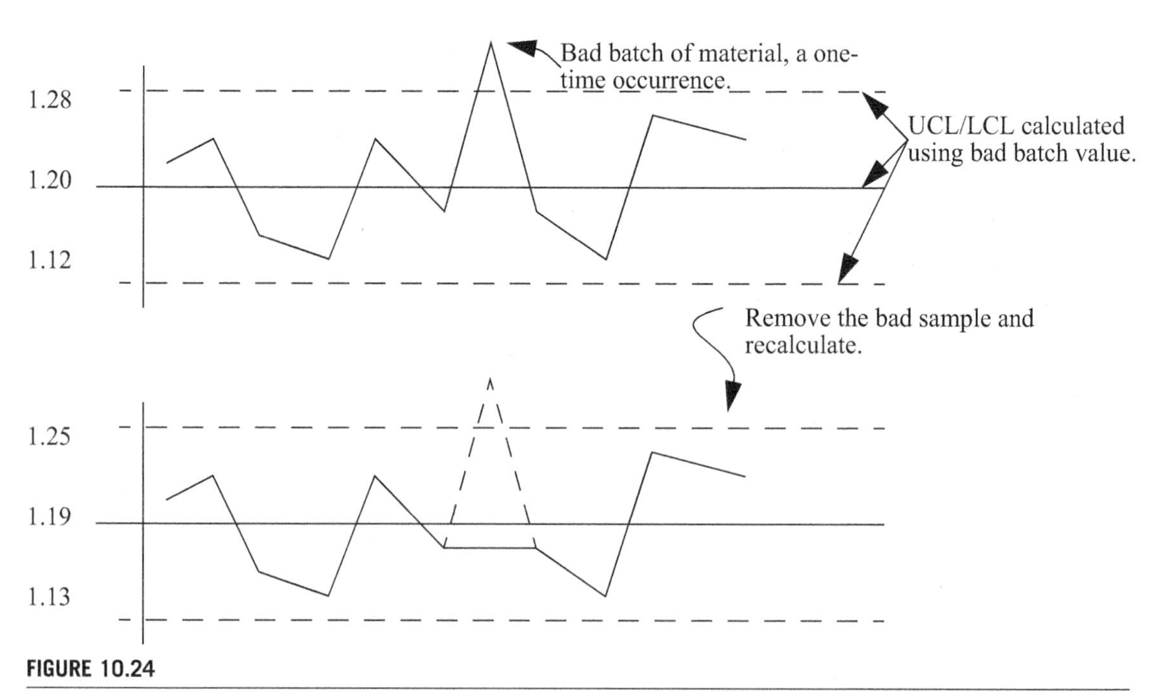

FIGURE 10.24

Recognizing and removing out-of-control cases. *LCL*, lower control limit; *UCL*, upper control limit.

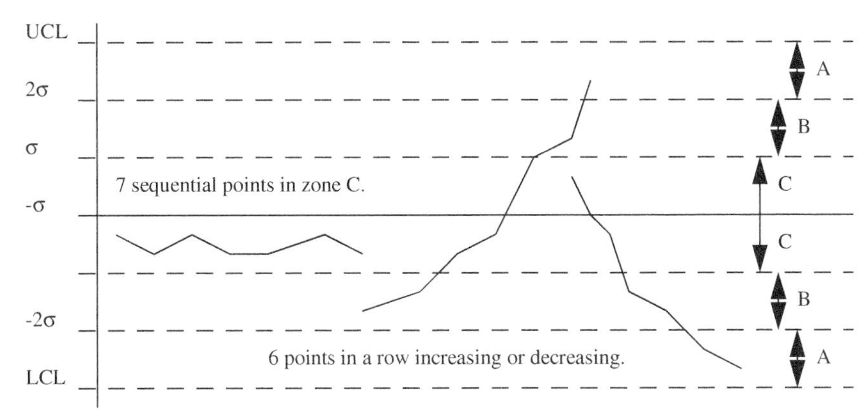

FIGURE 10.25

Example of patterns that are too consistent. *LCL*, lower control limit; *UCL*, upper control limit.

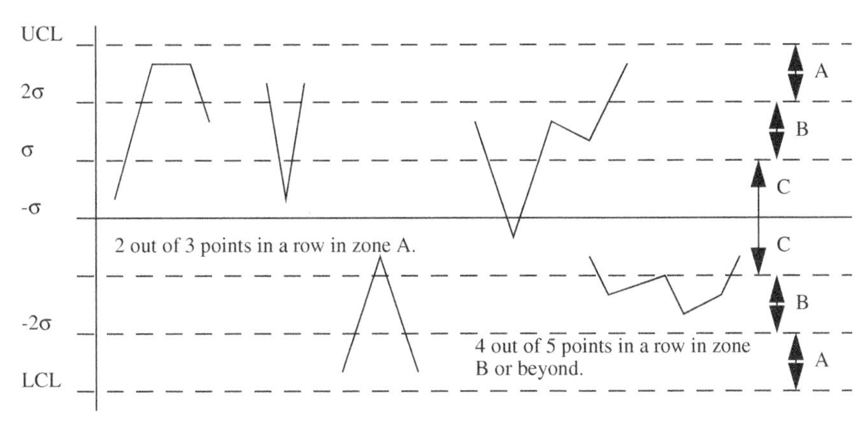

FIGURE 10.26

Too many points outside 1 standard deviation. *LCL*, lower control limit; *UCL*, upper control limit.

PROBLEMS

10.53 Describe SPC.

10.54 Why is the standard deviation important for process control?

10.55 What is the purpose of control limits in process monitoring?

10.56 What would happen if the SPC control limits were placed less than ± 3 standard deviations from the mean?

10.57 What factors can make a process out of control?

10.58 Draw the detailed Xbar chart for three data samples, of which sample 1 is 1.6, 1.3, 1.9, 1.9; sample 2 is 0.1, 2.5, 2.1, 3.3; and sample 3 is 2.7, 2.7, 4.9, 1.5.

Sample 1	Sample 2	Sample 3
1.6	0.1	2.7
1.3	2.5	2.7
1.9	2.1	4.9
1.9	3.3	1.5

10.59 What problems can be seen in this control chart?

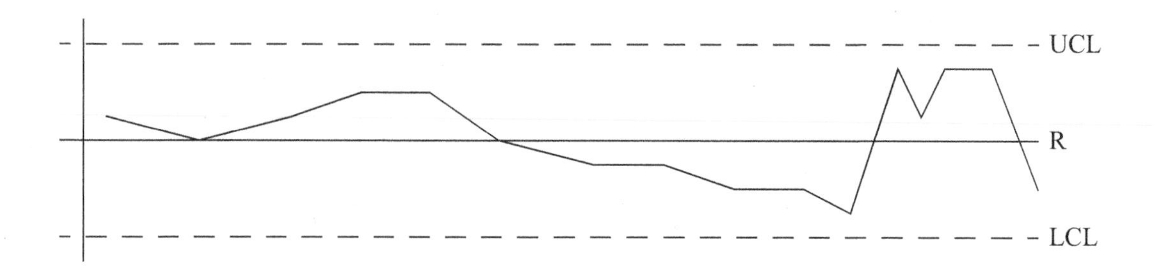

10.60 What problems can be seen in this control chart?

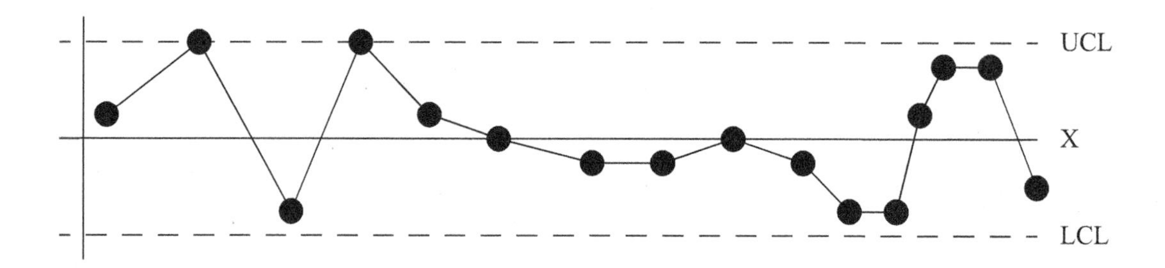

10.61 The data shown on the graph have been plotted from QC samples. Add the lines required to complete the Xbar chart. What tolerance is required to obtain three-sigma quality?

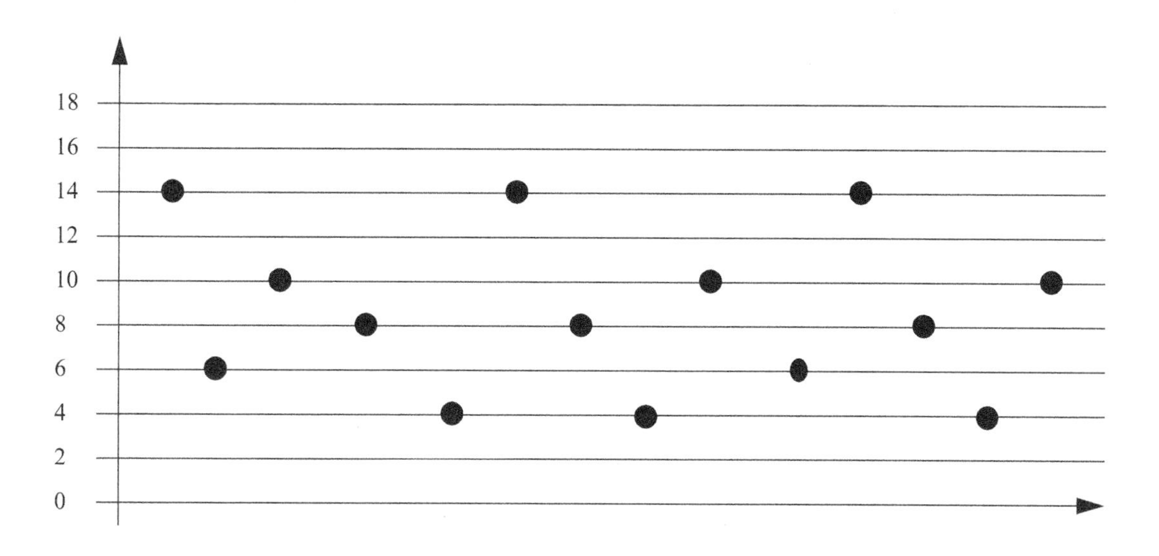

10.62 Draw the Xbar chart for the data in the table, using exact calculations. Then calculate the upper and lower control limits.

Sample 1	Sample 2	Sample 3
12.6	10.1	12.7
12.3	12.5	12.7
12.9	12.1	9.9
12.9	13.3	13.5
13.8	10.5	13.2
13.2	13.0	12.8
11.6	12.6	12.5

10.63 Four samples have been taken at the start of a new process run. However, one of the values, X1, was accidently erased after the calculations were done. Using the data shown here, find the missing value.

$\overline{X}_0 = 10.23$ $\qquad$ $UCL_{\bar{x}} = 12.54$

$\overline{X}_1 = ?$ $\qquad$ $LCL_{\bar{x}} = 9.26$

$\overline{X}_2 = 9.98$

$\overline{X}_3 = 11.75$

10.64 (a) Given the results from a designed experiment, as shown in the table, what are the main effects of A and B? (b) Draw a graph from the data in (a), and explain the significance of the effects.

Run	A	B	Samples
1	10	1	3.2, 9.8, 5.5
2	8	1	5.0, 6.7, 2.1
3	10	3	11.3, 7.2, 8.5
4	8	3	7.7, 6.0, 8.9

10.5.2 Parts inspection

SPC is used when you control the machines. When somebody else produces critical parts you still need to guarantee the parts meet specifications. The most common method is part or lot inspection. When parts arrive at your site you randomly collect some samples, measure the results, and then verify that they are statistically acceptable. It is normal that suppliers are trustworthy and parts are acceptable. However, there are times when a supplier provides parts that are unacceptable. When this happens the role of inspection changes from verification to enforcement.

Formal methods have been developed for accepting parts. These include a statistical process to estimate part quality and adjust sampling rigor when failure occurs. One general procedure is shown in Fig. 10.27. In general, a trusted supplier delivers parts and a random sample is taken, then the mean and standard deviation are calculated and compared with the tolerances. A common contractual requirement is that the part must lie between the tolerances with a variance of at least plus or minus six sigma (see the following subsection for more on six sigma). Suppliers earn a trusted status and have parts accepted easily, or sometimes with no inspection. When a batch does fail, the supplier moves to a less trusted status with greater scrutiny. In a very strict environment the parts will be removed from the plant immediately to avoid contaminating other batches of good parts. The rejection of the batch could require a formal response from a supplier or even a termination of a contract. A lenient approach is to accept it as is, or with a financial penalty. A middle-of-the-road solution is to have the supplier correct the issue and reinspect the parts.

When every incoming part is inspected it is called screening. This will be done for critical parts such as aircraft engine components. It may also be done when the supplier is not able to control the supply. For example, a rubber exporter may send batches that age during shipping. After delivery, the customer, a tire manufacturer, screens the batches and sells the substandard rubber to a playground equipment maker. It is worth noting that industries, such as automotive, are moving to 100% inspection on many components. Sensor and vision technologies make this relatively inexpensive.

Part inspection processes are not perfect and have the following advantages and disadvantages. It is critical for designers to understand this process when specifying parts. If specifications are difficult to produce it may create situations in which batches of parts are regularly rejected and slow production. As usual, looser tolerances will reduce these types of supply-chain problems:

Advantages
- Quality problems caused by suppliers can be identified.
- The inspection for trusted suppliers is reduced.
- Rejection of entire lots increases supplier quality incentives.

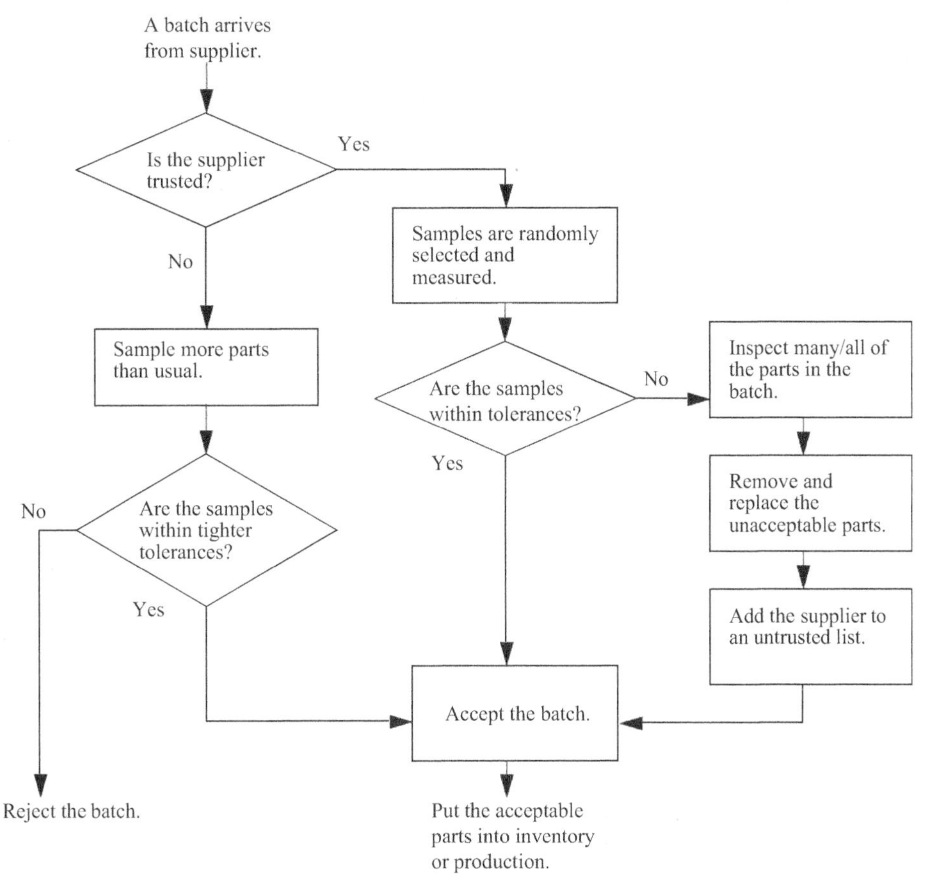

FIGURE 10.27

Incoming part inspection.

Disadvantages

- Good/bad lots may be rejected/accepted by poor sample selection.
- Planning, effort, and documentation are required.
- These samples describe only part of a lot.

PROBLEMS

10.65 What might happen if acceptance procedures were not used?

10.66 What are the benefits of trusting a supplier?

10.67 List three options if a batch of parts did not pass acceptance testing.

10.5.3 Six-sigma process capability

Consider a tolerance variation for a manufacturing process. The normal distribution will have a central mean based on the process setting. The width of the normal distribution can be described with the standard deviation. If we set the tolerance limits to the mean plus or minus six sigma, then only 3.4 parts in every million would fail. (Note: If the tolerance levels were at three sigma, there would be 66,807 failures per million parts.) The six-sigma philosophy is to establish a tolerance zone that encloses at least six standard deviations.

The basic six-sigma measure, C_p, is used when the process mean is very close to the tolerance mean (Fig. 10.28). The C_p value is 1 when the tolerance is set to ±3 standard deviations. A C_p of 2.0 would mean the tolerances are at ±6 standard deviations. Obviously a larger C_p value will produce more acceptable parts. It is generally recommended that C_p be greater than or equal to 1.5. The C_p value assumes that the process mean and tolerance mean are the same.

The process mean will rarely match the tolerance mean. For example, we may set a recipe tolerance at L = 195°C and U = 205°C, having a tolerance mean of 200°C. When the oven controls are set to 200°C the average oven temperature is 202°C. The difference between the mean and the upper

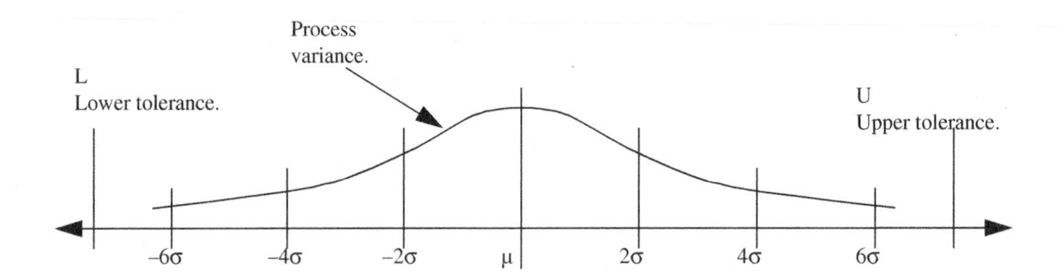

Where,

C_p = Inherent process capability.

$U - L$ = the difference between the upper and lower tolerance.

$6\sigma_0$ = the process capability. eqn. 9.2

$$C_p = \frac{U - L}{6\sigma} \quad \text{eqn. 9.3}$$

To interpret the *Cp* values note that higher than one is better.

C_p = 1 is marginal.

$C_p < 1$ is bad (not an acceptable process).

$C_p > 1$ is acceptable (1.33 is standard).

FIGURE 10.28

Using C_p to relate tolerances to machine capabilities.

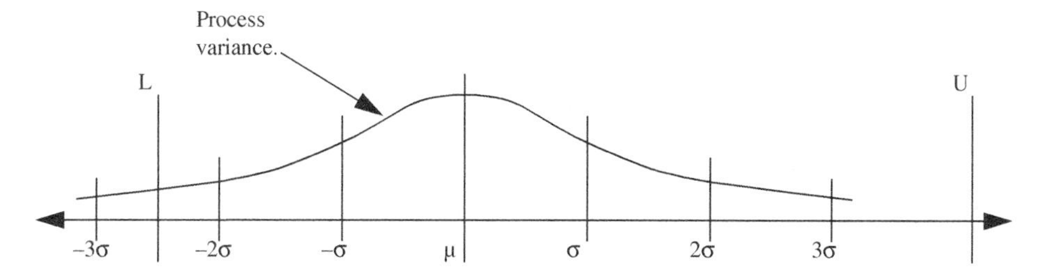

Where,

C_p = Process capability. C_{pk} = Process capability index

$$C_{pk} = \left(\frac{U-\mu}{3\sigma}\right) \text{ or } \left(\frac{\mu-L}{3\sigma}\right) \quad \text{Select the smaller of the two.} \qquad \text{eqn. 9.4}$$

To interpret the C_{pk} values note that higher than 2 is better.

If $C_p = 2\ C_{pk}$ then $\overline{X} = \mu$, and the process is always centered.

FIGURE 10.29

C_{pk} for uncentered process variation.

tolerance is 3°C, while the mean to the lower tolerance is 7°C. The C_{pk} compensates for asymmetry between process and tolerance means (Fig. 10.29). The C_{pk} calculation uses the tolerance that is the closest to the mean. The ratio is found by dividing by three sigma (not six sigma). When a C_{pk} value is greater than 2.0, the tolerances will contain plus or minus six sigma of variations, or 3.4 failures per million.

Fig. 10.30 is an analysis of a part in production. The part has a tolerance from 2.000 to 2.004. In production, the process has a mean of 2.0028 and a standard deviation of 0.0006. The C_p value is calculated and has an acceptable level of 1.11. A value of 0.667 is calculated for the C_{pk} value, well below 1.0. The poor C_{pk} value can be illustrated by the UCL, 2.0046, which is outside the upper tolerance of 2.004. As designed, this process would have many failures per thousand parts.

For example, given the control chart for a process, and a feature to be turned on the process, determine if the tolerances specified are reasonable.

PROBLEMS

10.68 What is process capability and how is it used?

10.69 What is acceptance sampling and when should it be used?

10.70 Describe the difference between C_p and the control limits.

10.71 Assume a process has a mean of 102 and a standard deviation of 1. What are the C_p and C_{pk} values for a tolerance of 100 ± 5?

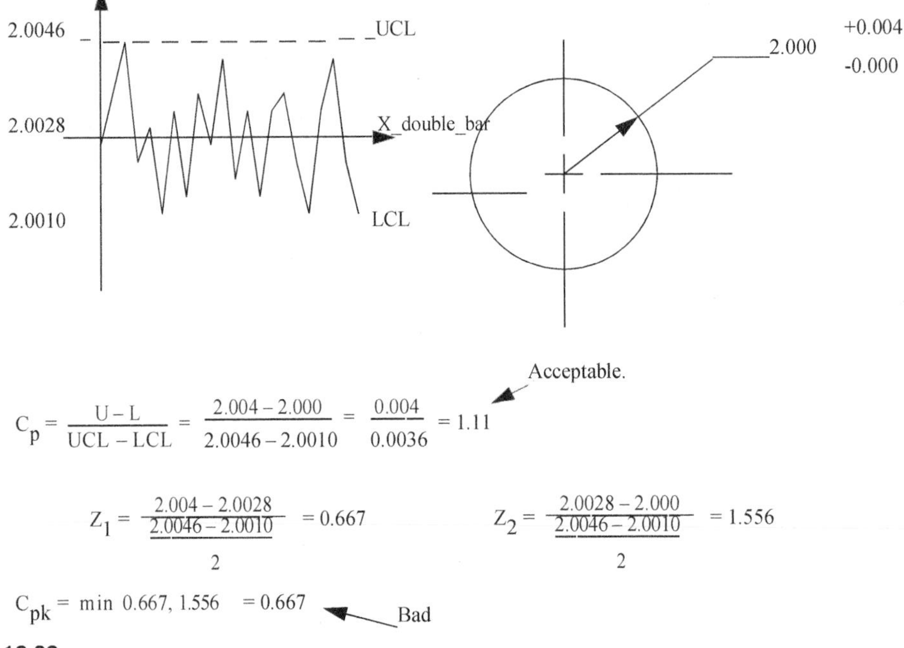

FIGURE 10.30

Using control charts to determine C_p and C_{pk}. *LCL*, lower control limit; *UCL*, upper control limit.

10.72 Assume a process has a mean of 102 and a standard deviation of 1. What tolerances would give a $C_p = 1.5$ and $C_{pk} = 1.5$?

10.73 What is the difference between C_p and C_{pk} ?

10.74 When selecting tolerances, what values should be used for fewer than 3.4 failures per million?

10.75 How can SPC data be used to calculate C_p and C_{pk} for a new design?

10.6 Parametric design and optimization

Equation-based system models provide excellent opportunities for design control and optimization. The model is normally oriented to some objective, such as cost, performance, weight, or life. These problems normally have objective and dependent variables. The problem-solving methods attempt to find the dependent control variable values that maximize, or minimize, the objective value. The dependent and objective variables often have constraints to keep their values within acceptable ranges. The skills to solve these problems are taught in algebra, calculus, and engineering courses:

- Objective: This is an output variable that should be minimized or maximized, often a cost. This is also called an output or dependent variable. An example is the cost of a cable.
- Objective functions: These are one or more equations that are used to calculate the objective value. An example is an equation that relates the cost of a cable to diameter, length, and material type.

Where,

P = Maximum system power
I = Maximum current
V = System supply voltage
C_H = Cost for a heat sink
C_R = Cost for a resistor
C_T = Cost for a transistor
C_{total} = The total cost

$$C_{total} = C_H + C_R + C_T \quad \text{eqn. 10.5}$$

$$P = IV \quad \text{eqn. 10.6}$$

$$C_H = 5.00P - 0.0005P^2 + 10.25 = 5.00IV - 0.0005I^2V^2 + 10.25 \quad \text{eqn. 10.7}$$

$$C_R = 0.50I^2 + 0.05P + 1.00 = 0.50I^2 + 0.05IV + 1.00 \quad \text{eqn. 10.8}$$

$$C_T = 0.79V + 0.09I + 1.24 \quad \text{eqn. 10.9}$$

$$\therefore C_{total} = 5.00IV - 0.0005I^2V^2 + 10.25 + 0.50I^2 + 0.05IV + 1.00 + 0.79V + 0.09I + 1.24$$

$$\therefore C_{total} = 5.05IV - 0.0005I^2V^2 + 0.50I^2 + 0.79V + 0.09I + 12.49 \quad \text{eqn. 10.10}$$

FIGURE 10.31

A sample cost model for a transistor–resistor–heat sink system.

- Independent variables: These are design factors that can be changed to control the design and, ultimately, the objective value. These may also be called control, independent, or input variables. Examples are a cable diameter, a cable length, and a cable material (steel or aluminum).
- Constraints: These are limits on values for inputs or outputs. For example, a cable diameter cannot be greater than 5 mm or less than 1 mm.

A sample system model is given in Fig. 10.31 for a transistor–resistor–heat sink system. Each of these components has a cost that is a function of the applied current (I), voltage (V), and power (P). Each of the three components has a cost that can be combined, shown in Eq. (10.10). If a designer specifies any three of these values, the third is calculated using Eq. (10.6). It is also possible for a designer to specify one or no value. If only one value is specified, the equation has one input and one output and can be solved using algebra or differential calculus. If none of the values is specified, there are two inputs and one output, requiring the use of more mature techniques such as linear programming and optimization.

Fig. 10.32 shows an example of single-variable optimization. In this example a value of 100 W was provided for P. This allows further simplification with I = 100/V or V = 100/I. In this case V was arbitrarily chosen for the decision variable. Eq. (10.10) was simplified, differentiated with respect to V, and set equal to 0. The roots of the third-order equation were found. In this case there is only one real root,

Where,

$$P = 100\text{ W} \qquad \text{Designer or customer selected.}$$

$$\therefore I = \frac{100}{V}$$

$$\therefore C_{total} = 5.05\left(\frac{100}{V}\right)V - 0.0005\left(\frac{100}{V}\right)^2 V^2 + 0.50\left(\frac{100}{V}\right)^2 + 0.79\ V + 0.09\left(\frac{100}{V}\right) + 12.49$$

$$\therefore C_{total} = \frac{5000}{V^2} + 0.79\ V + \frac{9}{V} + 512.49$$

Find the minima.

$$\frac{d}{dV}C_{total} = \frac{-2\,(5000)}{V^3} + 0.79 + \frac{-9.0}{V^2} = 0$$

$$\therefore -10{,}000 + 0.79\ V^3 + -9.0\,V = 0$$

$$V = 23.4\ ,\ -11.7 \pm j20.0 \qquad \text{Roots calculated using software.}$$

$$\therefore C_{total} = \frac{5000}{18.7\,(23.4)} + 0.79\,(23.4) + \frac{9.0}{23.4} + 512.49 = 542.79$$

$$\therefore C_{total} = \frac{5000}{18.7\,(24)} + 0.79\,(24) + \frac{9.0}{24} + 512.49 = 542.97$$

FIGURE 10.32

Single-variable optimization with calculus.

so the optimal voltage will be 23.8 V. The designer would probably round this up to 24 V, a standard voltage. If there were three real and positive roots the designer would have to calculate the costs for each to find the global optimum.

A constraint is added to a single variable optimization problem in Fig. 10.33. In this case the current must be greater than 10 A and the voltage is fixed at 24 V. In the first pass the derivative is used to calculate the minimum, which is −286 A, much less than 10 A. Therefore, the optimum cannot be reached, so the constraint of 10 will have to be used. A second value is calculated at I = 11 A as a test for the result and to provide guidance to the designer.

In the previous examples, one parameter was chosen to reduce the problem to a single variable. In a more complex example, I, P, and V will remain unknown. The objective is to find I and V values that minimize the device cost. The method shown in Fig. 10.34 shows the process of selecting a starting point and calculating the output. The variables are changed one at a time and the effect on the cost is observed. In this case the lowest cost comes for a device that does not use any current or voltage. This is an obvious but not very useful conclusion.

The previous example is revisited in Fig. 10.35 in which the new objective function is to minimize the cost per unit of power over a constrained design space of V = 0–10 and I = 0–10. As before, a starting location is selected and guesses are used to reduce the cost per unit of power. The result is that the lowest cost is at V = 10 and around I = 7. It is very important to note that this method is convenient but it can miss other optimum points.

Where,

$V = 24$ Designer specified voltage and constraint for the current.

$I \geq 10$

$$\therefore C_{total} = 5.05\,I\,(24) - 0.0005\,I^2(24)^2 + 0.50\,I^2 + 0.79\,(24) + 0.09\,I + 12.49$$

$$\therefore C_{total} = 121.29\,I + 0.212\,I^2 + 31.45$$

Find the minima.

$$\frac{d}{dI}C_{total} = 121.29 + 2(0.212)\,I = 0$$

$$I = \frac{-121.29}{2(0.212)} = \boxed{-286.06}$$

This current is less than 10 so it must be discarded.

The minima was below the constraint so we will use I = 10.

$$C_{total} = 121.29\,(10) + 0.212\,(10)^2 + 31.45 = 1265.55$$

To test the value try I = 11.

$$C_{total} = 121.29\,(11) + 0.212\,(11)^2 + 31.45 = 1391.29$$

FIGURE 10.33

A constrained single-variable optimization solution.

$$C_{total} = 5.05IV - 0.0005I^2V^2 + 0.50I^2 + 0.79V + 0.09I + 12.49$$

V	I	C_{total}
10	10	571.29
11	10	621.53
10	11	631.33
9	10	520.95
8	9	421.13
5	5	155.33
0	0	12.49

Stopping at 0 because the cost always increases with voltage and current. In other words there is no practical minimum.

FIGURE 10.34

Trial-and-error optimization to find the minimum cost.

$$\frac{C_{total}}{P} = \frac{5.05IV - 0.0005I^2V^2 + 0.50I^2 + 0.79V + 0.09I + 12.49}{IV}$$

V	I	C_{total}/P
0.1	0.1	1263.35
5	5	6.21
10	10	5.71
5	10	6.37
10	5	5.69
10	8	5.67
10	3	5.87
10	4	5.75
10	5.5	5.68
10	4.5	5.71
10	7	5.67

Use the design range from V = 0 to 10, and I = 0 to 10. Iterate by changing values based on observations.

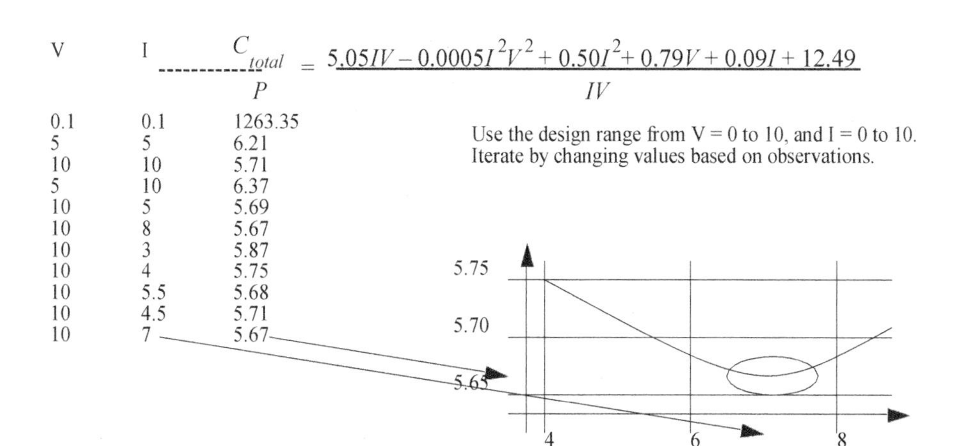

FIGURE 10.35

Optimum cost per power unit in a constrained design space.

Optimization techniques are used for systems that are complex, expensive, or produced in large volumes. The time required to develop the models, write the programs, and interpret the results can be small compared with the potential benefits.

PROBLEMS

10.76 Is an objective function required for optimization?

10.77 What is a minimum or maximum?

10.78 Give five examples of design constraints.

10.79 What would be a reasonable objective function for a product that should cost less that $10 and have a mass greater than 2.5 kg?

10.80 What is the difference between quality control and quality assurance?

10.81 If a batch of parts is rejected for SPC, does that mean it does not meet the tolerances?

10.82 SPC charts have upper and lower control limits that are three times the standard deviation from the mean. Why?

10.83 If the SPC UCL and LCL were increased to six sigma, would more parts be rejected? Explain.

10.84 Why does the actual production mean value differ from the specified dimension and tolerance?

10.85 Why would C_p and C_{pk} have different values?

10.86 The data in the table were measured over a 2-week period for a 1.000″ shaft with a tolerance of ±0.010″.

Date	Samples			
Nov. 1, 2021	1.0034	0.9999	0.9923	1.0093
Nov. 2, 2021	0.9997	1.0025	0.9993	0.9938
Nov. 3, 2021	1.0001	1.0009	0.9997	1.0079
Nov. 4, 2021	1.0064	0.9934	1.0034	1.0064
Nov. 5, 2021	0.9982	0.9987	0.9990	0.9957
Nov. 6, 2021	0.9946	1.0101	1.0000	0.9974
Nov. 7, 2021	1.0033	1.0011	1.0031	0.9935
Nov. 8, 2021	1.0086	0.9945	1.0045	1.0034
Nov. 9, 2021	0.9997	0.9969	1.0067	0.9972
Nov. 10, 2021	0.9912	1.0011	0.9998	0.9986
Nov. 11, 2021	1.0013	1.0031	0.9992	1.0054
Nov. 12, 2021	1.0027	1.0000	0.9976	1.0038
Nov. 13, 2021	1.0002	1.0002	0.9943	1.0001
Nov. 14, 2021	0.9956	1.0001	0.9965	0.9973

(a) Draw accurate Xbar control charts using graph paper or spreadsheet software.

(b) Determine C_p and C_{pk}.

(c) What would the tolerance have to be if we required six–sigma quality?

10.87 Draw a curve that shows the relationship between customer satisfaction and precision.

10.88 A distillation tower is used to separate recycled benzene chemical waste. The mean benzene purity is 96.5% with a standard deviation of 0.8%. What is the C_pp value if a customer orders (a) 97% ± 1%, (b) 95% ± 2%, (c) 96.5% ± 3%?

10.89 Find and summarize legal cases for the following situations. The summaries should include the defendants, a brief summary of the issue, the judgment, and the penalty.

(a) A company violating air-quality regulations

(b) An engineer criminally negligent for water pollution

(c) A company found criminally negligent for environmental contamination

10.90 Research and list the laws and regulations for your geographical area that deal with the use and disposal of potassium fluoride.

Further reading

Baxter, M., 1995. Product Design; Practical Methods for the Systematic Development of New Products. Chapman and Hall.

Basu, R., Wright, J.M., 2003. Quality Beyond Six Sigma. Butterworth-Heinemann.

Besterfield, D.H., 2008. Quality Control, eighth ed. Prentice Hall.

Dhillon, B.S., 1996. Engineering Design; A Modern Approach. Irwin.

Goldratt, E.M., Cox, J., 1984. The Goal: A Process of Ongoing Improvement. North River Press.

Harry, M.J., 1987. The Nature of Six Sigma Quality, first ed. Government Electronics Group, Motorola Inc., pp. 1–25.

MIL-STD-1472D, Military Standard: Human Engineering, Design Criteria for Military Systems, Equipment, and Facilities (14 Mar 1989).

Nordeen, D.L., June 1993. Total Quality Management in Industry. Automotive Engineering, pp. 35–41.

Psydek, T., 2003. The Six Sigma Project Planner: A Step-by-Step Guide to Leading a Six Sigma Project Through DMAIC. McGraw-Hill.

Toper, W.G., June 14, 1993. In: The ISO9000 Quality System Standards and Their Implication for Global Business, A Tutorial Presented at the Third Annual IIE Four-Chapter Conference at the Sheraton Falls View Hotel & Convention Centre, Niagara Falls, Ontario, Canada.

Truscott, W.T., 2003. Six Sigma: Continual Improvement for Businesses: A Practical Guide. Butterworth-Heinemann.

Ullman, D.G., 1997. The Mechanical Design Process. McGraw-Hill.

APPENDIX

Checklists

A.1 WORK BREAKDOWN STRUCTURE

During planning there is so much information that it is easy to overlook details. The following list is offered as an aid for planning design projects. The list is not exhaustive but focuses on common items found in project plans. Reading this section will not be productive, but skimming the list should help identify holes in plans.

Stakeholders

- Customers
    - Communication—frequency, types, format
    - Preferred—does the customer have a preferred supplier, information, materials, people, access?
    - History—find out about the customers, previous experiences and needed changes
- Staff
    - Skills—short supply, specialized knowledge, training
    - Available—other work, holidays, hiring, overtime, layoffs, part-time, contractors
    - Barriers—travel, safety, access, personal issues
    - Management—motivation, retasking
- In the company
    - Look at the organization charts for all of the people above you, below you, and to the sides, and for major business functions.
    - Ask other managers.
    - Talk to all of the people affected. Follow the money.
    - Follow the physical parts.
    - Develop a friendly and approachable relationship. Consider who controls or owns the spaces.
    - Consider who controls the supplies.
- External
    - Regulators, licensing, legal
    - Public, press
    - Standards and registration groups

Resources

- Equipment
    - Setup—tooling, programming, maintenance, supplies
    - Availability—scheduling, utilization

Supplies—special materials, consumables, waste
Accessories—cables, software, fixtures, tools, materials
Licenses—software and proprietary processes
Labor—people needed to operate and maintain the equipment
Facilities—power, water, space, safety

- Resources
  Information—similar projects for estimates, simultaneous projects for competing need
  Money—cash flow, rates, approval, purchasing
  External—renting, leasing, purchasing
  Facilities—offices, services, storage, production, laboratory
- Suppliers
  Process—lead time, process time, shipping, payment, meetings, contracts, bidding
  Type—raw, processed, commodity, special order, quality, alternatives
  Cost—minor, major, variable
  Strategic—dependable, capable, available, reputation, technical support, competitors
- Equipment
  Rental—purchase cost, rental or lease rates (hourly, daily, weekly, monthly, yearly), fees, taxes
  Purchase—costs for the equipment, software licenses
  Fees—license or regulation fees
  Maintenance—costs for regular upkeep or repair
  Consumables—paper, cutting tools, printed circuit board etchant, formaldehyde
  Basic labor requirements—operators, supervisors
- Materials
  Raw materials—shipping, form, testing
  Parts—shipping, storage, quality, handling
  Consumables and supplies—routine items, hidden items (e.g., pens and batteries)

Work tasks and schedules

- Tasks
  Deliverable—it should have only a single well-defined outcome
  Steps—a task can be broken into steps, but the subdeliverable result must be clear
  Sequence—there should be tasks or events that are before or after
  Parallel—when tasks can be done in parallel overall time can be saved
  Match—all low-level tasks should be part of high-level tasks or milestones
  Subtasks—reduce higher level tasks into subtasks
- Milestones: How much time does the design require? Based on experience and some calculations, a manager can set milestones for a design team. Typical tasks might include the following:
  Specifications—set as a clean start to the design process
  Concepts—approved by customers, generated as a quotation, verified by market study, verified by prototype/simulation
  Embodiment—preliminary detailed design with the selection of major components, layout, software, methods, processes
  Detailed—design of parts, hardware, software, frames, etc.
  Iteration—for designs that need a prototype for proof of concept

Preproduction—the design of production methods, tooling, testing equipment, process parameters, build testing
Ensure communication and start/stop points for the groups are clear.
The sequence of tasks can be critical.
Who needs to be involved?
Are there tasks or items that are push or pull in nature?
Focus on the start and end of tasks for the timeline (inputs and outputs) and process support (needed resources) for the task.

- Timing
  - Dates—start, end, internal reviews, due dates versus deadlines
  - Delays—suppliers, waiting for equipment, process time, work time, testing, legal, regulatory
  - Contingency—"slack time" is important when splitting tasks between multiple people, and even more important when using other people and resources you don't control (e.g., an outside supplier)
  - Undeniable—is it critical that you have an undeniable criterion for when a milestone or task is complete?
  - Milestones—methods for reviewing milestones or task completion include reviews, checklists, testing, and signed approvals
  - Pacing—plan out reasonable goals, not so far apart as to be frustrating, but not so close as to seem annoying; for major goals a week to a month is reasonable
  - Suppliers—hidden delays, vacations, and shutdowns
- People and resources
  - Booking—key resources should be available and booked
  - Leader—assign tasks to a leader
- Closing the project
  - Follow-up—after closure check on customer satisfaction
  - Resolution—a period of time when customers can report problems
  - Installation—work that must be done at the customer site
  - Maintenance—postcontract work
  - Acceptance—when a customer formally accepts the final results
  - Documentation—for maintenance, use, certifications, calibration
- Review for schedule problems
  - Gaps when no tasks are being done
  - Too many concurrent tasks
  - Too much/too little detail; charts can be broken into subcharts to isolate detail
  - Associate people and resources to tasks
  - Internal time constraints you can control versus external time constraints
  - For each task have an optimistic best case time, and a pessimistic worst-case time
  - Critical—are the tightest time constraints given buffer time?

Managing

- Organizing
  - Complexity—find a reasonable point between too much and not enough information
  - Approval—get approvals and support before moving forward

Allocate—consider planning one of the actions that needs to be planned: allocate time to do it
Experts—involve others in the planning process to get expertise you don't have, or to verify your thinking
Buy-in—ensure that all of the critical people care about project success
Approvals—Who needs to approve what project stages?
Escalation—What triggers management and customer involvement?
Objectives—for each stage the outcomes should be clearly communicated and agreed on

- Communication
  Written—plans take away the ambiguous details, and help everybody work together
  Visual—avoid using words to describe plans; they provide room for misunderstanding
  Individualize—plan details should be pared down to the critical for each person/group
  Follow-up—ensure details are received and agreed on
  Structure—find the structure and terminology for clients and suppliers and use their language
  Information—Who needs to be informed of project progress?
  Sharing—information should be shared with stakeholders to make them part of the process
  Considerate—too much correspondence will exhaust interest
  Ignored—simply sending a document does not mean it has been read
  Schedules—dates should be set and reminders sent a week and a day ahead
- Strategy
  Strategy—keep the plans simple and easy to implement
  Uncertainty—minimize risk and uncertainty first
  Plan B—have backup plans for risky items
  One choice—plan for the future: things that are done cannot be changed without new plans
  Arbitrary—decisions with no firm outcome should be approved early

Technical

- Expertise—what design functions are anticipated? This determines the types of roles on a design team. (These are listed in approximate order of responsibility and authority.) Note: All of these roles are important in a successful design. There should be people assigned to each explicitly:
  Marketing/product manager—makes major market/customer decisions
  Design engineer—makes major technical decisions and assesses results
  Manufacturing engineer—makes decisions about production of product
  Designer/engineer—does detailed design work based on major decisions
  Quality control engineer—evaluates quality problems and opportunities
  Materials specialist—selects materials
  Industrial designer—makes aesthetic decisions, typically an artist
  Drafter—completes drawings of parts
  Technician—builds, tests, evaluates product
  Vendor/supplier representative—a product manager from another company
- Design detail
  Volume—is the design for a small batch or mass production? Smaller batches don't require as much effort in refining the design. The design cost becomes a significant part of the final cost.

Incremental or revolutionary—Is the design an improvement or redesign of an existing product, or is it new? Product improvement/variation can be done with minimal technical effort. The amount of technical effort increases dramatically as we go to a new design.
Modules—How does the design naturally break into manageable parts? The team could have a constant membership in a static structure, or can have shifting responsibility and structure. An innovative design must allow more freedom and a dynamic structure. A well-defined design should use a clear structure and set of tasks.

Logistics and delivery

- Installation—rigging, electrical services, water services, other services, facilities need, safety equipment, calibration
- Shipping—time required, shipper available, special needs, multiple parts, legal, customs
- Qualification—testing, sample runs, adjustments
- Documentation—user, engineering, training, operators, legal, certification
- Delays—limited times or delays for receiving, shipping, installation
- International—shipping delays are longer, and the cost is higher, for longer distances

ESTIMATING

- Labor estimates
  - Durations—nominal, high, low, and worst-case times
  - Time off—breaks, vacations, training, meetings, sick days, education
  - Proficiency—skill levels, training
  - Commitment—overtime (reduced productivity), other jobs (task switching), effectiveness
  - Workers—employees may be able to provide opinions on estimated time and costs
  - Support—other staff that may not be seen, e.g., purchasing
  - Contract employees or temporary employees
- Suppliers
  - Hidden—costs for insurance, loss, surcharges, customs
  - Shipping—Who will pay the costs, and how much are they?
  - Subcontractors
- Knowledge
  - Delphi estimation—brief people together or separately. While separated, have them respond with estimates and opinions. Combine these, and ask new focused questions. Repeat as necessary. These can then be used to establish a range of estimates or possibilities.
  - Consultants
- Benchmark projects
  - Budgets
  - Time logs
  - Correspondence—talk to people about everything
- Data sources
  - Published standards
  - Bottom-up (roll-up)

Spreadsheet or project planning software
Use alternatives and average results
Analogies
Parametric scaling
Use experts
Include contingency reserves
Equipment schedules
Ask people who do the work
Schedules for competing projects
Do trial runs, pilot studies, prototypes, and so on.

A.2 PROJECT DETAILS

This list is designed to prompt thoughts about what could be included in a written project plan or report. When proposing a project it is essential to address more than the technical issues.
Timing and schedule

- Lists
    - Milestones: start and end
    - Major reviews, tests, and approvals
- Sequences
    - Timeline diagrams
    - Backup plans for risky tasks
    - Identification of the critical path
    - Identification of fallback paths for tangible risks
    - Slack time for failures
    - Competing work demands
- Risks
    - Tasks with a large variance
    - Tasks that may fail
    - Backup tasks

Deliverables for the sponsor

- Clarity
    - Delivery dates
    - Specifications
    - Delivery details
- Acceptance
    - Testable specifications
    - Tests for acceptance
    - Tests for performance bonuses
- Risk
    - Likelihood of failure on a deliverable and backup plan

Benefits and costs

- Strategic (sponsor and design team)
  - How the plan fits the business requirements and objectives
  - Specific reference to the mission and vision
  - Return on investment calculations
- Failure
  - Outcomes if the project fails
- Resource cost
  - How the project affects other priorities
  - Opportunity costs
  - Resources required
  - Potential losses
  - Use internal billing rates for people, equipment, and facilities
  - Recognize the cost for services and equipment not normally billed
- Resource benefits
  - Potential income and profits
  - Financial gain or cost reduction
  - Improved productivity
  - New capabilities or business opportunities
  - Regulatory compliance
  - Competitive advantage

Responsibility

- Authority
  - Spending limits before requiring approval
  - Permissible budget deviation (positive and negative)
  - Escalation procedures for budget and timing issues
  - Human resources
  - Resources from other departments
  - Change or work priorities
  - Levels of approval for project stages, changes, initiation, and completion
- Accountability
  - Costs or benefits of failure or success
- Administrative
  - Communication and reporting times and types
  - Budget tracking
  - Customer relations

Project resources

- Budget
  - A working budget with cash flow projections
  - Discounted cash flow for large projects over a year in duration
  - Detailed budgets for prototype, production, testing, or otherwise

Shipping costs in the budget
Detailed budget and purchasing plans
A budget listing each of the parts that must be purchased/acquired. Catalog pages and quotes can be used to validate the budget. In the final report, copies of receipts, or catalog pages will be required.

- Human
  - Who is doing which task?
  - Who needs to know?
  - Who needs to be consulted?
  - Who approves the final results?
  - Special work instructions
  - A clearly defined outcome
  - Available resources
  - Authority
  - Risks in staffing
  - Issues if the schedule changes
- Equipment and facilities
  - What equipment is needed?
  - What space is required?
  - Is anything in short supply?
  - Plans for limited resources such as computer numerical control machines or surface mounting equipment
  - Does anything new need to be purchased or rented?
  - Special requirements for parts such as a dark room, cold/heat, humidity, forklift, storage, etc.
  - Issues that may occur if the schedule changes
  - Special needs such as power (e.g., three-phase, 440 V AC), air (620 kPa 90 psi), fume hood, drains, water supply, and so on.

Communications and documentation

- Communications: when, between whom, format, approvals
  - Progress reports
  - Sponsor approvals
  - Invoices
  - Test results
  - Final approvals
- Design documents
  - Documentation and change procedures
  - Assessment, certification, and testing strategies
- In the absence of a formal method of initiating work (e.g., a purchase request) it is necessary to start project work phases with some sort of formal mechanisms; these can be in various forms:
  - A memo to another department
  - Written work instructions for an employee
  - An email to a colleague
  - A work request form

- Standard project plan documents
  Gantt chart: updated on a weekly basis and included with progress reports
  Budget: when changes are made, include an updated budget; the budget table should include descriptions, suppliers, quantity, price, and status
  Purchasing: status of ordered items should be indicated
  Testing: testing progress should be indicated, including any numerical results when available
  Bill of materials: detailed list of the required components

A.3 DESIGN DETAILS

This section offers a list of technical design details often found in reports. Naturally there are additional details that may be specific to a discipline and company. If an expected item on the list has been omitted, consider justifying the reason in the report.

Needs

- Assumptions
- Constraints
- Scope: What is and is not part of the project?

Concepts: describe the system at a higher level

- Drawing summary: selected isometric and assembly drawings
- System block diagrams
- Sketches

Specifications and performance levels

- Description of control scheme, such as the motion profile
- Schematics
- Calculations: free body diagrams (FBDs) and differential equations
- Project budget and bill of materials (BOMs)
- Weight inventory: itemized by each part of the design

Testing and prototyping

- Simulation results
- The tests that were done to describe the overall performance
- The results of formal tests should also be described
- A comparison of specified and actual performance

Machines

- Parts
  Assembly drawings
  Detailed part drawings
  Dimensioned with tolerances suitable for manufacturing
  Justification for tolerances less than 0.1 mm or 0.005 inch
  A detailed material list

- Strength
    - Force calculations for all critical members; FBDs required
    - Hand calculations for stress concentrations
    - Factor of safety calculations
    - Finite element analysis (FEA) verification of hand calculations
    - Material selection based on properties
    - A testing plan for part failure
- Fasteners
    - A list of fasteners
    - Verification of strength
    - Nuts, washers, bolts, heads, threads
    - Fastener torques
- Dynamics
    - Kinematics for moving mechanisms
    - FBDs and equations of motion
    - Dynamic forces
    - Fatigue from cyclic loading
    - Modes of vibration
    - Sound and vibration control
- Manufacturing
    - Welds
    - Process plans
    - Surface finishes

Structures and civil works

- Drawings
    - Topography
    - Soils composition
    - Geotechnical plans
    - Joints and welds to ASTM codes
- Services
    - Potable water, sewage, runoff
- Environmental
    - HVAC and airflow
- Thermal
    - Insulation and building envelope
    - HVAC
    - Lighting
- Safety

Electrical and industrial controls design

- Wiring and electrical design
    - Safety circuitry
    - Panel layout: component layout and conductor placement

Component lists and specifications
Peak and nominal current loads for fuses, service, and conductor sizes
Ladder wiring diagrams with wire numbering
- Controller
Controller CPU, input/output (IO), and module selection
Human–machine interface (HMI) designs and user model
Program design and ladder logic or similar
Communication layout
List of all IO points
- Fabrication
Grounding and shielding
Heat dissipation
Enclosures
Strain relief and conductor protection
Wire management
- Related
Pneumatic or hydraulic drawings
Power factor calculations

Feedback controls

- Basic design
Block diagram of the control system(s)
Block diagrams showing the system architecture
Expected inputs and responses
- Motion planning
Dynamic calculations
Motion profiles
Homing and calibration
- Analysis
Lumped parameter models of components
Numerical or hand simulation of the system responses
Root-locus and Nyquist plots for stability

Electronics

- Schematics and systems
Block diagrams
Analog circuits
Digital circuits
Links to specific and alternate circuit parts
- Printed circuit boards
Board layouts and minimum feature sizes
Signal propagation times
Cross talk and noise
Decoupling capacitors

Multiple layers for traces, ground planes, and vias
Double-sided boards
Accessible test points
Heat sinks

- Fabrication
    Connector size and placement
    Through-hole components
    Limits on trace size component spacing
    Plans to mount complex parts, including MLF, BGA, and very small surface-mount discrete parts
    Component tolerances and specifications
- Analysis
    Space and hand calculations for simulations
    Considerations for component lead and trace inductance and capacitance
    Thermal effects on components
    Controller design
    Specific parameters and tolerances for components
    Electromagnetic (EM) and radio-frequency (RF) emissions

Software design

- Software architecture
    System block diagram showing hardware and software
    Diagrams of software modules and layers
    Timing and state diagrams for sequential systems
    User models including objectives and assumptions
- Implementation
    Data structures
    Application programming interfaces (APIs) and detailed functional specifications
    Software modules and test harnesses/routines
    Special algorithms
    Error trapping, error recovery, fault tolerance, and fail safe
    Separate threads for graphical user interface, communications, and calculations
    Structured design methods across all modules
    Peripheral drivers, register construction, etc.
    Real-time process priority, preemption, and deadlock
- Analysis
    Scalability and order analysis
    Deterministic operation
    Testing and verification

Facilities and production

- Design for assembly (DFA)
    Bottom-up assembly
    Fewer operations
    Minimum tolerance stacks

Minimum tools and handling
Fixtures for complex assemblies
Easy-to-orient parts

- Parts and design for manufacturing (DFM)
  - Simplify the part, loosen tolerances, and combine manufacturing steps
  - Machine loading and unloading
  - Tooling purchasing
  - Tooling and equipment wear
  - Maintenance plans
  - Process plans
- Inventory
  - Bulk delivery to the first machine and direct shipping from the last
  - Effortless production with minimal setups
  - Storage plans
  - Shipping plans, including damage control
  - Unique part IDs
  - Supply-chain management
- Lines, jobs, or batches
  - Material handling plans, including belts, buffers, carts, pallets, hands carry, etc.
  - Material handling times
  - Metrology and quality control (QC) stations
  - Finishing, including painting, packaging, palletizing, and final inspection
  - Labeling
  - Operator requirements
  - Manual versus automatic methods
  - Physical layouts and the current space
- Analysis and planning
  - Lean manufacturing methods
  - Variability and C_p and C_{pk} for each process
  - Six-sigma quality limits
  - Quality control plans
  - Material requirements planning (MRP) analysis
  - Simulation
  - Noise, vibration, contaminants, and pollution
  - Scrap material cleaning and disposal
  - Work study for similar designs
  - Standard time estimation methods
  - Ergonomics

User or graphical interfaces

- User model
  - Various users, such as operator, consumer, maintenance
  - Expected sequences of operation and alternates
  - Size, color, language(s), terminology
  - Information output and user input at each stage

- Implementation
    - Look and feel of the screen: colors, graphics, layouts, buttons, touch, transitions
    - Screen sequences
    - An interaction plan: flowchart, state diagram, or script
    - A warning, error, panic, and fault plan
    - Accessibility features (sound, touch, etc.)
    - Undo and redo functions
    - User help
- Programming
    - Hardware interfaces to the main program, process, or controller
    - Shared memory and variables with other execution threads
    - Flags between the interface and other threads

Packages, enclosures, and aesthetics

- Appearance
    - Finish colors, patterns, and textures
    - Touch
    - Smell
    - Shape
    - Solid models rendered for appearance
    - A rapid prototyping (RP) model for look and feel
    - Consumer surveys of the look and feel
    - Customer preapproval
- Enclosure
    - Water, pressure, humidity tight
    - Heat or freezing resistant
    - Impact and vibration resistance
    - UV fading and tarnishing
    - Scuffing, scratching, and abrasion
- Implementation
    - Paint, powder coat, pigments in plastic, printed paper or film, brushed aluminum, etc.
    - Exposed surfaces and handling in production
    - Blister packs, twist ties, tape, elastic
    - Shipping protection: foam, spacers, packing materials

Delivery and commissioning

- Visual inspection
    - Verify that the machine meets internal and external safety codes, such as electrical codes (NEC), worker safety codes (e.g., OSHA).
    - Determine if all components are present.
- Mechanical installation
    - Physically locate the machine.
    - Connect to adjacent machines.
    - Connect water, air, and other required services.

- Electrical installation
    Connect grounds and power.
    Perform high-potential and ground-fault tests.
    Verify sensor inputs to the programmable logic controller (PLC).
- Functional tests
    Start the machine and test the emergency stops.
    Test for basic functionality.
- Process verification
    Run the machine and make adjustments to produce an acceptable product.
    Collect process capability data.
    Determine required maintenance procedures.
- Contract/specification verification
    Review the contract requirements and check off each one.
    Review the specification requirements and check off each one.
    Request that any noncompliant requirements are corrected.
- Put into production
    Start the process in the production environment and begin normal use.

A.4 MEETING DETAILS

The following list should be useful if you are planning a meeting. No meeting would be expected to use everything in the list; however, it should provide a reminder if anything is forgotten. It can be remarkably effective to imagine the meeting experience from beginning to end, for all involved, including support people. Planning is more important when guests are coming for the first time.

- Big picture
  - Purpose for the meeting
  - Meeting participants
  - Objective for the meeting: communication, networking, decisions
  - Date and location
- Schedule
  - Social and networking time
  - Break times
  - Breakfast, lunch, and dinner for long meetings and out-of-town participants
  - Avoid overlap with other meetings
  - Avoid holidays celebrated by any of the participants; prayer times may also be needed
  - Less than an hour, hours, full business day, or extended business day formats
  - Tours and demonstrations
- Meeting room
  - Enough table space for all attendees
  - Clear line of sight from all seats
  - Exhibit space for displays, parts, etc.
  - Screens, projectors, sound systems, etc.

- Power plugs for laptops
- Wireless or wired Internet
- Teleconferencing capabilities
- Water, coffee, and other refreshments
- White boards, markers, and erasers
- Location of bathrooms

- Staff
  - Staff to organize the meeting and welcome guests
  - Catering for lunches; remember vegetarian and nonpork food
  - AV and computer setup and help
  - Duplicate paper copies of handouts
  - Name tags if many new people are in the room
  - Provide recording equipment or minute-takers
  - Contact participants to coordinate and verify attendance
- Ensure all participants receive the following
  - Agenda including start and end times
  - Agenda with expected outcomes
  - List of participants
  - Request for available meeting times: use open days over a few weeks
  - Meeting location and directions
  - Parking details and passes if necessary
  - Local issues such as construction, events, etc.
- Traveling participants
  - Travel, lodging, and transportation details
  - Premeeting preparations
  - AV needs; e.g., bring a laptop or USB drive
  - International participants may require accommodations such as electrical converters, interpreters, letters, or documents for visas
  - Jet lag compensation
  - A meeting end time before airplane departures.

A.5 TRIZ CONTRADICTION CATEGORIES

Contradictions are a pair of items, one of which is to be maintained while the other is to be increased or decreased:

(1) Weight of moving object: gravity acts on a moving mass and causes forces on other components
(2) Weight of stationary object: gravity acts on a stationary mass, causing forces in support surfaces or components
(3) Length of moving object: one linear dimension of a moving object
(4) Length of stationary object: one linear dimension of a static object
(5) Area of moving object: a moving object in an area or the area of the moving object
(6) Area of stationary object: a static object in an area or the area of the object

(7) Volume of moving object: a moving object in a volume or the volume of the moving object
(8) Volume of stationary object: a static object in a volume or the volume of the static object
(9) Speed: a rate of action or velocity
(10) Force: as defined by physics
(11) Stress or pressure: as defined by physics
(12) Shape: the appearance or external geometry
(13) Stability of the object's composition: the micro and macro structures remain unchanged, including chemical, microstructure, wear, and assembly
(14) Strength: the point of failure
(15) Duration of action by a moving object: lifetime durability or length of an action
(16) Duration of action by a stationary object: lifetime durability or length of action
(17) Temperature: thermodynamic heat or energy
(18) Illumination intensity: anything to do with light
(19) Use of energy by moving object: as defined by physics
(20) Use of energy by stationary object: as defined by physics
(21) Power: as defined by physics
(22) Loss of energy: wasted or discarded energy
(23) Loss of substance: wasted or discarded mass
(24) Loss of information: wasted or discarded data, appearance, sensory feedback, ability
(25) Loss of time: the reduction of activity time
(26) Quantity of substance/the matter: useful properties of structural and consumable materials
(27) Reliability: ability to perform
(28) Measurement accuracy: the standard deviations of accuracy
(29) Manufacturing precision: the difference between desired and actual measurement average/mean
(30) External harm affects the object: ability to resist damage
(31) Object-generated harmful factors: ability to cause damage
(32) Ease of manufacture: effort and resources
(33) Ease of operation: effort-to-results ratio
(34) Ease of repair: effort to return a device to operation
(35) Adaptability or versatility: sensitivity, functions, responsiveness
(36) Device complexity: number of parallel and sequential operations in the system or required for operation
(37) Difficulty of detecting and measuring: indirect measurements, high noise-to-signal ratio, inadequate detection, complicated by other functions
(38) Extent of automation: self-monitoring and adjusting, less human effort
(39) Productivity: system activity or output levels

A.6 TRIZ DESIGN PRINCIPLES

These principles are used to change designs. Most TRIZ design suggestions will include a few of the following as options. A few of these will overlap or contradict one another. The terms are somewhat abstract so that they can be interpreted for a variety of applications. An analogous tool is used by fortune tellers so that a subject is able to interpret his or her situation into the answer. In this case, that effect is beneficial in generating new perceptions of design cases:

(1) Segmentation: more pieces
(2) Taking out: separating effects or components
(3) Local quality: overall differentiated materials, functions, structures
(4) Asymmetry: increase the number of unique geometries or specialize functions
(5) Merging: reduce unique components by combining or reusing geometries and functions
(6) Universality: the number of functions and/or specifications performed by the device
(7) Nested doll: telescoping or components combined for compressed storage
(8) Antiweight: buoyancy, lift, counterbalance weights and springs
(9) Preliminary antiaction: safety measures, selective protection or preparation
(10) Preliminary action: preuse preparation
(11) Beforehand cushioning: anticipate damage mitigation
(12) Equipotentiality: reduce the transfer or increase the similarity of energy levels, complexity, mass
(13) The other way round: reverse the effect, logic relationship, action/object, or common sense
(14) Spheroidality/curvature: use rounded geometries and actions
(15) Dynamics: add motion, articulation, adaptability
(16) Partial or excessive actions: accept less than optimal features or operations
(17) Another dimension: add a degree of freedom, another variable or function, another component, another layer
(18) Mechanical vibration: modify or use vibrations to perform or enhance a function
(19) Periodic action: modify or add a sinusoidal, cyclic, repeating function
(20) Continuity of useful action: replace partial or periodic functions with continuous effort, output, or operation
(21) Skipping: increase a parameter to bypass unwanted effects
(22) Blessing in disguise or "turn lemons into lemonade": turn a negative into a positive by reapplication, addition of a new function, or transformation into a useful form
(23) Feedback: monitor/measure system state and performance to adjust behavior
(24) Intermediary: add a temporary material, function, or operation
(25) Self-service: functions and materials reheal, adjust, or reuse themselves
(26) Copying: reproduction and repurposing to reduce repetition of effort
(27) Cheap short-living objects: reduce cost with inexpensive items, possibly reducing performance and increasing effort
(28) Mechanics substitution: change from one physics effect to another
(29) Pneumatics and hydraulics: use fluids or gases to distribute, store, or transfer energy
(30) Flexible shells and thin films: replace solids with shells; use surfaces as functional elements
(31) Porous materials: introduce holes or pores for storage, barriers, transmission, or weight reduction
(32) Color changes: use optical color or transparency to alter light reflection or transmission
(33) Homogeneity: decrease material types and property changes
(34) Discarding and recovering: components, materials, or functions are discarded after use; spent components are recycled or reused
(35) Parameter changes: change a material parameter to change the state of material properties, mechanics, and behavior

(36) Phase transitions: as materials change forms use the physical, or energy, property changes
(37) Thermal expansion: use the temperature-changing geometry of one or more materials
(38) Strong oxidants: oxidants can adjust combustion and reaction rates
(39) Inert atmosphere: inert substances act as insulators for chemical reactions; material properties can be adjusted with inert materials
(40) Composite materials: combine materials for new micro- and macroscopic properties

APPENDIX

Technical writing

B

B.1 INTRODUCTION

Simply put, writing is about the details; the words are secondary. A design report might include tables, drawings, part lists, calculations, procedures, source code, and schematics. These graphic and technical details are essential to make the report understandable; the text adds explanation and context. A technical report with only text is very difficult to read and requires substantial effort by the reader. Many people read textbooks by looking at the figures first and then read only if the figures are not clear. Readers value writing that provides the right details at the right time. Naturally, readers lose interest if the writer wanders off message, is not concise, and is not clear. For examples of good writing, find some textbooks that are highly regarded.

Consider how you read textbooks. As mentioned, most readers will look at the figures and details and then read text to resolve confusion; some "readers" avoid reading the text altogether, and focus only on the details. You should always keep this in mind when communicating: details first. Make it easy for the reader. Technical readers are less likely to read a report from beginning to end, though some will. The main approaches to reading technical documents are to (1) read everything, (2) skip to sections of interest and read only those, (3) skim and read selectively, or (4) skip or skim and look only at the figures. To write effectively for this audience you must assume that each section is self-contained and is easy to locate. Providing a visual cue, such as a title or figure, will help draw readers to the section. If the section relies on other knowledge, provide links or references to the other sections. In the first paragraph tell the reader what the section is about and how it concludes.

Some of the aspects of technical writing are listed below. The key principle is that busy professionals are paid to write and read the documents. Ideally they are written well the first time. They are clear, concise, and correct, and the needed information is readily available. The credibility of a report is based on the evidence it contains. This evidence then supports the conclusions drawn or message given by the author. The key in all written reports is that they can travel a long distance outside a company and become a formal record of commitments:

- The purpose of the writing is clear, including decisions, recommendations, and conclusions.
- The format of the writing is standard and well known.
- Somebody will use it.
- The report contains many details.
- The report may break various creative writing principles. Entertainment is not the primary purpose.
- The report can have legal implications or be required by law/regulation.

PROBLEM

B.1 Why should technical writing be as concise as possible?

B.2 REPORT AND DOCUMENT TYPES

Assume that everything you say or write is only "skimmed."

Routine business documents include memos (memorandums), meeting agendas, meeting minutes, and notices. Table B.1 shows a variety of other documents that are generated and distributed by groups. Naturally, this list is only a small sample of all of the different engineering functions and documents generated by engineers.

Every industry, company, and engineering profession has its own document types. For the purposes of brevity, broad categories are provided for some very specific engineering functions. This section discusses some documents used in various circumstances, but is by no means complete. (One could argue that the lists of departments should include facilities, research and development, information technology, and more.)

Engineering departments often receive documents that must be processed. Table B.2 shows a variety of documents that are received by the various departments.

Some of the standard documents are described in the following list. In some cases the format and content are specified by the customer, company, or profession. A common form is memos, short letters to let others know about changes. Given the free-form nature of memos, they often travel within departments for information and requests. Depending on the destination, these documents have varying levels and types of details. Letters of transmittal are similar, but for groups outside the company. For example, the accounting department wants to know about expenses and how to apply them to specific jobs. Managers want to know about the need for new resources and the progress of existing work. The customer primarily cares about the progress of their order and expected delivery dates:

- Memorandum: A memorandum, or memo, is an internal business communication or brief technical report designed to convey a business policy or technical information. The standard memo starts with the following header:
  Memorandum
  Date:
  From:
  To:
  Subject:
  cc:
- Letter of transmittal or notice: When a document is sent between companies, a letter of transmittal is often sent to describe the purpose of the document. A notice is similar to a memo, but it is directed to a customer for information purposes.
- Interim or progress report: A progress report provides details of project progress to a supervisor or customer. Teams are sometimes required to submit progress reports as often as weekly. These reports may include several elements divided into sections with a heading for each. Point form is common, but complete sentences should be used. Each section should include items completed since the last report, as well as current action items. If there is nothing to be said about a category use "no changes," "nothing done," or "complete" as appropriate.

Table B.1 Typical documents sent by design, manufacturing, and quality departments.

Destination department	Source departments		
	Design, research, development	**Manufacturing, fabrication, construction, assembly**	**Quality, testing**
Design	–	Engineer change requests Change notices	Designs of experiments Prototype laboratory reports
Manufacturing	Production methods Engineering change notices	–	–
Sales/marketing	Specifications and data sheets Market studies Life-cycle assessments	Production forecasts	Environmental audits Quality audits
Quality/testing	Quality requirements	–	–
Accounting/purchasing/finance	Material requirements Bills of materials for projects Employee work reports Purchase requests	Material orders Employee work reports Purchase requests	Purchase requests
Operations/facilities	–	–	–
Inventory/logistics/shipping	–	–	Material/part certifications
Human resources	Position descriptions Annual reviews	Position descriptions Annual reviews	Position descriptions Annual reviews
Manager/legal	Project proposals Project reports Executive briefs Invention disclosure forms Patent support information	Project proposals Project reports Production reports	Project proposals Test result summaries
Customer	Letters of transmittal	Letters of transmittal	Letters of transmittal
	Quotes (proposals)	Shipping details	Quality reports
	Manuals	–	Certifications
	Specifications	–	Laboratory reports
	Consulting	–	Consulting reports
	Technical and application notes	–	–
Other external	Letters of transmittal or notice Regulatory reports	Letters of transmittal or notice Patents	Letters of transmittal or notice Environmental reports Standards testing

Table B.2 Typical documents received by design, manufacturing, and quality departments.

Source department	Destination departments		
	Design, research, development	**Manufacturing, fabrication, construction, assembly**	**Quality, testing**
Design	–	–	–
Manufacturing	–	–	–
Sales/marketing	Specifications Prototypes and pictures Requests for quotes or proposals Design change requests	Orders or sales projections Order change requests	–
Quality/testing	–	–	–
Accounting/ purchasing/finance	Expense account summary invoices (for verification)	Expense account summary invoices (for verification)	Expense account summary invoices (for verification)
Operations/facilities	–	–	–
Inventory/logistics/ shipping	–	Bills of lading Packing lists	–
Human resources	–	–	–
Manager/legal	Procedure manuals	Procedure manuals	Procedure manuals
	Project proposals	Project proposals	Copies of contracts
	Project reports	Copies of contracts	–
	Executive briefs	–	–
	Copies of contracts	–	–
Customer	Letters of transmittal Requests for changes	Letters of transmittal Failure and defect reports	Letters of transmittal Certification reports Laboratory reports Consulting reports
Other external	Letters of transmittal or notice Standards and regulations presentations White papers Patents Studies	Letters of transmittal or notice Environmental policies Presentations/papers	Letters of transmittal or notice Regulatory policies Presentations/ papers Independent test reports

- Executive summary or brief: This is a condensed version of information to prompt or answer specific questions that a manager might have. These can be less than one page in length or a few pages for a major project.
- Study or consulting report: This is a report on a study or research done to investigate a topic of interest. The important results are captured. These reports often end with conclusions, suggestions, or recommendations.

- Certification and regulation: Government-mandated and voluntary standards often call for publicly filed reports on health products, environmental concerns, transportation safety, fire safety, electromagnetic emissions, an so on. These often come in defined formats and content. They normally require design, manufacturing, and testing details. The standards for health and aviation safety products are often noted as examples of complexity.
- Intellectual property
  - Patent: This is documentation that an idea was developed and disclosed by an inventor, giving him or her 20 years of legal protection.
  - Invention disclosure form: These are used to inform a company about something that might lead to a patent.
- Expense claims
  - Travel report: This describes interactions with the customer, knowledge learned, issues, etc.

Some business communications, such as reports and expense claims, are done using forms. These allow information to be provided in a condensed format that is tailored to an application. For routine and repetitive information, forms are easily constructed to obtain a complete set of information in a format that is easy to read. Although mundane, engineers may consider developing and using forms for items such as design change requests, part orders, quality reports, work orders, customer estimates, equipment schedules, sales engineering work sheets, quote check sheets, machine shop instructions, printed circuit board (PCB) design rules, test data reports, and on-site data commissioning. A paper or electronic form is laid out to collect the essential required, and other, information as necessary. Engineers designing systems that involve other people should consider developing forms to guide the flow of information.

PROBLEMS

B.2 List five documents for communicating between the accounting department and manufacturing.

B.3 What are five characteristics of a memo?

B.4 Why do businesses develop forms?

B.2.1 PROJECT DOCUMENTS

Project documents allow the developer or team to record all of the design decisions made during the course of the project. This report should also mention avenues not taken. Quite often the projects that we start will be handed off to others after a period of time. In many cases they will not have the opportunity to talk to us, or we may not have the time, so the project report serves as a well-known, central document that includes all relevant information. The project report should include the following:

- Project charter: a document to launch a project
  - Summary: title, initiating person, start and end dates, customer, crude labor, budget estimates
  - People: a list of key people and groups involved within the team
  - Scope and objectives: an outline of the project at a high level

  - Stakeholders/audience: a list of other major groups involved outside of the team
  - Primary stakeholders: management, related departments, unions, creditors, customers, suppliers, contractors, legal and regulatory bodies, employees, etc.
  - Secondary stakeholders: social/political groups, competitors, communities, public-service groups, professional groups, media, schools, hospitals, families, etc.
  - Timeline: major milestones and dates
  - Approvals: signed or equivalent permission to start
- Statement of work (SOW): defines the scope of the project
  - Summary of details
  - Approach: the expected path for the work
  - Strategies: priorities and fallbacks
  - Timeline
  - Scope: what is and is not part of the project
  - Assessment metrics
  - Processes for communication and agreement between customer and supplier
- Request for proposals; typical elements include the following:
  - SOW
  - Requirements for work
  - Deliverables
  - Access, services, equipment provided
  - Approvals required
  - Contract type
  - Major dates and schedule
  - Financial details—payments and hold-backs
  - Proposal format—approach, deliverables, schedule, budget, ability to do work
  - Evaluation criteria—lowest cost, ability to do work, etc.
- Quotations or contracts: formal documents that outline the work to be done and obligations; normally include delivery schedules, specifications, warranties, and payments

Details are critical in these documents. Some of the lists provided in Appendix A are thought starters for project and detailed design documents. Additional components will be needed for special design work, company requirements, and so on. For example, parts for aviation and medical markets have particularly complex design requirements and documentation.

Manuals and user documentation are normally required at the end of a project. These can be written by the engineers or by a technical writer. In either case the engineers are responsible for generating the technical content in a form that is suitable for the audience. Additional writing, typesetting, illustration, proofreading, and legal text may be added after. Typical examples of manuals and similar documents are shown in the following list. Sometimes documents, such as a quick-start guide, can be left until the end of the project. Others should be created as the project progresses, for example, documentation for a programming interface:

- User manuals
- Programming manuals
- Maintenance manuals
- Operator guides

- Training materials
- Safety labeling
- Design manuals
- Reference designs and design guides
- Tutorials and quick-start guides
- Specifications
- Marketing materials
- Certification paperwork for legal reasons
- Test results and qualifications

For the most part, these guides should be written at or a little below the target audience. Always assume that the documents will not be read front to back. Many people do not refer to these manuals until there is a problem and they need to find a clear answer. Well-written quick-start guides and manuals can avoid customer frustration, rejected products, and help-desk calls. Good practices to use when creating manuals include the following:

- Provide safety warnings in all places related to safety issues.
- Use pictures when possible.
- Include warnings and legal disclaimers.
- Provide revision numbers and dates.
- Use numbered steps for procedures.
- Provide checklists for inspections and maintenance.
- Include troubleshooting information (don't forget error codes).
- Keep it specific, complete, and easy to understand.
- Accuracy is key; all of these documents can result in liability issues if incorrect or incomplete.
- An index and detailed table of contents will make it easy to find details quickly.
- Employ a technical writer or graphic designer for public documents.
- Binding, for paper documents, and electronic formats are useful.
- A website is helpful when customer copies of documentation are lost.

Commissioning and acceptance reports are useful when a new design is being passed from a design team to a customer during a hand-off procedure. These reports document the testing results that show that the design satisfies all of the specifications and other standards. They are also used to document agreement between designers and customers. The commissioning or acceptance report should contain the following:

- Inspections: visual, mechanical, electrical
- Installation: mechanical, electrical, production, safety
- Testing: operates, meets specifications, reliable
- Other issues: deficiencies, maintenance, etc.

PROBLEMS

B.5 List five documents that might be delivered to a customer at the end of a project.

B.6 Why are project charters and statements of work used to start projects?

B.7 Give a reason a user manual might include a schematic or mechanical drawing.

B.2.2 TECHNICAL DOCUMENTS

Write reports, not mystery novels: give away the ending.

Design proposals, theses, and final reports have overlapping content but with different objectives (see Resource B.1). The design proposal is used to present all of the design details in a single document. Throughout the project the amount of design detail in the body increases. At the end of the project the test results section is completed. A great deal of attention is paid to the concepts, embodiments, budget, timeline, and other project planning details at the beginning of a project; however, these may be deemphasized or omitted in final reports.

Typical proposal and report elements are shown in the following list. It is worth noting that some organizations may not require a written report, but they do need the technical content of the report:

- Cover page
  - A title that allows the work to be easily identified
  - The name of the authors or originating company, department, class, etc.
  - Publication date or document tracking number
- Executive summary or abstract
  - A very concise overview of the work and outcomes
  - Unlike fiction, it should give away the end of the story
  - The reader should know what to expect if he or she reads the report
  - Many readers will get their information only from the summary, so make it count
- Table of contents: for documents with more than 10 pages
  - Should provide the major topical divisions for easy access
  - May include lists of tables and figures
- Nomenclature: for documents with extensive calculations
  - List all variables used in the report
  - The order is uppercase and then lowercase, alphabetically, with regular letters followed by Greek
  - Descriptions and units should be included
- Introduction
  - Provide the motivation and objectives for the work
  - Give essential details
  - Research should be documented with a literature review
- Design: these sections are included when design work has been done
  - Background: sponsor project needs
  - Specifications
  - Concepts
  - Embodiment
  - Detailed design
- Construction or fabrication
  - Equipment and materials required
  - Special instructions
  - Discuss equipment, tooling, components, testing, etc.
- Testing

    - Testing objectives or hypothesis
    - Experimental procedure
    - Data collection and observations
    - Basic data analysis and hypothesis testing and verification
    - Statement of hypothesis proof or qualification
- Conclusions and recommendations
    - Discuss testing results
    - Outline new knowledge and lessons learned
    - Recommend a future course of action
    - Discuss the fitness of the design or work with respect to the original motivation
    - Summarize the report content so that the reader can verify his or her knowledge
    - Repeat the significant results from the body of the report that the reader must know
    - Tables and graphics can be useful for effective presentation
- Appendices: for essential detail too large for the body of the report
    - Drawings
    - Schematics
    - Source code
    - Detailed budgets
    - Calculations
    - Extensive data

Resource B.1 A template for a design proposal or final report is included on this book's website: www.engineeringdesignprojects.com/home/content/communication-and-documentation.

Working notes and notebooks are maintained by most professionals throughout the day. In meetings they will keep track of who said what, what commitments were made, when things are due, problems, successes, and so on. At a minimum, these are used as reminders. In practical terms notes are used for generating reports and more. These are often done on pads of paper, in notebooks, or by computer. When used as a legal record for patent and liability reasons the process is formalized. Legally acceptable notes are written in pen in engineering notebooks with numbered pages stitched in to prevent removal. These are reviewed and notarized regularly. The notes in the books are meant to be added sequentially as the work is done, with dates and times included. An alternative is to take voice memos and have them typed later, a very common approach in the medical fields.

At times, professionals will present information to other professionals. One example is *white papers*, which are very similar to academic papers. These are done for technical audiences with the purpose of informing and educating. White papers are produced in large numbers by companies developing cutting-edge engineering tools and materials. Similar documents include design guides, reference designs, and data sheets. These are less about education and more about providing guidance to a knowledgeable design professional.

Engineers will occasionally create materials for audiences with less time to read or those who possess less technical knowledge. In such cases a presentation or poster may be used. A poster is a large printed format that conveys the key information visually so that a spectator can grasp the concept of the project at a glance and review the key concepts in under 1 min. The layout of the poster should be very visual, favoring figures, pictures, and graphs. Good practices for posters are described in the following list.

- Begin with a purpose and motivation for the work and a conclusion that refers back to the purpose.
- Describe the approach of the work.

- Acknowledge others who have contributed to the work.
- Use colors to make it attractive (avoid gaudy appearances).
- Use visual images to speed comprehension.
- Use a high-quality printing and mounting process. Glossy paper on foam core boards is standard.
- Use large fonts and condensed text. A few bullet-point sentences with 16- to 24-point fonts is recommended.
- The poster should be self-explanatory.
- Avoid trying to present too much detail.

Testing has an objective of proving some hypothesis. Engineering examples include proof that a design meets the specifications, or the statistical deviation of a material property. In school laboratories it is common to validate academic theory and investigate natural properties. In companies, tests are used to determine operation and customer needs. There are companies that exclusively deal with testing as an independent source of certification. Test reports are often documented using the following format:

- Purpose: a clear objective or hypothesis for the test work
- Background and theory: the basis for comparing the expected results
- Procedure: the experimental method
- Equipment: a list of the components and measurement equipment
- Results: measurements and observations from the test
- Analysis and discussion: the data are processed and compared with the background and theory
- Conclusion: refer to the purpose and analysis and state whether they agree, disagree, or are inconclusive
- Appendices: large volumes of results, calculations, or design work

Test results normally follow the scientific method with some objective. The results should be designed to conclusively prove the purpose/hypothesis. As a result, most test reports include calculations, numerical readings, graphs, tables, and so on. The outcome of the test should be summarized concisely, and, it is hoped, numerically, in the conclusions. Some of the common reasons for testing are:

- proving devices meet specifications
- determining parameters for a device or system
- establishing technical design limits
- proving a prototype or concept
- evaluating predictability

PROBLEMS

B.8 Why are abstracts used on large documents?

B.9 How should a design proposal and final report differ?

B.10 Does a report need figures if everything is described in text? Explain.

B.11 How are the report conclusions related to the purpose?

B.12 Are sections on testing or fabrication required in design proposals? Explain.

B.13 What section(s) might be added to a design proposal to present customer survey data and quality functional deployment?

B.3 DOCUMENT FORMATTING

The format of a document is used to organize and convey information in a consistent way. In technical documents the formatting ensures that the figures, equations, data, tables, and text are tied together in a consistent and logical manner. The obvious rule is to select a format and then apply it consistently. Some of the common variations in technical documents are described here, including section numbering, fonts, and references:

- Page numbering: Most software makes the process of numbering pages quite simple. Before the first page of the body of the document, the pages are numbered using roman numerals. For example page "i" may be the first page of the table of contents. The sequence starts again with Arabic numerals starting at "1" on the first page of the body. Sometimes in technical documents the pages are numbered by chapter, for example, "4−7" would be the seventh page in the fourth section. This can be helpful if you want to replace or break larger manuals into replaceable parts. When pages are blank they should still be numbered and contain the words "this page left blank intentionally." This will eliminate concerns about missing or misprinted pages.
- Front matter: The title page, contents listing, and any forward, preface, or introduction are all considered front matter (they appear at the front of the document). Copyrights are added to the beginning of many documents, including manuals and reports. The front matter may also include acknowledgments of technical sources, funding agencies, and technical assistance. Dedications are often common and are brief mentions of a personal nature of friends, family, colleagues, and the deceased.
- References and bibliographies: In technical work it is important to cite the sources used. The alternative is to cite no sources and then be responsible for justifying each item of data, equation, and design decision. When readers come across a reference they may want to read more to understand the technical details. Again, without a reference you are responsible for providing these details. A simple approach to references is to numerically list the author, title, publisher, and year. If it is a manual, list the company name and manual title. When Internet references are available provide hypertext links and the date you viewed it. Use reference numbers in the report body, for example, "[4]," to find the reference. There are other reference formats commonly used, and it may be necessary to change the format near the end of the project. A bibliography is a list of materials related to the work. Footnotes are references that are at the bottom of the page that refers to them.
- Appendices: When we have information that is needed to support a report, but is too bulky to include, one option is to add an appendix. When material is placed in an appendix, it must be summarized in the body of the report. The report should briefly summarize (usually a figure, graph, table, equation, or more) and then refer to the appendix. It is expected that the material summarized in the body will also appear in the appendices. Examples of appendices include the following:

- Sample calculations: These are redundant numerical calculations or a prolonged derivation of equations. The body of the report has a summary of key assumptions, sample calculations, and results. The calculations are often provided so that the reader may verify the work.
- Long tables of data: Tables of numerical data are often put in appendices. Typically a sample of the table is included in the body for discussion purposes. The additional data are often provided for the reader who has a use for them beyond the uses in the report.
- Program listings: Long listings of computer programs are often put in appendices. They are referenced in the body of the report near the algorithm, calculation, or method they implement. These listings are provided for readers who want to use the program.
- Multiple data graphs: Multiple sets of data graphs are often put in appendices and summarized in a report body. The graphs are provided so that the reader may use the graphs for verification or further analysis.
- Reviews of basic theory: These are often referenced in the body of the report for readers who may not have seen a topic previously. These are uncommon in student reports.

- Section numbering: The standard for larger documents is 1 Chapters, 1.1 Sections, and 1.1.2 Subsections. This convention is well understood and simplifies the relationship between the table of contents and the report sections. In short reports the numbers may be omitted, but a different heading style should be used for the three types of headings, to distinguish them from one another (e.g., all caps, title case, title case and italics, etc.).
- Page size: Many documents are no longer printed, but if they are, the standard paper sizes are US Letter and A4 Metric. If distributing a document in a fixed file format (such as PDF) the author should produce two versions for each paper size, or one version that can be reasonably printed on either. Distributing documents in an editable file format (such as DOC or ODF) allows easy resizing for different printers. For electronic formats such as web pages and help files, the text should be broken into smaller separate sections, typically a screen length in size.
- Fonts: Standard document fonts are normally 10 to 12 points. Bold, italic, and underlined fonts are used sparingly for emphasis. Larger fonts are often used for section headings and titles.
- Margins: Borders of 2.5 cm or 1 inch allow visual gaps to help the reader. In addition, most printers cannot print to the edge of a sheet. For left–right facing pages an extra gap is left by the binding to compensate for the visible area lost to the center crease. Often page numbers and special headers and footers lie outside the margins.

Exhibits is a broad term covering nontext elements of a report. These include items such as figures, equations, tables, drawings, and graphs. When presenting exhibits the general principles are as follows:

- All visual elements (e.g., figures and tables) should have a descriptive title and number.
- Every exhibit should be referred to, by number, in the text.
- Image resolutions should be high enough to be clear, typically 300 dpi or more. All important detail must be visible. Avoid pixelation.
- If there is too much detail, put it in an appendix.
- Photographs and drawings should be cropped to size and clearly visible.
- Screen captures should be clear and complete, but other detail is cropped out.
- Color exhibits can still be used if printed in black and white.

Like reports, figures are used throughout this book to illustrate concepts. Each figure should have a unique title that clearly and concisely describes what is shown. Nearby text refers to each figure by number and has a related discussion. Some of the general attributes of figures are as follows:

- Figure content can include drawings, schematics, graphs, charts, etc.
- A figure should be labeled underneath, sequentially, and given a brief title to distinguish it from other graphs, for example, "Figure 1: Voltage and currents for a 50-ohm resistor."
- In the body of the report the reference may be shortened to "Fig. B.1."
- The figures do not need to immediately follow the reference, but they should be kept in sequence. Often figures are moved to make the typesetting work out better.

Graphs and charts present data in standard formats, including line, bar, pie, and scatter. Given that the data are numerical in nature, there are a number of good practices, as summarized in the following list:

- If fitting a line/curve to the points, indicate the method used (e.g., linear regression).
- Try not to use more than five curves on the same graph.
- Use legends that can be seen in black and white.
- Clearly label units and scales on each axis.
- Label axes with descriptive terms, for example, "Hardness (RHC scale)" instead of "RHC."
- Scale the curve to make good use of open spaces on the graph.

$$\overset{+}{\rightarrow} \quad \sum F_x = -T_1 \sin 60^\circ + F_R \sin\theta_R = 0 \qquad \text{eqn. B.1}$$

$$+\uparrow \quad \sum F_y = -T_1 - T_1 \cos(60^\circ) + F_R \cos\theta_R = 0 \qquad \text{eqn. B.2}$$

substitute B.1 into B.2.

$$\therefore F_R = \frac{T_1 \sin 60^\circ}{\sin\theta_R} = \frac{T_1 + T_1 \cos 60^\circ}{\cos\theta_R}$$

$$\therefore \frac{\sin 60^\circ}{1 + \cos 60^\circ} = \frac{\sin\theta_R}{\cos\theta_R} = \tan\theta_R$$

$$\therefore \tan\theta_R = \frac{0.866}{1 + 0.5}$$

$$\boxed{\therefore \theta_R = 30^\circ}$$

$$98 \sin 60^\circ = F_R \sin 30^\circ$$

$$\boxed{\therefore F_R = 170N}$$

FIGURE B.1

Sample calculation to resolve force components.

- Avoid overly busy graphs.
- Titles should indicate clearly and distinctly why the content of the figure is significant.
- Points should be drawn and connected with straight (or no) lines if experimental.
- Smooth lines are drawn for functions or fitted curves. If a curve has been fitted, the fitting method should be described. For example, Least Squares Linear Regresion.
- If using graphing software, don't put a title on the graph.

Sketches are hand-drawn or created using simple drawing software. Unlike drawings, these are not meant to be geometrically accurate. Sketches are normally used in the early stages of designs to show concepts. They will also be used for illustration in detailed design work including free body diagrams, conductor placement, and flow patterns.

Engineering drawings provide detailed geometries for solid parts. Applications of these range from part placement on circuit boards, piers for bridges, tooling geometry, bioreactor piping, and pistons. The conventions for drawings are very well understood although there are some variations between disciplines and applications. Some of the general rules for technical drawings are the following:

General requirements

- Use a title block with the top—front—side views distributed normally. Isometric views are shown at an angle.
- Complete the title block with part name, client, date, designer name, dimensions, material, and default tolerances.

Orthographic views

- There should be three views unless axial symmetry allows fewer.
- The front view should be the most descriptive.
- Blind holes made by drilling must have a drill point shown.
- All parts must be manufacturable.
- Avoid shaded drawings unless looking for interference or providing an aesthetic view.

Schematics and PCBs

- Use multiple drawings for each functional system.
- Show where data buses enter and leave a drawing.
- Show connectors and test points.
- Label all components and show polarities and orientations for parts.
- Add special labeling where critical.
- Position parts to minimize overlap on traces.
- If crossing conductors touch, use a black dot to show electrical contact.

Dimensioning

- The location and size of each feature must be clearly defined.
- Critical assembly dimensions must be directly readable and not require addition.
- Holes that form patterns must be dimensioned relative to one another and relative to a major feature.
- Smaller dimensions should be closer to the part.
- Chained dimensions must be aligned.

- Hole sizes and dimensions should be on the profile view.
- Arcs/circles more than 180 degrees are sized by diameter, otherwise the radius is used.
- Redundant dimensions should be eliminated.

Tolerancing

- Tolerances must be reasonable for manufacturing capabilities.
- Tolerances must ensure proper assembly and operation at maximum/minimum material conditions.
- Mating parts should not have identical dimensions, they should be free running or press fits.
- Smaller tolerances should be used for mating parts.
- A general part tolerance should be defined for the part, and smaller tolerances indicated for critical dimensions, to reduce clutter.

Tables present information that can be structured into a small number of categories. These allow details to be presented in a compact form that is easy to read. General rules of form for tables are (1) a numbered descriptive title is shown above, (2) row and column headings provide adequate descriptions and suitable units, (3) the table data should be readable, and (4) the table should be described, or called out, in nearby text. These principles are illustrated in Table B.3

Calculations and equations are required to justify design work. These should follow the conventions of the discipline. For example, Fig. B.1 shows a set of calculations for a slip-tip problem from a statics course, with summed forces in Eqs. (B.1) and (B.2). Some of the rules for documenting calculations are as follows:

- When presenting equations, use a good equation editor and watch to make sure fine details like subscripts are visible.
- Number equations that are referred to in the text.
- Box in equations of great significance.
- Left justify equations, or center all equations by the equals sign.
- Express results in engineering notation.
- Use subscripts consistently.
- Highlight final results with a box, equation number, bold font, or equivalent.
- Define variables before they are used. This can be with a nomenclature page after the table of contents.
- When possible, italicize variables.
- Keep solutions in variable form until the end of the problem, then substitute numbers if required.

In engineering work numbers are important. When representing numbers it is best to use engineering notation in which all exponents are factors of 3 (i.e., ..., −6, −3, 0, 3, 6, ...). The rules of

Table B.3 A comparison of toy vehicle properties.

Mirror description	Mass (kg)	Color	Shape	Material
Car	3	Red	Rectangular	Die cast
Truck	6	Blue	Long	Polypropylene
Motorcycle	2	Green	Small	Aluminum

significant figures should be observed when using numbers, but it is better to use variables and substitute numbers as the last step of a calculation. Some of the rules for engineering numbers follow:

- Put a space between numbers and units.
- Verify that units match the numerical results.
- Radians are one of the units that may not observe normal conventions.
- Use engineering notation (move exponents three places) so that units are always in standard powers of micro, milli, kilo, mega, giga, and so on. Avoid number formats such as "0.00000456" that include too many leading zeros.
- Use significant figures to round the numbers into meaningful values. For example, stating a length of 0.345432 inch for a dimension measured with a ruler is ridiculous.
- Units are always required.
- Take care to distinguish frequencies stated in Hertz versus radians/s; don't use "cycles/sec."
- Include a "0" before a leading decimal point, such as 0.5; not just .5.

PROBLEMS

B.14 What is an exhibit?

B.15 How are appendices used to improve the readability of reports?

B.16 Why are equations numbered?

B.17 Should every table and figure be mentioned in the text? Explain.

B.18 Why are references useful in technical reports?

B.19 Why should a zero be placed before a decimal point when a number is less than 1?

B.4 TECHNICAL STYLE, GRAMMAR, AND SYNTAX

Consider the writing style you appreciate.

Technical writing is different from writing for entertainment or persuasion. This is not to say that technical writing cannot be entertaining or persuasive, but the primary goal is to document and describe. This leads to the three C's of *clarity*, *conciseness*, and *completeness*. Clarity is important when dealing with complicated topics, but this is easily lost with vague text. Correctness avoids problems with mixed messages or simply incorrect details. Conciseness is critical to keep the discussion focused and easier to absorb. When writing is complete, it will provide the details needed for understanding.

To begin with the obvious, fundamental spelling and grammar are important. Spelling and grammar mistakes are a source of confusion and misunderstanding. For example, a point form sentence that reads "• Increase the resistors" might mean "• Increase the resistance," or it may mean "• Add more resistors." Correctness comes in two forms. One is the basic construction of the language, syntax, and grammar. The other is the technical content. The following list indicates some of the general problems encountered when writing technical reports, along with some strategies for fixing these problems. An excellent reference for this type of writing is Strunk and White (2000).

- Basic spelling: A document should always be checked for spelling. Considering that utilities for checking spelling are available in most software and operating systems, this is expected. Be aware that "spell checkers" will only point out misspelled words, not words used inappropriately, so you should also proofread.
- Technical spelling: Many technical terms are not in the dictionaries used for checking spelling. You may add these terms to the dictionary or visually verify. Be very careful when using the "autoreplace" options in software.
- Basic grammar: "Grammar checkers" can be used to look for obvious problems. Using simple sentence structures will reduce problems and speed the writing process. Grammar-checking software should not be used as a replacement for proofreading.
- Technical grammar: Normally grammar-checking software will reject text written in passive voice, but the software can often be reconfigured. This software will also be confused by the interchangeable use of nouns and verbs common in technical English, such as "input."
- Jargon and acronyms: A number of technical terms and acronyms have been developed for efficiency and clarity. Examples include DMM, HTTP, kitted, parted, and so on. All acronyms should be defined when first used.

An author has many choices concerning how to write a sentence and paragraph. This style choice is a function of the words and structures used to communicate a message. In technical writing, this is mainly a function of precision. Determine what you need to say and then express it clearly. Adding unnecessary content and complication only creates barriers to the rate and depth of reader comprehension. The guidelines shown in the following list should lead to better technical writing.

- Don't find creative ways to say technical things. Many students have been taught that they should not repeat themselves and instead should find multiple ways to say things. When this is done in technical documents, it leads to confusion. Authors should use precise terms (as many times as needed) and avoid trying to generate creative word choices. For example, we could increase confusion by describing translation also as motion, movement, sliding, displacing, and so on.
- Keep it simple. In an attempt to increase the "prestige" of their documents many authors will use uncommon or pretentious words. This often leads to confusion and should be avoided. In some cases, when authors are unsure, they will respond by making their writing style more complex, but most readers recognize this. For example, "Electronic computer-based digital readings can provide a highly accurate data source to improve the quality of the ascertained data" could be replaced with "Computer-based data collection is more accurate."
- Clear, concise, and complete (the three C's): In some courses, students may have been required to write reports with a minimum number of words. This requirement may have encouraged students to increase their verbiage. However, readers appreciate shorter documents that get to the point. For example, "Readings of the pressure, as the probe was ascending up the chimney toward the top, were taken" is better put "Pressure probe readings were taken at multiple chimney heights." Also, it is better to break complex ideas into smaller pieces.
- There is no great opening paragraph. Many student authors spent a large amount of time on the opening paragraph to set the tone for the report. All too often the longer a student tries to write the

opening paragraph, the worse it becomes. In most cases, these opening paragraphs can be deleted entirely from the document without any negative impact. Ironically, the writing of these students often improves once they get beyond the first paragraph, but often they have already lost the interest of their readers.

- Transitions are not that important. Students are often coached to create clean transitions between sentences and paragraphs. As a result they often add unnecessary sentences and words to make these transitions. Words that are warning signs include "also" and "then." Standard technical documents have standard structural forms that provide the major transitions for readers.
- Don't keep the "good stuff" to the end. Many student authors try to write their reports so that there is a "climax." It can be very frustrating for a technical reader to have to read 90% of a report before he or she encounters some discussion of the results. A technical report is not a mystery novel.
- Saying it more than once is acceptable. Most student authors feel that it is unacceptable to state a fact more than once. In truth, you want to state facts as many times as necessary to make a technical point. In the case of very important details, they will be stated in the abstract, the introduction, the discussion, and the conclusion sections.
- Colloquialisms: Avoid informal language in technical reports. Use of informal language such as "cookin' with gas" will look unprofessional, confuse some readers, and easily date the material.
- Repetition: Early writing instruction often encourages writers to find interesting descriptions and variations. As a writing tool this does help students explore the language. On the other hand, a professional has a collection of technical words and phrases with specific meanings. Even at the risk of seeming repetitive, these should always be used the same way each time. Consider the example of an author who describes a specific screw with the following variations: threaded shaft, slot head, threaded fastener, M20, retaining screw, screw, etc.

Table B.4 lists a number of examples of reasonable replacements for complex phrases. When editing or writing, the default should be the simpler form. The more complicated forms should be used only if there is a specific reason. This list is not exhaustive, and each dialect of English has unique phrases that have developed over time; they may be accepted in one region, but make no sense in another (see Resource B.2).

At the paragraph level and above there are strategic issues that influence the effectiveness of a document:

- Reading sequence: How can the document be read?
  - Linear: Read from beginning to end in a fixed sequence (reports and proposals).
  - Nonlinear: Read from beginning to end in a variety of sequences.
  - Random access: Small sections of the documents are read as needed.
- The message
  - State the objective or outcome of the work and repeatedly address it while writing.
  - Use summaries to restate important points, and put them in context.
  - Consider the big picture: overview, repeat while adding detail, summarize.
  - Interleave visual and written content.
  - All statements should be justified; avoid personal opinions or "gut feels."
- Negative statements may be necessary
  - Communicate issues clearly without vague language to soften the impact.
  - Reduce surprises later.

Table B.4 A plain English translation chart.

Good	Bad
Was or is	It became obvious that
	Came in at
	In order to be
	Needed to be
	Needed to be used
	Decided to be
	So as to
	Can be located
	Found to have
	Found through
	It was found that
	Implementation of
	Important
	Precise
	Exact
	Perfect
	Noted to be
	Involved
	Allowed for it
	Was found to be
Reviewed	Was looked through
With	Along with
Selected	Decided on
Measured, calculated	Found
	The wearing of
	Needed to
Measured	Read
Chose	Optimized
Parallax error	Human error
Damper	Dampener (makes things wet)
Resistor	Resister (someone who resists)
Build, calculate, write	Create
Axle	Axel
Illustrates	Represents

    - It sends a message that more support/resources may be needed.
    - It encourages trust.
- Technical depth and completeness
    - Provide the level of detail suitable for the audience, or provide references if needed.
    - Follow the problem-solving approach normally used by the audience.
    - Ask the question: Could somebody understand my work if I were not here to answer questions?
    - If you were to restart your work, would your report help you save time and effort?
    - Consider that many of the readers do not have English as a first language or do not know many of the phrases you do.
- Procedure
    - Proofread as you write; it will be easier to correct.
    - There are no rewards for flowery, creative, and poetic language.
    - The main purpose of the text is to clarify (not present) the details.
    - "Textbook rule"—Write in the style you prefer to read in such as textbooks, manuals, or guides.
    - The 90/10 rule—90% preparation, 10% writing at the end—is good.
    - 90% writing + 90% preparation is not a good method.

Resource B.2 Technical Writing Standard ASD-STE100 (www.asd-ste100.org/).

PROBLEMS

B.20 Should a technical report keep the reader's interest by finding interesting variations for names and operations?

B.21 Why is it better to state the outcome at the beginning of a report? The alternative is to save it to the end of the report, to surprise the reader.

B.22 What do the three C's mean?

B.23 When is past tense, or passive voice, useful in technical writing?

B.24 What are three advantages and three disadvantages of using jargon and acronyms?

B.5 WRITING PROCESS

A good report can be described in one sentence. An effective technical writer will not write text until the other work is done. A poor writer will begin to write first and then fill in details as needed. Outlines are the key to organization. Simple outlines are sets of bullet points that can be rearranged until they make sense. Technical outlines also include calculations, specifications, drawings, sketches, test results, and much more. In other words, a poor writer will rush to write, a good writer will do all of the background work first. An effective procedure for writing engineering reports is outlined in the following list. This procedure leaves writing to a later stage, when you know what you need to say. The key to this approach is to do the technical work first. Make a point of creating figures, tables, and calculations as you go. Point form or short notes can be used to capture information that will be used when writing later.

(1) Background work (90%)
    (a) Develop a single-sentence description of the purpose for the report.
    (b) Define the goals for the project clearly in bullet point form.
    (c) Plan and do the work as normal. Regardless of what the report entails, this will often include creating sketches, drawings, graphs, or charts of collected data, pictures, etc.
    (d) As work continues on the project, add notes and figures.
    (e) The document outline is the technical exhibits and point form text, in a logical order. The content should tell the story by itself before the text is added.
(2) Structuring (5%)
    (a) Do the analysis (preferably on computer) of the data and results. These should be organized into a logical sequence.
    (b) Review the results to ensure they make sense and follow a logical flow. If necessary, add figures to help clarify. Write figure and table captions that describe the materials that will be included in your report.
    (c) Review the materials to verify that they make sense without the text.
    (d) Within the required sections, write bullet form notes to lay out the document.
    (e) Use your notes, and other records, to add bullet point information.
(3) Writing (5%)
    (a) When the project is complete, convert the bullet point form to full text.
    (b) Verify that the report conforms to guidelines.
    (c) Proofread and edit.

The writing process has inertia. The first paragraph always seems to take too much time. After the first paragraph is written the process becomes easier and faster. Writer's block normally occurs when the next writing steps are not clear. This will happen when we are not sure what we need to say, why we need to say it, or how we need to say it. This can be caused by a lack of understanding of the topic or viewpoint. Some simple strategies are provided in the following list for some of the common issues:

- If you are unsure what to say, stop, step away from the writing, figure it out, and then start again.
- Knowledge: If you are not sure what you are writing about you should spend time clarifying your knowledge before returning to writing. Reorganizing the material often helps to create clarity.
- Lack of knowledge: Current knowledge is based on fundamental knowledge discovered and used before. This means that no matter how simple something apparently is, it has more layers of knowledge than could be known by any one person. If you don't know everything, you should define what you do and don't know.
- Skip that great opening paragraph. It is quite acceptable to start by writing central sections of a report. Many authors will write the abstract, introduction, and conclusions last.
- Your report doesn't need to sound impressive. Simply write what you mean to say. If you are having trouble saying it, skip it and come back later, or leave it out if you can.
- If you feel like you are babbling, then consider adding a figure or other exhibit.

The single largest mistake that engineers make is to start writing before assembling the technical content.

Given that the end product is the written word, it is obvious to start there. The writing in a report is analogous to a new house that is finished with paint and carpet. Behind the finish materials there is technical framing and service work that holds it together. To start reports with writing is akin to trying to build a house by painting and carpeting first. A good professional will prepare the plans and background materials first, and then the finishing touches of writing hold it together.

PROBLEMS

B.25 Explain how the exhibits in a report are like a rough draft of an essay.

B.26 If you are writing and you feel confused, what should you do?

B.27 When somebody says you will write a report, does that mean that you will spend most of the time writing?

INSTRUCTOR PROBLEMS

B.28 Is writing text the most important step in writing a report?

B.29 What are the three C's of communication?

B.30 Describe reasonable audience expectations for the following: (a) a software user manual; (b) an automobile mechanics manual; (c) a project proposal; (d) test certifications for a medical device

B.31 Explain how the purpose of work should be described and answered at the start and end of a technical document.

B.32 Why are references important in technical documents?

B.33 Consider a document that includes both exhibits and text. How should they be related?

Reference

Strunk, W., White, E.B., 2000. The Elements of Style, fourth ed. Pearson.

Further reading

Bryan, W.J. (Ed.), 1996. Management Skills Handbook; A Practical, Comprehensive Guide to Management. American Society of Mechanical Engineers, Region V.
Dhillon, B.S., 1996. Engineering Design; A Modern Approach. Irwin.
Heldman, K., 2009. PMP: Project Management Professional Exam Study Guide, fifth ed. Sybex.
Heerkens, G.R., 2002. Project Management. McGraw-Hill.
Pritchard, C., 2004. The Project Management Communications Toolkit. Artech House.
Ullman, D.G., 1997. The Mechanical Design Process. McGraw-Hill.
Wysocki, R.K., 2004. Project Management Process Improvement. Artech House.
Zambruski, M.S., 2009. A Standard for Enterprise Project Management. CRC Press.

APPENDIX

Accreditation requirements mapping

There are topics that are difficult to address in curricula that are based heavily in mathematics, science, and technology. These topics are sometimes addressed in multiple courses, but the assessment process is often left for the capstone/thesis/senior project course.

This appendix provides tables that relate book chapters to accreditation standards, including those of Australia, Canada, the United Kingdom, and the United States. The tables are a planning tool for mapping learning outcomes assessment and book chapters. The matrix values indicate:

- H, a high level of coverage;
- L, limited coverage;
- M, a medium level of coverage;
- empty, no significant coverage.

C.1 UNITED STATES OF AMERICA

Table C.1 Sample curriculum mapping for ABET EAC programs.

	Chapters									
ABET EAC criterion (USA)	**1. Design projects**	**2. Planning and managing projects**	**3. Customer requirements and specifications**	**4. Concepts and technical specifications**	**5. People and teams**	**6. Decision-making**	**7. Finance, budgets, purchasing, and bidding**	**8. Reliabilityandsystemdesign**	**9. Communication, meetings, and presentations**	**10. General design topics**
1. An ability to identify, formulate, and solve complex engineering problems by applying principles of engineering, science, and mathematics		L					L	H		H
2. An ability to apply engineering design to produce solutions that meet specified needs with consideration of public health, safety, and welfare, as well as global, cultural, social, environmental, and economic factors	L		H	H	L					H
3. An ability to communicate effectively with a range of audiences		L	L		H				H	
4. An ability to recognize ethical and professional responsibilities in engineering situations and make informed judgments, which must consider the impact of engineering solutions in global, economic, environmental, and societal contexts	L	L		H	H	H	L			H
5. An ability to function effectively on a team whose members together provide leadership, create a collaborative and inclusive environment, establish goals, plan tasks, and meet objectives				L	H				H	

Table C.1 Sample curriculum mapping for ABET EAC programs.—cont'd

	Chapters									
ABET EAC criterion (USA)	**1. Design projects**	**2. Planning and managing projects**	**3. Customer requirements and specifications**	**4. Concepts and technical specifications**	**5. People and teams**	**6. Decision-making**	**7. Finance, budgets, purchasing, and bidding**	**8. Reliabilityandsystemdesign**	**9. Communication, meetings, and presentations**	**10. General design topics**
6. An ability to develop and conduct appropriate experimentation, analyze and interpret data, and use engineering judgment to draw conclusions				L		H	L	L		H
7. An ability to acquire and apply new knowledge as needed, using appropriate learning strategies		H	H	H		L				

EAC, *Engineering Accreditation Commission.*

Table C.2 Sample curriculum mapping for ABET ETAC programs.

	Chapters									
ABET ETAC criterion (USA)	**1. Design projects**	**2. Planning and managing projects**	**3. Customerrequirements and specifications**	**4. Concepts and technical specifications**	**5. People and teams**	**6. Decision-making**	**7. Finance, budgets, purchasing, and bidding**	**8. Reliability and system design**	**9. Communication, meetings, and presentations**	**10. General design topics**
1. An ability to apply knowledge, techniques, skills, and modern tools of mathematics, science, engineering, and technology to solve well-defined engineering problems appropriate to the discipline		L				L	H	H		H
2. An ability to design solutions for well-defined technical problems and assist with the engineering design of systems, components, or processes appropriate to the discipline	H	H	H	H		H		H		H
3. An ability to apply written, oral, and graphical communication in well-defined technical and nontechnical environments, and an ability to identify and use appropriate technical literature					L		L		H	L
4. An ability to conduct standard tests, measurements, and experiments and to analyze and interpret the results		L	H	L		L		H		H
5. An ability to function effectively as a member of a technical team					H	L			H	

ETAC, *Engineering Technology Accreditation Commission.*

C.2 CANADA

Table C.3 Sample curriculum mapping for CEAB programs.

	Chapters									
CEAB criterion (Canada)	**1. Design projects**	**2. Planning and managing projects**	**3. Customer requirements and specifications**	**4. Concepts and technical specifications**	**5. People and teams**	**6. Decision-making**	**7. Finance, budgets, purchasing, and bidding**	**8. Reliability and system design**	**9. Communication, meetings, and presentations**	**10. General design topics**
3.1.1 A knowledge base for engineering: demonstrated competence in university-level mathematics, natural sciences, engineering fundamentals, and specialized engineering knowledge appropriate to the program		L				L		H		H
3.1.2 Problem analysis: an ability to use appropriate knowledge and skills to identify, formulate, analyze, and solve complex engineering problems to reach substantiated conclusions	L		H	H			L	H		H
3.1.3 Investigation: an ability to conduct investigations of complex problems by methods that include appropriate experiments, analysis and interpretation of data, and synthesis of information to reach valid conclusions			L	L				L		H
3.1.4 Design: an ability to design solutions for complex, open-ended engineering problems and to design systems, components, or processes that meet specified needs with appropriate attention to health and safety risks, applicable standards, and economic, environmental, cultural, and societal considerations	H	H	H	H			H	H		H
3.1.5 Use of engineering tools: an ability to create, select, apply, adapt, and extend appropriate techniques, resources, and modern engineering tools to a range of engineering activities, from simple to complex, with an understanding of the associated limitations		H	H	H			H			H

Continued

Table C.3 Sample curriculum mapping for CEAB programs.—cont'd

	Chapters									
CEAB criterion (Canada)	**1. Design projects**	**2. Planning and managing projects**	**3. Customer requirements and specifications**	**4. Concepts and technical specifications**	**5. People and teams**	**6. Decision-making**	**7. Finance, budgets, purchasing, and bidding**	**8. Reliability and system design**	**9. Communication, meetings, and presentations**	**10. General design topics**
3.1.6 Individual and team work: an ability to work effectively as a member and leader in teams, preferably in a multidisciplinary setting					H	L			H	
3.1.7 Communication skills: an ability to communicate complex engineering concepts within the profession and with society at large. Such ability includes reading, writing, speaking, and listening, and the ability to comprehend and write effective reports and design documentation, and to give and effectively respond to clear instructions						L			H	
3.1.8 Professionalism: an understanding of the roles and responsibilities of the professional engineer in society, especially the primary role of protection of the public and the public interest					H	H			H	
3.1.9 Impact of engineering on society and the environment: an ability to analyze social and environmental aspects of engineering activities. Such ability includes an understanding of the interactions that engineering has with the economic, social, health, safety, legal, and cultural aspects of society, the uncertainties in the prediction of such interactions, and the concepts of sustainable design and development and environmental stewardship	H	H	H	H	H		L		L	H
3.1.10 Ethics and equity: an ability to apply professional ethics, accountability, and equity					H				L	

Table C.3 Sample curriculum mapping for CEAB programs.—cont'd

CEAB criterion (Canada)	Chapters									
	1. Design projects	2. Planning and managing projects	3. Customer requirements and specifications	4. Concepts and technical specifications	5. People and teams	6. Decision-making	7. Finance, budgets, purchasing, and bidding	8. Reliability and system design	9. Communication, meetings, and presentations	10. General design topics
3.1.11 Economics and project management: an ability to appropriately incorporate economics and business practices, including project, risk, and change management, into the practice of engineering and to understand their limitations			H			H	H	H		
3.1.12 Lifelong learning: an ability to identify and address their own educational needs in a changing world in ways sufficient to maintain their competence and to allow them to contribute to the advancement of knowledge			L	L			L		L	H

CEAB, *Canadian Engineering Accreditation Board.*

C.3 AUSTRALIA

Table C.4 Sample curriculum mapping for Australian engineering programs.

Stage 1 Competency standard for professional engineer (Australia)	Chapters									
	1. Design projects	2. Planning and managing projects	3. Customer requirements and specifications	4. Concepts and technical specifications	5. People and teams	6. Decision-making	7. Finance, budgets, purchasing, and bidding	8. Reliability and system design	9. Communication, meetings, and presentations	10. General design topics
1.1 Comprehensive, theory-based understanding of the underpinning natural and physical sciences and the engineering fundamentals applicable to the engineering discipline										
1.2. Conceptual understanding of the mathematics, numerical analysis, statistics, and computer and information sciences that underpin the engineering discipline										
1.3. In-depth understanding of specialist bodies of knowledge within the engineering discipline								H		H
1.4. Discernment of knowledge development and research directions within the engineering discipline	L	H	H	H		L			L	
1.5. Knowledge of contextual factors impacting the engineering discipline	L		H	H		L	H	L		H
1.6. Understanding of the scope, principles, norms, accountabilities, and bounds of contemporary engineering practice in the specific discipline	H		H	H			L			H
2.1. Application of established engineering methods to complex engineering problem solving							L	H		H
2.2. Fluent application of engineering techniques, tools, and resources		L					L	H		H
2.3. Application of systematic engineering synthesis and design processes	H		H	H		L	L	H		H

Table C.4 Sample curriculum mapping for Australian engineering programs.—cont'd

	Chapters									
Stage 1 Competency standard for professional engineer (Australia)	**1. Design projects**	**2. Planning and managing projects**	**3. Customer requirements and specifications**	**4. Concepts and technical specifications**	**5. People and teams**	**6. Decision-making**	**7. Finance, budgets, purchasing, and bidding**	**8. Reliability and system design**	**9. Communication, meetings, and presentations**	**10. General design topics**
2.4. Application of systematic approaches to the conduct and management of engineering projects	H	H	L	L	L	H	L		L	
3.1. Ethical conduct and professional accountability					H					
3.2. Effective oral and written communication in professional and lay domains					H				H	
3.3. Creative, innovative, and proactive demeanor	H		H	H	H					
3.4. Professional use and management of information		L			L	L	L		L	
3.5. Orderly management of self, and professional conduct					H	H			L	
3.6. Effective team membership and team leadership					H				H	

C.4 UNITED KINGDOM

Table C.5 Sample curriculum mapping for UK CEng programs.

	Chapters									
UK CEng accreditation criterion by the Engineering Council	**1. Design projects**	**2. Planning and managing projects**	**3. Customer requirements and specifications**	**4. Concepts and technical specifications**	**5. People and teams**	**6. Decision-making**	**7. Finance, budgets, purchasing, and bidding**	**8. Reliability and system design**	**9. Communication, meetings, and presentations**	**10. General design topics**
Knowledge and understanding of scientific principles and methodology necessary to underpin their education in their engineering discipline, to enable appreciation of its scientific and engineering context, and to support their understanding of historical, current, and future developments and technologies										
Knowledge and understanding of mathematical principles necessary to underpin their education in their engineering discipline and to enable them to apply mathematical methods, tools, and notations proficiently in the analysis and solution of engineering problems								M		
Ability to apply and integrate knowledge and understanding of other engineering disciplines to support study of their own engineering discipline		L		L		L		M		H
Understanding of engineering principles and the ability to apply them to analyze key engineering processes			M	M				H		H
Ability to identify, classify, and describe the performance of systems and components through the use of analytical methods and modeling techniques			H	H		L		H		H
Ability to apply quantitative methods and computer software relevant to their engineering discipline to solve engineering problems		M						M		M
Understanding of and ability to apply a systems approach to engineering problems	H		M	H				H		H

Table C.5 Sample curriculum mapping for UK CEng programs.—cont'd

	Chapters									
UK CEng accreditation criterion by the Engineering Council	**1. Design projects**	**2. Planning and managing projects**	**3. Customer requirements and specifications**	**4. Concepts and technical specifications**	**5. People and teams**	**6. Decision-making**	**7. Finance, budgets, purchasing, and bidding**	**8. Reliability and system design**	**9. Communication, meetings, and presentations**	**10. General design topics**
Ability to investigate and define a problem and identify constraints, including environmental and sustainability limitations, health and safety, and risk assessment issues	H	H	H	H			M			H
Understanding of customer and user needs and the importance of considerations such as aesthetics			H	H						L
Ability to identify and manage cost drivers							H			
Ability to use creativity to establish innovative solutions				H						M
Ability to ensure fitness for purpose for all aspects of the problem, including production, operation, maintenance, and disposal			H	H		M	H	H		H
Ability to manage the design process and evaluate outcomes	H	H	H	H		H			H	H
Knowledge and understanding of commercial and economic context of engineering processes		M	H	M			H			H
Knowledge of management techniques that may be used to achieve engineering objectives within that context	H	H			H	H	H		H	
Understanding of the requirement for engineering activities to promote sustainable development			M				M			M
Awareness of the framework of relevant legal requirements governing engineering activities, including personnel, health, safety, and risk (including environmental risk) issues			H				M			H
Understanding of the need for a high level of professional and ethical conduct in engineering					H	M				

Continued

Table C.5 Sample curriculum mapping for UK CEng programs.—cont'd

	Chapters									
UK CEng accreditation criterion by the Engineering Council	**1. Design projects**	**2. Planning and managing projects**	**3. Customer requirements and specifications**	**4. Concepts and technical specifications**	**5. People and teams**	**6. Decision-making**	**7. Finance, budgets, purchasing, and bidding**	**8. Reliability and system design**	**9. Communication, meetings, and presentations**	**10. General design topics**
Knowledge of characteristics of particular materials, equipment, processes, or products										
Understanding of contexts in which engineering knowledge can be applied (e.g., operations and management, technology development, etc.)	H	H	H	H		H	M	H	M	
Understanding of the use of technical literature and other information sources			H	H		H		M	H	M
Awareness of the nature of intellectual property and contractual issues			H	H			M		M	
Understanding of appropriate codes of practice and industry standards			H	H						M
Awareness of quality issues								H		M
Ability to work with technical uncertainty			H	H	M	H		H		

CEng, *chartered engineer.*

Index

'Note: Page numbers followed by "f" indicate figures and "t" indicate tables.'

D

E

Q

R

S